RESIDUE REVIEWS

VOLUME 32

The Triazine Herbicides

SINGLE PESTICIDE VOLUME:

THE TRIAZINE HERBICIDES

RESIDUE REVIEWS

Residues of Pesticides and Other
Foreign Chemicals in Foods and Feeds

RÜCKSTANDS-BERICHTE

Rückstände von Pestiziden und anderen
Fremdstoffen in Nahrungs- und Futtermitteln

Editor
FRANCIS A. GUNTHER

Assistant Editor
JANE DAVIES GUNTHER

Riverside, California

VOLUME 32

SPRINGER-VERLAG
BERLIN • HEIDELBERG • NEW YORK
1970

ISBN 978-1-4615-8466-7   ISBN 978-1-4615-8464-3 (eBook)
DOI 10.1007/978-1-4615-8464-3

© 1970 by Springer-Verlag New York Inc.
Softcover reprint of the hardcover 1st edition 1970
Library of Congress Catalog Card Number 62–18595.

Title No. 7851

Preface

More and more biologists, chemists, pharmacologists, toxicologists, governmental agencies, and "food control" (regulatory) officials around the world are finding it increasingly difficult to keep abreast of the technical literature in the pesticide field; indeed, many libraries do not have even a small proportion of the journals and other sources that now regularly contain research, development, and application information about all aspects of modern chemical pest control. As a result, a very large number of requests has come to RESIDUE RE-VIEWS to publish detailed digests of information on single pesticide chemicals so that the interested person in any part of the world could easily be brought up to date with all available important information without having to search probably several hundred literature sources, many of them obscure or simply not available except in very large libraries. The service and convenience rendered the readership by such a series of volumes on major individual pesticide chemicals would therefore be considerable.

Type and scope of coverage in this series of single-pesticide volumes will of course vary with available information. The coverage should be as complete as possible, however, to be of maximum value to all interested individuals, industries, research institutions, and governmental agencies concerned with the continuing production of an adequately large yet safe food supply for the world. Among the topics bracketed for a single pesticide should ideally be:

 I. Introduction
 II. History of development and use, including alternate names around the world, patent information
 III. Chemistry, manufacture, and stability
   a) Synthesis
   b) Chemical and physical properties
   c) Commercial synthesis and composition of commercial product
   d) Formulations
   e) Storage stability of formulated products
   f) Compatability with other materials
   g) Photodecomposition
   h) Stability of parent compound in water, animal and plant tissues, and soils
   i) Other known chemical and biochemical reactions
 IV. Pharmacology and toxicology
   a) Acute, chronic, dermal, and inhalation toxicities

  b) Metabolic pathways and products in animals, plants, and other
     soils
  c) Other important effects on mammals
  d) Effects on wildlife in controlled tests
V. Biological properties and uses, world coverage
  a) Pests and crops involved
  b) Performance
  c) Methods of application
  d) Dosages and formulations
  e) Any secondary effects (e.g., phytotoxicity with insecticides or
     fungicides)
VI. Tolerances, world coverage
VII. Residues encountered in foodstuffs in developmental programs and in
     practice, with dosages, timing, methods of application, persistence
     curves, half-lives, etc
  a) Raw agricultural commodities
  b) Processed food products
  c) Residue removal by washing, processing, etc
  d) Metabolites or other alteration products as residues
VIII. Analytical methods, with full details of recommended methods
  a) Assay of technical grade product
  b) Assay of formulated products
  c) Residue methods for foodstuffs, waters, soils, including recoveries,
     minimum detectabilities, sensitivities, precision, etc. both in the
     presence of and in the absence of substrate extractives
IX. Incidence in, and effects on, environment
  a) Incidence in the environment
  b) Persistence in the environment
  c) Soil-compound interactions (soil flora and fauna)
  d) Stability in soils, water, and soil water
  e) Effects on fish and other wildlife, pollinating insects
  f) Involvement in the food chain
X. Discussion and conclusions

Individual chapters or sections may be credited to particular
persons as authors, of course, but it is suggested that a single person be
responsible for tone, style, and acceptable RESIDUE REVIEWS
format of the total manuscript.

As with other volumes of RESIDUE REVIEWS, manuscripts
are normally contributed by invitation, but in English only. Preliminary communication with the editor is necessary before volun-
teered reviews are submitted in manuscript form.

Department of Entomology   Francis A. Gunther
University of California
Riverside, California
June 2, 1969

Foreword

The s-triazine herbicides are widely used for preemergence and postemergence weed control for industrial and agricultural purposes. Most are applied to the soil and their major absorption by plants is through the roots. The fate of s-triazines in the soil is one of the most important aspects concerning their widespread use. Their fate determines the amount of weed control obtained, selectivity, persistence, effect on soil organisms, soil and water contamination, and crop rotation. Herbicides must survive in the soil environment long enough to kill the undesired weeds. If their soil persistence is too long they interfere with the use of the land for growth of other crop plants.

All organic compounds added to the soil must ultimately decompose or be altered to become part of the soil complex. The nature and speed of the alteration or loss are determined by the intricate interaction of the chemical and physical properties of the herbicide and the chemical, physical, edaphic, and biological properties of the soil. The long-term use of herbicides is contingent upon knowledge of their interactions with soils.

The s-triazine herbicides are well suited for research concerning their interaction with soils. They are used in several crops and also in non-crop areas under a wide range of climatic and soil conditions. A large number of herbicidal s-triazines are possible through altering substitution at the 2-, 4-, and 6-positions on the ring. Different substitutions result in a wide range of physical, chemical, and biological properties of the compounds. Names, structures, uses, and properties of marketed s-triazine herbicides are shown in the following Reference Tables I and II which were compiled by *Geigy Agricultural Chemicals* personnel.

The papers presented in this special volume of RESIDUE REVIEWS are the outgrowth of an international symposium held on February 17–19, 1969, at the *University of California, Riverside,* especially for this publication. Personnel of *Geigy Agricultural Chemicals* and the editor of RESIDUE REVIEWS conceived the idea of holding the symposium to bring together all available information concerning s-triazine-soil interactions. The *University of California College of Biological and Agricultural Sciences* agreed to co-sponsor the symposium. A grant-in-aid was given by Geigy to the University to cover expenses of the majority of the participants and the expenses of the symposium.

Reference Table I. *Names, structures, uses, and properties[a] of marketed triazine herbicides*

Structure and chemical denomination	Code number	Common name	Trademarks	Main uses	m.p.(°C.)	Solubility (p.p.m.)		Vapor pressure (mm. Hg 293°K)	pK	Dipole moment (Debyes)
						Water	Organic solvents			
2-Chloro-4,6-bis-ethylamino-s-triazine	G27692	Simazine	PRIMATOL S® Europe GESATOP® Europe PRINCEP® U.S.A.	General weed control Corn, grapes, small grains, fruit trees, forestry, alfalfa	223–5	5	Sl. sol. Chloroform 900 n-Pentane 3 Ac. ester 1,200	6.1×10^{-9}	1.65	Solubility too limited
2-Chloro-4-ethylamino-6-isopropylamino-s-triazine	G30027	Atrazine	PRIMATOL A® Europe GESAPRIM® Europe AATREX® U.S.A.	General weed control Corn, sorghum, grapes (in special cases), fruit trees, pineapples, sugar cane	173–5	33	Sol. Ac. ester 28,000	3.0×10^{-7}	1.68	4.6_2
2-Chloro-4,6-bis-isopropylamino-s-triazine	G30028	Propazine	GESAMIL® Europe MILOGARD® U.S.A.	Sorghum, carrots, other umbelliferae	212–4	8.6	Sl. sol.	2.9×10^{-8}	1.85	4.5_2

Structure	Chemical name	Code	Common name	Trade name	Country	Use	m.p.		Sol. acetone			
C_2H_5NH, Cl, $NH\text{-tert.}C_4H_9$ triazine	2-Chloro-4-ethylamino-6-tert.-butylamino-s-triazine	GS13529	—	PRAMITOL M 50®	Europe	General weed control	177–9	8.5	Sol. acetone	1.12×10^{-8}	—	—
C_2H_5NH, Cl, $N(C_2H_5)_2$ triazine	2-Chloro-4-ethylamino-6-diethylamino-s-triazine	G27901	Trietazine	GESAFLOC®	Japan	Chrysanthemum	102–4	20	Very sol.	—	1.88	4.7$_7$
SCH_3, C_2H_5NH, NHC_2H_5 triazine	2-Methylthio-4,6-bis-ethylamino-s-triazine	G32911	Simetryne	GY-BON®	Japan	Rice	81–2.5	450	Very sol.	7.1×10^{-7}	—	—
SCH_3, CH_3NH, $NHiC_3H_7$ triazine	2-Methylthio-4-n-methylamino-6-isopropylamino-s-triazine	G34360	Desmetryne	SEMERON®	Europe	Cruciferous crops	84–6	580	Very sol.	1.0×10^{-6}	3.08	—

ix

Reference Table I. (*Continued*)

Structure and chemical denomination	Code number	Common name	Trademarks	Main uses	m.p.(°C.)	Solubility (p.p.m.)		Vapor pressure (mm. Hg 298°K)	pK	Dipole moment (Debyes)
						Water	Organic solvents			
2-Methylthio-4-ethylamino-6-isopropylamino-s-triazine	G34162	Ametryne	GESAPAX® Europe No trademark in U.S.A.	Sugar cane, pineapples	84–6	193	Very sol.	8.4×10^{-7}	3.12	3.1_5
2-Methylthio-4,6-bis-isopropylamino-s-triazine	G34161	Prometryne	GESAGARD® Europe CAPAROL® U.S.A.	Cotton, celery, legumes, seed grasses, carrots, rice, sunflowers	118–20	48	Very sol.	10×10^{-6}	3.05	3.5_4
2-Methylthio-4-isopropylamino-6-γ-methoxypropylamino-s-triazine	G36393	Methoprotryne	GESARAN® Europe	Small grains, flax	68–70	320	Very sol.	—	3.03	—

GS14260	Terbutryne	IGRAN®	Europe & U.S.A.	Small grains	104–5	58	Very sol.	9.6×10^{-7}	—	—
G32293	Atratone	GESATAMIN®	Africa	Sisal, coffee, sugar cane, general weed control	94–6	1654	Very sol.	2.9×10^{-6}	4.20[b]	—
G31435	Prometone	PRIMATOL O® PRAMITOL®	Europe U.S.A.	General weed control	91–2	750	Very sol.	2.3×10^{-6}	4.3	2.94

Structures and chemical names:

- SCH₃ ... C_2H_5NH ... NH–tert. C_4H_9
  2-Methylthio-4-ethylamino-6-tert.-butylamino-s-triazine

- OCH₃ ... C_2H_5NH ... $NHiC_3H_7$
  2-Methoxy-4-ethylamino-6-isopropylamino-s-triazine

- OCH₃ ... iC_3H_7NH ... $NHiC_3H_7$
  2-Methoxy-4,6-bis-isopropylamino-s-triazine

[a] Physical data from the Analytical Laboratories, *J. R. Geigy S. A.*, Basle/Switzerland.
[b] Weber, J. B. Spectrochimica Acta *23A*, 458 (1967).

Reference Table II. *Code numbers, structures, and properties of some triazine derivatives*

Code number	Common name	Substitution on triazine ring at positions			pK Value	Solubility in water (p.p.m. at 20–25°C.)
		2	4	6		
G11348	—	SCH₃	NHiC₃H₇	N(C₂H₅)₂	4.43	—
G25804	Chlorazine	Cl	N(C₂H₅)₂	N(C₂H₅)₂	1.74	9
G27692	Simazine	Cl	NHC₂H₅	NHC₂H₅	1.65	5
G27901	Trietazine	Cl	NHC₂H₅	N(C₂H₅)₂	1.88	20
G30026	Norazine	Cl	NHCH₃	NHiC₃H₇	—	260
G30027	Atrazine	Cl	NHC₂H₅	NHiC₃H₇	1.68	33
G30028	Propazine	Cl	NHiC₃H₇	NHiC₃H₇	1.85	8.6
G30031	Ipazine	Cl	NHiC₃H₇	N(C₂H₅)₂	1.85	40
G30044	Simetone	OCH₃	NHC₂H₅	NHC₂H₅	4.17	3,200
G31430	—	OCH₃	N(C₂H₅)₂	N(C₂H₅)₂	4.76	—
G31432	—	OCH₃	NHC₂H₅	N(C₂H₅)₂	4.51	40
G31435	Prometone	OCH₃	NHiC₃H₇	NHiC₃H₇	4.28	750
G31717	Ipatone	OCH₃	NHiC₃H₇	N(C₂H₅)₂	4.54	100
G32292	Noratone	OCH₃	NHCH₃	NHiC₃H₇	4.15	3,500
G32293	Atratone	OCH₃	NHC₂H₅	NHiC₃H₇	4.20	1,654
G32911	Simetryne	SCH₃	NHC₂H₅	NHC₂H₅	—	450
G34161	Prometryne	SCH₃	NHiC₃H₇	NHiC₂H₇	4.05	48
G34162	Ametryne	SCH₃	NHC₂H₅	NHiC₃H₇	—	193
G34360	Desmetryne	SCH₃	NHCH₃	NHiC₃H₇	—	580
G36393	Methoprotryne	SCH₃	NHiC₃H₇	NH(CH₂)₃OCH₃	—	320
GS14254	—	OCH₃	NHC₂H₅	NHsC₄H₉	—	620
GS14260	Terbutryne	SCH₃	NHC₂H₅	NHtC₄H₉	—	58

Over 150 scientists met at Riverside, California, for the three-day conference. Sixteen papers were presented covering every aspect of s-triazine soil interactions. Substantial discussion of each subject occurred. The manuscripts are being published in this issue of RESIDUE REVIEWS. The discussions as such are not included, but each author evaluated the discussion of his subject and incorporated pertinent information into his final manuscript.

The symposium and the resulting publication represent a unique cooperative effort by an industrial organization and a public agency working together for the consolidation and dissemination of knowledge and to determine where more knowledge is needed. It is hoped that this endeavor sets a precedent for future conferences with the same high ideals.

The symposium was planned and executed by two committees:

Dr. Homer LeBaron, Dr. James David, and Dr. Peter Dubach from *Geigy Agricultural Chemicals*, and Dr. Joe Goodin, Dr. James Martin, and Dr. Lowell Jordan, of the *University of California*. The success of the symposium resulted from the excellent presentations by the speakers and the lively participation of the scientists who attended.

University of California   Lowell S. Jordan
Riverside, California
June 2, 1969

Table of Contents

Contents

History of the development of triazine herbicides

By

E. KNÜSLI*

Technology does not like historical reflections. The technological world asks for results and for progress. Achievement is important, not, in general, the road which leads to achievement. What was new yesterday is routine today, and what is described as revolutionary today may be considered antiquated tomorrow. Our age is a daily challenge, beloved or hated, depending on where or how we stand.

However, all who are engaged in a biological sector of modern technology are influenced by a further field of power which opposes an uncontrolled rushing forward and this force may simply be defined as nature or natural life. An engineer who is asked to build a bridge will stand one day in front of his construction, tested, safe, ready for use, with no question left open. He will, of course, direct his ideas and dreams immediately to the next and even more perfect bridge, but what he had made before was concluded, final in itself. The man or better the community of men dealing with nature can hardly dream of ever reaching this point; they may add more and more to the picture of understanding by diligence, perseverance and skill, but only in very limited fields will they arrive at the point where they can say now we can stop, we know everything, or—even in a more restrictive sense— we know everything we should know.

The attendants to the Triazine Symposium were kind enough to answer in the affirmative an appeal for establishing the state of our knowledge in one small, special sector of man's confrontation with nature, in particular with the fight against undesired, and for desired, plant growth; and again, not for consideration of the whole field, but of the sector "Herbicides"; not the whole sector "Herbicides," but the subsector "Triazine Herbicides," and even not the whole subsector "Triazine Herbicides," but one important aspect of them only: their behaviour in soil. The immodesty to invite for such a detailed study may be explained and, we hope, excused by the complexity of the many factors which interfere and by the principle that it may be more useful and more profitable to discuss one aspect thoroughly and as exhaustively as possible than to deal superficially with a multitude of facets. It can-

---

* J. R. Geigy S.A., Basle.

1

not be stressed enough that the same principle is also a necessity for all basic research done in laboratories, in greenhouses, and in the field.

At this moment when we try to review the current status of our knowledge and to recognize where we have still important loopholes in our knowledge, it may be justified, despite the reservations made initially, to recall quickly the situation of the field of herbicides as a whole when the first triazine derivatives left the chemical laboratory to be screened as potential herbicides. In 1952 research people, chemists and biologists, when asked to look for new contributions had a relatively easy job to become acquainted with the status of the art. They then made a literature search of their own; had they waited one year more, they could have used the excellent second edition of the "Textbook and Manual of Weed Control" by W. W. ROBBINS, A. S. CRAFTS, and R. N. RAYNOR which appeared in 1953 (London: McGraw Hill). When leaving out the inorganic chemicals, the old phenols and cresols and trichloroacetic acid, the catalogue of chemicals was quite limited. 2,4-D and MCPA already had a glorious and victorious 12-year old career behind them and were, with respect to the tonnage produced and used, apparently out of reach for successive developments. With not much more than four additional pages the representative textbook mentioned could pay tribute to the other organic compounds: IPC, Crag I, Crag II, endothal, maleic hydrazide, N-naphthylphthalamic acid, and the newborn CMU. The usefulness of aminotriazole was already deposed in a patent application, but had not yet reached the stage of general public knowledge. From the Chapter "Selective Herbicides" the following sentence is worth quoting: "Recently there was developed a wide interest in what is termed 'preemergence' treatment of weeds." .... sixteen years ago!

The pace of the research done in industry for originating new active compounds has increased tremendously since the early fifties. When then roughly ten firms had herbicidally active compounds of their own research in the marketing or developing stage, the number may approach 40 today. When looking at the patents reported in 1968, more than 100 companies claimed, mentioned, or pretended herbicidal activities of chemicals. Despite these impressive figures, there is no doubt that only companies of a certain size can take the risks and the uncertainties of the developing of new active principles.

An examination of the herbicides used today shows a picture which is characterized by the fact that activity is not limited to a narrow range of chemically different structures. In this respect the field is clearly distinct from the synthetic insecticides where nearly all compounds used nowadays belong to four chemical groups only: halogenated hydrocarbons, esters of phosphoric or thiophosphoric acids, carbamates, and, less important, thiocyanates.

But, despite the variety of active structures already known, today—as in the early fifties—research workers are still looking for leads to new herbicides. It is tempting to check backwards which leads could have induced a research group to consider triazines as potential herbicides. Four possibilities may be cited, but it has to be emphasized that it looks much too obvious when doing it in a retrospective way.

One lead may have been the knowledge of maleic hydrazide:

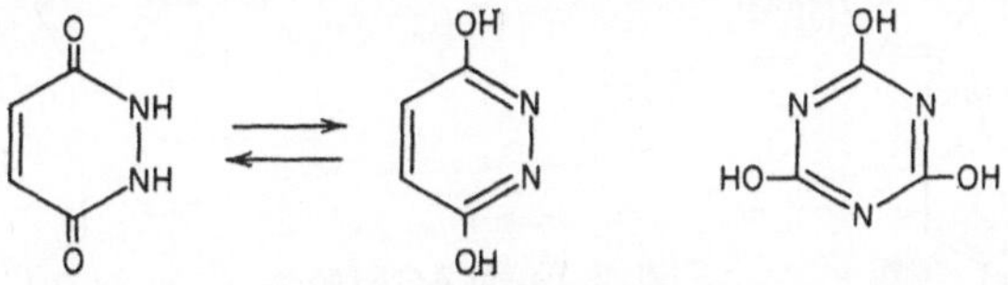

Maleic hydrazide in its enolized form could have induced a chemist to consider also the quite analogous cyanuric acid and its derivatives.

A joking and gambling with atoms and molecules could have led one to the triazine ring by starting from the structure of aminotriazole:

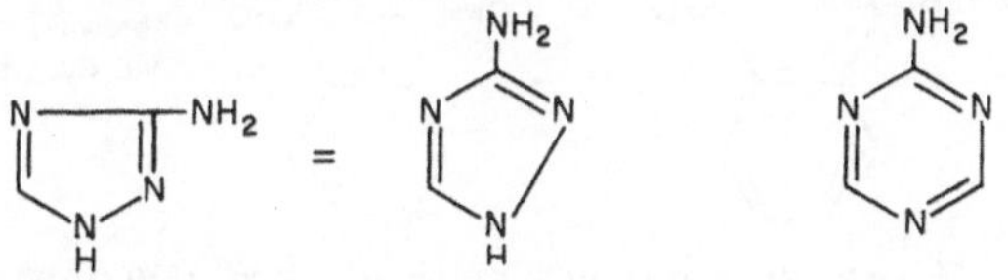

A further lead could have been hidden behind the antimitotic effects of triazines bearing the ethyleneimino radical, reported in connection with chemotherapeutic objectives:

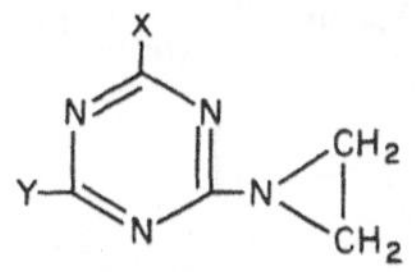

although just these pharmacological effects may have caused mental reservations or at least established even from the beginning the necessity of preparing analogues free of such properties.

Finally a further working hypothesis could be induced by certain parallelisms recognized in other fields between analogous structures of ureas and triazines. This parallelism may be illustrated by one example each out of the field of pharmaceuticals and dyestuffs:

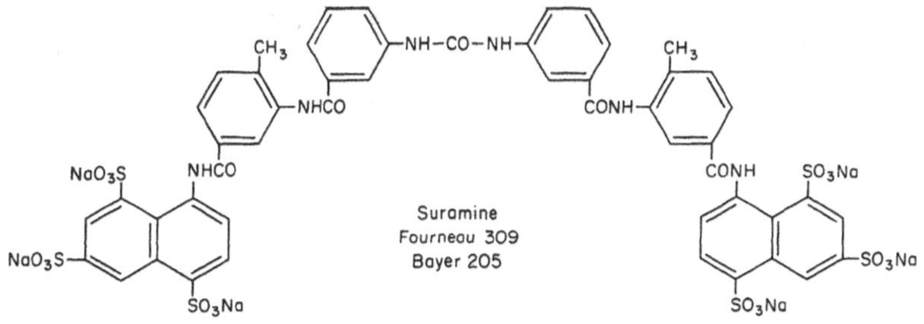

Suramine
Fourneau 309
Bayer 205

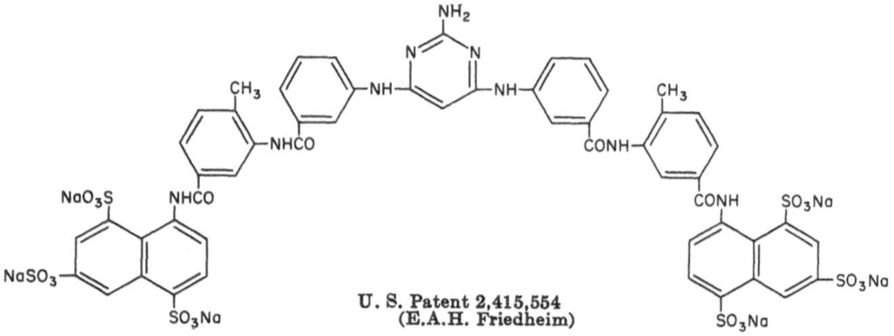

U. S. Patent 2,415,554
(E.A.H. Friedheim)

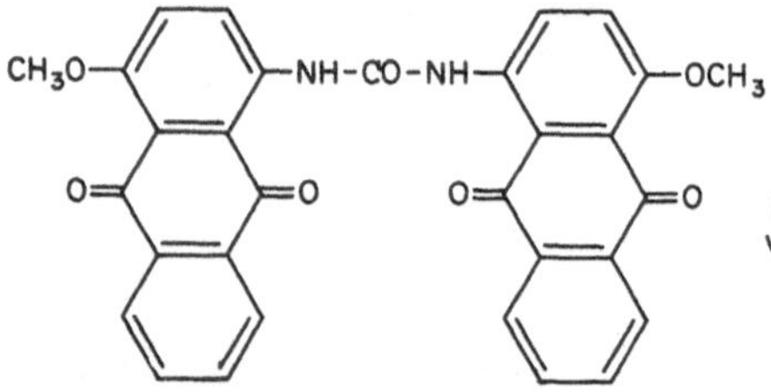

brown-red
vat dyestuff

M. Battegay and D. Bernhardt. Bl. (4) **33**, 1510 (1923)

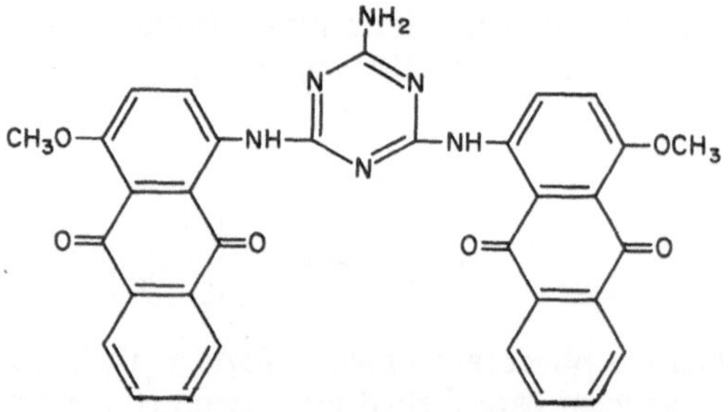

Cibanone Red G

M. Matter. Diss. ETA Zürich (1936)

The first relates to antimalarials, the second to vat-dyestuffs.

It is known now that in the early fifties, at different places, work was underway following at least three of the above-mentioned leads. In all cases the first step along the leads or the working hypothesis did not disclose compounds of dramatic activities. None of the following structures:

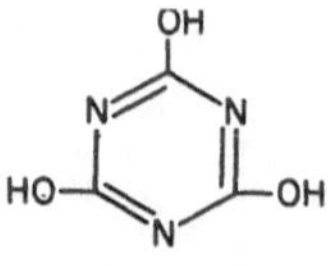

Cyanuric acid

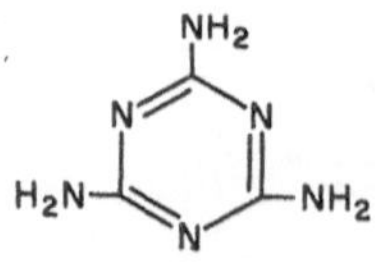

Melamine

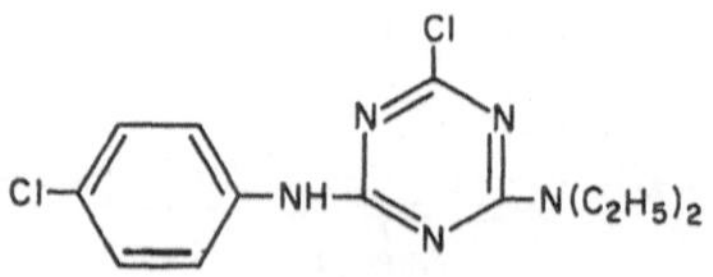

G 27902

have activities about which the chemists or biologists could have been enthused. Our Swiss group had been dealing with the urea-analogy working hypothesis. Besides the strict analogues they synthesized and screened further variations along the triazine pattern and achieved the following compounds:

G 25804
Chlorazine

and

G 27692
Simazine

This type of structure disclosed outstanding herbicidal activity with, at the same time, remarkable selectivities for important crop plants.

Subsequently intense studies circling around the triazine model were undertaken by our group and in the recent past also by third parties. It has been recognized that mainly by suitable variation of the N-alkyl groups and replacement of the chlorine atom by alkoxy or alkylmercapto groups, useful compounds with in part surprisingly different properties can be obtained. The compounds which have reached the marketing level up to now are enumerated and briefly described in Reference Table I of the Foreword. They differ, of course, in their importance.

Five representatives belong to the group of 2-chloro-4,6-diamino-s-triazines, two are 2-methoxy-4,6-diamino-s-triazines, and six are 2-methylthio-4,6-diamino-s-triazines. A review of the biological characteristics of the more important compounds of this list is given by A. GAST in his paper in this volume.

Besides the work along the chemical lines the main part of the workload was the identification of the sectors where the compounds have their practical value and in the determination of the data for the respective labels, in the working out of analytical methods, and, subsequently, of the residue pictures, in the studies on the toxicological behaviour of the compounds, of the metabolism in plants, in animals, and in soil, and in studies for the elucidation of the mechanism of the interference with plant life.

The biochemical, the metabolism, and the analytical aspects are covered by recent review papers. With respect to the mode of action it may be briefly summarized that the triazine herbicides are potent inhibitors of the Hill reaction, resembling the phenylureas. The inhibition of photosynthesis is the most obvious explanation for their interference with plant growth. Besides this line there are also effects on light-independent reactions. Particularly a positive influence of the triazines on the uptake of nutrients by the roots was noticed, and changes in the nitrogen metabolism seem to occur, too. These facts have been known for quite a time, but now the detailed picture of the biochemical reactions which lead to the above-mentioned phenomena becomes more complete. Most valuable information could be obtained by examining more closely the so-called side-effects such as deeper green colour and stimulated growth of treated surviving plants. It has been recognized that triazines interfere with the complex of biochemical reactions around phytokinin and indolylacetic acid. This leads to an influencing of the rate of RN- and protein synthesis and, via the structural protein, to a positive or negative affect on chloroplasts and chlorophyll. The selectivity of the triazine herbicides is mainly correlated with different abilities of the plant species to metabolize the herbicide and to translocate it to the vital centres of the cells. Up to now three major degradative pathways are evident when s-triazines are confronted with plants, animals, or soil: hydrolysis of the substituent at carbon atom 2—be it chlorine, methoxy, or methylthio, a stepwise N-dealkylation at the carbon atoms 4 and 6, and the splitting of the triazine ring. In the case of hydrolysis of the methylthio-group a previous oxidation to the sulfoxy and sulfono radical occurs.

The mentioned review papers show that it is imperative that the originators endow their products with a certain portion of basic information. But they also show that no aspect can get a satisfactory level of knowledge without the interest and cooperation of many research people outside of industry. Triazines had from the very beginning the luck to encounter the enthusiasm and devotion of a great number of scientists at universities and experimental stations, and this should be recognized at this place with deep gratitude.

It would be rewarding if a review like the one at this symposium was not only constructive with respect to its objective *per se*, but also in a more general sense, namely for a more impartial judgement of the contribution made by herbicides and pesticides in general. The benefit of a pharmaceutical, the product of another line of chemical endeavours, is directly evident to its end-user. The pleasure caused by dyestuffs in their manifold application can be directly felt and is obvious to everybody. The services of the progress of mechanical science and communication are easily recognized and appreciated. The progress made through herbicides and pesticides is often—too often— honoured by the reproach that they and their promotors poison mankind. The consumer of food in abundance and good quality has no direct evidence of the positive interference of them. On the contrary, he is rather inclined to follow tendencies which over-emphasize their shortcomings and their limitations, and deny the positive aspects. Nobody better than people and institutions engaged actively in the fight against weeds and other pests know that the means created by modern chemistry cannot be used undeliberately and uncritically. They are, on the other side, convinced of their inestimable value and harmlessness when applied intelligently and properly. If it is accepted that one major goal in this whole picture is the adequate instruction and information of the end-user then the need for a better knowledge of the fundamentals is expressed simultaneously. May the following papers be a further attempt to show the food-consumer—the final link of the food-chain—the seriousness of so many workers involved in such endeavours.

Summary

The situation in the field of chemical weed control is recalled, as it was in 1952—the time when the first triazine derivatives were synthesized and screened in the laboratories of J. R. Geigy S. A., Basle. Besides the inorganic salts, the old phenols, cresols and TCA, the catalogue of chemical week killers was quite limited: 2,4-D, MCPA, IPC, Crag I, Crag II, endothal, maleic hydrazide, N-naphthylphthalamic acid, newborn CMU, and aminotriazole deposed in a patent application. As to the application technique, the preemergence treatment of weeds just started to find interest. Reference to the increased pace of research in chemical weed control is made: then, about ten firms had herbicidally active compounds of their own research in the marketing or developing stage—this number approaches 40 today. In 1968 more than 100 companies applied for the patenting of chemicals for which herbicidal activity was pretended.

It is described how in 1952, leads to synthetic activities in the field of triazines might have originated—the Geigy group speculating on analogies between ureas and triazines and finding only poor herbicidal activity with the most logical analogue derivatives to the urea herbicides. Further variations of the structure lead to success, however.

The names, structures, physical properties, and uses of the presently marketed Geigy triazine herbicides are disclosed. Briefly, reference is made to efforts and results in the fields of biological evaluation, development of analytical methods, toxicology, metabolism in animals, plants and microorganisms, and mode of action, including side effects and soil interaction.

The help obtained by public and private institutions in this vast area of "Getting insight into the products" is recognized with gratitude. Also, the hope is expressed that, as a side effect of this symposium a more impartial attitude in the assessment of the benefits and shortcomings of agro-chemicals in general may emerge.

Résumé[*]

Historique de l'évolution des herbicides du groupe des triazines

On rapelle la situation dans le domaine de la lutte chimique contre les mauvaises herbes tellequ'elle se présentait en 1952, lorsque les

---

[*] Traduit par l'auteur.

premiers dérivés des triazines étaient synthétisés et examinés dans les
laboratoires de la maison J. R. Geigy S. A. à Bâle. Outre les sels
inorganiques, les anciens phénols, les crésols et le TCA, le catalogue
des herbicides chimiques était très limité et ne comprenait que le
2,4-D, le MCPA, l'IPC, le Crag I, le Crag II, l'Endothal, l'hydrazide
maléique, l'acide N-naphtylphtalamique, le CMU récemment découvert et l'aminotriazole qui venait d'être breveté. En ce qui concerne les
techniques d'application, le traitement en préémergence des mauvaises
herbes commençait à attirer l'intérêt. On signale le développement
rapide de la recherche dans le domaine de la lutte chimique contre
les mauvaises herbes. Au début des années 50 environ 10 entreprises
seulement développaient ou présentaient sur les marchés des composés
herbicides résultant de leur propre recherche; ce nombre s'élève
aujourd'hui à environ 40. En 1968, plus de 100 compagnies ont présenté
des demandes de brevets pour des produits chimiques prétendant
posséder une activité herbicide.

Différentes considérations auraient pu conduire les chimistes à des
synthèses dans le domaine des triazines, quant au groupe Geigy celui-ci
a spéculé sur l'analogie existant entre les urées et les triazines. Les
dérivés ressemblant le plus aus herbicides dérivant de l'urée ne présentaient cependant qu'une faible activité herbicide, mais les variations
ultérieures de la structure initiale ont conduit au succès.

Les noms, les structures, les propriétés physiques et les usages des
herbicides du groupe des triazines commercialisés actuellement par
Geigy sont énumérés et l'on résume les efforts et les résultats concernant l'évaluation biologique, l'évolution des méthodes analytiques,
les problèmes toxicologiques, le métabolisme chez l'animal, les plantes
et les microorganismes et finalement le mode d'action.

L'aide apportée par des institutions officielles et privées pour mieux
connaître les aspects multiples des produits est vivement appréciée.
On espère aussi qu'une attitude plus objective dans l'estimation des
avantages et des inconvénients des produits agrochimiques en général
résulte en tant qu'effet secondaire de ce Symposium.

Zusammenfassung*

Entwicklungsgeschichte der
Triazin-Unkrautvertilgungsmittel

Es wird die Situation auf dem Gebiete der chemischen Unkrautbekämpfung im Jahre 1952 in Erinnerung gerufen, als die ersten
Triazin-Derivate in den Laboratorien der J. R. Geigy A. G. in Basel
synthetisiert und geprüft wurden. Ausser den anorganischen Salzen,
den alten Phenolen, Kresolen und dem TCA führte der Katalog der

---

* Übersetzt vom Autor.

chemischen Mittel für die Bekämpfung von Unkräutern nur wenige
Präparate auf: 2,4-D, MCPA, IPC, Crag I, Crag II, Endothal,
Maleinsäurehydrazid, N-Napththylphthalaminsäure, das neu ent-
deckte CMU und das gerade zum Patent angemeldete Aminotriazol.
Die Vorauflaufbehandlung zeichnete sich eben als neue Anwendungs-
technik ab. Es folgt ein Hinweis auf den enormen Auftrieb, den die
Forschung auf dem Gebiete der chemischen Unkrautbekämpfungs
nahm: anfangs der Fünfzigerjahre hatten ungefähr 10 Firmen
Unkrautbekämpfungsmittel aus ihrer eigenen Forschung auf dem
Markt oder in der Entwicklung; heute gegen 40. 1968 reichten mehr
als 100 Firmen Patentanmeldungen ein für Präparate mit angeblicher
unkrautbekämpfender Wirkung.

Es werden verschiedene Arbeitshypothesen zitiert, die Anstoss zur
Prüfung von Triazinderivaten hätten geben können; Geigy vermutete
Ähnlichkeit zwischen de Harnstoffen und den Triazinen. Während die
den herbiziden Harnstoffen am nächsten verwandten Analogen nur
eine dürftige Unkrautbekämpfungswirkung aufwiesen, haben die
weiteren Abwandlungen der Struktur zum Erfolg geführt.

Die Namen, Strukturen, physikalischen Eigenschaften und An-
wendungsbereiche der sich gegenwärtig auf dem Markte befindlichen
Geigy-Triazin-Unkrautbekämpfungsmittel werden erläutert. Des weiteren wird kurz auf die Anstrengungen und Resultate auf den Sektoren
der biologischen Auswertung, der Entwicklung analytischer Methoden,
der Abklärung der Toxikologie, des Stoffwechsels in Tieren, Pflanzen
und Mikroorganismen, der Wirkungsweise hingewiesen.

Die Hilfe, die durch öffentliche und private Institutionen auf
diesem Gebiete zur besseren Kenntnis der vielfältigen Aspekte der
Produkte geleistet wurde und noch immer geleistet wird, wird dankbar
anerkannt. Ferner wird der Hoffnung Ausdruck gegeben, dass dieses
Symposium zu einer im allgemeinen unvoreingenommeneren Haltung
in der Abwägung von Vorteilen und Unzulänglichkeiten der Agro-
chemikalien beiträgt.

References

DELLEY, K., K. FRIEDRICH, B. KARLHUBER, G. SZEKELY, and K. STAMMBACH:
    The identification and determination of various triazine herbicides in bio-
    logical materials. Z. für analyt. Chem. **228**, 23 (1967).
EBERT, E., and P. W. MUELLER: Aspects biochimiques des herbicides à base de
    triazines. Experientia **24**, 1 (1968).
GYSIN, H., and E. KNUESLI: Chemistry and herbicidal properties of triazine
    derivatives. Adv. Pest Control Research **3**, 289 (1960).
KNUESLI, E., D. BERRER, G. DUPUIS, and H. O. ESSER: s-Triazines. In P. C.
    Kearney and D. D. Kaufman (eds.): The degradation of triazine herbicides.
    New York: Dekker (1969).

Use and performance of triazine herbicides on major crops and major weeds throughout the world

By

A. GAST*

When in 1952 our test work started with triazines as possible active ingredients for weedkillers, nobody would have thought that the way was thus paved for a development which could finally be a reason to organize a symposium at Riverside in 1969.

In December 1952 we had found a compound which seemed to be interesting enough to be followed further, namely G25804 (chlorazine). A first communication was published by GAST *et al.* in 1955 and, at practically the same time, ANTOGNINI and DAY (1955) read a paper on "Experimental Compound 444 as a Pre- and Post-emergence Herbicide" at the Southern Weed Control Conference at St. Petersburg, Florida. This compound was reported to be tolerated by cotton, sweet corn, and snap-beans at a rate of 12 lb./acre applied pre-emergent; eight to 12 lb./acre produced a good weed control. A next publication by GAST *et al.* (1956) mentioned among other compounds, G27692, (simazine) and G27901 (trietazine). In this publication the high tolerance was especially mentioned.

Table I provides a good comparison between these early triazines and shows distinctly the superiority of G27692 over G25804. Intensive field testing and comparing experiments between several triazines proved clearly that G27692 was the most interesting compound. The long-lasting herbicidal effect and the extremely high selectivity made our decision easy to concentrate our efforts on it. For a certain period G27901 took some of our interest too, as in former experiments this compound seemed to be more selective than G27692 on potatoes and tobacco. After some years of intensive field testing in the above crops, we came to the conclusion to drop it, as the safety margin was not high enough at herbicidally-active dosage rates. Trietazine has, however, found a very limited and quite specific use in chrysanthemums in Japan. The observation that deep-rooted perennials were insufficiently controlled by G27692 was an indication for a strong adsorption of the

---

* *J. R. Geigy S.A.*, Basle.

11

Table I. *Total fresh weight of weeds after*
103 days (Gast *et al.* 1956)

Product	kg. a.i./ha	Fresh weight (g.)
G27692	1	360
	2.5	45
	5	120
G27901	1	4870
	2.5	3490
	5	1680
G28279	1	5525
	2.5	5370
	5	2480
G25804	1	4643
	2.5	4545
	5	2390
CMU	1	4860
	2.5	3265
	5	1105
Untreated I	—	4555
II	—	5040
III	—	4114

chemical in the superficial soil layers and a hint for further testing in deep-rooting crops.

Thus the important lines for further testing of G27692 were already given:

1. Based on the long residual effect, tests as an industrial herbicide were indicated.
2. The outstanding tolerance of corn had to be investigated and tests about the possibilities in practical use in this crop were of predominant importance.
3. A large amount of practical tests in fruit orchards, vineyards, etc. was indicated.
4. Considering the long-lasting effect and its positive and negative aspects for practical use, research on the influence of soil composition on the activity was one of the main concerns in our early work.

From the group of the chlorotriazines simazine (G27692), atrazine (G30027) and propazine (G30028) met with special consideration in corn, all these compounds showing an extraordinary high degree of tolerance to the crop. When atrazine became triazine number one in corn finally, we had to consider that simazine as well as propazine have some practical disadvantages compared with atrazine. Both compounds are much less soluble in water and depend therefore to a high degree on sufficient rainfall for the development of their herbicidal effect; they have no activity on the leaves of weeds and work consequently as pre-emergence herbicides only. The herbicidal effect of the compounds is therefore highly dependent on soil conditions. Atrazine, on the other hand, shows a tendency to create selections of relatively tolerant grasses, such as *Panicum*, *Setaria*, and similar species. In this respect propazine behaves similarly to atrazine. The slightly higher tolerance to the afore mentioned grasses is sufficient as to allow the practical use of atrazine and propazine in sorghum.

Propazine was also used in a limited extent in umbelliferous crops, *i.e.*, celery and carrots. These plants show a distinctly higher tolerance to propazine than to atrazine or simazine (GAST 1959) In perennial woody crops, as grapevines, apples, pears and other fruit trees, forest and ornamental nurseries, and reforestation, simazine and atrazine are recommended herbicides. Whether preference should be given to simazine or atrazine depends upon several factors: susceptibility of the crop, soil, meteorological conditions, and weeds to be controlled. In general atrazine shows a lower safety margin in these crops,. this being mainly due to the possibility of leaching in deeper soil layers.

Continuous use of simazine in fruit orchards or vineyards may cause selections of resistant weed vegetations, mainly consisting of *Cirsium*, *Convolvulus*, *Potentilla*, and *Ranunculus*. Where this takes place, alternative treatments with the leaf- and soil active atrazine should be applied in order to interrupt or stop such a dangerous development. In tropical countries, with good rainfall conditions, simazine is a leading product for tea, rubber, coffee, and sisal.

The group of methoxytriazines is characterized by a relatively high water solubility and leaf and soil activity and, on the other hand, very reduced selectivities. Consequently these compounds, especially prometone and atratone, are used in practice as industrial weed killers and prometone, in special recommendations, as a brush-killer. It is of interest that also compounds of such aggressive properties show signs of selectivities. Prometone, for instance, selects *Galium aparine* at dosage rates which are able to kill trees, and it is typical for methoxytriazine plots that after a relatively short time a relict vegetation of *Oxalis* species may develop.

The highest variability in practical use is shown by the methylthiotriazines. Compared with chloro- or methoxytriazines, their activity in the soil lasts during a relatively short period under temperate climatic conditions, a fact which makes it possible to apply them to less tolerant crops. In general these compounds prove to be very active on the leaves, a fact which makes applications on just emerged weeds especially effective. Some of the methylthiocompounds are quite specific in their phytotoxicity or their tolerance, respectively. Good examples are shown by G36393 and prometryne, the former being highly effective against

Matricaria, whereas the latter is more active against *Vicia*. According to the region concerned, *Matricaria* and *Vicia* are, however, important weeds in small grains.

G36393, alone or in combination with simazine, applied post-emergent, safely controls grasses such as *Alopecurus myosuroides*, and *Apera spica venti* in winter-sown small grains.

GS14260 is the first triazine compound which can be applied pre-emergent in winter cereals to fight against grasses and annual broadleaves.

G34360 (desmetryne) takes advantage of a gap in the spectrum of activity of the methylthiotriazines, most of them being not very active against cruciferous plants. Desmetryne may thus be applied in kale and other cruciferous crops.

Reviewing the recommendations for methylthiocompounds, one finds prometryne to be the most polyvalent compound which, depending on local circumstances, can be applied in cotton, peas, beans, soybeans, sunflower, leek, onions, parsnip, artichokes, pepper, rice, carrots, celery, broad beans, and peanuts. In combination with simazine, we find it in potatoes or, as a postdirected treatment combined with atrazine, against atrazine-tolerant grasses in corn.

Ametryne is, due to its high activity, in practical use only in sugar-cane and pineapples.

A problem which may arise with methylthiotriazines, particularly with prometryne or desmetryne, is the danger of increased phytotoxicity under high temperature conditions, due to the relatively high vapour pressures of these compounds, compared with other triazine herbicides. Therefore, special attention must be given to the evaporation of warm soil surfaces, in warm climates (GAST 1962).

Simetryne, a relatively old compound, is again tested in paddy rice, as it seems to be better tolerated than prometryne.

So far, we did not deal in particular with the place of triazines in industrial weed control. The long-lasting effect of the chloro- or methoxytriazines is a property which of course is extremely useful for industrial applications. The already mentioned shortcoming of creating selections of resistant relict vegetation may be reduced or overcome with combinations of quite a lot of herbicides. Among the most important components for combined products, aminotriazole, 2,4-D, 2,4,5-T, or methylthiotriazines have to be mentioned.

A major responsible factor for the activity of triazines—the soil composition—became evident in our very first experiments carried out in this direction in 1956. In a soil which was very rich in organic matter (40 percent Witzwil), the activity was very low, as well with respect to the initial effect as with the period of herbicidal effect. We assumed then, that adsorption and microbial breakdown are responsible for this phenomenon. Another type of soil containing high amounts of clay but no organic matter reduced the persistence, too. To similar conclusions came BURSCHEL in 1961.

The increasing practical use of simazine and, later on, of other triazines, and thereby the increasing amount of gathered information, either information which confirmed our first findings, or information which did not fit into our picture, confirmed that the interactions between soil and our herbicides are of dominant importance. More knowledge about these interactions is necessary as well in order to give the farmer a correct recommendation of the products and to obtain good performance, as well as to explain damage cases.

Summary

Of the triazine herbicide derivatives synthesized since 1952, three major classifications have originated: the chloro-, methoxy-, and methylthiotriazines.

The first chlorotriazines that exhibited satisfactory herbicidal activity were chlorazine (G25804), simazine (G27692), and trietazine (G27901). A table is presented comparing the activity of these herbicides. Extensive testing clearly indicated the superiority of simazine. The long lasting herbicidal effect and high crop selectivity of simazine were reasons enough to concentrate research efforts on this triazine. Emphasis was placed on the development of simazine as a herbicide for corn, perennial fruit crops, and industrial weed control.

A later group of chlorotriazines, atrazine (G30027) and propazine (G30028), were given special consideration, along with simazine, as a corn herbicide. Because of certain advantages, namely greater water solubility and post-emergence activity, atrazine became the prime candidate for corn. Simazine found a significant place in perennial crops and propazine in sorghum.

The methoxytriazines are characterized by relatively high water solubility, foliar activity, and low crop selectivity. Consequently, two methoxytriazines, atratone (G32293) and prometone (G31435) are used as industrial herbicides.

The group of triazine derivatives showing the greatest amount of variability are the methylthiotriazines. Compared to the other triazines, the methylthiotriazines have shorter periods of soil activity and can be used in a greater variety of crops. They also are very active post-emergence herbicides. Numerous methylthiotriazines have been developed and usage ranges from crucifer crops to pineapples. Of these herbicides, prometryne (G34161) is the most versatile. Combinations of the methylthiotriazines with simazine and atrazine are used on a commercial basis. G36393 is combined with simazine for post-emergence weed control in winter-sown cereals. Prometryne is combined with simazine for potato weed control and with atrazine for corn.

The wide commercial usage of the triazine herbicides is influenced by their variability in activity. It is evident that soil interaction plays a dominant role in determining this activity.

Résumé*

Revue des usages et des effets à l'échelon mondial des herbicides du groupe des triazines sur les principales cultures et mauvaises herbes.

Trois groupes principaux de triazines: les chloro-, méthoxy- et méthylthiotriazines ont trouvé leur origine parmi les herbicides dérivés des triazines synthétisés à partir de 1952.

Les premières chlorotriazines ayant manifesté une activité herbicide satisfaisante ont été la chlorazine (G 25804), la simazine (G 27692) et la triétazine (G 27901). Un tableau comparatif concernant l'activité de ces herbicides est présenté. Des essais extensifs ont montré la supériorité de la simazine. L'effet herbicide persistant et la forte sélectivité de ce produit à l'égard des cultures motivaient une concentration des travaux de recherches sur cette triazine. L'accent fut placé sur la mise en valeur de l'effet herbicide du produit dans le cas du maïs, des cultures fruitières et dans la lutte contre les mauvaises herbes à l'échelon industriel.

Une attention spéciale a été accordée à l'atrazine (G 30027) et à la propazine (G 30028) faisant partie d'un groupe plus récent de chlorotriazines, ainsi qu'à la simazine comme herbicide pour le maïs. En raison de certains avantages, en particulier de la plus grande solubilité dans l'eau et l'activité en post-émergence, l'atrazine devint l'herbicide le plus prometteur pour le maïs. La simazine trouva une place de choix dans les cultures fruitières et la propazine pour le sorgho.

Les méthoxytriazines sont caractérisées par une solubilité relativement élevée dans l'eau, une activité sur le feuillage et une faible sélectivité à l'égard des cultures. Pour cette raison, deux méthoxytriazines, l'atratone (G 32293) et la prométone (G 31435) sont utilisées comme herbicides industriels.

Le groupe des triazines présentant la plus grande variété est celui des méthylthiotriazines. Comparées aux autres triazines, les méthylthiotriazines ont une durée d'activité plus courte dans le sol et peuvent être utilisées sur une plus large gamme de cultures. Elles sont très actives comme herbicides en post-émergence. De nombreuses méthylthiotriazines ont été synthétisées et leurs usages couvrent de nombreuses applications, des crucifères aux ananas. Parmi ces herbicides, la prométryne (G 34161) est la plus polyvalente. Des combinaisons de méthylthiotriazines avec la simazine et l'atrazine sont utilisées commercialement. Le composé G 36393 est combiné à la simazine pour

---

* Traduit par l'auteur.

la lutte en postémergence contre les mauvaises herbes dans les céréales
semées l'hiver. La prométryne est combinée à la simazine pour la
lutte contre les mauvaises herbes dans les pommes de terre et à
l'atrazine pour le maïs. L'utilisation commerciale très vaste des herbicides du groupe des triazines est influencée par la variabilité de leur
activité. Il est évident que le sol joue un rôle dominant dans la détermination de cette activité.

Zusammenfassung*

Uebersicht über Gebrauch und Wirkung von Triazinherbiziden in den wichtigsten Kulturen

Unter den herbiziden Triazinverbindungen, die seit 1952 untersucht worden sind, lassen sich drei Hauptgruppen unterscheiden:
Chlor-, Methoxy- und Methylthiotriazine.

Die ersten geprüften Chlortriazine, die eine ausreichende herbizide
Wirkung aufwiesen, waren Chlorazin (G 25804), Simazin (G 27692)
und Trietazin (G 27901). Eine Tabelle ermöglicht den Vergleich der
Wirkung dieser Verbindungen. Eine hervorragende Wirkungsdauer
und herausstechende Selektivitäten auf Kulturpflanzen waren die
Gründe für die intensive Versuchs- und Entwicklungsarbeit für
Simazin als Herbizid in Mais, Obstkulturen und als Industrieherbizid.

Atrazin (G 30027) und Propazin (G 30028) wurden im Vergleich
zu Simazin speziell in Mais intensiv untersucht. Gewisse Vorteile von
Atrazin, namentlich eine höhere Wasserlöslichkeit und Blattaktivität
gaben den Ausschlag, dass diese Verbindung vor Simazin zum wichtigsten Herbizid in Mais wurde. Propazin fand Einsatzmöglichkeiten
in Sorghum und Umbelliferenkulturen (z.B. Sellerie und Karotten).

Methoxytriazine sind ausgezeichnet durch relativ hohe Wasser-
löslichkeiten, hohe Blattaktivität und geringe Selektivitäten, so finden
Atraton, (G 32293) und Prometon (G 31435) Verwendung als
Industrieherbizide.

Die Gruppe der Methylthiotriazine zeigt die höchste Variabilität
in ihren Einsatzmöglichkeiten; sie weisen relativ kurze Wirkungs-
zeiten auf und wirken via Blatt und Boden.

Während Prometryn in einem weiten Sortiment von Kulturpflanzen
verwendnet werden kann, ist der Gebrauch von Ametryn auf Ananas
und Zuckerrohr beschränkt. G 36393 allein oder in Kombination mit
Simazin wird in Getreide, speziell gegen grasartige Unkräuter eingesetzt. Prometryn dient mit Atrazin kombiniert zur Bekämpfung von
Gräsern in Mais, während eine Kombination mit Simazin als Kartoffel-
herbizid Verwendung findet. GS14260 ist das erste Triazinherbizid,
dam im Vorauflaufverfahren im Getreidebau eingesetzt werden kann.

Der praktische Einsatz der Triazinherbizide wird weitgehend von
den Bodenqualitäten beeinflusst.

---

*Übersetzt vom Autor.

References

ANTOGNINI, J., and B. E. DAY: Experimental compound 444 as a pre- and post-emergence herbicide. S. Weed Control Conf., St. Petersburg, Fla. (1955).

BURSCHEL, P.: Untersuchungen über das Verhalten von Simazin im Boden. Weed Research 1, 131 (1961).

GAST, A.: Neuere Triazine. Mededelingen van de Landbouwhogeschool en de Opzoekingsstations van de Staat te Gent **XXIV**, 857 (1959).

—— Beiträge zur Kenntnis des Verhaltens von Triazinen im Boden. Mededelingen van de Landbouwhogeschool en de Opzoekingsstations van de Staat te Gent **XXVII**, 1252 (1962).

——, E. KNUESLI, and H. GYSIN: Ueber Pflanzenwachstumsregulatoren. 1. Mitteilung: Chlorazin, eine phytotoxisch wirksame Substanz. Experientia **11**, 107 (1955).

—— —— —— Ueber Pflanzenwachstumsregulatoren. 2. Mitteilung: Ueber weitere phytotoxische Triazine. Experientia **12**, 146 (1956).

Introduction to triazine-soil interactions

By
Peter Dubach[*]

Contents

I. Introduction

The more man's technology develops, the more important it becomes to study its influence on the environment: on the soil, water, and air, and on the flora and fauna. This is mainly a matter of public concern, questioning whether or not human habitats will not be changed too much by technology to allow a healthy and worthwhile living. On the other hand, technology itself has to cope with the elements of the environment and their influence on technological installations or operations therefore must be considered, too. Indeed, today's studies on interactions between environment and technology constitute a substantial part of the total worldwide research efforts.

The technology of controlling weeds by chemical means instead of by expensive and tiresome mechanical and manual methods is of rather recent origin, especially the sophisticated use of the preventive application of herbicides on the still weed-free soil. From the very beginning of the use of such pre-emergence herbicides it was recognized that the nature of the soil greatly influences the general and selective phytocidal activity of these chemicals. Recognizing also the fact that the soil, as

---

[*] Research Department, Agro-Chemicals, *J. R. Geigy S.A.*, Basle.

the producer of man's food, is a very vital element of our environment, interactions between herbicides and soil have immediately found considerable worldwide attention.

After about 15 years of practical experience with the triazines, which are among the first and most widely used soil-applied herbicides, a symposium specifically oriented to review the interactions of this class of compound with the soil is certainly justified, although many of the problems are not yet solved. However, research in the field of triazine-soil interaction has now reached a point where we can sit together and consolidate our findings, formulate more clearly the still open questions, and discuss appropriate ways and means to make progress in this field.

A prerequisite in the study of any interaction is a thorough knowledge of the specific nature of the participants, *i.e.*, the triazine herbicides on one side and the soil on the other side.

II. The triazines

The triazines are particularly suited for study due to the fact that the biologically active derivatives can vary quite extensively in the nature of their substitution. They are weakly basic herbicides and their physicochemical properties are very variable, even within the restricted, most interesting group of the marketed derivatives:

Solubility in water from five p.p.m. (simazine) to 1,800 p.p.m. (atratone) pK from 1.65 (simazine) to 4.20 (atratone)

It is hoped that as a result of this symposium the physico-chemical properties of the triazines may be ranked more clearly in relation to their importance in regulating soil interactions.

Some of the physico-chemical properties along with chemical structure, main use, trademark, common name, and code number are listed in Reference Tables I and II of the Foreword. The wide range of derivatives offers a fascinating opportunity to study extensively relationships between structure and behavior in the soil. In this connection it should perhaps be stressed that soil interaction studies with biologically less interesting derivatives are necessary and fruitful if these derivatives are members of a structurally related series of compounds. It may lead to a better understanding of the interactions, which again would help to make a better choice amongst similarly interesting derivatives. Having thus discussed the triazines in a few words, we should now focus briefly on the other partner of the interplay.

III. The soil

Using different soils, it has been shown that, under controlled greenhouse conditions, herbicide activity was reasonably well correlated with soil characteristics such as organic matter content and general adsorptive behavior. However, when herbicide activity with these soils was investigated in the field, the correlation was not good and it was concluded that the influence of climate was as important or more important than that of soil type (HANCE et al. 1968).

Naturally we all know how dramatically the selective and nonselective phytocidal activity of herbicides applied to the same soil may change following an increase or decrease in temperature or precipitation. Not only the absolute level of precipitation, temperature, or radiation over a set time period is of importance, but very often the sequence of changes of these climatic factors within that period are of even greater consequence. Significant differences are noted if the herbicide is applied on the soil in a dry spell followed by a rainy period or vice-versa. We all know about these climatic influences, but we probably agree that we are as yet far away from assessing them in a quantitative manner. Whenever we try to extrapolate greenhouse results into the field, this has to be done with due care: we must be aware of the soil in a more embracing manner, recognizing it as a dynamic, complex, multiphase system, as the interface where earth's litho-, aqua-, atmo- and biosphere intermingle and interact. The soil contains numerous organic, inorganic, and complex constituents mostly of poorly known chemical structure. Quoting the late Professor DEUEL (1960) "the chemical architecture of the soil has to be considered; the soil is heterogenous, it contains gaseous, liquid, solid and living phases, and most important— it is very rich in surfaces. In the soil there are areas of a high degree of disorder as in gases and liquids, and areas of a high degree of order as in crystals, but a large and characteristic part of the soil is something intermediate. The chemist encounters still more difficulties, if he is studying the soil as a dynamic system. Mechanical translations and chemical transformations are continuously taking place in the soil. The inorganic compounds are derived mainly from the parent material which is present from the very beginning of soil formation; whereas the organic compounds are derived from biotic material which is being continuously replenished. The soil is a special kind of chemical as well as biochemical laboratory in which inorganic and organic compounds are synthesized, both inside and outside of organisms."

IV. Triazine-soil interactions

After having summarized very briefly some of the properties of the partners of the interaction let us try to describe this interaction play itself. We may classify the interactions taking place between the triazines and the soil into three major groups of processes:

1. Distribution of the triazines into the various phases of the open soil system, eventually beyond the area which is important for plant growth.

2. Chemical and biochemical transformations of the triazines in the various phases of the soil system.
3. Transformation of the various phases of the soil system due to the biological activity of the triazines.

After having briefly described the soil system and the principal groups of interactions, it is easy to understand that the triazines are less active in the soil than in a simple system like a hydroponic solution; this observed drop in activity is a very important feature of herbicide behavior, usually called inactivation. Let's now discuss these three principal groups of interactions.

a) Distribution of the triazines in the open soil system

We must expect that the rate of distribution of triazines depends on the degree of its dispersion in the spraying mixture; this dispersion will be quite heterogenous. That part of the triazine which is molecularly or nearly molecularly dispersed in the spraying mixture distributes easily and rapidly into the liquid soil phase and begins from here to redistribute into the gaseous phase and the surfaces of the solid soil phase. The non-molecularly dispersed solid, or undissolved main part of the herbicide, less readily distributes molecularly into the liquid (dissolution) or gaseous phase (sublimation); it then redistributes into the solid soil phase, too. A direct molecular distribution of the solid triazine onto the surfaces of the solid soil phase, theoretically at least, seems not impossible. As the practical application rates of triazines are extremely low in relation to the size of the available surfaces, distribution onto these surfaces, i.e., adsorption, is expected to lead the whole process to a complete disappearance of solid triazine, even of extremely sparingly soluble derivatives. Just to what extent the dissolution process might thus be speeded up would be very interesting to know. Eventually, and perhaps more quickly than one would assume, all the applied herbicide must be expected to be either in an adsorbed, dissolved, or gaseous state.

However, this distribution process in the soil system never reaches equilibrium, because: a) the system is an open, and dynamic one as described before and it is exposed to solar heat irradiation and to the force of gravity. There will always be locations with very distinct differences in concentration of herbicide. b) The herbicide distributes continuously into the biophase, i.e., is taken up by plants and microorganisms. c) The herbicide is exposed to chemical and biochemical transformation or degradation reactions.

Principally the shiftings of the herbicide from one soil phase into the other are reversible. However, the equilibrium (e.g., for the adsorption—desorption process) might be very much on one side. Also, this equilibrium must be expected to be quite variable with respect to the involvement of different constituents of the solid soil phase, even within restricted groups of related constituents such as sesquioxides, clays, humic substances, proteins, and polysaccharides, these groups still being rather heterogenous.

Whenever irreversibility of the distribution process is observed, then transformation of the herbicide, cageing, or formation of covalent linkages with soil constituents must be anticipated.

Distribution of the triazine from the various soil phases, primarily from the gaseous and liquid phases, into the adjoining atmosphere, leads to a loss of active ingredient in the soil. Naturally this volatilization process is expected and has been found to be more pronounced for derivatives with less adsorptive properties and with higher vapor pressures.

As differences in the selective phytocidal activity of herbicides, especially within a group of related derivatives such as chloro-, methylthio-, or methoxytriazines, might also be a result of favorable positioning of the herbicide in the soil, its vertical distribution, or leaching is a process of utmost importance. For readily adsorbed herbicides and in soils of higher adsorptive capacity, this process will be governed by the extent of adsorption. At what level of adsorptive capacity of the soil the water solubility of the herbicide becomes more important than its adsorptivity would be very interesting to know.

We have so far not yet discussed the distribution of the herbicide into the biosphere, i.e., the uptake by plants and microorganisms, Whereas the microbial aspect is specifically covered by this symposium the aspect of uptake and metabolisms of triazines by plants is not discussed, being a topic to fill another symposium. A few words on this topic may be appropriate. We generally are ready to accept that biological activity is displayed by that part of the chemical, which is in the dissolved or gaseous state. For plant nutrients, the theory of a direct uptake from the adsorbed state by contact exchange has been put forward (JENNY and OVERSTREET 1939) and it may be good to consider it also for herbicides. From what we said earlier, we must expect the availability of adsorbed triazines to be quite variable as the adsorption onto different types of soil constituents is considered. A herbicide may experimentally be found adsorbed and yet be available to the plant (HARRIS 1966). Fundamentally, availability of the adsorbed triazine through the liquid soil phase is given by the reversibility of the adsorption-desorption process, the rate of the desorption process being important for the display of biological activity. Last but not least, let's not forget that to some degree there certainly exist differences in the ability of plants (and microorganisms) to take up chemicals, especially, perhaps, from the adsorbed state, although these differences might be smaller than one would be ready to expect for selective herbicides (GYSIN and KNÜSLI 1960).

b) Transformation or degradation processes*

It is most interesting to know the nature, the rate, and the products of these transformation reactions. We seem to have adequate proof that degradation processes take place as well in the biophase as in the abiotic phases of the soil system, the relative importance of the various phases in this respect still being at stake. Knowledge is especially lacking with what happens with the triazines in the adsorbed state. Theoretically, they may be protected from or made even more liable to chemical or biochemical attack.

The persistence of a herbicide in the critical soil zone, another important feature of herbicide behavior, is given by the rate of the distribution and degradation reactions. It should not be too short to allow for adequate length of weed control nor should it be too long to cause damage to succeeding susceptible crops, or to pollute the aquasphere or atmosphere of our environment. Due to the rather slow rate of distribution and the inherent degradability of the triazines, the danger of pollution with these herbicides is virtually non-existent. However, damage to succeeding susceptible crops can and does occasionally occur, but can be avoided by proper application and selection of products or mixtures amongst the triazine herbicides. This special topic is dealt with in a specific paper.

c) Transformation of soil system due to biological activity of triazines

The quantities of active ingredient of a triazine required for practical weed control are so small, usually much less than one g./sq. m., that a direct influence on the soil system can be excluded. However, indirect influences by the impact on the biophase could well be noticeable. The attempted attack on a narrow segment of the biophase, the weeds, by itself is bound to produce a remarkable set of side reactions, at least during the time of the activity of the herbicide. The removal of weeds means more water and nutrients for the crop, and for the microorganisms a decrease of their organic energy supply; it also means a change in the exposure of the soil to atmospheric conditions producing a change in the microclimate near the soil surface. Primarily the removal of weeds means a change in soil cultivation practices. Together with a decrease of the soils total root mass the herbicide treatment must necessarily lead to a change in soil structure, as compared to a mechanically weeded soil. In fact, agronomy has for the first time the tools in hand to carefully study whether the effects of cultivation other than removing weeds are beneficial or detrimental to soil fertility. The trends of such studies being such as to show cultivation practices *per se* to have no or even a negative influence on soil fertility (Henin 1968).

---

* See Knüsli et al. (1969).

Due to the above-described readily observable side effects of the elimination of weeds, the absolute and relative number of microorganisms will be apt to be different in a treated soil. The real questions then are: does such a change influence yields and how long, if detrimental, would such an effect last? Generally such effects with herbicides, including the triazines have not been detrimental to yield and not lasting (UPCHURCH 1966).

Under given ecological conditions the soil system is in a very stable dynamic equilibrium, e.g., it is practically impossible to have the organic matter content permanently increased by the addition of organic materials; likewise, the constant elimination of weeds will not necessarily lead to a decrease in the organic matter content of the soil (DUBACH 1968).

V. Other features of use

Having thus described in a broad and very general way soil herbicide interaction processes we can now recognize inactivation, positioning, volatilization as it may damage the crop, persistence, and impact on soil fertility as the most important features for the safe, efficient, and economical use of selective and non-selective soil-applied herbicides. These critical features of herbicide behavior in soil are intimately interrelated and resultants of the various described processes each of which is discussed extensively in the following contributions in this volume. The greatest attention, however, has been obtained by the process of adsorption as it controls all the critical features of soil-triazine interactions.

The ultimate goal of soil-herbicide interaction studies is to define and to assess the most important properties of the interacting partners in order to control the performance of herbicides applied to the soil: a) by selecting from a group of related triazines the best available herbicide for the particular use, with a hope to be able to rely much more on physico-chemical properties in the future; and b) by constantly improving application techniques; i.e., adjusting formulation, placement, time, and rate in the best possible way to the specific conditions of the environment, the soil, and climatic factors.

Summary

Investigations in the field of triazine-herbicide-soil interactions are placed within the frame of ever increasing worldwide environmental studies to understand the impact of technology on man's environment and to determine the factors which are most important for its safe, efficient and economical use.

General properties of the partners of the interaction, i.e., of the triazines and of the soil are briefly mentioned and a special point is made

that the soil is an open, dynamic, multiphase system, resulting from the
interaction of earth's litho-, aqua-, atmo-, and biosphere.

Herbicide-soil interactions are described in a general way dividing
the various simultaneous and related processes into *three major groups:*
(1) distribution of the herbicide in the soil system, (2) transformation
or degradation of the herbicide, and (3) transformation of the soil
system.

Adsorption, leaching, degradation, volatilization, and impact on the
biophase of the soil, the well-known individual processes, are mentioned
as the factors which influence the various features responsible for safe,
efficient, and economical performance of soil-applied selective and non-
selective herbicides, such as: inactivation, positioning and persistence
of the herbicide in the soil, crop damage by volatilization, and impact on
soil fertility.

The ultimate goal of triazine-soil studies is to define and to assess
the most important properties of the partners in order to control the
interactions by making the best choice within a narrow group of related
biologically active triazines, and by constantly improving application
techniques in terms of formulations, placement, timing and rates.

In this connection, an attempt is made to present a few generalizing
statements and to focus on possible controversial states of knowledge.

Résumé*

Introduction à l'étude des actions réciproques
entre le sol et les triazines

Des recherches dans le domaine des actions réciproques entre le sol
et les herbicides du groupe des triazines sont placées dans le cadre des
études mondiales toujours croissantes sur le milieu, afin de comprendre
l'incidence de la technologie sur le milieu ambiant de l'homme et de
déterminer les facteurs les plus importants pour son usage inoffensif,
efficace et économique.

Les propriétés générales des agents exercant des actions réciproques,
c-à-d des triazines et du sol sont brièvement indiqées et il est montré, en
particulier, que le sol est un système ouvert, dynamique, comprenant de
multiples phases et résultant de l'action réciproque de la lithosphère, de
l'hydrosphère, de l'atmosphère et de la biosphère.

Les actions réciproques des herbicides et du sol sont décrites de façon
générale, en séparant les différents processus simultanés et dépendents
en *trois groupes principaux:* (1) la distribution de l'herbicide dans le sol,
(2) la transformation ou la dégradation de l'herbicide, et (3) la transformation du sol.

L'adsorption, la filtration, la dégradation, la volatilisation et

---

* Traduit par S. DORMAL-VAN DEN BRUEL.

l'incidence sur la biosphère—*qui sont des processus individuels* bien connus—sont mentionnés comme étant les facteurs influençant les différentes *caractéristiques* responsables de l'innocuité, de l'efficacité et de l'usage économique des herbicides sélectifs et non sélectifs appliqués au sol, à savoir: inactivation, localisation et persistance de l'herbicide dans le sol, dégâts aux cultures par volatilisation et incidence sur la fertilité du sol.

Le but final des études "triazines-sol" est de définir et d'établir les propriétés les plus importantes des agents en présence, afin de *vérifier les actions réciproques:* 1) en faisant le meilleur choix dans un groupe restreint de triazines apparentées, biologiquement actives, et 2) en améliorant constamment les techniques d'application exprimées en formulations, placement, moment d'application et doses d'emploi.

Une tentative est faite à cet égard pour présenter quelques conclusions générales et attirer l'attention sur des états de connaissance pouvant donner lieu à des controverses.

Zusammenfassung*

Einführung in die Triazin-Boden-Interaktionen

Untersuchungen auf dem Gebiet der Interaktion zwischen Triazin-Herbiziden und dem Boden werden in den Rahmen der weltweit immer mehr zunehmenden Umwelt-Studien gestellt. Solche Studien sollen einerseits den Einfluss der Technik auf die Umwelt untersuchen, andererseits sollen sie zeigen, wie die Technik unter verschiedenster Umwelt gefahrlos, wirkungsvoll und wirtschaftlich eingesetzt werden kann.

Es werden kurz die allgemeinen Eigenschaften der Partner dieser Interaktion, d.h. der Triazine und des Bodens, erwähnt. Dabei wird hervorgehoben, dass der Boden ein offenes, dynamisches Mehrphasen-System ist und die Resultante der Wechselwirkungen zwischen der Litho-, Aqua-, Atmo- und Biosphäre der Erde darstellt.

Die Interaktionen zwischen Herbiziden und dem Boden werden in einer allgemeinen Weise beschrieben, die die verschiedenen, gleichzeitig laufenden und von einander abhängigen Prozesse in drei Hauptgruppen einteilt: (1) Verteilung des Herbizids im Bodensystem, (2) Veränderung oder Abbau des Herbizids, und (3) Veränderung des Bodensystems.

Es werden die gut bekannten, individuellen Prozesse, wie: Adsorption, Auswaschung, Abbau, Verdampfung und Einfluss auf die Biosphäre des Bodens erwähnt und als bestimmende Faktoren der für

---

* Übersetzt vom Autor.

den gefahrlosen, wirkungsvollen und wirtschaftlichen Einsatz von Bodenherbiziden wichtigen Charakteristika, wie: Inaktivierung, Lokalisierung, Wirkungsdauer, Schädigung der Kultur durch Verdampfung und Beeinflussung der Bodenfruchtbarkeit, dargestellt.

Es ist das Ziel von Triazin-Boden-Untersuchungen, die wichtigsten Eigenschaften der Partner zu erkennen und zu gewichten, um die Interaktion unter Kontrolle zu bringen: (1) Indem die beste Wahl innerhalb einer engen Gruppe verwandter biologisch wirksamer Triazine getroffen wird, und (2) indem die Anwendungstechnik (Formulierung, Dosis, Plazierung und Wahl des Applikationszeitpunktes) dauernd verbessert wird. In diesem Zusammenhang wird ein Versuch gemacht, einige allgemeine Feststellungen vorzulegen und auf mögliche gegensätzliche Kenntnissachlagen hinzuweisen.

References

DEUEL, H.: Interactions between inorganic and organic soil constituents. Trans. 7th Internat. Congress Soil Sc. **1**, 38 (1960).

DUBACH, P.: Unpublished data, *J. R. Geigy AG*, Basel. (H-Bericht No. 550) (1968).

GYSIN, H., and E. KNÜSLI: Chemistry and herbicidal properties of triazine derivatives. Adv. Pest Control Research **3**, 289 (1960).

HANCE, R. J., S. D. HOCOMBE, and J. HOLROYD: The phytotoxicity of some herbicides in field and pot experiments in relation to soil properties. Weed Research **8**, 136 (1968).

HARRIS, C. I.: Adsorption, movement, and phytotoxicity of monuron and s-triazine herbicides in soil. Weeds **14**, 6 (1966).

HENIN, M. S.: La culture sans labour. C.R. Acad. Agr. France **54**, 126 (1968).

JENNY, H., and R. OVERSTREET: Surface migration of ions and contact exchange. J. Phys. Chem. **43**, 1185 (1939).

KNÜSLI, E., D. BEERER, G. DUPUIS, and H. ESSER: s-Triazines. In P. C. Kearney and D. D. Kaufman (eds.): Degradation of Herbicides, New York: Dekker (1969).

UPCHURCH, R. P.: Behavior of herbicides in soil. Residue Reviews **16**, 46 (1966).

Factors influencing the adsorption, desorption, and movement of pesticides in soil

By

GEORGE W. BAILEY* and JOE L. WHITE**

Contents

* Southeast Water Laboratory, Federal Water Pollution Control Administration, *U.S. Department of the Interior*, Athens, Georgia.
** Department of Agronomy, *Purdue University*, Lafayette, Indiana. Journal Paper No. 3656 of *Purdue University* Agricultural Experiment Station.

I. Introduction

Seven factors are known to influence the fate and behavior of pesticides in soil systems: (1) chemical decomposition, (2) photochemical decomposition, (3) microbial decomposition, (4) volatilization, (5) movement, (6) plant or organism uptake, and (7) adsorption. The phenomenon of adsorption-desorption directly or indirectly influences the magnitude of the effect of the other six factors. Adsorption, therefore, appears to be one of the major factors affecting the interactions occurring between pesticides and soil colloids.

There have been several review articles recently concerned with the overall behavior of insecticides (Eno 1958, Edwards 1964), herbicides (Ennis 1954, Aldrich 1953, Newman and Downing 1958, Sheets and Danielson 1960, Woodford and Sagar 1960, Hartley 1961 and 1964, Upchurch 1964, Holly 1965, Sheets 1964, Kearney et al. 1965, Alexander 1966, Dustman and Stickle 1966, Kaufman 1966, Martin 1966) and soil fumigants (Newhall 1946, Taylor 1951, McBeth 1954, Dieter 1959, Goring 1962, Domsch 1964) in soil systems. These articles discuss the general role of adsorption as it influences the fate and behavior of pesticides in soil. The specific topic of adsorption and desorption of organic pesticides by soil colloids and/or clay minerals has been recently reviewed (Bailey and White 1964, Kunze 1966). The movement of pesticides in soils has been reviewed by LeGrand (1966) from a hydrological viewpoint and by Hartley (1961 and 1964) and Bailey (1966) from a physical-chemical viewpoint, with particular emphasis on the role of adsorption on pesticide movement in soils. The nature of clay-organic complexes has been treated by MacEwan (1962), Calvet (1963), and Greenland (1965).

In a previous review article (Bailey and White 1964), it was shown that such factors as soil or colloid type, physical-chemical nature of the pesticide, soil reaction, temperature, nature of the saturating cation on the colloid exchange sites, and nature of the formulation directly influence the adsorption-desorption of pesticides by soil systems. These topics and their role in adsorption will be re-examined in light of current literature. In addition, such topics as (1) mathematical models describing adsorption processes, (2) detailed examination of adsorption mechanisms, (3) the nature and role of surface forces and surface acidity in pesticide-soil colloid interactions, (4) role of spectroscopy in elucidation of pesticide-soil colloid complexes, and (5) movement of pesticides through and off of soil surfaces are treated. For sake of convenience the topics (1) adsorption and desorption of pesticides by soil colloids and (2) leaching and movement of pesticides in, through, and over soil surfaces will be treated separately.

A portion of the literature to be cited concerns the reactivity of soil constituents with organic compounds that are not currently recognized as pesticides. However, in general, pesticides are only organic compounds with known toxicological properties. Therefore, any knowledge available on the interaction of any family or type of organic compounds with soils or soil constituents may shed some light on the fundamental behavior of organics in soils and sediments. The literature for this article was reviewed through December 1969.

II. Nature and properties of soil colloids

The chemical and physical properties of soils are influenced strongly by soil constituents which have high specific surfaces or highly reactive surfaces. Since high specific surface is associated with small particle size, the colloidal fraction of the soil will be the dominant factor in influencing interactions between pesticide molecules and the soil.

The colloidal constituents of soils may be divided into the organic fraction and the mineral fraction. The humic colloid fraction has not been completely characterized, but it appears that much of the reactivity of this fraction is embodied in the fraction designated "humic acid." The mineral fraction is composed of crystalline clay minerals and crystalline and amorphous oxides and hydroxides. Table I gives the cation exchange capacity and specific surface values for selected soil constituents.

Table I. *Selected physical properties of soil constituents*
(after BAILEY and WHITE 1964)

Soil constituent	Physical property	
	Cation exchange capacity (meq./100 g.)	Surface area (sq. m./g.)
Organic matter	200 to 400	500 to 800
Vermiculite	100 to 150	600 to 800
Montmorillonite	80 to 150	600 to 800
Dioctahedral vermiculite	10 to 150	50 to 800
Illite	10 to 40	65 to 100
Chlorite	10 to 40	25 to 40
Kaolinite	3 to 15	7 to 30
Oxides and hydroxides	2 to 6	100 to 800

Humic acid has been described by VAN DIJK (1966) as being globular, polydisperse, and irregular polycondensate. Humic acids are polybasic acids with at least two kinds of acid groups, *i.e.*, carboxyl and phenolic hydroxyl groups. The cation exchange capacity of humic acid is higher than that of clay minerals, being of the order of 200 to 400 meq./ 100 g. Functional groups such as carboxyl, amino, phenolic hydroxyl, and alcoholic hydroxyl, in addition to directly affecting the adsorption of cationic and anionic pesticides by humic acid, may also provide sites for hydrogen bonding interactions with the pesticide molecules. Because of the very great complexity of humic acid and the experimental difficulties in applying spectroscopic techniques to the study of interactions between two groups of organic compounds of considerable complexity, relatively little information on the mechanism of adsorption of pesticides by organic matter is available.

As first suggested by PAULING (1930a and b) the fundamental structural units of clay minerals in the layer silicate group are (a) sheets of tetrahedra consisting of oxygen around the small Si^{4+}, Al^{3+}, and less frequently Fe^{3+} cations, and (b) sheets of octahedral groups of O and OH around the cations Al^{3+}, Mg^{2+}, Fe^{2+}, Fe^{3+}, and occasionally other ions such as Li^+, Cr^{2+}, Zn^{2+}, etc. The tetrahedral sheets consist of Si—O or Al—O tetrahedra connected at three corners in the same plane, giving a two-dimensional network of hexagonal rings in the ideal case. The fourth unlinked oxygen corners point in the same direction. The octahedral sheets consist of hydroxyl ions in two planes, above or below a plane of aluminum or magnesium ions which are octahedrally coordinated by the hydroxyls. RADOSLOVICH (1960) has shown that there may be considerable distortion in both the tetrahedral and octahedral sheets. These two layers combine in such a way that the unlinked oxygens of the tetrahedra replace two-thirds of the hydroxyls in one plane of the octahedral layer. The remaining hydroxyls in this plane are at the centers of the hexagons formed by the oxygens of the tetrahedra. These tetrahedral and octahedral sheets are firmly bound together to form layers that are very stable; forces between the layers are usually weaker as evidenced by the pronounced basal cleavage shown by the layer silicates. Larger ions such as Na^+, K^+, and Ca^{2+}, as well as layers of water molecules, may occur between the layers of certain layer silicates.

The layer lattice silicates may be divided into two main structural groups on the basis of the ratio of tetrahedral sheets to octahedral sheets in the unit layer. A further division is made on the basis of the cation content of the octahedral layer, either two cations per half unit cell (dioctahedral) or three cations per half unit cell (trioctahedral). First, the kaolin group is an example of a 1:1 structure (Fig. 1), as it is made up of one sheet of tetrahedrally coordinated cations with one sheet of octahedrally coordinated cations. The thickness of a single 1:1 layer is about 7.2A. The 1:1 layer silicate group includes kaolinite, dickite, nacrite, serpentine minerals, and halloysite. The other type of structure, which is the basic layer of micas, chlorites, pyrophyllite, talc, vermiculite, and the montmorillonite minerals is the 2:1 type (Fig. 2), for it is made up by combination of two tetrahedrally coordinated sheets of cations, one on either side of an octahedrally coordinated sheet. The thickness of a single 2:1 layer is about 9.6A. In general, the 1:1 type layer silicates are electrically neutral or possess a very low negative

Fig. 1. Geometrical model of kaolinite structure (1:1 type clay mineral) showing the manner in which tetrahedral and octahedral layers are organized

charge. The 2:1 layers often carry a negative charge due to isomorphous substitutions in which Si^{4+} in tetrahedral positions is replaced by Al^{3+}, or Mg^{2+} replaces Al^{3+} in octahedral sites. These negative charges are balanced by positively charged ions or groups of atoms which occur between successive 2:1 layers. The layer height for the minerals composed of 2:1 layers depends on the size of the positively-charged interlayer group. In the micas, K^+ ions usually balance the charge on the 2:1 layers and the thickness of a mica layer is about 10A; chlorite is essentially a 2:1 layer plus an octahedral layer of the type found in brucite, giving a layer thickness of about 14A. In the vermiculites, moderately hydrated cations such as Mg are found between 2:1 layers and the height of a vermiculite layer is about 14A. In the montmorillonite minerals the balancing cations are even more highly hydrated and the layer height depends on the specific nature of the cation and the humidity.

Because of the extensive occurrence and stability of the dioctahedral micas many of the 2:1 minerals in soils are the result of weathering actions on these micas. Under moderately intense weathering conditions these micas may be altered to vermiculite, montmorillonite, or to a chlorite-like mineral with hydroxy-aluminum interlayers.

The expanding 2:1 minerals, such as montmorillonite and vermiculite, have a high cation exchange capacity and a high surface area (see Table I); these properties give rise to coulombic forces and van der Waals forces and consequently impart a very considerable adsorption capacity. Non-expanding clay minerals, such as illite or mica, kaolinite, and chlorite, because of their low cation exchange capacity and low surface area do not have as large an adsorption capacity as montmoril-

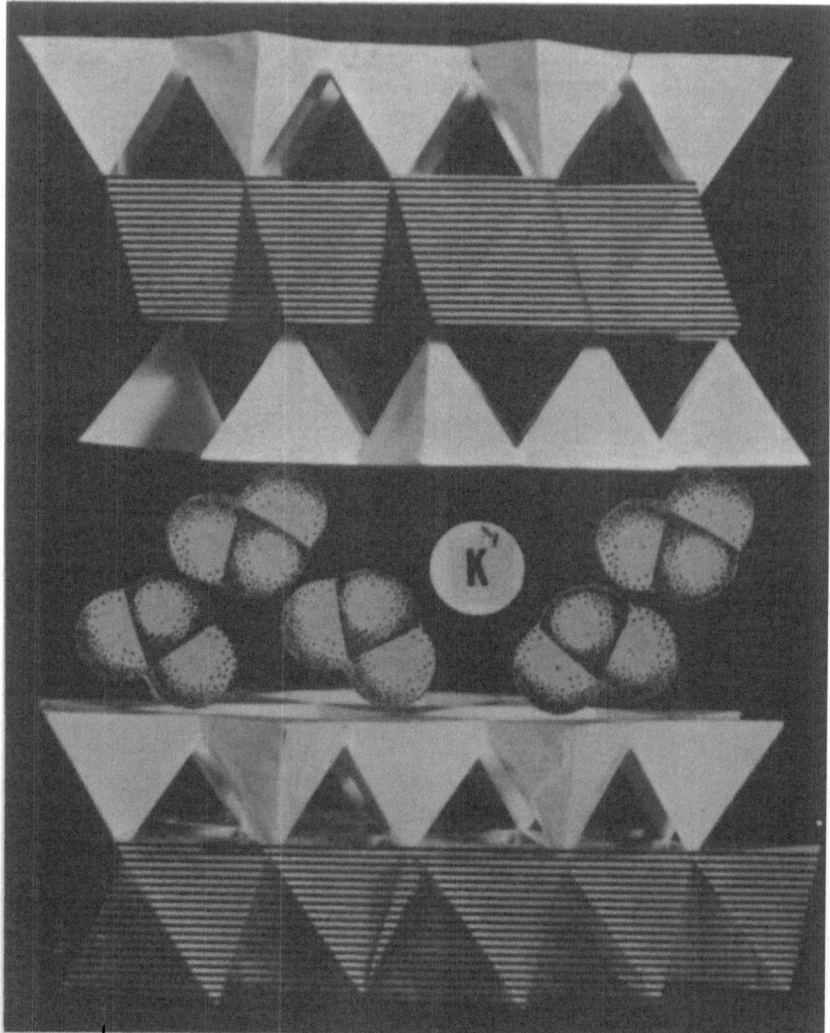

Fig. 2. Geometrical model of montmorillonite structure (2:1 type clay mineral);
a potassium ion and water molecules are shown schematically in the space
between adjacent plates of the mineral

lonite and vermiculite. The adsorption of various herbicides by montmorillonite was considerably greater than with illite or kaolinite
(FRISSEL 1961). Similar observations have been reported by many
others (HILL 1956, YUEN and HILTON 1962).

Crystalline and amorphous oxides and hydroxides of silica, iron, and
aluminum occur in soils as separate phases as well as coatings on surfaces of layer lattice silicates. Very little research on the physical and

chemical properties of these materials as they occur in soils has been done. Some of the crystalline materials may have very low surface areas, whereas some of the amorphous materials such as allophane may have large surface area (500 sq. m./g.) and be positively charged (AOMINE and OTSUKA 1968).

In summary, the interaction of pesticides with soils takes place at the surfaces of the soil constituents. In the case of the organic fraction of the soil, these interactions involve functional groups such as carboxyl, amino, and phenolic hydroxyl. The surfaces of the mineral fraction consist of oxygen and/or hydroxyl groups and are charged or uncharged. Table II summarizes the properties of the surfaces of the mineral

Table II. *Nature of surfaces of soil minerals* (after FRIPIAT 1965)

Hydroxylic surfaces	(a) electrically charged: Kaolinitic clays, amorphous alumino-silicates and mixed gels, montmorillonite and mica edges
	(b) electrically neutral: Silica gels, hydrated aluminas
Oxygen surfaces	(a) electrically charged: Montmorillonite and mica internal surfaces
	(b) electrically neutral: Dehydrated pure silica gels, aluminas, etc.

fraction. The nature of the mineral surface which is further modified as a result of the interactions within the clay surface-cation-water system system will be discussed later.

III. Adsorption models

a) *Langmuir adsorption equation*

The Langmuir adsorption equation (LANGMUIR 1916 and 1918) was initially derived for the adsorption of gases by solids, the derivation being based upon three assumptions: (1) energy of adsorption is constant and independent of the extent of surface coverage (*i.e.*, a homogeneous surface), (2) adsorption is on localized sites and there is no interaction between adsorbate molecules, and (3) maximum adsorption possible is that of a complete monomolecular layer.

With regard to the adsorption of gases by solids, the Langmuir adsorption equation may be written:

$$\theta = \frac{bp}{1 + bp}$$

where θ is the fraction of the surface covered by adsorbed molecules, p is the pressure, and b is a constant. The constant b is related to the free energy of adsorption (BRUNAUER *et al.* 1967).

The belief has become widespread that the Langmuir equation

applies only to adsorption on an energetically uniform surface where there are no interactions between adsorbed molecules. This is not true according to Brunauer *et al.* (1967). These authors state that the most important factor in the constant b is the heat of adsorption. Since most surfaces are energetically heterogeneous (this would certainly apply to the surfaces of soil components) and at low coverages (*i.e.*, small values of theta) adsorption takes place on the sites possessing the highest energies, thus a plot of the heat of adsorption against θ is ordinarily a decreasing function. Lateral interaction energy between physically-adsorbed molecules increases the heat of adsorption. The greater the number of adsorbate molecules on the surface, the greater the contribution to the heat of adsorption. Therefore, a plot of lateral interaction energies against surface coverage is an increasing function. These two opposing effects, in certain cases, compensate for each other, making the heat of adsorption approximately constant. Therefore, due to these compensating factors, the Langmuir equation is obeyed.

The Langmuir adsorption equation may also be written in terms of concentration, in the form:

$$\frac{x}{m} = \frac{K_1 K_2 C}{1 + K_2 C}$$

where x/m is the amount adsorbed per unit amount of adsorbent, K_1 and K_2 are constants for the system, and C is the equilibrium concentration in solution.

At low concentrations, it is possible to neglect the product $K_2 C$ in comparison with unity. The equation then reduces to:

$$\frac{x}{m} = K_1 C$$

In this situation, the amount adsorbed becomes directly proportional to the concentration. At high concentrations, it is impossible to neglect unity in comparison with $K_2 C$ and the equation becomes:

$$\frac{x}{m} = \frac{K_1}{K_2}$$

The amount adsorbed becomes independent of concentration at intermediate concentration, the resultant expression being of the type

$$\frac{x}{m} = KC^{1/n}$$

where $1/n$ lies between zero and one. This is the classical or Freundlich adsorption equation.

In order for the Langmuir equation to be considered applicable to a given set of data, not only a straight line plot must be obtained but upon evaluation, the constant b must be a reasonable value.

The Langmuir equation is only useful when multilayer adsorption or catalytic condensation phenomena are not involved. This generally is the case for chemisorption and physical adsorption of gases at the critical temperature and frequently for adsorption from solution.

For a more detailed and theoretical discussion of this adsorption equation, the reader is referred to other sources (ADAMSON 1960, GREGG 1961, OSIPOW 1962, HAYWARD and TRAPNELL 1964, KIPLING 1965).

b) Freundlich adsorption equation

It was observed early in adsorption studies that much of the experimental data did not obey the Langmuir isotherm. FREUNDLICH (1926) found that the adsorption isotherms for many dilute solutions could be fitted by the equation:

$$\frac{x}{m} = KC^{1/n}$$

where x is the weight of adsorbate taken up by a weight m of solid, and C is the concentration of the solution at equilibrium; K and n are constants. The above equation, with concentration replaced by pressure, was also used by FREUNDLICH and others to describe the isotherm for adsorption of gases by solids. It was originally an empirical equation without a theoretical foundation.

KIPLING (1965) describes a derivation of the Freundlich equation involving the combination of an expression for the free energy of the surface with the Gibbs adsorption equation. This derivation is based on an approximation which is only applicable to dilute solutions.

Although it has been noted previously that the Freundlich adsorption equation could be considered similar to the Langmuir adsorption equation at intermediate concentrations, the Freundlich isotherm is often obeyed by systems which do not obey the Langmuir isotherm. For this reason it should not be regarded simply as an approximate form of the Langmuir isotherm.

The Freundlich isotherm implies that the heat of adsorption decreases logarithmically as the fraction of the surface covered (θ) increases. The Langmuir isotherm equation implies that the heat of adsorption is independent of surface coverage. This is based on a model which assumes a uniform surface. In contrast to this, the Freundlich isotherm is derived on the assumption that the decrease in heat of adsorption with increasing surface concentration or coverage is due to surface heterogeneity.

The heterogeneity of the adsorbent remains the unknown factor in all investigations of adsorption from solution. The presence of this factor nullifies all efforts to treat either the Langmuir or the Freundlich equations as anything more than empirical descriptions.

c) Brunauer, Emmett, and Teller (BET) adsorption theory

The Langmuir adsorption equation was only applicable to mono-layer adsorption. The theory of multimolecular adsorption was developed by Brunauer, Emmett, and Teller (1938) and extended in applicability by Brunauer, Deming, Deming, and Teller (BDDT) (1946). This theory is an extension of the Langmuir theory for mono-layer adsorption and is widely used for determining the surface area of solids. The theory and equation will describe the five types of adsorption isotherms: (1) monolayer adsorption (Langmuir type adsorption isotherm), (2) monolayer adsorption (sigmoid-shaped), (3) multilayer adsorption, (4) monolayer plus capillary condensation, and (5) multilayer plus capillary condensation.

The following fundamental assumptions were made in derivation of the BET-BDDT equations:

1. Adsorption occurs on fixed adsorption sites.
2. The area of the total adsorbing surface is measured in terms of the cross-sectional layer of the adsorbate molecules, and that this cross-sectional area is assumed to be the same as that in the liquid substance.
3. The first layer of adsorbed molecules is held to the adsorbent by a force which is related to the average heat of adsorption of the first layer and to the temperature.
4. That the adsorptive forces that bind the first layer of molecules to the adsorbate do not reach much beyond this layer so that the heat of adsorption of the second and all successive layers will be approximately that of a liquefaction (condensation of a gas).

One form of the BET-BDDT equation is:

$$\frac{P}{V(P_0 - P)} = \frac{1}{V_m C} + \frac{C-1}{V_m C} \cdot \frac{P}{P_0}$$

where P is the equilibrium pressure at which a volume, V, of a gas is adsorbed; P_0 is the liquefaction or saturation pressure, V_m is the volume of gas adsorbed when the entire surface is covered with a complete monolayer, C is a constant related to the heat of adsorption. The constant C is equal to:

$$C = \frac{b_2 a_1}{b_1 a_2} e \frac{E_1 - E_2}{RT}$$

where $b_2 a_1 / b_1 a_2$ is the ratio of the evaporation-condensation coefficient for the adsorbed layers (often, this ratio has a value near unity), E_1 is the average heat of adsorption of the first adsorbed layer, E_2 is the heat of liquefaction, R = molar gas constant, and T is the temperature.

The equation is valid when a straight line is obtained upon plotting

$$\frac{P}{V(P_0 - P)} \quad versus \quad \frac{P}{P_0}$$

over the region in which this equation holds. From the slope and intercept values for V_m and C can be calculated. This equation has been used for studying the adsorption of pesticides with reasonably high vapor pressures (JURINAK 1957).

d) Gibbs adsorption isotherm

The general form of the Gibbs adsorption isotherm for the adsorption of a soluble substance at an interface is given by OSIPOW (1962) as:

$$\Gamma = -2 \frac{a}{RT} \left(\frac{\partial \gamma}{\partial a} \right)_T$$

where Γ is the surface excess of solute in moles/cm², γ is the surface tension in dynes/cm., a is the activity of the solute in moles, T is the absolute temperature, and R is the molar gas constant. BANGHAM (1937) showed this equation could be applied to the adsorption of a gas on a solid. The final expression which he derived was

$$\gamma_s - \gamma_{sf} = \frac{RT}{V\epsilon} \int_0^p \frac{v}{p} \, dp$$

where p is the equilibrium pressure of the gas, v is the volume of gas adsorbed/g. of solid, V is the molar volume of the gas, ϵ is the area/g. of solid, and γ_s and γ_{sf} are the surface tensions of a clean solid surface and the solid with an adsorbed gas film, respectively.

While no experimental procedure has been devised for the determination of the surface tension of a clean solid surface, the equation can be used to calculate the decrease of the free surface energy resulting from the adsorption of a gas or vapor. This can be done by plotting v/p versus p and calculating the area under the curve.

KIPLING (1965) has shown the relation between the Freundlich equation and the Gibbs adsorption equation in the following manner. If σ_0 ergs/sq. cm. is the free surface energy of the surface in contact with the pure solvent, and σ_1 that of the surface covered with a monolayer of solute, then σ, the free surface energy of the surface when a fraction θ is covered with solute is given by

$$\sigma = \sigma_0 - (\sigma_0 - \sigma_1)\theta$$

Now

$$\theta = \left(\frac{x}{m} \right) \Big/ \left(\frac{x}{m} \right)_m$$

where $\left(\dfrac{x}{m} \right)_m$ is the corresponding monolayer capacity.

Hence:

$$\sigma = \sigma_0 - \frac{(\sigma_0 - \sigma_1)\dfrac{x}{m}}{\left(\dfrac{x}{m}\right)_m}$$

Recalling the approximate expression for the Gibbs surface excess

$$\Gamma = -\frac{c}{RT} \cdot \frac{d\sigma}{dc}$$

and equating x/m with the Gibbs surface excess, Γ, for dilute solutions, then integration and replacement of the quantity

$$\frac{RT\left(\dfrac{x}{m}\right)_m}{\sigma_0 - \sigma_1} \quad \text{by} \quad \frac{1}{n}$$

gives $\dfrac{x}{m} = KC^{1/n}$ which is the usual form of the Freundlich equation.

IV. Factors influencing adsorption and desorption

a) Physico-chemical character of the adsorbent

The properties of the adsorbent which influence its behavior in interactions with the adsorbate are primarily related to the area and configuration of the surface, and to the magnitude, distribution, and intensity of the electrical field at the surface.

Since adsorption reactions involve interactions at surfaces, one of the most important properties of adsorbents is their surface area. The 1:1 minerals, primarily those of the kaolin group, because of their low cation exchange capacity and low surface area, have very limited adsorption capacities. Because virtually all adsorption is on external surfaces and the extent of these surfaces is so limited, it has not been possible to experimentally observe the interactions on these surfaces of the 1:1 clays by infrared spectroscopy without resorting to intersalation techniques.

The 2:1 minerals which can expand, such as montmorillonite and vermiculite, have very high cation exchange capacities and surface areas. These values are from ten to 100 times greater than those for kaolinite. As a result, the extent of adsorption is sufficiently great to be readily measured and in many cases the spectra of the adsorbed molecules may be observed (Russell et al. 1968 a and b).

The non-expanding 2:1 minerals, such as illite and chlorite, are intermediate in adsorption capacities between the 1:1 and expanding 2:1 clay minerals.

Colloidal particles are characterized by a surface density of charge. This may be estimated by dividing the number of charges per unit weight by the specific surface area (FRIPIAT 1965) and expressed in electrons/sq. mμ (sq. mμ = 100 sq. A). Clay minerals and amorphous materials may be classified into low- (0.1 electron/sq. mμ) and high-charged (one electron/sq. mμ) groups. Silica, alumina, and ferric or ferrous hydroxide as pure compounds have a low density of charge, while clay minerals and mixed silica, alumina, and ferric or ferrous gels have a high density of charge. The mean value of the surface densities of charge calculated for a large number of clay minerals, soil clays, and amorphous mixed silica-alumina gels range around 1.4 electrons/sq. mμ (FRIPIAT 1957). This observation is very important since it demonstrates that surfaces bearing electrical charges do not accommodate a variable or indefinite number of negative sites. It is of interest to observe that the surface density of charge of kaolinite may be higher than that of montmorillonite. In most cases, the total charge and surface area are more important factors than surface density of charge in adsorption of pesticides.

The manner in which the surface charge density can be of importance in understanding the adsorption of pesticides has been illustrated in a very elegant manner by WEED and WEBER (1968). The compounds used in this study were the divalent organic cations paraquat[1] and diquat; paraquat charge centers are about seven A apart, where the diquat charge centers are 3.5 A apart. It was observed that at low surface charge densities (montmorillonite) paraquat was preferentially adsorbed over diquat. With increase in surface charge density (vermiculites, expanded micas) the adsorption sites are more closely spaced and diquat was preferentially adsorbed due to the shorter separation of its charge centers (3.5 A).

The total charge on the surface of minerals is largely due to the extent to which isomorphous substitution occurred in the genesis of the mineral. The intensity of the electrical field at the surface may be affected by the location of the isomorphous substitutions. In the case of layer lattice silicates, isomorphous substitutions may occur in either the octahedral layer or the tetrahedral layer. It was suggested by WEAR and WHITE (1951) that the proportion of charges originating in the tetrahedral positions should be related to the fixation of potassium by 2:1 minerals due to the stronger attractive forces at the clay surface. Recent infrared data (J. L. WHITE, unpublished data) indicates that the stretching frequency for the ammonium ion adsorbed on montmorillonite (Wyoming bentonite) occurs at 3,285 cm.$^{-1}$, while the stretching frequency for ammonium in an expanded dioctahedral mica with a high proportion of tetrahedral charge occurs at 3,240 cm.$^{-1}$ This decrease in

---

[1] The chemical name of this pesticide and all others cited in the text are given in Table VI.

frequency of 45 cm.$^{-1}$ indicates that the ammonium ion is bonded more strongly at the surface of the mineral having a higher tetrahedral charge. It may be postulated that a similar effect might be observed in the case of pesticides that are cationic in nature, such as paraquat and diquat. The location of the isomorphous substitutions might also influence adsorption of pesticides indirectly through differences in degrees of dissociation of exchange cations and water molecules as a result of the strength of the attractive force at the clay surface.

The adsorbent may influence adsorption through its effect on the orientation of the adsorbate. This influence may be exerted through a combination of two factors: (1) surface configurations such as the holes in the distorted hexagonal network of surface oxygens in the 2:1 minerals—which provide positions for "keying" of molecules and impose limitations on orientation of the adsorbate, and (2) the attractive forces between adjacent 2:1 layers may be sufficiently great to cause the adsorbate to be oriented in such a manner as to give a minimum interlayer distance, or to limit adsorption to one layer.

Information concerning the orientation of molecules adsorbed on mineral surfaces has been primarily derived from x-ray diffraction studies of interlayer complexes formed between montmorillonite or vermiculite and organic compounds (GREENE-KELLY 1955a and b). GREENE-KELLY (1955a and b) found that saturated and unsaturated ring compounds formed two main types of complexes, depending on whether the rings lie parallel to (001) of montmorillonite (type-A arrangement), or perpendicular to (001) (type-B arrangement). Type-A complexes with flat aromatic rings give d_{001} of approximately 12.5 A, but with puckered, saturated rings give d_{001} of about 13.5 A. The latter value is consistent with the spacing developed by single layers of extended aliphatic chains ($d_{001} = 13.2$ to 13.6 A). Type-B complexes have d_{001} of 15.0 to 15.4 A with aromatic compounds such as aniline, methyl aniline, nitrobenzene, pyridine, isoquinoline, benzidine, and approximately 14.6 to 14.8 A with saturated ring compounds such as cyclohexanol, piperidine, and tetrahydrofuran.

GREENLAND and QUIRK (1962) found that the adsorbed 1-n-alkyl pyridinium cations normally lie along the surface of montmorillonite, adsorption of larger amounts of the bigger ions giving rise to two- and three-layer complexes. In some instances, cetyl pyridinium bromide adsorption resulted in the molecules standing up on the clay surface. BRINDLEY and THOMPSON (1966) concluded that chain molecules terminating in —OH, —COOH, and —NH₂ readily form complexes with the molecules.

WEED and WEBER (1968) report d_{001} spacings of 12.8 to 12.9 A for montmorillonite saturated with paraquat and diquat, indicating a one-layer complex with the rings of the organic cations parallel to the clay surface. CRUZ et al. (1968) have reported both one- (13.3 A) and two-layer (17.6 A) complexes of the protonated hydroxypropazine cation with montmorillonite. Infrared pleochroism studies of the 17.6 A complex indicate the cation makes an angle of about 60° with the clay surface.

b) Physico-chemical character of adsorbate

The following properties determine the role of the adsorbate in adsorption and desorption of organics by soil colloids (GREENLAND 1965, BAILEY *et al.* 1968): (1) chemical character, shape, and configuration, (2) acidity or basicity of the molecule (pKa or pKb), (3) water solubility, (4) charge distribution on the organic cation, (5) polarity, (6) molecular size, and (7) polarizability. There are many ways in which each of these properties of the adsorbate may be manifested in the overall adsorption reaction. An attempt will be made to separate each of these factors and discuss them separately, but it should be realized that more than one of these factors may be, and probably are, operating simultaneously.

1. Chemical character, shape, and configuration of the adsorbate.— Four factors determine the role of the chemical character of the adsorbate in the overall adsorption reaction: (1) nature of the functional group, (2) nature of the substituting groups, (3) position of the substituting group with respect to the functional group, and (4) presence and magnitude of unsaturation in the molecule. Since the majority of pesticides are aromatic or heterocyclic in character, the discussion will be limited to these types of compounds.

The nature of the functional group determines: (1) whether a compound is acidic, basic, or amphoteric in nature, (2) ability to undergo hydrogen bonding, and (3) ability to undergo coordinate covalent bonding (*i.e.*, to form a chelate structure with a transition metal ion). The nature of the substitution on the ring as well as the position of the substituting group enhances or lessens each of the above phenomena. The principal ways in which substituents can affect the pKa values are: (1) inductive and field effects, (2) resonance effects (particularly important in *para*-positions of aromatic molecules, (3) steric factors (bulky groups around a basic center; steric hindrance to solvation, steric inhibition of resonance in neutral molecules), (4) tautomerism (pKa different for enol *versus* keto form), (5) solvation effects, (6) internal hydrogen bonding, and (7) stereoisomerism (different pKa values for pairs of geometrical or conformation isomers). The nature and position of the substituting group may affect the ability of the molecule to undergo intermolecular hydrogen bonding (*i.e.*, hydrogen bonding to the silicate surface) by promoting intramolecular hydrogen bonding. Spectroscopic data cited by PAULING (1967) indicate that the presence of a substituting group capable of hydrogen bonding positioned *ortho* with respect to a functional group capable of hydrogen bonding resulted in strong intramolecular hydrogen bonding, while similarly, if the same substituting group were *meta* or *para* to the same functional group, intramolecular hydrogen bonding would be weak or absent.

The nature and position of the functional groups and the position and length of the alkyl portion of the molecule determines the lyophobic to lyophilic balance of the molecule. The exact nature of the balance will determine the relative affinity the molecules will have for polar and for nonpolar adsorbents.

The extent of interlamellar expansion of montmorillonite is different between ionic and polar types of molecules (GRIM 1953). With ionic species, the degree of interlamellar expansion increases with increasing number of carbon atoms, while the converse occurs in the case of neutral polar type adsorbates. For large ions, two layers are required to neutralize the charge on the clay; while in the case of polar adsorbates, the number of layers is determined by the energy of adsorption which decreases as the nonpolar part of the molecules becomes larger.

Molecular configuration and shape are important in determining if there can be rotation about a double bond which, in turn, may influence the orientation of molecules at the colloidal surface (either external or interlamellar surface). The type of molecular orientation may, in turn, determine the contribution of each type of bonding mechanism and thus the overall adsorption potential. Molecular orientation is important in determining: (1) if the geometrical considerations are such that adsorption could occur in high surface charge density clays (e.g., vermiculite), (2) orientation of the molecule with respect to the force fields emanating from the clay surface, and (3) orientation of functional groups capable of hydrogen bonding with the silicate surface.

The influence of aromatic character on adsorption through the possible formation of various resonance structures is illustrated by the work of LAILACH et al. (1968). On the basis of pKa, the authors predicted the magnitude of adsorption by montmorillonite of the following compounds should increase in the order: hypoxanthine (pKa 1.98), 5-amino-6-methyluracil [AMU] (pKa 3.28), and adenine (pKa 4.2). However, the sequence was found to be: AMU, hypoxanthine, adenine. AMU, both as a neutral molecule and in the protonated form, is a cation and is non-aromatic whereas hypoxanthine and adenine in both states are aromatic. Through resonance stabilization, the adsorption potential of the two aromatic compounds may have been lowered, which decreased the magnitude of adsorption.

BAILEY et al. (1968) in a study on the effect of functional group, nature, and position of ring substitution on the magnitude of adsorption of a wide variety of compounds, concluded that the chemical character of the molecule effects retention by colloidal system in three different ways: (1) determines if the molecule is fundamentally acidic or basic in character and the relative acidic or basic strength, (2) affects the water solubility of the molecule, and (3) determines the relative importance to (1) and (2) of van der Waals type forces.

LAMBERT (1967) has mathematically derived a fundamental relationship between adsorption by soils and chemical structure of certain classes of chemicals. This relationship is based upon extra-thermodynamic linear free energy approximations (*i.e.*, standard free energies are additive functions of a molecular structure) and uses parachor as approximate measure of a molecular volume of the chemical under consideration. This relationship utilizes an equilibrium concentration in soil; organic matter is considered the absorbing media. This development is a result of an earlier paper by this author and his collaborators (LAMBERT *et al.* 1965) describing the movement and sorption of certain types of chemicals in the soil based upon chromatographic theory. In using this relationship, it should be kept in mind what some of the assumptions are in its development: (1) that the soil organic matter is the major or dominant adsorbing media in the soil and (2) that the adsorbate or solute is an uncharged molecule. The latter assumption means that ion exchange and proton transfer adsorption mechanisms are not operative.

2. Dissociation constant.—The pKa of a compound indicates the degree of acidity or basicity that a compound will exhibit and therefore should be very important in determining both the extent of adsorption and the ease of desorption by colloidal systems.

FRISSEL (1961) studied the adsorption of such herbicide acids as 2,4-D and 2,4,5-T and noted that negative adsorption occurred until the pH of the clay-water system approached the pKa for the particular compound. Thereafter, positive adsorption commenced and increased gradually as the pH of the system decreased. Amphoteric antibiotics, such as terramycin and auromycin and the basic antibiotic, steptomycin, were studied by GOTTLIEB and coworkers (SIMINOFF and GOTTLIEB 1951, GOTTLIEB and SIMINOFF 1952, MARTIN and GOTTLIEB 1952, GOTTLIEB *et al.* 1952, MARTIN and GOTTLIEB 1955) and found to be strongly adsorbed by bentonite or illite while the neutral or acidic antibiotics were adsorbed to a lesser extent. Similarly, PINCK *et al.* (1961 a) found that the amphoteric antibiotics were adsorbed substantially more by various clay minerals than were the basic antibiotics and that the neutral or acidic antibiotics were hardly adsorbed. Such clay minerals as vermiculite, illite, or kaolinite were incapable of adsorbing either acidic or neutral antibiotics, while montmorillonite was capable of adsorbing both types. The release of such antibiotics was also affected by their degree of acidity or basicity. PINCK *et al.* (1961 b) showed that the compounds of the amphoteric group of antibiotics were released from all the clay minerals, the maximum amount being released from kaolinite and the least by montmorillonite. Not one of the four basic antibiotics was released from vermiculite or montmorillonite; one was partially released from illite and two were released to some degree from kaolinite.

The effect of pKa on the adsorption of a basic compound of widely varying chemical character by montmorillonite has been studied by BAILEY et al. (1968). These investigators studied, by means of adsorption isotherms, the adsorption of such basic compounds as s-triazines, substituted ureas, phenylcarbamates, anilides, and anilines on montmorillonite. They concluded that the major factor governing the magnitude of adsorption by different chemical families basic in character is the dissociation constant of the adsorbate. TALBERT and FLETCHALL (1965) studied the adsorption of five s-triazines by 25 different soil types, the extent of adsorption being expressed by means of distribution coefficients, K_d (amount adsorbed at equilibrium/concentration remaining in equilibrium solution). The degree of adsorption based on the averaged K_d values (in parentheses) was propazine (2.0) < atrazine (2.7) < simazine (3.7) < prometone (7.8) < prometryne (9.1). Similar results were reported by HARRIS (1966) for the same series of s-triazines using the Freundlich "n" value as a measure of adsorbability. The methylmercapto- and methoxy-derivatives are much more basic in chemical character (pKa 4.2 to 4.8) than are the chloro derivatives (pKa 1.5 to 2.0) which probably accounts for the observed differences in adsorption.

From a consideration of the pKa of the s-triazines (WEBER 1966, BAILEY and WHITE 1965, *Weed Society* 1967), the basicity of the s-triazines regardless of their groups on the 4 and 6 positions are in the order hydroxy > methoxy > methylmercapto > chloro. WEBER (1966) noted for the methoxy-analogues that basicity increased with increasing number of carbons on the amino alkyl groups. In both cases, increasing basicity can be explained in terms of the presence of more nucleophilic groups on the molecule.

The basicity of the desorbing agent appears to be important in the overall adsorption-desorption reaction, especially if the adsorbed material is basic in nature. RAMAN and MORTLAND (1968) have studied charge transfer reactions at clay mineral surfaces by infrared spectroscopy. They concluded that proton transfer from the protonated species adsorbed on the clay surface to an uncharged molecule at the clay surface was dependent upon the relative basicities of the two interacting compounds as well as the relative concentrations or activities of the reactants and the products. If the results of JOHNSON and RUMON (1965) are applicable to charge transfer reactions in clay-organic systems, then there may have to be a pKa differential between the desorbing uncharged basic molecules and that of the protonated specie (conjugate acid) on the clay surface. These workers found that in the case of a 1:1 pyridine-benzoic acid complex that a critical Δ pKa (difference between the pKa of the base and the acid) was necessary for proton transfer to occur. Proton transfer occurred between the acid and base as evidenced by the appearance of an $N—H^+$ band as seen by infrared spectroscopy when the Δ pKa was about 3.75 or greater.

3. Water solubility.—There appears to be a relationship between water solubility and extent of adsorption, but only within a family of compounds. BAILEY et al. (1968) compared the extent of adsorption of a wide variety of basic herbicides onto sodium and hydrogen montmorillonite utilizing the Freundlich K value as a criterion of adsorbability (Table III). The data show the following trends: (1) direct relationship

Table III. *Effect of water solubility on the magnitude of adsorption of various organic compounds by Na- and H-montmorillonite using the Freundlich "K" and "n" values as a criterion of adsorption* (after BAILEY et al. 1968)

Adsorbate	Water solubility (p.p.m.)	Adsorbent			
		Na-montmorillonite		H-montmorillonite	
		K	n	K	n
simetone	3,200	2,200	3.23	—[b]	—[b]
atratone	1,800	400	2.08	—[b]	—[b]
prometone	750	150	1.56	—[b]	—[b]
trietazine	20	58	1.00	—[b]	—[b]
propazine	8	18	0.89	—[b]	—[b]
atrazine	70	15	1.18	—[b]	—[b]
3-phenylurea	Sl. sol.	28[d]	2.33	330	1.18
monuron	230	24[d]	2.08	100	1.02
diuron	42	23[d]	1.08	70	0.95
fenuron	2,900	14[d]	1.00	115	1.22
aniline	34,000	130	1.11	1,300	1.24
CIPC	108	27	0.93	30	1.09
propanil	500	16	0.90	65	1.70
IPC	20–32	NA[a]	NA	30	1.56
dicryl	8–9	NA	NA	30	1.61
solan	8–9	NA	NA	58	5.88
2,4,5-T	238	—[c]	—[c]	105	2.38
2,4-D	725	—[c]	—[c]	—[c]	—[c]
phenoxyacetic acid	12,000	—[c]	—[c]	—[c]	—[c]
benzoic acid	2,700	—[c]	—[c]	—[c]	—[c]
amiben	700	—[c]	—[c]	—[c]	—[c]
picloram	430	—[c]	—[c]	37	1.28

[a] NA = not adsorbed.
[b] No value due to complete adsorption.
[c] No value, negative adsorption.
[d] Questionable conformity to Freundlich equation.

exists between water solubility and adsorbability for the s-triazines on Na-montmorillonite (with the exception of atrazine), (2) for the substituted urea, a direct relationship existed between water solubility and adsorbability for adsorption onto the hydrogen clay—no great deal of difference existed in the case of the Na clay, and (3) no such relationship existed in the case of the aniline, phenylcarbamate, anilide, and amide groupings. On the basis of these findings, BAILEY et al. (1968) concluded that within a chemical family or within an analogue series basic in chemical character, the magnitude of adsorption is related to and governed by the degree of water solubility. HARRIS and WARREN (1964) studied the adsorption of diquat, CIPC, DNBP, and atrazine on various types of inorganic and organic adsorbents. No relationship was found between water solubility and adsorption. This is what would be expected since these herbicides do not belong to the same chemical family.

The direct relationship between water solubility and adsorbability for the substituted ureas is in direct opposition to results reported in the literature. WOLF et al. (1958) found that the degree of adsorption of four substituted ureas (fenuron, monuron, diuron, and neburon) by soils to be inversely related to the order of their water solubilities. These same four substituted ureas were found to have a comparative adsorption ratio inversely related to their solubilities. Studies with Hawaiian sugar cane soils by YUEN and HILTON (1962) and HILTON and YUEN (1963) indicated that in all cases there was greater adsorption of diuron than monuron. The solubility of monuron is five times that of diuron. FRISSEL (1961) reported greater adsorption onto clay minerals of diuron over monuron from pH 8 to pH 3. However, ASHTON (1961) found that the order of lateral movement of the substituted ureas and the order of their water solubilities were the same.

TALBERT and FLETCHALL (1965) reported that there was greater adsorption of the methoxy- and methylmercapto-derivatives than there was of the chloro-s-triazine derivatives, indicating there was a direct relationship between water solubility and adsorbability. However, it should be recalled that not only does the water solubility increase between the chloro-, the methoxy-, and the methylmercapto-derivatives, but also that the methoxy- and methylmercapto-derivatives are more basic than are the chloro-derivatives. HANCE (1965 a) studied the adsorption by soils of ureas and their derivatives (urea, fenuron, methylurea, phenylurea, monuron, monolinuron, diuron, linuron, neburon, and chloroxuron) and found no relationship between adsorption and water solubility.

For a particular family of adsorbates, several factors may be interacting in determining whether or not there is a direct relationship or not between water solubility and adsorbability. These may include such factors as surface acidity and the relative polarity of the adsorbent.

Adsorption studies onto carbon (a relatively nonpolar or lipophilic adsorbent) suggests that adsorbability is inversely proportional to water solubility. LEOPOLD et al. (1960) found an inverse relationship between water solubility and the extent of adsorption of 17 various chlorinated phenoxy herbicides by activated carbon. Adsorption was also independent of pH from pH of 2.2 to 8. Increased chlorination of the ring resulted in a decrease in water solubility and an increase in degree of adsorption.

WARD and UPCHURCH (1965) studied the adsorption of 52 structurally related N-phenylcarbamates, acetanilides, and anilines from a mixed solvent system (2 percent ethanol, 98 percent water) onto three model organic adsorbents, nylon, cellulose triacetate, and cellulose. Statistical analysis indicated an inverse relationship between solubility and adsorption accounted for 60 percent of the total variation in adsorption. The authors attributed the remaining 40 percent (approximately) of the adsorption variation not accounted for by solubility to specific steric or electronic differences among the chemicals. These authors also proposed that the preferred adsorption mechanisms of the amido compounds from aqueous solution is via the adsorbate's imino hydrogen and the adsorbent's carbonyl oxygen. These differences, the authors indicated, would influence the adsorption bonding sites to a different extent and thereby affect the amount of adsorption.

WARD and HOLLY (1966) studied the adsorption by alkyl s-triazines possessing chloro-, methoxy-, and methylmercapto-groups in the 2-position from a mixed solvent system onto nylon and cellulose acetate. They found a linear relationship between the amount of adsorption and the degree of partitioning between cyclohexane (nonpolar solvent) and water (polar solvent) by the s-triazines having common substituents. Correlation between the partition coefficient was much better than between the water solubilities of the triazines as a measure of adsorbability onto model organic substrates. It was further found that the order of decreasing adsorbability was a function of the electronegativity in the 2-position, this being methylmercapto > chloro > methoxy > hydroxy when the 4- and 6-substituents are the same. These authors propose that the adsorption sites of the s-triazines are the electropositive substituents in the two position and those of the solids are the carbonyl oxygen; adsorption was proposed to result from dipole-dipole interactions. These authors also pointed out little attention has been given to adsorption as it relates to the hydrophilic-lipophilic (hydrophobic) balance of the solute. This would appear to be particularly important in the adsorption by organic matter or by natural clay-organic complexes. It was also found there was positive adsorption of all the s-triazines by nylon and cellulose acetate but none on cellulose indicating that triazines can be adsorbed by natural amido and carbonyl-type adsorbents.

4. Charge distribution on the organic cation.—The distribution of the electrostatic charge on an organic cation has a very drastic influence on the adsorption and desorption of the compound by different clay minerals. In competitive adsorption studies, paraquat was preferentially adsorbed over diquat (both are divalent organic cations) by montmorillonite, kaolinite, and by the external surfaces of vermiculite, while diquat was preferentially adsorbed over paraquat on the internal surface of vermiculite and the external surface of nonexpanded mica (Weed and Weber 1968, Philen 1968). Weed and Weber (1968) found that at a surface charge density of about 8×10^4 esu/sq. cm. adsorption of paraquat and diquat on external surfaces were nearly equal. The difference in adsorptive behavior between the two compounds was attributed to differences in the separation of the electrostatic charge. At lower surface charge densities, the more widely spaced electrostatic charge centers on paraquat (7 A) can more effectively counter the adsorbent charge, while at high surface charge densities the adsorption sites are closer together; diquat can more effectively counter these sites due to shorter separation of charge centers (3.5 A).

Both electrostatic charge distribution and surface charge density appear to be important factors influencing the desorption of organic cations. Paraquat was much more effective in displacing diquat from montmorillonite than was diquat displacing paraquat [87 percent diquat desorbed by paraquat versus 15 percent paraquat desorbed by diquat (Weed and Weber 1968)]. Similarly, paraquat was harder to displace from montmorillonite than diquat by inorganic salt solutions.

The site of adsorption also greatly influences the ability to desorb the two bipyridinium cations. Approximately 80 percent of both diquat and paraquat adsorbed on kaolinite could be displaced with $1N$ $BaCl_2$. However, a total of only five percent of each compound adsorbed onto montmorillonite was removed by similar strength salt solutions (Weber and Weed 1968). In the case of montmorillonite, the amount of paraquat and diquat removed by the barium ion was thought to have come from the exterior surfaces, that adsorbed on internal surfaces not being replaced.

5. Molecular size.—Molecular size will have an effect on the magnitude of adsorption and on the ability of organics to be adsorbed on internal surfaces by high layer charge clay minerals, e.g., vermiculite. If a large ion can cover more than one electrostatic exchange site (e.g., exchange site area for montmorillonite is approximately 80 A), the result will be an overall decrease in the adsorption capacity of the mineral (barring additional adsorption by some other mechanism such as physical adsorption). Hendricks (1941) found the large alkaloid molecules, brucine and codiene, neutralized less of the acid sites on montmorillonite than the smaller molecules, aniline or benzidine, even though the former are stronger bases than the latter. This was attributed to the covering of more than one exchange site by the large-sized molecule. If the minimum dimension of a molecule is greater than five A, it would be expected that such a molecule would not be adsorbed on internal surfaces of vermiculite.

Parachor is a measure of molecular size. McEwan (1948) stated that the number of monomolecular layers adsorbed on internal surfaces by montmorillonite in contact with an excess of the liquid appeared to increase with the function $\frac{u}{[P]}$ where u is the molecular dipole moment and $[P]$ is the parachor. WEBER et al. (1965) reported that the herbicide with the largest parachor was adsorbed to the greatest extent onto charcoal at pH 6.

Molecular size may be a factor in the movement of herbicides (OGLE and WARREN 1954, COGGINS and CRAFT 1959) and has been shown to be a prime factor in the vapor phase movement of soil fumigants (HENSON and NEX 1953, GORING 1957).

Overall gross size and chain length are important in at least three different ways: (1) with regard to Traub's rule, which states "For a non-electrolyte within a homologous series, adsorption onto nonpolar adsorbents increases as molecular weight increases.", (2) adsorption may be prevented or lessened due to an adverse molecular configuration (i.e., due to steric hindrance), and (3) the overall van der Waals energy of adsorption will increase with molecular size, especially with increase in the number of double and triple bonds. CRAFT (1961) and HANCE (1965 a) noted that both increasing the number of chloro-substituents on the phenyl ring (which lowers the water solubility of the compound) and lengthening the N-alkyl chain leads to greater adsorption of the substituted ureas by soils. GEISSBÜHLER et al. (1963) found that the addition of a second aromatic ring caused an even more pronounced effect than increasing the number of chloro-substituents or lengthening the side chain on the adsorption of substituted urea by soils.

The size of the adsorbed organic ion is important to the ease by which it can be desorbed. HENDRICKS (1941) found that larger organic ions adsorbed onto montmorillonite are difficult or impossible to replace with smaller ions. He attributed this to the presence of van der Waals forces in addition to coulombic forces.

The work of LAILACH et al. (1968) shows the influence of molecular size on adsorption by montmorillonite. Caffeine was found to be strongly adsorbed onto montmorillonite saturated with various inorganic cations at a pH greater than 7. This is unexpected since the molecule is weakly basic (pKa 0.61) and is non-aromatic in nature. Strong adsorption was attributed to the large molecular size of this compound compared to the other molecules studied. A similar explanation was given for the strong adsorption of 7-methyl- and 9-methyladenine.

Unfavorable molecular configuration with respect to interlamellar adsorption may override the enhancing effect of molecular size on adsorption. Various nucleosides were less tightly bound than their purines or bypyrimidines, the reason being attributed to the non-planar character of the nucleoside molecules. The high degree of adsorption which is expected from their large molecular size is affected by their non-planar form.

6. Polarity.—The properties, dipole moment and dielectric constant, give a measure of the influence of polarity on adsorption. Since the major portion of the surface area in expansible clays is internal in nature, the total adsorption capacity is mainly a function of interlamellar surface area. The ability of an organic to be adsorbed in the interlamellar region is of great importance. In order for a compound to be adsorbed in interlayer positions, sufficient energy must result from the complex formation to equal or exceed the forces binding the individual plates together. The exact magnitude of the forces binding the plates together is a function of the layer charge or the surface charge density, the nature of the exchangeable cation present, and the nature of the solvating agent present. Not all organics are adsorbed interlamellarly by montmorillonite and vermiculite. McEWAN (1948) found from x-ray diffraction analysis that n-heptane and n-hexane are not interlamellarly adsorbed by montmorillonite, while GREENE-KELLY (1955) found that benzene in the absence of water is also not absorbed in the interlamellar space of montmorillonite. This would indicate that the polarity of a neutral molecule is very important in determining if interlamellar adsorption can occur. The presence of polar groups on a molecule enables the clay sheets to be separated from each other and diminishes the clay minerals-exchangeable ion electric field. BARSHAD (1952) concluded that the force of attraction between the molecule adsorbed and the surface of montmorillonite is directly proportional to the dipole moment of the adsorbate and inversely proportional to the dielectric constant of the adsorbate.

The polarity of the molecule will also influence the degree to which it is solvated in solution, which, in turn, will influence the overall energy required for adsorption to occur. The relative polarity of the solute, the adsorbent, and the solvent would appear quite important in determining the extent of the overall adsorption reaction. If the adsorbate is nonpolar (e.g., DDT) and the adsorbent (e.g., clay minerals) is less polar than the solvent, then it would seem that the solute would exhibit greater preference for the clay than for the water. This is based on the old adage "like adsorbs like."

7. Polarizability.—The polarizability of a molecule is a measure of the ease by which the positive and negative charges in a molecule can be displaced with respect to each other in the presence of an electric field. Polarizability is made up of two parts, a nuclear or atom polarizability and an electron polarizability. Polarizability increases with an increase in the mobility of valence electrons, and thus polarizability would increase with increasing numbers of conjugate double bonds or with increasing distance of the valence electrons from the nucleus (shielding effect). It would be expected that where no permanent dipoles exist in molecules, that those molecules with higher strength of polarizability would be more strongly adsorbed. The magnitude of adsorption by a polar adsorbent should be influenced by the degree of polarizability of the molecule. The nature of the exchangeable cation would also affect the adsorption of a polarizable molecule; the greater the polarizing power of the cation, the greater the extent of adsorption. The polarizing power of an inorganic cation usually increases with valence charge and is inversely related to the size of the cation.

c) Soil reaction

This property of the clay-water system influences the properties of both the adsorbent and adsorbate. The pH of the soil solution will determine the degree of dissociation or association of adsorbate, the exact extent being a function of the actual value of the pKa. Therefore, whether a compound is present in the molecular, the cationic, or anionic form can affect the extent and magnitude of adsorption and the strength by which it is held, since the energy of adsorption may be vastly different between dissociated and the associated form. FRISSEL and BOLT (1962) found that the adsorption of herbicides of widely different molecular structures increased as the pH was decreased, the pH where maximum or near minimum adsorption occurred being a function of the particular compound and the adsorbent. BAILEY et al. (1968) found that regardless of the chemical character of the adsorbate (Table III), adsorption occurred to the greatest extent on the highly acid hydrogen-montmorillonite (pH 3.35) compared to the near neutral sodium-montmorillonite (pH 6.8). The same authors conclude that the magnitude of adsorption of organic compounds with widely different chemical character is governed by three factors: (1) pH of the clay system, (2) water solubility, and (3) the dissociation constant of the adsorbate. These workers also found that the adsorption of acidic-type compounds was dependent upon the pH of the suspension, while the adsorption of a basic compound was dependent upon the surface acidity. BAILEY et al. (1968) found negative adsorption for 2,4,5-T, 2,4-D, benzoic acid, amiben, and picloram when the adsorbent was sodium montmorillonite (pH 6.8). In the case of the hydrogen-montmorillonite (pH 3.35), picloram and 2,4,5-T were positively adsorbed, phenoxyacid exhibited both positive and negative adsorption, while 2,4-D, benzoic acid, and amiben either were not adsorbed or were negatively adsorbed. FRISSEL and BOLT (1962) reported positive adsorption of 2,4-D at pH's below 4, but negative adsorption by both montmorillonite and illite above pH 4. There appears to be slightly more driving force (i.e., a more highly negative partial molar free energy) for the adsorption of picloram than for phenoxyacetic acid on hydrogen-montmorillonite (BAILEY et al. 1968). BAILEY et al. (1968) reported a change in the pH of the supernatant liquid on the adsorption of adsorbates of widely-different chemical character. For all the chloro-s-triazine analogues, the phenylcarbamates, aniline, amides, anilides, phenylalkanoic acids, and benzoic acids, the pH of the supernatant following the adsorption reaction with sodium-montmorillonite was lower than that of the initial solution. The pH was higher for the hydrogen-montmorillonite systems. The above trend was reversed for atratone, prometone, and fenuron. In the case of simatone and picloram, there was no appreciable increased pH of the solution. In general, the magnitude of the pH change was concentration-dependent—the higher the initial concentration of the organic in solution, the greater the pH change. This suggests that during the reaction of the above adsorbates with sodium-montmorillonite that protons are released during the adsorption reaction, which lowers the pH of the system; while in the case of the reaction with the hydrogen-montmorillonite, one of two things is occurring—protons either are being removed from the system making it more alkaline, or hydroxyl groups are being generated during the adsorption reaction making the system more alkaline (*i.e.*, increasing the pH).

BAILEY *et al.* (1968) studied the effect of pH on the adsorption of a wide variety of chemically basic and acidic compounds by montmorillonite and concluded that: (1) maximum retention of basic compounds would be expected to occur when the surface acidity was at least one to two pH units lower than the lowest dissociation constant of the molecule, (2) adsorption of basic compounds will be principally due to van der Waals forces where the surface acidity is more than two pH units lower than the dissociation constant, (3) positive adsorption of acidic compounds will commence when the pH of the bulk solution is approximately one to 1.5 pH units above the dissociation constant of the acid, and (4) the surface acidity of montmorillonite appears to be three to four pH units lower than the pH of the bulk solution. TALBERT and FLETCHALL (1965) found that an increase in pH resulted in the decreased adsorption of simazine and atrazine by soils. In a similar fashion, McGLAMERY and SLIFE (1965) found that atrazine adsorption onto soil and humic acid isolated from leonardite increased as pH decreased, but adsorption was only slightly affected by the parameters, temperature and concentration of adsorbate.

WEBER (1966) in an adsorption study of 13 related s-triazines by montmorillonite found that maximum adsorption of all the compounds occurred at a pH in the vicinity of the dissociation constant of each compound. A further lowering of the pH resulted in some desorption of each of the adsorbed triazines. He attributed this to the competition of hydrogen ions for exchange sites at low pH levels. This result is directly opposite that found by FRISSEL and BOLT (1962) for the adsorption of chloro-s-triazines. They found that about three to four pH units above the dissociation constant adsorption started to increase as pH decreased down to pH one; the amount of adsorption was essentially exponential in nature. WEAVER (1947) reported that the adsorption of 2,4-D on a hydrogen-saturated cation exchanger was nearly twice as great as the pH 2.5 or below as at pH 3.3. This would be expected since at a pH of one unit lower than the pKa (pKa of 2,4-D = 3.31), 90 percent of the compound would be associated, adsorption of the undissociated molecule probably would occur via hydrogen bonding to the exchange resin.

However, the work of LEOPOLD et al. (1960) and GEISSBÜHLER et al. (1963) indicates that the pH dependence of adsorption does not universally apply to all adsorbents and adsorbates. From an adsorption study of various chlorinated phenoxyacetic acid derivatives, LEOPOLD et al. (1960) found that the adsorption of 2,4-D on activated carbon was pH independent in the range of pH 2.2 to pH 8. This would suggest that adsorption is due to van der Waals rather than coulombic forces. GEISSBÜHLER et al. (1963) found that the magnitude of adsorption of chloroxuron by different soils was essentially pH independent. The phenylureas are in general very weak bases (pKa of monuron and diuron are in the order of −1 and −2) and adsorption mechanisms other than proton transfer probably are operative. KIM and WEED (1968) indicate, based on infrared spectroscopy data, that there is no protonation of such phenylureas as fenuron, monuron, cyluron, and norea by montmorillonite. These authors indicated that bonding occurred through the carbonyl group of the organic either by hydrogen bonding through water of hydration of the cation or by ion dipole bonding.

In an adsorption study of simazine with 18 soils from the southeastern United States, NEARPASS (1965) found that the extent of adsorption was not significantly correlated with soil pH (in fact, it was negatively correlated), but was highly significantly correlated with titratable acidity. This author also found that simazine penetrated to a greater depth in limed soils than in unlimed soils. NEARPASS (1967), studying the effect of the predominating cation on the adsorption of simazine and atrazine by soils, found that the adsorption of simazine and atrazine is governed largely by the hydrogen ion activity relationship which occurred between the solution and the solid phase of the soil. At equal concentrations, atrazine was less strongly adsorbed than simazine.

The adsorption of purines, pyrimidines, and their nucleosides by Na-, Li-, Ca-, and Mn-montmorillonite over a pH range of 2 to 12 was found to be pH dependent, greater adsorption occurring as the pH decreased (LAILACH et al. 1968). Only thymine, uracil, and their nucleosides were not adsorbed even at pH of 2. Adsorption was postulated to occur primarily as a cation exchange reaction (mechanism 10 or mechanism 8, Table IV) under acid conditions, however, no spectroscopic data were offered to substantiate their postulation.

d) Surface acidity

It has been recognized for some time that the activity of protons in the bulk suspension (i.e., as measured by pH) and the activity of pro-

Table IV. *Possible adsorption mechanisms[a] for the various organic com-*

Adsorbate family	Physical adsorption	R—N—H . . . . . O—Clay	>C—H . . . . . O—Clay	>C=O . . . M^{r+}—Clay	R—C—O . . . R . . HO—Clay
	1[b]	2[b]	3[b]	4[b]	5[b]
s-Triazine					
Chloro-derivatives	A	Ab	Aa(?), Ab	—	—
Methoxy-derivatives	A	Ab	Aa(?), Ab	Ab	—
Substituted Ureas	A	Ab	—	Ab	—
Aniline	A	Ab	—	—	—
Phenylcarbamates					
IPC[c]	Aa, NNa	Ab	—	Ab, NNa	Ab
CIPC	A	Ab	—	Ab	Ab
Anilide					
dicryl[c]	Ab, NNa	Ab	A	Ab, NNa	—
propanil	A	Ab	Ab	Ab	—
Amide solan[c]	Aa, NNa	Ab	Ab	Ab, NNa	—
Phenylalkanoic acids					
Phenoxyacetic					
2,4,5-T	Aa, NNa	—	Aa(?), Ab	Aa, NNa	—
2,4-D	N	—	N	N	N
Benzoic acid					
benzoic acid	N	—	—	N	—
amiben	N	N	—	N	—
Picolinic acid					
picloram[c]	Aa, NNa	Ab, NNa	—	Ab. NNa	—

[a] A = Adsorption mechanism applicable in both acidic and neutral systems.
　Aa = Adsorption mechanism applicable when acidity of the system is such that pH ≤ pK + 2.
　Ab = Adsorption mechanism applicable when acidity of system is such that pH > pK + 2.
　— = Appropriate group not present.
　N = No adsorption in either acidic or neutral systems.
　NNa = No adsorption in the Na-montmorillonite system (pH 6.8).
[b] Adsorption mechanism number.
[c] pKa values unavailable, postulated adsorption mechanism(s) based upon: occurrence of adsorp-

tons at or in close proximity to the colloidal surface (*i.e.*, the acidity in the interfacial region) may differ drastically. The term "surface acidity" as applied to soil systems is the acidity at or in close proximity to the colloidal surface and reflects the ability of the system to act both as a Bronsted acid and a Lewis acid. This is a composite term which reflects both the total number of acid sites and their relative degree of acidity.

Surface acidity is probably the most important property of the soil or colloidal system in determining the extent and nature of adsorption and desorption of basic organic compounds as well as determining if acid-catalyzed chemical degradation occurs.

The nature and properties of water in the interfacial region differ from those of bulk water. Low (1962) recently reviewed the main properties of adsorbed water on clays; the properties of the clay-water-ice system has been discussed recently by ANDERSON (1967). Nuclear magnetic resonance (NMR) studies by PICKETT and LEMCOE (1959), DUCROS and DUPONT (1962), HECHT et al. (1966) have shown that water adsorbed by montmorillonite has a higher degree of dissociation *pounds adsorbed by Na- and H-montmorillonite* (after BAILEY *et al.* 1968)

$>C=O \cdots O \cdots$ H / H $\cdots M^{z+}$ – Clay 6[b]	$-\overset{O}{\overset{\|}{C}}-OH \cdots$ $\cdots O-$Clay 7[b]	B + [H+ – Clay] → (HB+ – Clay) 8[b]	$M^{z+}(H_2O)$ – Clay + B → [(M^{z+} + OH⁻) – Clay] + (HB+ – Clay) 9[b]	B + $H_3O^+_{(soln)}$ → HB+ HB+ + M^{z+} Clay → HB – Clay + $M^{z+}_{(soln)}$ 10[b]
—	—	Aa	Ab	Ab
Aa(?), Ab	—	Aa	Ab	Ab
Aa(?), Ab	—	Aa	Ab	Ab
—	—	Aa	Ab	Ab
Aa(?), NNa	—	Aa, NNa	NNa	Ab
Aa(?), Ab	—	Aa	Ab	Ab
Aa(?), NNa	—	Aa, NNa	Ab	Ab
Aa(?), Ab	—	Aa	Ab	Ab
Aa(?), NNa	—	Aa, NNa	Ab	Ab
Aa(?), N / N	Aa, NNa / N	—	—	—
N / N	Aa, NNa / N	N	N	Ab
Aa(?), Ab	Aa, NNa	Aa, NNa	Ab, NNa	Ab

tion by the H-montmorillonite or nature of the functional group(s) present.

than that of normal water. The NMR studies of DUCROS and DUPONT (1962) showed that the degree of dissociation of adsorbed water of montmorillonite is at least 1,000 times higher than that in liquid water. Electrical conductivity studies by FRIPIAT *et al.* (1965) on montmorillonite indicate that the electrical conductivity on clays is mainly protonic; protons originating from the dissociation of water molecules which is enhanced by surface electrical fields. The degree of dissociation is of the order of magnitude of 10^{-2}, the dissociation constant being 10^6 times greater than that of bulk water. HECHT *et al.* (1966) studied the protons of water adsorbed by montmorillonite, vermiculite, beidellite, saponite, and sepiolite in the temperature range of 4° to 500° K and indicated that there is proton delocalization in the adsorbed water; the importance of the process of proton delocalization varied with the crystalline state of the mineral.

DUCROS and DUPONT (1962) considered that the constitutional protons involved in the octahedral layer are included to some extent in this proton delocalization process. FRIPIAT *et al.* (1967) have shown that structural type proton delocalization does occur in micas. The

residual water on montmorillonite is bound very tightly as evidenced
by the fact that it is not possible to dehydrate montmorillonite without
destroying the structure (MERING 1946, FRIPIAT et al. 1960).

There is overwhelming evidence, mainly from infrared studies,
pointing to the fact that there is protonation of compounds basic in
chemical character both by clays where hydrogen and aluminum are
the predominant exchangeable cation and also those saturated with
transition, alkaline, and alkali metal cations.

By means of infrared spectroscopy, MORTLAND et al. (1963) were
able to show that NH_3, chemisorbed by montmorillonite and vermiculite saturated with various metal cations, existed as NH_4^+. The existence
of NH_4^+ was thought to be a result of the interaction of NH_3 with protons dissociated from residual water on the exchangeable cations and
the interlamellar silicate surfaces. SWOBODA and KUNZE (1968) found
that the ability of various montmorillonites to adsorb weakly basic
amines and pyridine was related to the pK_b of the organic and the position of isomorphous substitution. Montmorillonite with tetrahedrally-
located substitutions were able to adsorb weaker bases ($pK_b > 11.4$)
than montmorillonites with a lower degree of substitution emanating
from the tetrahedral layer (reactive with bases of pK_b of 8.8 or less but
not with pK_b 9.6 or 9.4). Translated in terms of pKa, the montmorillonites were able to react with bases having pKa values in the range of
approximately 2.6 to 5.2. The lower the pKa, the greater the acid
strength and the lower the basicity on the molecule.

The importance of this adsorption mechanism and its total contribution to the overall order of magnitude of adsorption for mont-
morillonite-type clay minerals would depend upon the pKa of the
adsorbate and the origin of the negative charge in the alumino-silicate.
FARMER and MORTLAND (1966), using infrared spectroscopy, found
that dehydration of magnesium-montmorillonite previously reacted
with pyridine, resulted in the formation of a pyridine-pyridin-
ium montmorillonite complex. This reaction was reversible in
nature; the reaction was not noted when calcium or copper were the
exchangeable cations. Proton transfer and cation formation were also
observed when coordinated pyridine was displaced by water. In both
instances a mechanism similar to that given in no. 9, Table IV was
probably operative.

The infrared studies of TAHOUN and MORTLAND (1966) showed that
amides could be protonated by both hydrogen and aluminum-saturated
montmorillonites, the protonation occurring on the carbonyl group of
the amide molecule. The reaction was reversible since upon removal of
water, the cation reverted back to molecular form. The source of the
proton in the case of aluminum is due to partial dissociation of the water
hydrating the aluminum ion forming the hydroxy-aluminum polymers.
MORTLAND (1966) reported that urea was protonated when reacted with
hydrogen, iron, and aluminum montmorillonite but formed hemisalts

when urea was present in excess of the number of available protons. The fully-protonated urea disappeared upon the dehydration; the original protonated condition could be re-established by rehydration. The importance of water in the formation of the protonated form was noted by these authors. Protonation was concluded to occur on the carbonyl group rather than on the amino group of the urea molecule.

The basicity of the adsorbed organic determines if the proton can be removed upon dehydration; ammonium- or ethylammonium-saturated montmorillonite retained their protonated form upon heating or vacuum, due to their greater basicity (MORTLAND 1966). The residual water in magnesium and calcium montmorillonite is sufficiently dissociated to protonate ammonia (MORTLAND 1963), and ethyl amine (FARMER and MORTLAND 1965), but it is not sufficiently acidic to protonate urea (MORTLAND 1966) and certain amide molecules (TAHOUN and MORTLAND 1966 a). FRIPIAT et al. (1962) found that protonation of the amines to form alkylammonium cations could occur in the presence of sodium and calcium montmorillonite as well as with hydrogen-aluminum clay systems. RUSSELL (1965), using infrared spectroscopy, found that reaction of gaseous NH_3 with montmorillonite and saponite resulted in the formation of ammonium cations and metal hydroxides in the innerlayer spaces, which was attributed to the reaction of ammonia gas with hydrated cations (mechanism 9, Table IV). RUSSELL et al. (1968 a) found that the 3-aminotriazole molecule became protonated upon adsorption on the surfaces of montmorillonite to produce a 3-aminotriazolium cation. A mechanism similar to no. 9, Table IV was given by the authors as an explanation for proton formation and transfer.

McATEE (1962) found that octadecylamine was adsorbed onto sodium montmorillonite in such a fashion that about 29 meq. of the exchangeable ion, sodium, was replaced. The mechanism by which this occurred was postulated to be the protonation of the base due to hydrolysis of water followed by proton transfer to the base. Since this experiment was carried out using a clay slurry and high pressure filtration, protonation might have occurred by the dissociation of the water in the interfacial region (mechanism 9, Table IV). By means of infrared spectroscopy, SWOBODA and KUNZE (1964) found that pyridine could be adsorbed on montmorillonite by both physical and chemisorption; introduction of water vapor resulted in the desorption of the physically-adsorbed pyridine, but did not remove the chemically-adsorbed pyridine. Pyridine was adsorbed in the protonated form by both the magnesium- and the hydrogen-saturated clays. Infrared studies reported by RUSSELL et al. (1968 b), CRUZ et al. (1968), and BAILEY et al. (1970) clearly show that various s-triazines are protonated on montmorillonitic surfaces in the presence of a variety of different cations. These investigators have also postulated that protonation occurs by means of mechanism 9, Table IV.

The nature of the cation appears to be of great importance in the ability of the interfacial region to protonate organic bases and also on the magnitude of protonation and adsorption. Russell et al. (1968 a) found that protonation of 3-aminotriazole increased in the order of calcium < magnesium < aluminum and attributed this to the increased polarizing power of the cation from calcium to magnesium to aluminum. These same authors also reported that protonation occurred when the exchangeable cations were sodium, ammonium, copper, and nickel.

Anderson (1968) in a recent review article speculated that the properties of the interfacial water in a clay-water-ice system would change as a function of water content; thus, surface acidity, which is a property of the interfacial region, would be some function of water content. Mortland (1968) determined, using infrared spectroscopy, that the amount of NH_4^+ sorbed by montmorillonite upon addition of NH_3 gas was inversely related to the water content of the clay, and the amount of NH_4^+ formed was considered a measure of the surface acidity of the clay.

Surface properties of aluminosilicates with particular emphasis on surface acidity has been recently reviewed (Fripiat 1960 and 1964).

Summary of recent investigations clearly indicates that the protonation of organics in the interfacial region of clays is a function of basicity of the molecule, the nature of the exchangeable cation, water content of the system, and origin of negative charge in the aluminosilicate.

The role of the interfacial region is important not only in determining the adsorption mechanism and the energy by which the adsorbate is held, it is vitally important in determining if the adsorbed organic is degraded. This is of vital importance in determining the persistence and ultimate toxicity of the molecule in that (1) the degradation product may be more or less toxic than the original compound, (2) that it may be more or less tightly bound than the original compound, and (3) its water solubility may either be greater than or less than the original compound, which will affect its leaching and movement into the groundwater.

Rice and Osugi (1918) found that reaction of cane sugar with soils resulted in the inversion of the sugar. These authors concluded that (1) the inversion activity of soils was chiefly a property of the insoluble portion of the soil since none or very little inverting power was found in water extracts from soils, (2) inversion did not continue in sugar extracts after soil was removed, (3) inversion increased with increasing amounts of soil in contact with the sugar solution while no measurable change in the hydrogen ion concentration of the extract was found, and (4) greater inversion occurred by shaking soils with the sugar solution than by allowing the mixtures to stand quietly. McAuliffe and Coleman (1955) pointed out the catalytic role of acid clays. They found per unit of hydrogen ion activity as deduced from pH measurements, that the clay was 2.5 to 4.5 times more effective a catalyst for the inversion of sucrose and on the hydrolysis of ethyl acetate than was hydrochloric acid. The nature of the ion has a great importance on the catalytic role of the interfacial region. MCAULIFFE and COLEMAN (1955) found on the basis of total acidity that the aluminum-clays had a lower catalytic activity than hydrogen-clays. DAVIES and THOMAS (1952) found when compared on an equivalent basis that hydrogen-saturated exchange resins were more effective in the hydrolysis of esters than was hydrochloric acid. No correlation was found to exist between potentiometric hydrogen ion activities in clays or exchange resins suspension and their efficiencies in the catalytic hydrolysis of ethyl acetate or an inversion of sucrose (MCAULIFFE and COLEMAN 1955).

The low temperature transformations of a wide variety of organic compounds on a number of clay minerals has been attributed to the high surface acidity of the clay. FOWKES et al. (1960) reported the catalytic decomposition of such insecticides as DDT, aldrin, toxaphene, chlordane, dieldrin, heptachlor, Aramite, and endrin occur in wettable powders and dusts as a result of the catalytic action of the clay surface. Reaction rates of these insecticides in the presence of such clay minerals as Wyoming bentonite, attapulgite, and kaolinite increase with the increasing acidity of the clay surface. Treatment of the clays with such basic substances as urea or hexamethylenetetramine resulted in a diminished rate of decomposition apparently due to a reduction in surface acidity.

ROSENFIELD and VAN VALKENBURG (1965) report that Ronnel undergoes an acid-catalyzed molecular rearrangement reaction when adsorbed on bentonite. It has been reported (RUSSELL et al. 1968 b, CRUZ et al. 1968, BAILEY et al. 1970) that certain s-triazines, in addition to being protonated in the interfacial region of clays, are degraded to the keto-form of the hydroxy derivative. Coordination complexes have also been reported to be degraded in the interfacial region. Cobalt (III) hexammine and cobalt (III) chloropentammine cations adsorbed onto montmorillonite decomposed upon dehydration with evolution of NH_3 gas resulting in NH_4^+ formation, and the production of cobalt (II) hydroxide (CHAUSSIDON et al. 1962, FRIPIAT and HELSEN 1965).

FRIPIAT and HELSEN (1964), using ultraviolet spectroscopy, showed the transformation of triphenylcarbinol into triphenylcarbonium ion and attributed this to the acid character of the hydration water of montmorillonite. CHAUSSIDON and CALVET (1965) observed the decomposition of aliphatic amines adsorbed on montmorillonite and their transformation into hydrocarbons at temperatures below 100°C. FRIPIAT et al. (1966) report that dehydration of glycine and alanine montmorillonite complexes produced secondary amide linkages.

Attempts to measure the surface acidity of the interfacial region have been made through the use of Hammett indicators. WALLING

(1950) and Benesi (1956) have used Hammett indicators as a qualitative measure of the relative proton donating power of dry silicate surfaces in a nonaqueous system. The Hammett acidity function, H_0, is given by the formula:

$$H_0 = pK_{HB^+} - \log \frac{C_{BH^+}}{C_B}$$

where pK_{HB^+} is the log of the dissociation constant of the conjugate acid, C_{HB^+} is concentration of conjugate acid, and C_B is concentration of neutral base. The surface acidity, H_0, of 1 to 0.2μ hydrogen- and sodium-montmorillonite (source Wyoming bentonite) dried at 110°C. has been reported to be -5.6 to -8.2 and $+4.0$, respectively (Bailey et al. 1968). Of course, the H_0 value for dried clays and the surface acidity in the interfacial region would not be expected to be identical.

The exact surface acidity for clay systems still is not known. The work of McLaren and Esterman 1957, Harter and Ahlrich 1967, and calculations (Hartley and Roe 1940) indicate the surface acidity to be at least two pH units (100X) more acid than the suspension pH. Bailey et al. (1968) concluded from an adsorption study of adsorbates of widely differing chemical character and basicity that the surface acidity of montmorillonite is three to four pH units (1,000X to 10,000X) lower than the pH of the bulk solution. The surface acidity values obtained by Harter and Ahlrich (1967) by means of infrared spectroscopy have been questioned by Mortland (1967) and McLaren and Seaman (1968).

Bernstein (1960) derived an equation relating internal pH (surface acidity) of a partly neutralized clay and external pH (pH of the bulk solution), this equation being:

$$p\bar{H} = pH - 7 + \tfrac{1}{2}pKa$$

where $p\bar{H}$ is the internal pH or surface acidity, pH is the external pH, and pKa is the dissociation constant of the exchange group on the clay. The dissociation constant of the exchange group for montmorillonite as given by the author is about 10^{-4} and probably an order of magnitude lower for illite. From this equation, it would appear that the surface acidity would be about four units lower than the external pH for montmorillonite.

e) Temperature

Since the pesticide formulations are polycomponent systems, the influence of temperature on the behavior of these materials upon addition to the soil system may not always be predictable on the basis of generalizations from two- or three-component systems.

Adsorption processes are exothermic and desorption processes are endothermic in nature, and an increase in temperature would normally be expected to reduce adsorption and favor the desorption process. This corresponds to a weakening of the attractive forces between the solute and the solid surface (and between adjacent adsorbed solute molecules) with increasing temperature, and corresponding increase in solubility of the solute in the solvent. HARRIS and WARREN (1964) reported exceptions to this generalization; they reported the adsorption of simazine, atrazine, and monuron by bentonite was greater at 0° than at 50°C.

Exchange reactions tend to be temperature independent; HARRIS and WARREN (1964) found that diquat was completely adsorbed at 0° and at 50°C.

Temperature may influence adsorption through its effects on solubility and vapor pressure. Generally speaking, an increase in temperature leads to decreased adsorption; however, there are exceptions in which the effect of temperature on solubility is such that increased adsorption occurs at higher temperatures; i.e., FREED et al. (1962) found increased adsorption of EPTC at higher temperatures.

f) Electrical potential of clay surface

The electrical field arising from the charge-balancing cations is considered to be responsible for the various surface phenomena observed in clays, zeolites, and other alumino-silicates. The behavior of these materials in dilute aqueous solutions has been explained by the double layer theory. It is now becoming apparent that soils are more nearly solid-state systems with a limited moisture content and most of the effects observed are in the domain of surface chemistry.

FRIPIAT (1968) has recently reviewed work on the magnitude of surface fields of alumino-silicate surfaces as related to the transformation of adsorbed molecules in soil colloids.

Theoretical estimates of the field arising from ionic surfaces have been made for calcium X-zeolites by PICKERT et al. (1964) and for sodium and calcium X-zeolites in more detail by DEMPSEY (1967). Direct measurements of the surface fields in Y-zeolites have been made by WHITE et al. (1967) through a study of the perturbation of OH groups in decationated Y-zeolite by physically-adsorbed gases. From these measurements and calculations it was concluded that an electrical field of the order of magnitude of 10^8 to 10^9 volts/cm. is acting on OH groups present at ionic alumino-silicate surfaces.

From the heat of immersion of a variety of solids of different surface polarity in n-butyl derivatives having a range of dipole moments, ZETTLEMOYER (1965) has estimated the surface force field emanating from the solid. These are shown in Table V.

FRIPIAT (1968) has suggested that when a polarizable molecule having a polarizability equal to 10^{24} cm.^{3} is submitted to an electrical field of 10^8 volts/cm. (0.33 × 10^{-6} e.s.u.) the induced dipole in the molecule amounts to 0.33 Debye unit (10^{-18} e.s.u.); i.e., a significant

Table V. *Surface area and field strength of selected adsorbents*
(after Zettlemoyer 1965)

Solid	B.E.T. Area (sq. m./g.)	Field strength (e.s.u./sq. cm. $\times 10^{-5}$)
Al_2O_3	0.4	1.9
SiO_2 (Aerosil)	120	1.1
Graphon	95	0
Teflon	9	0
Carbon black	120	0.7
Y-zeolite	900	3.3[a]

[a] From Fripiat (1968).

increase compared to the usual range of permanent dipoles. This electrical interaction results in a weakening of some chemical bonds. Mortland et al. (1963) have shown that the water molecules at monolayer coverage on a montmorillonite surface have a degree of dissociation higher than in bulk water. Fripiat et al. (1965) obtained from thermodynamic and electrical conductivity measurements a value of about 10^{-2} for this degree of dissociation. Various chemical reactions induced by this high acidity have been reported and are summarized below:

(a) decomposition of $Co(NH_3)_6^{3+}$ into N_2, $Co(OH)_2$, NH_3 and NH_4^+ (Chaussidon et al. 1962, Fripiat and Helsen 1966),

(b) formation of triphenylcarbonium ion from triphenylcarbinol (Fripiat et al. 1964),

(c) decomposition of amines (Calvet and Chaussidon 1965),

(d) protonation of pyridine (Farmer and Mortland 1966, Swoboda and Kunze 1964),

(e) formation of 3-aminotriazolium cation from 3-aminotriazole (Russell et al. 1968 a),

(f) formation of protonated hydroxypropazine and hydroxyatrazine from propazine and atrazine, respectively (Russell et al. 1968 b),

(g) peptide linkages from glycine and from alanine (Fripiat et al. 1966), and

(h) protonation of amines (Farmer and Mortland 1965).

g) Nature of formulation

As previously indicated (Bailey and White 1964), the nature of the formulation may have an effect on the relative adsorption, desorption, and availability of pesticides. It has been reported that mont-morillonite (LAW and KUNZE 1966) and kaolinite (BARBARO and HUNTER 1967) adsorb surfactants. In both cases, the extent of adsorption was a function of the nature of the surfactant, the cationic surfactant being adsorbed to the greatest extent. In the case of montmorillonite (LAW and KUNZE 1966), the cationic surfactants were adsorbed to the greatest extent compared to anionic or non-ionic surfactants (both cationic and nonionic compounds were held in the internal surfaces of montmorillonite and tended to form double layers if the concentration was sufficiently high). Since surfactants are present in the pesticide formulation, these may result in certain cases in competition between the pesticide and the surfactant for adsorption sites, which would affect the movement and bioactivity of the particular pesticide.

V. Mechanisms of adsorption

Several mechanisms or combination of mechanisms can be postulated for adsorption of organic compounds by alumino-silicates. Some of these are as follows: (1) physical adsorption—adsorption due to van der Waals forces (a summation of dipole-dipole interactions, dipole-induced dipole interactions, and induced dipole-induced dipole interactions, ion dipole interactions), (2) hydrogen bonding, (3) coordination complexes, and (4) chemical adsorption.

For the purpose of this discussion, hydrogen bonding, physical adsorption, and chemical adsorption will be taken as three distinct and separate mechanisms. Normally, there is a question of whether to classify hydrogen bonding as physical adsorption or chemical adsorption.

Possible adsorption mechanisms for various compounds were given in Table IV. This table will be used in the ensuing discussion on adsorption mechanisms as it applies to various groups of compounds. A coordination complex is a type of chemical adsorption but will be discussed separately to gain a better understanding of the overall adsorption processes that occur in soil.

It is realized that not all mechanisms occur simultaneously; however, two or more may occur simultaneously depending on the nature of the functional group and the acidity of the system.

a) Physical adsorption

Physical adsorption involves van der Waals forces which result from short range dipole-dipole interactions of several kinds. The dispersion interaction (induced dipole-induced dipole) appears to be the most important factor in determining the van der Waals forces for simple molecules. Due to the complex nature of the soil colloid-cation-water system, it is not feasible to treat the individual interactions which are responsible for van der Waals forces.

The role of van der Waals forces in the adsorption of neutral polar and non-polar organic molecules by montmorillonite was demonstrated by the research of BRADLEY (1945), MACEWAN (1958), and GREENE-KELLY (1955 a). The compounds used in these studies included substances such as ethylene glycol, glycerol, and saturated and unsaturated ring compounds. Since these compounds are adsorbed in interlayer positions in montmorillonite and vermiculite the exchangeable cation may have a significant effect on adsorption (BRINDLEY 1966).

Van der Waals forces are also involved in the adsorption of organic cations. The importance of the shape of the cations in determining the extent of van der Waals interaction has been shown by the work of GREENLAND et al. (1965) on the adsorption of some amino acids and peptide cations by montmorillonite. Since van der Waals forces are known to decay rapidly with distance, their contribution to the adsorption energy would be greatest for those ions which are in closest contact with the surface, or enable close contact to be maintained with the adjacent adsorbed ions. Thus, the small and spherically shaped methyl-ammonium ions fulfil this requirement whereas the bulky and irregularly shaped tetra-n-propyl- and tetra-n-butylammonium ions provide less intimate contact with the clay surface and hence the adsorption energy is less than for straight chain monoalkyl-ammonium ions of comparable molecular weight.

BRINDLEY and THOMPSON (1966) have compared the d_{001} spacings of the neutral and cationic forms of 1,3-di-4-pyridylpropane, 4-phenyl-propylpyridine, 1,3-di-4-piperidylpropane, and 1-(N-hydroxyethyl-4-piperidyl) propane and shown that in most cases the orientation of the molecule is unchanged in the interlayer space of a synthetic Ca-fluor-hectorite when the molecule is converted to the cation. There is a decrease of 0.15 to 0.35 A in the spacing of the complex when the molec-ular form of the compounds is replaced by the cationic form. In the case of 4-phenylpropylpiperidine the neutral molecules develop a double layer complex and the cationic forms develop a single layer complex. Complexes with neutral amines and with cationic amines likewise give, respectively, double-layer and single-layer complexes (BRINDLEY 1965).

Due to the rather large size of many of the pesticide molecules, the interlayer adsorption energy for the neutral molecules may be so high that adsorption will occur only on external surfaces of expanding clay minerals.

b) Chemical adsorption

Chemical adsorption by soils and soil constituents can occur by at least four different mechanisms: (1) ion exchange, (2) protonation at the silicate surface or colloidal surface by reaction of the base with the hydronium ion on the exchange site (mechanism 8, Table IV), (3) protonation in the solution phase with subsequent adsorption of the organic molecule via ion exchange (mechanism 10, Table IV), and (4) in systems having water of hydration, protonation by reaction with the dissociated protons from residual water present on the surface or in coordination with the exchangeable cation (mechanism 9, Table IV).

1. Ion exchange.—The fact that ion exchange occurs with organic cations is very well documented (SMITH 1934, GIESEKING 1939, HENDRICKS 1941, GRIM et al. 1947, JORDAN 1949, REAY and BARRER 1957, COWAN and WHITE 1958, McATEE 1959, KURITENKO and MIKHALYUK 1959, DODD and RAY 1960, BLACK and VAN WINKLE 1961, SUTHERLAND and MacEWAN 1961, McATEE 1962, WEISS 1963, KUL'CHITSKII 1961, GREENLAND and QUIRK 1962, KINTER and DIAMOND 1963, ROWLAND and WEISS 1963, KESAREE et al. 1962, BROOKS 1964, McATEE and HACHMAN 1964, GREENLAND 1965, FARQUI et al. 1967, BODENHEIMER and HELLER 1968). As indicated, the adsorbate may ultimately reside in the interfacial region as a cation due to either protonation with the hydronium ion on the exchange site, or to protonation with the organic base in solution with subsequent adsorption by ion exchange, or thirdly, due to protonation or ionization of the basic organic compound from protons furnished by the dissociation of water at the surface of the clay and/or that hydrating the exchangeable cations (i.e., proton transfer). With regard to the behavior and persistence of pesticides in the soil, it is most important to know the stoichiometry of the ion exchange reaction and the ease of replaceability of the adsorbed organic cation. Also of vital interest is the adsorption capacity or the amount that can be adsorbed (i.e., is adsorption limited to the cation exchange capacity or can an amount in excess of the ion exchange capacity be adsorbed; and, in turn, what is the ease of replacement of this excess material?). This latter case would be important under soil sterilization conditions.

The nature of the inorganic exchangeable ion appears to have an effect on the ease and stoichiometry of ion exchange reactions. McATEE (1962) found that an amine salt (octadecylamine) and a quaternary ammonium salt (dimethylbenzylaurylammonium chloride) replaced sodium stoichiometrically from montmorillonite, while a greater than stoichiometric amount of each organic cation was required to replace calcium and magnesium. The ease of replaceability of the organic cations by inorganic cations was in the order of magnesium > calcium > sodium. McATEE (1959) also indicated that the size of the organic cation influences replaceability since the quaternary ammonium salt replaced both sodium and calcium easier than the amine salt.

There have been numerous reports in the literature of adsorption in excess of the cation exchange capacity by adsorbates of widely-different chemical character (GRIM et al. 1947, COWAN and WHITE 1958,

FRISSEL 1961, BERGMANN and O'KONSKI 1963, FARMER and MORTLAND 1965, MORTLAND 1966, TAHOUN and MORTLAND 1966, BAILEY et al. 1968, BODENHEIMER and HELLER 1968). The adsorption of such bases as ethylamine (FARMER and MORTLAND 1965), urea (MORTLAND 1966), and amides (TAHOUN and MORTLAND 1966) by montmorillonite in excess of the cation exchange capacity (CEC) has been explained by the formation of "hemi" salts where two basic molecules share a single proton. COWAN and WHITE (1958) found in a study of the adsorption of chloro-hydrates of n-primary aliphatic amines by montmorillonite that adsorption in excess of the CEC was related to the chain length. The hydrochloride salt of octylamine was adsorbed in excess of the CEC, but the amount of the hydrochloride salt of heptylamine adsorbed was in proportion to the exchange capacity. The authors believed that the excessive quantity of amine retained was due to the intervention of a complexation reaction:

$$RNH_3^+ - clay + RNH_3Cl \rightarrow clay - RNH_3^+(RNH_3Cl),$$
$$clay - RNH_3^+(RNH_3Cl) \rightarrow clay - RNH_3^+(RNH_2) + HCl.$$

For the above mechanism to be operative, a decrease in pH of the reaction solution must occur. This has been observed by GRIM et al. (1947), who reported that the amount of dodecylamine acetate and ethyl dimethyl octadecenyl ammonium bromide adsorbed was in the excess of the CEC of montmorillonite. The methylene blue cation has been found to be adsorbed by montmorillonite in excess of the CEC (FRISSEL 1961, BERGMANN and O'KONSKI 1963). The nature of the exchangeable cation present appeared to determine the extent of the adsorption of methylene blue in excess of the CEC (FRISSEL 1961). BERGMANN and O'KONSKI (1963) found that the adsorption isotherm of the reaction of methylene blue cation and Na-montmorillonite could be expressed by the equation.

$$\frac{x}{m} = k_1 + k_2C^{1/n}$$

where x/m = the amount adsorbed, k_1 = value for the cation exchange capacity of the clay, and $k_2C^{1/n}$ = amount adsorbed due to physical adsorption (Freundlich adsorption equation). This equation describes two different adsorption processes: ion exchange and physical adsorption.

The type of the clay mineral affects the amount in excess of the CEC that can be adsorbed. BLACK and VAN WINKLE (1961) found that montmorillonite could adsorb 110 percent in excess of the cation exchange capacity of n-butyl ammonium and n-dodecyl ammonium acetate, while sodium vermiculite could only adsorb these two cations 50 percent in excess of the cation exchange capacity. These authors reported that washing removed that quantity which was adsorbed in excess of CEC and attributed the excess adsorbed to physical adsorp-

tion. It has also been found by HENDRICKS (1941) and by KINTER and
DIAMOND (1961) that large organic cations are adsorbed in quantities
less than the cation exchange capacity, and this has been attributed to
steric interferences (*i.e.*, covering up of more than one exchange site
by such a large molecule).

The nature of the clay mineral appears to have an effect on the
exchangeability of one organic cation by another. MCATEE (1959)
found that the displacement of the dimethylbenzylauryl cation by a
quaternary ammonium ion was greater for hectorite than Wyoming
bentonite and attributed this to the difference in the origin of negative
charge, isomorphic substitution occurring in the tetrahedral layer of
Wyoming bentonite and the octahedral layer in hectorite. The dissociation constant of the base and its cationic size affects the replaceability of one organic cation by another (MCATEE 1962). The greater
the basic strength of the amine, the better replacing agent it was; the
larger the chain length of the exchanger, the less effective it was as an
exchanger. Presumably, this effect of basicity is due to the ability of
charge or proton transfer to occur. GREENLAND and QUIRK (1960)
found that ethyl, butyl, and octyl pyridinium bromides were adsorbed
up to the cation exchange capacity of montmorillonite, while dodecyl
and cetyl pyridinium bromides were adsorbed at twice or three times
the CEC, respectively.

Two pesticides which are applied as organic cations are diquat and
paraquat. Diquat and paraquat have been reported by WEBER and
WEED (1968) to adsorb to approximately the cation exchange capacity of sodium-montmorillonite and sodium-kaolinite. KNIGHT and
TOMLINSON (1967) reported that paraquat was not adsorbed up to the
CEC capacity of soil as measured by inorganic cation displacement
techniques. The nature of the adsorbent has a great affect on the extent
of adsorption and the ease of desorption of the bipyridilium compounds.
KNIGHT and TOMLINSON (1967) found that removal of soil-organic
matter did not change the amount of paraquat adsorbed drastically,
and concluded that paraquat adsorption is a property of clay minerals,
particularly the presence of expansible clay minerals. WEBER and
WEED (1968) reported that 80 percent of both paraquat and diquat
could be displaced from a kaolinite clay with 1M barium chloride
solution, while only five percent of each compound was removed from
montmorillonite. WEBER *et al.* (1965) reported that 100 percent of
paraquat adsorbed onto sodium- and hydrogen-saturated amberlite
cation exchange resins could be replaced by extraction with 1M sodium
chloride and barium chloride solutions. No adsorption, positive or negative, of diquat or paraquat was reported to occur on IRA-411 anion
exchange resin (WEBER *et al.* 1965). KNIGHT and TOMLINSON (1967)
also showed that paraquat was preferentially adsorbed by soils in
competition with solutions of $0.1N$ to $2.0N$ with respect to the ammonium ion. These authors indicated that paraquat was very tightly held

by the soil, and that the majority of the adsorbed material is herbicidally inactive.

GREENLAND (1965) has suggested that two factors of importance in the adsorption of organic ions are (1) the large number of possible points of contact between ion and adsorbent would lead to a large change in the entropy of the system, and (2) the occurrence of specific adsorption sites, which depend upon a particular molecular characteristic of the cation and substrate. KNIGHT and TOMLINSON (1967) indicate that soil has heterogeneous adsorption sites and that paraquat is adsorbed with different energies of adsorption, but the majority of the sites bind paraquat very tightly. The difficulty in displacing large organic cations from mineral surfaces has been attributed to a combination of electrostatic forces, and van der Waals adsorption forces (HENDRICKS 1941, WEBER et al. 1965, KNIGHT and TOMLINSON 1967, WEBER and WEED 1968). X-ray diffraction data indicates that both diquat and paraquat are oriented with the plane of the diphenyl ring parallel to the surface of the lattice (WEBER et al. 1965). The d_{001} spacing at maximum saturation has been found to be essentially independent of moisture content (BAILEY, unpublished data). The expansion of the lattice with respect to the dimensions of paraquat is less than the normal van der Waals thickness of the ion (WEBER et al. 1965, KNIGHT and TOMLINSON 1967).

2. Protonation.—As indicated earlier, adsorption may occur due to protonation at or near the surface. This is a very important adsorption mechanism, especially in regard to the adsorption of pesticides basic in chemical character. Evidence for this phenomenon will be cited and this mechanism will be discussed in great detail in the section on surface acidity. The various mechanisms by which a basic molecule could be protonated and adsorbed are given in Table IV.

c) Hydrogen bonding

It is difficult to classify hydrogen bonding as physical adsorption or chemisorption. HADZI et al. (1968) have suggested that there is a parallel between hydrogen bonding and protonation. Protonation may be considered as a full charge transfer from the base (electron donor) to the acid (proton) electron acceptor—the hydrogen bond is a partial charge transfer. Low (1961) has suggested that hydrogen bonding may occur at clay surfaces between the surface oxygens and protons of adsorbed molecules, such as water, due to the distortion of the lone-pair electrons of surface oxygen atoms by the protons of the adsorbed molecules.

For those organic compounds possessing a basic chemical character and containing an N—H group, adsorption could occur by formation of a hydrogen bond between the amino group and the oxygen of the clay surface (mechanism 2, Table IV). This would be a prime mechanism for the adsorption of the molecular form of the basic organic compounds. The adsorption of diethyl amine (MacEwan 1948), and other aliphatic amines (MacEwan 1948, Jordan *et al.* 1950, Cowan and White 1958) by montmorillonite has been postulated to occur by such a mechanism.

The importance of functional groups in influencing bonding to the surface of clays has been considered by Brindley and Thompson (1966). They concluded that chain molecules terminating in —OH, in —COOH, and in —NH₂ readily form complexes with montmorillonite, the molecules standing at steep angles to the (001) plane of the montmorillonite whereas similar molecules terminating in —Cl and —Br do not.

d) Coordination

Coordination compounds or metal complexes are compounds that contain a central atom or ion, usually a metal surrounded by a cluster of ions or molecules. The complex is formed by the donation of electron pairs by the ligand and the acceptance by the metal (usually a transition metal ion) resulting in the filling or partial filling of inner "d" orbitals (*i.e.*, a coordinate covalent bond). The number of ligands that can be grouped around a particular ion (*i.e.*, the coordination number) is a function of the particular metal and the configuration of the ligand. The ligand therefore is a Lewis base, the metal a Lewis acid. The term "innersphere" and "outersphere" coordination are often used. In innersphere coordination the ligand is directly coordinated to the metal, while in outersphere coordination the ligand is bonded to, in the case of clay-water systems, the water of hydration surrounding the cation. For a greater insight into the nature and properties of coordination compounds, the reader is referred to Basolo and Johnson (1964) or Jones (1964).

Recent evidence indicates that the coordination type of bonding may be quite important in determining the fate and behavior of pesticides in soil. Ashton (1963) deduced from chromatographic evidence that amitrole formed complexes in solution with nickel, cobalt, and copper ions, but did not react with magnesium, manganese, ferric, or ferrous iron. Sund (1956) similarly showed that amitrole formed "complexes" with the transition metal ions—nickel, cobalt, and iron—as well as with the alkaline earth metal, magnesium. Presumably, these were coordination complexes.

Various investigators employing infrared spectroscopy indicate that certain ligands form coordination complexes with various metals on clays (Mortland 1966, Farmer and Mortland 1965, Mortland and Meggitt 1966, Dowdy and Mortland 1967).

Farmer and Mortland (1966) present infrared data which indicated that pyridine can coordinate directly to the copper ions present on the exchange sites of montmorillonite (innersphere coordination) and also coordinate indirectly to the copper ion through the water of hydration (outersphere coordination). Mortland (1966) presents infrared data which indicates that urea is held onto Cu (II)-, Mn (II)-, and Ni (II)-montmorillonite by means of a coordinate covalent bond (mechanism 4, Table IV). Bonding occurs through the carbonyl group rather than the amino group. Infrared studies by Mortland and Meggitt (1966) on EPTC-montmorillonite complexes revealed a decrease in the carbonyl stretching and an increase in the CN stretching frequencies. In light of the definition of the coordinate compounds given above, the bonding of the carbonyl groups to the transition metal ions studied was probably covalent bonding (mechanism 4, Table IV). The EPTC-montmorillonite complex was stable against atmospheric humidity, but EPTC was completely displaced when the complex was immersed in water.

Aliphatic amines appear to form coordination complexes with Cu-montmorillonite (Farmer and Mortland 1965); the resultant complex is quite strong—the amine could not be displaced by water vapor.

Alcohols also form coordinate complexes with transition metals on clay (Dowdy and Mortland 1967).

The herbicide amitrole forms coordinate complexes with Ni- and Cu-saturated montmorillonite (Russell et al. 1968 a). Earlier, Ercegovich and Frear (1964) postulated that there might be "complex" formed between soil clays and amitrole. Tahoun and Mortland (1966) have shown by means of infrared spectroscopy that primary, secondary, and tertiary amides can coordinate on the surface of montmorillonite saturated with certain metal ions. The strength of the bond was a function of the nature of the exchangeable cation and the order of the amide. For a given ion, the bond strength is in the order of tertiary > secondary > primary. For an amide order, the bond strength in decreasing order is transition metal (e.g., copper) > alkaline earth > alkali metal ion.

The total amount of the ligand that can be complexed is a function of the coordination number of the metal ion. By means of chemical analysis Farmer and Mortland (1965) calculated that in the pyridine-Cu-montmorillonite complex there were approximately four pyridine molecules surrounding each copper ion, indicating a square planar configuration.

The nature of the transition metal forming the coordinate complex on the clay may determine whether or not the organic will be degraded. Farmer and Mortland (1965) and Mortland (1966) reported observations which indicated that pyridine and urea, respectively, underwent degradation. In the case of pyridine, it was suggested that one of the degradation products was the pyridine-free radical, the formation and possible further reaction of the free radical being catalyzed by the copper ion.

VI. Role of spectroscopy in adsorption studies

The infrared spectra of molecules in the adsorbed state seem to offer at present the most satisfactory means of obtaining detailed information about the nature of the adsorptive bond. The infrared spectra of physically adsorbed molecules are similar to the spectra of the gaseous, liquid, or dissolved states. Minor differences in band positions may indeed be observed in physical adsorption, but they are of the same magnitude as the differences between the various unadsorbed states (gas, liquid, or solution of the adsorbate); no new bands are found that cannot be attributed to vibrations already known from infrared or Raman bands of the gaseous or liquid states. For chemisorbed molecules the infrared spectra differ markedly from those of the unadsorbed state.

Polar molecules have long been known to substitute for water in the interlayer space of expanding layer silicates, but only by the application of infrared methods has it been possible to get direct evidence of the mechanism of adsorption. In studying the adsorption of ammonia (RUSSELL 1965), ethylamine (FARMER and MORTLAND 1965), pyridine (FARMER and MORTLAND 1966), nitrobenzene, and benzoic acid (YARIV et al. 1966) several different mechanisms of adsorption have been distinguished, in all of which the exchangeable cation plays a predominant role.

Infrared techniques have been used to establish the acidic nature of residual water on clays (MORTLAND et al. 1963) and in providing evidence for the coordination of EPTC to metal cations through the carbonyl and nitrogen groups on the molecule (MORTLAND and MEGGITT 1966). Protonation of 3-aminotriazole (RUSSELL et al. 1968 a) and protonation and hydrolysis of s-triazines on montmorillonite have been established through infrared studies (RUSSELL et al. 1968 b, CRUZ et al. 1968).

Although the amount of surface exposed by kaolinite is too small for infrared techniques to detect organic compounds on external surfaces, LEDOUX and WHITE (1966 a and b) have utilized intercalation complexes of kaolinite to obtain information on hydrogen bonding of compounds such as urea, formamide, hydrazine, and potassium acetate. Studies of these expanded kaolinite systems have provided evidence for the nature of the interaction of organic compounds with uncharged hydroxyl surfaces and oxygen surfaces.

SERRATOSA (1966) has reported infrared studies of the orientation of pyridine in montmorillonite and vermiculite and shown that the pyridinium ion is parallel to the montmorillonite surface but occupies a vertical position in vermiculite. CRUZ et al. (1968) have made infrared pleochroism studies of the 17.6 A complex of protonated hydroxypropazine with montmorillonite and found that the triazine makes an angle of about 60° with the montmorillonite surface.

The usefulness of infrared spectroscopy in studying adsorbed species in many systems, including alumino-silicates and layer-lattice silicates, has been thoroughly documented and discussed recently by LITTLE (1966) and HAIR (1967).

By a combination of infrared, ultraviolet, and nuclear magnetic resonance spectroscopy, the following information regarding the nature of the organo-clay complex can be obtained: (1) elucidation of bonding mechanisms, (2) molecular orientation, (3) acidity of residual water on clays, (4) interfacial degradation or transformation, (5) tautomeric and resonance form present, and (6) site of protonation on molecule.

VII. Leaching and movement of pesticides in soils

The fact that adsorption has a great influence on the nature and extent of the leaching and movement of pesticides is commonly realized. However, the relationship between adsorption and movement is not fully understood. UPCHURCH and PIERCE (1957 and 1958) indicated that at least two steps are involved in the leachability of a herbicide: (1) entrance of the compound into solution and (2) adsorption of the compound to soil particles. Entrance of the pesticide into solution can take place either from the dissolution of the pesticide present in particulate form or from the desorption of pesticide present on colloidal surfaces. In addition to these factors, HARTLEY (1964) stated that the leaching of herbicides into soil is influenced by the moisture level of the soil at the time of application and the evapo-transpiration ratio.

The factors which appear to affect overall pesticide movement most are (1) adsorption, (2) physical properties of the soil, and (3) climatic conditions. The ensuing discussion will show how each of these influence pesticide movement and their interrelationship.

a) Role of physical properties of soil on pesticide movement

Numerous investigators studying a wide variety of pesticides have clearly shown that pesticides are leached to a greater degree in the light-textured soils than in heavier-textured soils (HANKS 1946, LINDER 1952, SMITH and ENNIS 1953, OGLE and WARREN 1956, SHERBURNE and FREED 1954, HERNANDEZ and WARREN 1950, HURTT et al. 1958, GANTZ 1960, STROUBE and BONDORENKO 1960, COMES et al. 1961, HATTRUP and MUZIK 1961, RODGERS 1962, RAUSER and SWITZER 1963, GEISSBÜHLER et al. 1963, HARRIS 1964, DUBEY and FREEMAN 1964, DONALDSON and FOY 1965, KEARNEY et al. 1965, HARRIS 1966, SHAHIED and ANDREWS 1966, KEYS and FRIESEN 1968). This implies that differences in soil texture affect the movement of pesticides. Soil texture changes generally affect a resultant change in soil structure; therefore, both soil texture and structure should affect the movement and leachability of pesticides.

The four principal means for pesticide transport within soils are (1) diffusion in the airspace of the soil, (2) diffusion in soil water, (3) downward flowing water, and (4) upward moving water.

Movement by diffusion through the soil and airspace would be important for those pesticides with a high vapor pressure—one class of compounds being the soil fumigants. Soil porosity has been found to be one of the most important factors affecting the diffusion of the soil fumigants (HANSON and NEX 1953). GORING (1957) has pointed out that pore size as well as the airspace continuity is important in clay soils for the movement of vapor. Under certain conditions in soils of high clay content, molecular diffusion may be restricted due to the presence of pores with prohibitively small diameters. HEMWALL (1962) concluded from the mathematical treatment of fumigant movement in soils that one of the optimum soil properties to maximize fumigant efficacy would be a low, continuous airspace. Movement by air diffusion is probably more important than downward water flow for the transport of those pesticides possessing a high volatility. Similarly, air diffusion will also play a dominant role in the eventual loss of pesticides from the soil by volatilization.

Calculations by HARTLEY (1961) indicated that several years would be necessary for as little as one percent of the concentration of surface-applied herbicides to migrate by diffusion to a depth of only two feet in a moist soil. It appears that percolating water is the principal means of movement of the relatively non-volatile pesticides, and that diffusion in soil-water is important only for transport over very small distances. Mathematical treatment is reported by LINDSTROM et al. (1968) for the diffusion of 2,4-D in saturated soils. The diffusion coefficient for 2,4-D acid in nine soils was generally found to be inversely related to soil texture.

Upward movement of pesticides may be a factor, especially in those areas of irrigation and where high evapo-transpiration ratios are prevalent and may affect the overall movement and persistence of certain pesticides in the soil. HARRIS (1964) reported the upward movement of dicamba and diphenamid occurred under sub-irrigation conditions and also when surface-applied water was allowed to evaporate freely from the soil surface. Presumably, the upward movement was due to the pesticide being dissolved in upward-flowing capillary water. HARRIS (1964) concluded that weather conditions may be important in addition to total rainfall in determining the overall movement of herbicides in soil. ASHTON (1961) and JORDAN et al. (1963) demonstrated that lateral movement of herbicides results from the lateral movement of capillary water under furrow irrigation. COMES et al. (1961) presented evidence for the upward movement of endothal in sub-irrigated soil. PHILLIPS (1964) reported the upward movement of phenols in soils.

Pore size and pore-size distribution affect the rate in which water enters and moves through the soil. This will affect the extent of the band spreading of the downward-moving pesticide front. The nature of this concentration profile band as it reaches the groundwater could be quite important biologically. For example, if there are no irreversible physico-chemical reactions between a solute and the components of the system, then linear isotherms with Gaussian-type concentration peaks (piston flow type breakthrough curve) would be expected. However, for the same pesticides under the same conditions of concentration of full velocity, if there are irreversible physical-chemical reactions or if the band spreads due solely to such process as dispersion, the band or concentration profile will be skewed, the degree of skewing depending upon mechanisms responsible for the interactions. In the first instance cited, essentially, the initial concentration would reach the subsurface drainage outlet at a given time, while in the latter case, depending upon the extent of band spreading, a very substantially lower amount of the pesticide would reach the watercourse at a given time. In the first case, the concentration may be high enough (even if only a portion of the originally-applied concentration moves) that it cannot be assimilated, and therefore an undesirable effect on certain segments of the biota population would result. In the other case, the pesticide concentration may be of such small magnitude as to be innocuous to the biota present.

The rate of water movement should also influence the nature of the equilibrium between the pesticide in solution and that adsorbed onto the colloidal surface. From both a mass action and a solubility viewpoint, this would appear to favor increased desorption.

The presence of an impervious layer horizon influences the rate of water movement and would probably increase the extent of both desorption and pesticide movement.

The miscible displacement method has been used to follow solute movement in soils as well as to determine the microscopic flow velocity distribution of water as it moves through the soil. The usefulness of dyes (COREY 1968), ^{2}H, ^{3}H, and ^{18}O (COREY and HORTON 1968) as tracers to investigate water movement through soils has been studied. BIGGAR and NIELSEN (1962) have used this method of miscible displacement to study the behavior of tracers in different porous media and at different moisture content. The latter authors explain the distributional differences of the tracers measuring the effluent on the basis of the relative effects of pore geometry, diffusion rates, and such physical-chemical phenomena as adsorption and ion exchange. They found that changes in water content of the system changed the outlet arrival time of the tracer (position of the breakthrough curve) and the concentration distribution of the tracer (shape of the breakthrough curve). The holdback of the two tracers was three to four times greater in the unsaturated than in the saturated media. This would indicate that the water content of the soil would greatly effect the movement of the pesticide if it behaved similarly to the inorganic tracers. HARTLEY (1964) cites results from leaching studies of the soluble organic dye from sand columns under saturated conditions at two high rates of flow and with free drainage. There was much more band spreading in the case of free drainage than at the highest flow rate. In a similar study, however, HARTLEY (1964) found that with a sand-plaster mixture, there was a marked delay in leaching out the dye when the dye was applied in solution to a previously dried-out column.

NIELSON and BIGGAR (1962) examined the use of several theoretical models in describing miscible displacements in such porous media as glass beads, sieved aggregates, and soils. It was shown that mathematical descriptions which do not include individual mechanism such as adsorption or ion exchange are unsatisfactory. The work of DAVISON and SANTELMANN (1968) showed fluometuron to be as mobile as the chloride ion under both high and low flow rates. Low rates of water were shown to affect the flow of fluometuron through soils as evidenced by differing shapes of the distribution curves. In systems where adsorption was minimized, a less total volume of water was required to displace fluometuron than in the case where a material which exhibited a high adsorption for the herbicide.

Attempts have been made to use chromatographic theory to describe movement of pesticides in soils. FRISSEL and POELSTRA (1967) have theoretically evaluated the different approaches used for the chromatographic transport of ions or various types of solute in solution through soils. KING and McCARTY (1968) developed a chromatographic theory to predict pesticide movement in soils. The degradation of the pesticide during the time it resided in the soil was taken into account by use of appropriate kinetic equations. LAMBERT et al. (1965) developed a slotted tube technique to test the applicability of the chromatographic theory to movement of pesticides through soils.

The resultant movement and distribution of the pesticides in the soil as related to the physical-chemical properties of the soil appear to be the function of the relative effects of (1) reversible adsorption (molecular and ionic), (2) chemical reaction (chelation and precipitation), (3) irreversible adsorption (exclusion of solute by solute-solid interaction or velocity distribution with velocities near zero), (4) pore geometry (microscopic flow velocity distribution), (5) entrance of pesticides into solution, (6) amount and average flow velocities, and (7) moisture content of soil at time of application.

b) Role of climatic forces on pesticide movement

The total amount of rainfall or irrigation water received, the intensity (water flux), and frequency of received water all appear to effect movement of pesticides in soils. These factors should be important in facilitating the entrance of pesticides into solution whether the pesticide be present in particulate or adsorbed form.

For a given compound on a given soil type, the greater the amount and the frequency of rainfall, the greater the amount of pesticide leached, the deeper the pesticide is leached in the profile. Within limits, the converse appears to hold true for rainfall intensity. Pesticides are leached in greater amounts and to a greater depth under the lower rainfall intensities. Burnside et al. (1961), in a field leaching study of simazine, found that the depth to which simazine moved was not great (essentially all remained in the top six inches), but increased with the total amount of water applied. In the leaching study with monuron, Upchurch and Pierce (1957) found that an increase in the total amount of water increased the depth of movement. In desorption studies of monuron and diuron from Hawaiian sugarcane soils, Yuen and Hilton (1962) and Hilton and Yuen (1966) found that as the soil-water ratios decreased (greater volume of liquid to solid phase), greater recoveries of the two adsorbed substituted ureas were effected. A linear increase in percent recovery of the adsorbed herbicide with increase in solution volume was found by Hilton and Yuen (1966) when herbicide concentrations were plotted as a semilog function-versus-solution volume. Sherburne and Freed (1954) compared the movement of monuron to the movement of compounds in chromatography and concluded that at the depth of the highest monuron concentration in the soil columns was a function of the amount of water at the surface. Donaldson and Foy (1965) related differences in the movement of 2,3,6-TBA, dicamba, and amiben through soils to differences in the rate of water movement through soils; the greater the extent of water movement, the greater the leaching of these three herbicides.

The applied dosage level and total quantity of water affect the extent of adsorption. Sheets (1964) reported that when both monuron and diuron were applied in increased concentrations to a sandy loam soil, herbicide movement took place more rapidly. The movement of DNBP was related to water applied; the greater the amount of water, the greater the downward movement (Davis and Selman 1954).

With regard to the effect of rainfall intensity on pesticide movement, Upchurch and Pierce (1958) found that simulated rainfall varying from $1/16$ inch to four inches/application had little influence on monuron removed from the top two inches. In the two to eight inch zone, greater accumulation of monuron occurred with low rainfall intensities. Comparing the parameters of rainfall intensity and frequency, these same authors found that less intense application resulted in greater removal of the herbicide from the upper surface than did less frequent applications. About one half of the frequency effect was attributed to the fact that less frequent herbicide application permitted more moisture to evaporate from the soil surface, thereby making less moisture available to percolate through the soil. Harris (1964) reported that movement of dicamba and diphenamid was greater with either five or ten inches of water when the water was applied in 0.25-inch increments rather than in one-inch increments. RODGERS (1962) reported for all four s-triazines, atrazine, simazine, atratone, and ipazine, in three different soils that higher rates of simulated rainfall resulted in leaching to a depth of one to two inches greater than did the lower rate. DAVIDSON *et al.* (1968) found that the rate at which fluometuron and diuron moved through a water-saturated glass bead and uniformly packed soil column was a function of the water flux or average pore velocity.

Deviations from the above-stated trends have been found and are mainly attributed to soil type and structure differences and the effect of these parameters on gross water movement. The evapo-transpiration ratio will affect the rate of leaching between rains; evaporation of water from the surface would cause the surface few inches of the soil to be less well-extracted of chemicals than are other horizons farther down the profile. Low moisture content in the surface would result in diminution of pesticide movement in two different ways: (1) lowering the total water solubility of the pesticide, and (2) enhancing the ability of the pesticide to compete for adsorption sites. With regard to the former for a unit concentration of pesticide dissolved in a unit volume of water surrounding an adsorbent, if this volume water decreases (due to a decreasing moisture content), the pesticide concentration/unit volume would then increase. Therefore, under certain moisture content, the solubility of the pesticide in question would be exceeded and crystallization would occur. With regard to the latter process at low moisture levels, the proportionate number of water molecules present to compete for an adsorption site is small, and less polar pesticide molecules may compete more favorably for adsorption sites. At higher moisture contents, the pesticide cannot compete favorably, and the extent of adsorption is reduced.

The movement of pesticides by overland flow is equally important as movement of the pesticide downward in the profile, especially as it affects water pollution. Two questions need to be answered in this regard: (1) Is pesticide transport mainly in the liquid phase? (2) Or, is it initially adsorbed on particulate matter and then transported "piggyback" to the watercourse? TRICHELL *et al.* (1968) found that the loss of 2,4,5-T, dicamba, and picloram in runoff water was influenced by cultural practice, nature and rate of application, and the slope of the field. When loss was determined 24 hours after application, loss of dicamba and picloram was greater from sod than from fallow plots, while 2,4,5-T losses were about equal. The amount of picloram lost in the simulated rainfall of 0.5 inch varied with the degree of application, the percentage loss was the same. BARNETT *et al.* (1966) studied the loss of 2,4-D in the washoff from cultivated fallow land as a function of formulation of rainfall intensities and storm duration using the simulated rainfall technique. Concentrations of 2,4-D in the washoff positively correlated with the rate applied, and were found to be greatest in the earlier part of storm, and to decrease with storm duration. The iso, octyl, butyl, ether, and ester formulations were found to be far more susceptible to removal by washoff than was the amine salt. Soil bioassays indicated that most of the 2,4-D remained in the surface three inches of the soil.

Summary

The literature is reviewed pertaining to: (1) adsorption and desorption of organic pesticides by soil colloids, (2) movement of pesticides through soils and off of soil surfaces, and (3) physical-chemical properties of soil constituents which influence pesticide adsorption and desorption. The adsorption theories of Freundlich, Langmuir, Gibbs, and Brunauer, Emmett, and Teller (BET) are briefly treated. The effects and interactions of such factors as physico-chemical nature of the adsorbent, physico-chemical nature of the pesticide, soil reaction, surface acidity, nature of the saturating cation on the colloid exchange site, temperature and nature of the formulation on adsorption and desorption are discussed in detail. The applicability of various adsorption mechanisms (physical and chemical adsorption; hydrogen and coordinate bonding) to a variety of pesticide-soil complexes is examined. The role of surface forces and of surface acidity on the character of the pesticide-soil complex and the applicability of spectroscopy to the elucidation of the chemical character of such complexes is discussed. Movement of pesticides in soil and off of soil surfaces is treated in terms of the following three factors: (1) adsorption, (2) physical properties of the soil, and (3) climate.

Résumé*

Facteurs influençant l'adsorption, la désorption et le mouvement des pesticides dans le sol

On a examiné la littérature relative aux sujets suivants: (1) adsorption et désorption des pesticides organiques par les colloïdes du sol; (2) mouvement des pesticides au travers des sols et à la surface des sols; et (3) propriétés physico-chimiques des constituants du sol influençant l'adsorption et la désorption des pesticides. Les théories de Freundlich, Langmuir, Gibbs et Brunauer, Emmett et Teller (BET) concernant l'adsorption sont brièvement traitées. Les effets et actions réciproques

---

* Traduit par S. DORMAL-VAN DEN BRUEL.

de facteurs tels que la nature physicochimique de l'adsorbant, la nature
physico-chimique du pesticide, la réaction du sol, l'acidité superficielle,
la nature du cation saturant sur le radical échangeur des colloïdes, la
température et la nature de la formulation en relation avec l'adsorption
et la désorption sont discutés en détail. L'applicabilité de différents
mécanismes d'adsorption (adsorption physique et chimique; liaisons
hydrogène et de coordination) à une variété de complexes "pesticides-
sol" est examinée. Le rôle des tensions superficielles et de l'acidité
superficielle sur le caractère du complexe "pesticide-sol" et l'applicabilité de la spectroscopie à l'élucidation du caractère chimique de tels
complexes sont discutés. Le mouvement des pesticides dans le sol et à
la surface du sol est exprimé par les trois facteurs suivants: (1) adsorption, (2) propriétés physiques du sol, et (3) climat.

Zusammenfassung*

Faktoren, die die Adsorption, Desorption und Tätigkeit der Unkrautvertilgungsmittel im Boden beeinflussen

Die zugehörige Literatur zu: (1) Adsorption und Desorption von
organischen Pestiziden durch Bodenkolloide; (2) Tätigkeit von Pestiziden im Boden und von Bodenoberflächen entfernt; und (3) physika-
lisch-chemische Eigenschaften von Bodenbestandteilen, die die Adsorption und Desorption von Pestiziden beeinflussen, ist besprochen worden.
Die Adsorptionstheorien von Freundlich, Langmuir, Gibbs, Brunauer,
Emmett und Teller (BET) werden kurz behandelt. Die Wirkungen und
Wechselwirkungen solcher Faktoren, wie die physiko-chemische Natur
der adsorbierenden Substanz, die physiko-chemische Natur der
Pestizide, Boden-Reaktion, Oberflächensäuregehalt, Natur des sätti-
genden Kations an der Kolloidaustauschstelle, Temperatur und Natur
der Formulierung in Adsorption und Desorption, werden in Einzel-
heiten diskutiert. Die Anwendungsmöglichkeit verschiedener Adsorp-
tions-Mechanismen (physikalische und chemische Adsorption; Wasser-
stoff- und gleichartige Bindung) auf eine Vielzahl von Pestizid-Boden-
Verbindungen wird geprüft. Die Rolle der Oberflächenkräfte und des
Oberflächensäuregehaltes in Bezug auf den Charakter des Pestizid-
Boden-Komplexes, und die Anwendungsmöglichkeit der Spektroskopie
zur Aufklärung des chemischen Charakters solcher Verbindungen
werden diskutiert. Die Tätigkeit von Pestiziden im Boden und von der
Bodenoberfläche entfernt wird im Sinne der folgenden drei Faktoren
behandelt: (1) Adsorption, (2) physikalische Eigenschaften des
Bodens und (3) Klima.

---

* Übersetzt von M. Düsch.

Table VI. *Common and chemical names of organic pesticides referred to in text*

Common name	Chemical name
aldrin	1,2,3,4,10,10-hexachloro-1,4,4a,5,8,8a-hexahydro-1,4-*endo,exo*-5,8-dimethanonaphthalene
amiben	3-amino-2,5-dichlorobenzoic acid
amitrole	3-amino-1,2,4-triazole
Aramite	2-(*p*-tert-butylphenoxy)isopropyl 2-chloroethyl sulfite
atratone	2 methoxy-4-ethylamino-6-isopropylamino-*s*-triazine
atrazine	2-chloro-4-ethylamino-6-isopropylamino-*s*-triazine
CDEC	2-chloroalkyl diethyldithiocarbamate
chlordane	1,2,4,5,6,7,8,8-octachloro-3a,4,7,7a tetrahydro-4,7-methanoindane
chloroxuron	*N′*-4-(4-chlorophenoxy)phenyl-*N,N*-dimethylurea
CIPC (chloropropham)	*m*-chloroisopropyl ester, carbanilic acid
cycluron	3-cyclooctyl-1,1-dimethylurea
DDT	2,2-bis(*p*-chlorophenyl)-1,1,1-trichloroethane
dicamba	2-methoxy-3,6-dichlorobenzoic acid
2,4-D	2,4-dichlorophenoxyacetic acid
dicryl	3′,4′-dichloro-2-methyl-acrylanilide
dieldrin	1,2,3,4,10,10-hexachloro-6,7-epoxy-1,4,4a,5,6,7,8,8a-octahydro-1,4-*endo, exo*-5,8-dimethanonaphthalene
diquat	6,7-dihydrodipyrido[1,2-a:2′,1′-c] = pyrazidinium salt
diphenamid	*N,N*-dimethyl-2,2-diphenylacetamide
diuron	3-(3,4-dichlorophenyl)-1,1-dimethylurea
DNBP	4,6-dinitro-*o*-sec-butylphenol
endrin	1,2,3,4,10,10-hexachloro-6,7-epoxy-1,4,4a,5,6,7,8,8a-octahydro-1,4-*endo,endo*-5,8-dimethanonaphthalene
fenuron	3-phenyl-1,1-dimethylurea
fluometuron	3-(*m*-trifluoromethylphenyl)-1,1-dimethylurea
heptachlor	1,4,5,6,7,8,8-heptachloro-3a,4,7,7a-tetrahydro-4,7-methanoindene
hxdroxyatrazine	2-hydroxy-4-ethylamino-6-isopropylamino-*s*-triazine
hydroxypropazine	2-hydroxy-4,6-bis(isopropylamino)-*s*-triazine
ipazine	2-chloro-4-dimethylamino-6-isopropylamino-*s*-triazine
IPC (propham)	isopropyl ester, carbanilic acid
linuron	3-(3,4-dichlorophenyl)-1-methoxy-1-methylurea
monolinuron	3-(4-chlorophenyl)-1-methoxy-1-methylurea
monuron	3-(p-chlorophenyl)-1,1-dimethylurea
neburon	1-butyl-3-(3,4-dichlorophenyl)-1-methylurea
norea	3-(hexahydro-4,7-methanoindan-5-yl)-1,1-dimethylurea
paraquat	1,1′-dimethyl-4,4′-bipyridinium salt
picloram	4-amino-3,5,6-trichloro-picolinic acid
prometone	2,4-bis(isopropylamino)-6-methoxy-*s*-triazine
prometryne	2,4-bis(isopropylamino)-6-methylmercapto-*s*-triazine
propanil	3′,4′-dichloro-propionanilide
propazine	2-chloro-4,6-bis(isopropylamino)-*s*-triazine
Ronnel	*o,o*-dimethyl-*o*-2,4,5-trichlorophenyl-phosphorothioate
simazine	2-chloro-4,6-bis(ethylamino)-*s*-triazine
simetone	2,4-bis(ethylamino)-6-methoxy-*s*-triazine
Solan	3′-chloro-2-methyl-p-valerotoluidide
2,3,6-TBA	2,3,6-trichlorobenzoic acid
2,4,5-T	2,4,5-trichlorophenoxyacetic acid
trietazine	2-chloro-4-diethylamino-6-ethylamino-*s*-triazine
toxaphene	a chlorinated camphene containing 67 to 69 percent chlorine

References

ADAMSON, A. W.: Physical chemistry of surfaces. New York: Interscience (1960).

ALEXANDER, M.: Biodegradation of pesticides. In: Pesticides and their effects on soils and water. Wisconsin, Soil Sci. Soc. Amer., Inc., ASA special publ. 8, 78 (1966).

ALDRICH, R. J.: Herbicides: Residues in soil. J. Agr. Food Chem. 1, 257 (1953).

ANDERSON, D. M.: Phase interface between ice and silicate surfaces. J. Colloid Interfac. Sci. 25, 174 (1967).

—— Phase boundary water in frozen soils. In symposium: Freezing and thawing. Soil Sci. Soc. Amer. Meeting, New Orleans (1968).

AOMINE, S., and H. OTSUKA: Surface of soil allophanic clays. Trans. 9th Internat. Congress Soil Sci. 1, 731 (1968).

ASHTON, F. M.: Movement of herbicides in soil with simulated furrow irrigation. Weeds 9, 612 (1961).

—— Fate of amitrole in soil. Weeds 11, 167 (1963).

BAILEY, G. W.: Entry of biocides into watercourses. Proc. Symposium on Agricultural Waste Waters. California: Water Resources Center Report 10, 94 (1966).

——, and J. L. WHITE: Review of adsorption and desorption of organic pesticides by soil colloids, with implications concerning pesticide bioactivity. J. Agr. Food Chem. 12, 324 (1964).

——, J. L. WHITE: Herbicides: A compilation of their physical, chemical, and biological properties. Residue Reviews 10, 97 (1965).

——, W. R. PAYNE, JR., and K. W. BUXTON: A spectroscopic study of s-triazine-montmorillonite interactions. In preparation (1970).

——, J. L. WHITE, and T. ROTHBERG: Adsorption of organic herbicides by montmorillonite: Role of pH and chemical character of adsorbate. Soil Sci. Soc. Amer. Proc. 32, 222 (1968).

BANGHAM, D. H.: The Gibbs adsorption equation and adsorption on solids. Trans. Faraday Soc. 33, 805 (1937).

BARBARO, R. D., and J. V. HUNTER: Surfactural adsorption on several homoionic forms of kaolin. Water Research 1, 157 (1967).

BARNETT, A. P., E. W. HAUSER, A. W. WHITE, and J. H. HOLLADAY: Loss of 2,4-D in washoff from cultivated fallow land. Weeds 15, 133 (1967).

BARSHAD, I.: Factors affecting the interlayer expansion of vermiculite and montmorillonite with organic substances. Soil. Sci. Soc. Amer. Proc. 16, 176 (1952).

BASOLO, F., and R. C. JOHNSON: Coordination chemistry: The chemistry of metal complexes. New York: Benjamin (1964).

BENESI, H. A.: Acidity of catalyst surfaces. I. Acid strength from colors of adsorbed indicator. J. Amer. Chem. Soc. 78, 5490 (1956).

BERNSTEIN, F.: Distribution of water and electrolyte between homoionic clays and saturating NaCl solution. Clays and Clay Minerals 8, 122 (1960).

BERGMANN, K., and C. T. O'KONSKI: A spectroscopic study of methylene blue monomer, dimer, and complexes with montmorillonite. J. Phys. Chem. 67, 2169 (1963).

BIGGAR, J. W., and D. R. NIELSON: Miscible displacement. II. Behavior of tracers. Soil Sci. Soc. Amer. Proc. 26, 125 (1962).

BIKERMAN, J. J.: Surface chemistry: Theory and application. New York: Academic Press (1958).

BLACK, J. L., and M. VAN WINKLE: Sorption of n-butylammonium and n-dodecyl-ammonium acetate by sodium montmorillonite and sodium vermiculite. J. Chem. Eng. Data 6, 557 (1961).

BODENHEIMER, W., and L. HELLER: Sorption of methylene blue by montmorillonite saturated with different cations. Israel J. Chem. 6, 307 (1968).

BRADLEY, W. F.: Molecular associations between montmorillonite and some polyfunctional organic liquids. J. Amer. Chem. Soc. 67, 975 (1945).

BRINDLEY, G. W.: Clay-organic studies. X. Complexes of primary amines with montmorillonite and vermiculite. Clay Minerals **6**, 91 (1965).
—— Ethylene glycol and glycerol complexes of smectites and vermiculites. Clay Minerals **6**, 237 (1966).
——, and T. D. THOMPSON: Clay organic studies. XI. Complexes of benzene, pyridine, piperidine and 1,3-substituted propanes with a synthetic Ca-fluorhectorite. Clay Minerals **6**, 345 (1966).
BROOKS, C. S.: Mechanism of methylene blue dye adsorption on siliceous minerals Kolloid Zh. **199**, 31 (1964).
BRUNAUER, S., L. E. COPELAND, and D. L. KANTRO: The Langmuir and BET theories in the solid-gas interface. E. A. Flood, ed.: The solid-gas interface New York: Dekker (1967).
——, L. S. DEMING, W. E. DEMING, and E. TELLER: On a theory of the van der Waals adsorption of gas. J. Amer. Chem. Soc. **62**, 1723 (1940).
——, P. H. EMMETT, and E. TELLER: Adsorption of gases in multimolecular layers. J. Amer. Chem. Soc. **60**, 309 (1938).
BURNSIDE, O. C., E. L. SCHMIDT, and R. BEHRENS: Dissipation of simazine from the soil. Weeds **9**, 477 (1961).
CALVET, R.: Mise au pont bibliographiques les complexes organques des argiles. Ann. Agron. **14**, 31 (1963).
CHAUSSIDON, J., and R. CALVET: Evolution of amine cations adsorbed on montmorillonite with dehydration of the mineral. J. Phys. Chem. **69**, 2265 (1965).
—— ——, J. HELSEN, and J. J. FRIPIAT: Catalytic decomposition of cobalt (III) hexammine cations on the surface of montmorillonite. Nature **196**, 161 (1962).
COGGINS, C. W., JR., and A. S. CRAFTS: Substituted urea herbicides: Their electrophoretic behavior and the influence of clay colloid in nutrient solution on their phytotoxicity. Weeds **7**, 349 (1959).
COMES, R. D., D. W. DOHMONT, and H. P. ALLEY: Movement and persistence of endothal (3,6-endoxohexahydropthalic acid) as influenced by soil texture, temperature, and moisture levels. J. Amer. Soc. Sugar-beet Technol. **11**, 287 (1961).
COREY, J. C.: Evaluation of dyes for tracing water movement in acid soils. Soil Sci. **106**, 182 (1968).
——, and J. H. HORTON: Movement of water tagged with ^2H, ^3H, and ^{18}O through acidic kaolinitic soil. Soil Sci. Soc. Amer. Proc. **32**, 471 (1968).
COWAN, C. T., and D. WHITE: The mechanism of exchange reactions occurring between sodium montmorillonite and various n-primary aliphatic amine salts. Trans. Faraday Soc. **54**, 691 (1958).
CRAFT, A. S.: The chemistry and mode of action of herbicides. New York: Interscience (1961).
CRUZ, M., J. L. WHITE, and J. D. RUSSELL: Montmorillonite-s-triazine interactions. Israel J. Chem. **6**, 315 (1968).
DAVIDSON, J. M., and P. W. SANTELMANN: Displacement of fluometuron and diuron through saturated glass beads and soil. Weed Sci. **16**, 544 (1968).
——, C. E. RIECK, and P. W. SANTELMANN: Influence of water flux and porous material on the movement of selected herbicides. Soil Sci. Amer. Proc. **32**, 629 (1968).
DAVIES, C. W., and G. G. THOMAS: Ion-exchange resins as catalysts in the hydrolysis of esters. J. Chem. Soc. **2**, 1607 (1952).
DAVIS, F. L., and F. T. SELMAN: Effects of water upon the movement of dinitro weed killers in soils. Weeds **3**, 11 (1954).
DEMPSEY, E.: Calculation on model zeolite crystals. Proc. Soc. Chem. Ind., Meeting on Zeolites (preprint) (1967).
DIETER, C. E.: Factors affecting results with soil fumigants. Plant Disease Reporter Suppl. **227**, 98 (1954).
DODD, C. G., and R. SATYABRATA: Semiquinone cation adsorption on montmorillonite as a function of surface acidity. Clays and Clay Minerals **8**, 237 (1960).

DOEHLER, R. W., and W. A. YOUNG: Some conditions affecting the adsorption of quinoline by clay minerals in aqueous solution. Clays and Clay Minerals 9, 468 (1962).
DOMSCH, K. H.: Soil fungicides. Ann. Rev. Phytopathol. 2, 293 (1964).
DONALDSON, T. W., and C. L. FOY: The phytotoxicity and persistence in soils of benzoic acid herbicides. Weeds 13, 195 (1965).
DOWDY, R. H., and M. M. MORTLAND: Alcohol-water interactions on montmorillonite surfaces. II. Ethylene glycol. Soil Sci. 105, 36 (1967).
DUCROS, P., and M. DUPONT: Etude par resonance magnetique nucleire des protons dans les argiles. Compt. Rend. 254, 1409 (1962).
DUBEY, H. D., and J. F. FREEMAN: Influence of soil properties and microbial activity on the phytotoxicity of linuron and diphenamid. Soil Sci. 97, 334 (1964).
DUSTMAN, E. H., and L. F. STICKEL: Pesticide residues in the ecosystem. In: Pesticides and their effects on soils and water. Wisconsin, Soil Sci. Soc. Amer., Inc., ASA special publ. 8, 109 (1966).
EDWARDS, C. A.: Insecticide residues in soils. Residue Reviews 13, 83 (1964).
ENNIS, W. B., JR.: Some soil and weather factors influencing usage of pre-emergence herbicides. Proc. Soil Sci. Soc. Florida 14, 130 (1954).
ENO, C. F.: Insecticides and the soil. J. Agr. Food Chem. 6, 348 (1958).
ERCEGOVICH, C. D., and D. E. H. FREAR: The fate of 3-amino-1,2,4-triazole in soils. J. Agr. Food Chem. 12, 26 (1964).
FARMER, V. C., and M. M. MORTLAND: An infrared study of complexes of ethylamine with ethylammonium and copper ions in montmorillonite. J. Phys. Chem. 69, 683 (1965).
—— —— An infrared study of the coordination of pyridine and water to exchangeable cations in montmorillonite and saponite. J. Chem. Soc. (London) A, 344 (1966).
FARQUI, F. A., S. OKUDA, and W. O. WILLIAMSON: Chemisorption of methylene blue by kaolinite. Clay Minerals 7, 19 (1967).
FOWKES, F. M., H. A. BENESI, L. B. RYLAND, W. M. SAWYER, K. D. DETLING, E. S. LOEFFLER, F. B. FOLCKEMER, M. R. JOHNSON, and Y. P. SUN: Clay-catalyzed decomposition of insecticides. J. Agr. Food Chem. 8, 203 (1960).
FREED, V. H., J. VERNETTI, and M. MONTGOMERY: The soil behavior of herbicides as influenced by their physical properties. Proc. W. Weed Control Conf. 19, 21 (1962).
FREUNDLICH, H.: Colloid and capillary chemistry. London: Methuen (1926).
FRIPIAT, J. J.: Surface properties of clays and gels. Trans. Internat. Congress Soil Sci. 7, 502 (1960).
—— Surface properties of alumino-silicates. Clays and Clay Minerals 12, 327 (1963).
—— Surface chemistry and soil science. In E. G. Hallsworth and D. V. Crawford (ed): Experimental pedology, p. 3. London: Butterworths (1965).
—— Surface field and transformation of adsorbed molecules in soil colloids. Trans. 9th Internat. Congress Soil Sci. 1, 679 (1968).
——, and J. HELSEN: Kinetics of decomposition of cobalt coordination complexes on montmorillonite surfaces. Clays and Clay Minerals 14, 163 (1965).
——, J. CHAUSSIDON, and R. TOUILLAUX: Study of dehydration of montmorillonite and vermiculite by infrared spectroscopy. J. Phys. Chem. 64, 1234 (1960).
——, J. HELSEN, and L. VIELVOYE: Formation de radicaux libres sur les surface des montmorillonites. Bull. Groupe Fr. Argiles 15, 3 (1964).
——, P. ROUXHET, and H. JACOBS: Proton delocalization in micas. Amer. Mineral. 50, 1937 (1965).
——, A. SERVAIS, and A. LEONARD: Etude de l'adsorption des amines par les montmorillonites. III. La nature de la liaison amine-montmorillonite. Bull. Soc. Chim. France, p. 635 (1962).
——, P. CLOOS, B. CALICIS, and K. MAKAY: Adsorption of amino-acids and peptides

by montmorillonite. II. Identification of adsorbed species and decay products by infrared spectroscopy. Internat. Clay Conf. (Jerusalem), p. 233 (1966).

Frissel, M. J.: The adsorption of some organic compounds, especially herbicides, on clay minerals. Verslag Landbouwk. Onderzoek **76**, 3 (1961).

——, and G. H. Bolt: Interactions between certain ionizable organic compounds (herbicides) and clay minerals. Soil Sci. **94**, 284 (1962).

——, and P. Poelstra: Chromatographic transport through soils. Plant and Soil XXVI **26**, 285 (1967).

Gantz, R. L.: Persistence and movement of CDAA and CDEC in soils and the tolerance of corn seedlings to these herbicides. Weeds **8**, 599 (1960).

Geissbühler, H., C. Haselbach, and H. Aebi: The fate of N'-(chlorophenoxy)-phenyl-N,N-dimethyl urea (C-1938) in soils and plants. I. Adsorption and leaching in different soils. Weed Research **3**, 140 (1963).

Gieseking, J. E.: Cation exchange on montmorillonite. Soil Sci. **47**, 1 (1939).

Goring, C. A. I.: Factors influencing diffusion and nematode control by soil fumigants. Dow Information Bull. no. 110 (1957).

—— Theory and principles of soil fumigation. Adv. Pest. Control Research **5**, 47 (1962).

Gottlieb, D., and P. Siminoff: The production and role of antibiotics in the soil. II. Chloromycetin. Phytopathol. **42**, 91 (1952).

—— ——, and M. M. Martin: The production and role of antibiotics in soil. IV. Actidione and clavacin. Phytopathol. **42**, 493 (1952).

Gray, R. A., and A. J. Weierich: Leaching of five thiocarbamate herbicides in soils. Weed Sci. **16**, 77 (1968).

Greene-Kelly, R.: Sorption of aromatic organic compounds by montmorillonite. I. Orientation studies. Trans. Faraday Soc. **51**, 412 (1955 a).

—— Sorption of aromatic organic compounds by montmorillonite. II. Packing studies with pyridine. Trans. Faraday Soc. **51**, 425 (1955 b).

Greenland, D. J.: Interaction between clays and organic compounds in soils. I. Mechanisms of interaction between clays and defined organic compounds. Soil Fertilizer **28**, 415 (1965).

——, and J. P. Quirk: The adsorption of alkyl-pyridinium bromides by montmorillonite. Clay and Clay Minerals **9**, 484 (1962).

——, R. H. Laby, and J. P. Quirk: Adsorption of amino-acids and peptides by montmorillonite and illite. I. Cation exchange and proton transfer. Trans. Faraday Soc. **61**, 2013 (1965).

Gregg, S. J.: The surface chemistry of solids, 2ed. New York: Reinhold (1961).

Grim, R. E.: Clay mineralogy. New York: McGraw-Hill (1953).

——, W. H. Allaway, and F. L. Cuthbert: Reactions of different clay minerals with organic cations. J. Amer. Ceram. Soc. **30**, 137 (1947).

Hadzi, D., C. Klofutar, and S. Oblak: Hydrogen bonding in some adducts of oxygen bases with acids. Part IV. Basicity in hydrogen bonding and in ionization. J. Chem. Soc. (A) **1968**, 905 (1968).

Hair, M. L.: Infrared spectroscopy in surface chemistry. New York: Dekker (1967).

Hance, R. J.: The adsorption of urea and some of its derivatives by a variety of soils. Weed Research **5**, 98 (1965 a).

—— Observations on the relationship between the adsorption of diuron and the nature of the adsorbent. Weed Research **5**, 108 (1965 b).

Hanks, R. W.: Removal of 2,4-dichlorophenoxyacetic acid and its calcium salt from six different soils by leaching. Bot. Gaz. **108**, 186 (1946).

Hanson, W. J., and R. W. Nex: Diffusion of ethylenedibromide in soils. Soil Sci. **76**, 209 (1953).

Harris, C. I.: Movement of dicamba and diphenamid in soils. Weeds **12**, 112 (1964).

—— Adsorption, movement, and phytotoxicity of monuron and s-triazine herbicides in soil. Weeds **14**, 6 (1966).

——, and G. F. Warren: Adsorption and desorption of herbicides by soil. Weeds **12**, 120 (1964).

HARTER, R. D., and J. L. AHLRICHS: Determination of clay surface acidity by infrared spectroscopy. Soil Sci. Soc. Amer. Proc. **31**, 30 (1967).

HARTLEY, G. S.: Physico-chemical aspects of the availability of herbicides in soils. In K. K. Woodford and G. R. Sagar (eds.): Herbicides in the soil. New York: Blackwells (1961).

—— Herbicide behavior in the soil. I. Physical factors and action through the soil. In L. J. Audus (ed.): The physiology and biochemistry of herbicides. New York: Academic Press (1964).

——, and J. W. ROE: Ionic concentration at interfaces. Trans. Faraday Soc. **36**, 101 (1940).

HATTRUP, A. R., and T. J. MUZIK: Effect of various environmental factors on the persistence and movement of 2,3,6-trichlorophenylacetic acid in soil. Research Progress Rept. Weed Control Conf. (1961).

HAYWARD, D. O., and B. M. W. TRAPNELL: Chemisorption, 2 ed. Washington: Butterworth (1964).

HECHT, A. M., M. DUPONT, and P. DUCROS: Etude des phénomènes de transport de l'eau adsorbée dans certains mineraux argileux par la resonance magnetique nucléaire. Bull. Soc. franc. Min. Crist. **89**, 6 (1966).

HEMWALL, J. B.: Theoretical consideration of soil fumigation. Phytopathol. **52**, 1108 (1962).

HENDRICKS, S. B.: Base exchange of the clay mineral montmorillonite for organic cations and its dependence upon adsorption due to van der Waals forces. J. Phys. Chem. **45**, 65 (1941).

Herbicide Handbook, *Weed Society of America.* New York: Humphrey Press (1967).

HERNANDEZ, T. P., and G. F. WARREN: Some factors affecting the rate of inactivation and leaching of 2,4-D in different soils. Proc. Amer. Soc. Hort. Sci. **56**, 287 (1950).

HILL, G. D.: Soil factors and herbicide action. Abstr. Weed Soc. Amer., p. 42 (1956).

HILTON, H. W., and Q. H. YUEN: Adsorption of several pre-emergence herbicides by Hawaiian sugarcane soils. J. Agr. Food Chem. **11**, 230 (1963).

HOLLY, K.: The disappearance of herbicides from soil. Weed Abstr. **14**, 268 (1965).

HURTT, W., J. A. MEADE, and P. W. SANTELMANN: The effect of various factors on the movement of CIPC in certain soils. Weeds **6**, 425 (1958).

JOHNSON, S. L., and K. A. RUMON: Infrared spectra of solid 1:1 pyridine-benzoic acid complexes; the nature of the hydrogen bond as a function of the acid-base level in the complex. J. Phys. Chem. **69**, 74 (1965).

JONES, M. M.: Elementary coordination chemistry. New Jersey: Prentice-Hall (1964).

JORDAN, J. W.: Alteration of the properties of bentonites by reaction with amines. Mineral. Mag. **28**, 598 (1949).

——, B. J. HOOK, and C. M. FINALYSON: Organophilic bentonites. II. Organic liquid gels. J. Phys. Chem. **54**, 1196 (1950).

JORDAN, L. S., B. E. DAY, and W. A. CLERX: Effect of incorporation and method of irrigation on pre-emergence herbicides. Weeds **11**, 157 (1963).

JURINAK, J. J., and D. H. VOLMAN: Application of the Brunauer, Emmett, and Teller equation to ethylene dibromide adsorption by soils. Soil Sci. **83**, 487 (1957).

KAUFMAN, D. D.: Structure of pesticides and decomposition by microorganisms. In: Pesticides and their effects on soils and water. Wisconsin Soil Sci. Soc. Amer., Inc. ASA Special Publ. **8**, 85 (1966).

KEARNEY, P. C., C. I. HARRIS, D. D. KAUFMAN, and T. J. SHEETS: Behavior and fate chlorinated aliphatic acids in soils. Adv. Pest Control Research **6**, 1 (1965).

KESAREE, M., T. DEMIREL, and E. ROSAUER: X-ray diffraction studies of quaternary ammonium treated montmorillonite. Proc. Iowa Acad. Sci. **69**, 384 (1960).

KEYS, C. H., and H. A. FRIESEN: Persistence of picloram activity in soil. Weeds **16**, 341 (1968).

KING, P. H., and P. L. McCARTY: A chromatographic model for predicting pesticide migration in soils. Soil Sci. **106**, 248 (1968).

KINTER, E. B., and S. DIAMOND: Characterization of montmorillonite saturated with short-chain amine cations. II. Interlayer surface coverage by the amine cations. Clays and Clay Minerals **10**, 174 (1963).

KIM, J. T., and S. B. WEED: Adsorption of substituted urea derivatives by montmorillonite. Amer. Soc. Agron. Abstr., p. 90 (1968).

KIPLING, J. J.: Adsorption from solutions of non-electrolytes, p. 215. London: Academic Press (1965).

KNIGHT, B. A. G., and T. E. TOMLINSON: The interaction of paraquat (1:1'-dimethyl 4:4'-dipyridylium dichloride) with mineral soils. J. Soil Sci. **18**, 233 (1967).

KUL'CHITSKII, L. I.: A spectrophotometric study of the adsorption of methylene blue by highly dispersed aluminosilicate. Kolloid Zh. **23**, 63 (1961).

KUNZE, G. W.: Pesticides and clay minerals. In: Pesticides and their effect on soils. Wisconsin Soil Sci. Soc. of Amer., Inc. ASA Special Publ. **8**, 49 (1966).

KURITENKO, O. D., and R. V. MIKHALYUK: Adsorption of aliphatic amines on bentonite from aqueous solutions. Kolloid Zh. **20**, 181 (1959).

LAILACH, G. E., T. D. THOMPSON, and G. W. BRINDLEY: Absorption of pyridines purines, and nucleosides by Li-, Na-, Mg-, and Ca-montmorillonite (clay-organic studies XII). Clays and Clay Minerals **16**, 275 (1968).

—— —— —— Absorption of pyrimidines, purines, and nucleosides by Co-, Ni-, Cu-, and Fe(III) montmorillonite (clay organic studies XIII). Clays and Clay Minerals **16**, 295 (1968).

LAMBERT, S. M.: Functional relationship between sorption in soil and chemical structure. J. Agr. Food Chem. **50**, 572 (1967).

——, P. E. PORTER, and R. H. SCHIEFERSTEIN: Movement and sorption of chemicals applied to the soil. Weeds **13**, 185 (1965).

LANGMUIR, I.: The constitution and fundamental properties of solids and liquids. J. Amer. Chem. Soc. **38**, 2221 (1916).

—— The adsorption of gases on plane surfaces of glass, mica, and platinum. J. Amer. Chem. Soc. **40**, 1361 (1918).

LAW, J. P., JR., and G. W. KUNZE: Reactions of surfactants with montmorillonite: Adsorption mechanisms. Soil Sci. Soc. Amer. Proc. **30**, 321 (1966).

LEDOUX, R. L., and J. L. WHITE: Infrared studies of hydrogen bonding interaction between kaolinite surfaces and intercalated potassium acetate, hydrazine, formamide, and urea. J. Colloid Interface Sci. **21**, 127 (1966 a).

—— —— Infrared studies of hydrogen bonding of organic compounds on oxygen and hydroxyl surfaces of layer-lattice silicates. Proc. Internat. Clay Conf. 1966. (Jerusalem) **1**, 361 (1966 b).

LEGRAND, H. E.: Movement of pesticides in the soil. In: Pesticides and their effects on soils and water. Wisconsin. Soil Sci. Soc. Amer., Inc. ASA Special Publ. **8**, 71 (1966).

LEOPOLD, A. C., K. P. VAN SCHAIK, and N. NEAL: Molecular structure and herbicide adsorption. Weeds **8**, 48 (1960).

LINDER, P. J.: Movement and persistence of herbicides following their application to the soil surface. Proc. N.E. Weed Control Conf. **7**, (1952).

LINDSTROM, F. T., L. BOERSMA, and H. GARDINER: 2,4-D diffusion in saturated soils: A mathematical theory. Soil Sci. **106**, 107 (1968).

LITTLE, L. H.: Infrared spectra of adsorbed species. London: Academic Press (1966).

LOW, P. F.: Physical chemistry of clay-water interaction. Adv. Agron. **13**, 269 (1961).

MacEwan, D. M. C.: Complexes of clays with organic compounds. I. Trans. Faraday Soc. **44**, 349 (1948).
—— Interlamellar reactions of clays and other substances. Clays and Clay Minerals **11**, 431 (1962).
Martin, J. P.: Influence of pesticides on soil microbes and soil properties. In: Pesticides and their effects on soils and water. Wisconsin Soil Sci. Soc. Amer., Inc. ASA Special Publ. **8**, 95 (1966).
——, and P. F. Pratt: Fumigants, fungicides and the soil. J. Agr. Food. Chem. **6**, 345 (1958).
—— ——, C. F. Eno, A. S. Newman, and C. R. Downing: What pesticides do to soils. J. Agr. Food Chem. **6**, 344 (1958).
Martin, M., and D. Gottlieb: The production and role of antibiotics in the soil. III. Terramycin and aureomycin. Phytopathol. **42**, 294 (1952).
—— —— The production and role of antibiotics in soil. V. Antibacterial activity of five antibiotics in the presence of soil. Phytopathol. **45**, 407 (1955).
McAuliffe, C., and N. T. Coleman: H-ion catalysis by acid clays and exchange resins. Soil Sci. Soc. Amer. **19**, 156 (1955).
McAtee, J. L.: Inorganic-organic cation exchange on montmorillonite. Amer. Mineral. **44**, 1230 (1959).
—— Organic cation exchange on montmorillonite as observed by ultraviolet analysis. Clays and Clay Minerals **10**, 153 (1962).
——, and J. R. Hachman: Exchange equilibria on montmorillonite involving organic cations. Amer. Mineral. **49**, 1569 (1964).
McBeth, C. W.: Some practical aspects of soil fumigation. Plant Disease Reporter Suppl. **227**, 95 (1954).
McGlamery, M. D., and F. W. Slife: The adsorption and desorption of atrazine as affected by pH, temperature and concentration. Weeds **14**, 237 (1966).
McLaren, A. D., and E. F. Esterman: Influence of pH on the activity of chymotrypsin at a solid-liquid interface. Arch. Biochem. Biophys. **68**, 157 (1957).
——, and G. V. F. Seaman: Letters to the editor: Concerning the surface pH of clays. Soil Sci. Soc. Amer. Proc. **32**, 127 (1968).
Mering, J.: On the hydration of montmorillonite. Trans. Faraday Soc. **42B**, 205 (1946).
Mortland, M. M.: Urea complexes with montmorillonite: An infrared absorption study. Clay Minerals **6**, 143 (1966).
—— Letters to the editor. Soil Sci. Soc. Amer. Proc. **31**, 578 (1967).
—— Protonation of compounds at clay mineral surfaces. Trans. 9th Internat. Congress Soil Sci. **1**, 691 (1968).
——, and W. F. Meggitt: Interaction of ethyl-N,N-di-*n*-propyl-thiocarbamate (EPTC) with montmorillonite. J. Agr. Food Chem. **14**, 126 (1966).
——, J. J. Fripiat, J. Chaussidon, and J. Uytterhoeven: Interaction between ammonia and the expanding lattices of montmorillonite and vermiculite. J. Phys. Chem. **67**, 248 (1963).
Nearpass, D. C.: Effects of soil acidity on the adsorption, penetration, and persistence of simazine. Weeds **13**, 341 (1965).
—— Effect of the predominating cation on the adsorption of simazine and atrazine by Bayboro clay soil. Soil Sci. **103**, 177 (1967).
Newhall, A. G.: Volatile soil fumigants for plant disease control. Soil Sci. **61**, 67 (1948).
Newman, A. S., and C. R. Downing: Herbicides and the soil. J. Agr. Food Chem. **6**, 351 (1958).
Nielson, D. R., and J. W. Biggar: Miscible displacement. III. Theoretical considerations. Soil Sci. Soc. Amer. Proc. **26**, 216 (1962).
Nose-Kazuo, S., and K. Fukunaga: Pentachlorophenol (PCP) adsorption on soil. II. Effects of humus, pH, and exchangeable cations upon PCP adsorption of soil. J. Soil Sci. and Manure Japan **34**, 291 (1963).

OGLE, R. E., and G. F. WARREN: Fate and activity of herbicides in soils. Weeds **3**, 257 (1956).

OSIPOW, L. I.: Surface chemistry: Theory and industrial applications. American Chemical Society Monograph Series No. 153, pp. 16 and 24 New York: Reinhold (1962).

PAULING, L.: The structure of micas and related minerals. Proc. Nat. Acad. Sci. **16**, 123 (1930 a).

—— The structure of the chlorites. Proc. Nat. Acad. Sci. **16**, 578 (1930 b).

—— The nature of the chemical bond, 3rd ed., p. 487. New York: Cornell University Press (1967).

PHILEN, O. D.: Determination of surface charge density of layer silicates by competitive adsorption of organic cations. Abstr. Amer. Soc. Agron., p. 150 (1968).

PHILLIPS, F. T.: The aqueous transport of water-soluble nematocides through soils. III. Natural factors modifying the chromatographic leaching of phenol through soil. J. Sci. Food Agr. **15**, 458 (1964).

PICKERT, P. E., J. A. RABO, E. DEMPSEY, and V. SCHOMAKER: Zeolite cations with strong electrostatic fields as carboniogenic catalytic centers. Proc. Internat. Congress Catalysis, 3rd, Amsterdam (1964) **1**, 714 (1965).

PICKETT, A. G., and M. M. LEMCOE: An investigation of shear strength of the clay water system by radio-frequency spectroscopy. J. Geophys. Research **64**, 1579 (1959).

PINCK, L. A., W. F. HOLTON, and F. E. ALLISON: Antibiotics in soils. I. Physico-chemical studies of antibiotic-clay complexes. Soil Sci. **91**, 22 (1961 a).

——, D. A. SOULIDES, and F. E. ALLISON: Antibiotics in soils. II. Extent and mechanism of release. Soil Sci. **91**, 94 (1961 b).

RADOSLOVICH, E. W.: The structure of muscovite $KAl_2(Si_3Al)O_{10}(OH)_2$. Acta Cryst. **13**, 919 (1960).

RAMAN, R. V., and M. M. MORTLAND: Proton transfer reactions at clay mineral surfaces. Soil Sci. Soc. Amer. Proc. **33**, 313 (1969).

RAUSER, W. E., and C. M. SWITZER: Effects of leaching on the persistence of amiben toxicity in various soils. Hormolog **4**, 13 (1963).

REAY, J. S. S., and R. M. BARRER: Sorption and intercalation by methylammonium montmorillonites. Trans. Faraday Soc. **53**, 1253 (1957).

RICE, F. E., and S. OSUGI: The inversion of cane sugar by soils and allied substances and the nature of soil acidity. Soil Sci. **5**, 333 (1918).

RODGERS, E. G.: Leaching of four triazines in three soils as influenced by various frequencies and rates of simulated rainfall. Proc. S. Weed Control Conf. **15**, 268 (1962).

ROSENFIELD, C., and W. VAN VALKENBURG: Decomposition of (O,O-dimethyl-O-2,4,5-trichlorophenyl) phosphorothioate (Ronnel) adsorbed on bentonite and other clays. J. Agr. Food Chem. **13**, 68 (1965).

ROWLAND, R. L., and E. J. WEISS: Bentonite-methylamine complexes. Clays and Clay Minerals **10**, 460 (1963).

RUSSELL, J. D.: Infra-red study of the reactions of ammonia with montmorillonite and saponite. Trans. Faraday Soc. **61**, 2284 (1965).

——, M. I. CRUZ, and J. L. WHITE: The adsorption of 3-aminotriazole by montmorillonite. J. Agr. Food Chem. **16**, 21 (1968 a).

—— —— ——, G. W. BAILEY, W. R. PAYNE, JR., J. D. POPE, JR., and J. I. TEASLEY: Mode of chemical degradation of s-triazines by montmorillonite. Science **160**, 1340 (1968 b).

SERRATOSA, J. M.: Infrared analysis of the orientation of pyridine molecules in clay complexes. Clays and Clay Minerals **14**, 385 (1966).

SHAHIED, S., and H. ANDREWS: Leaching of trifluralin, linuron, prometryne, and cotoran in soil columns. Proc. S. Weed Control Conf. **19**, 522 (1966).

SHEETS, T. J.: Review of disappearance of substituted urea herbicides from soil. J. Agr. Food Chem. **12**, 30 (1964).

——, and L. L. DANIELSON: Herbicides in soil: The nature and fate of chemicals applied to soils, plants and animals. *U.S. Department of Agriculture* ARS 20-9, 170 (1960).

SHERBURNE, H. R., and V. H. FREED: Adsorption of 3(p-chlorophenyl)-1,-dimethylurea as a function of soil constituents. J. Agr. Food Chem. 2, 937 (1954).

SIMINOFF, P., and D. GOTTLIEB: The production and role of antibiotics in the soils. I. The fate of streptomycin. Phytopathol. 41, 420 (1951).

SMITH, C. R.: Base exchange reactions of bentonite and salts of organic bases. J. Amer. Chem. Soc. 56, 1561 (1934).

SMITH, R. J., JR., and W. B. ENNIS, JR.: Studies on the downward movement of 2,4-D and 3-chloro-IPC in soils. Proc. S. Weed Control Conf. 6, 63 (1953).

STROUBE, E. W., and D. P. BONDARENKO: Persistence and distribution of simazine applied in the field. Proc. N.C. Weed Control Conf. 17, 40 (1960).

SUND, K. A.: Residual activity of 3-amino-1,2,4-triazole in soils. J. Agr. Food Chem. 4, 57 (1956).

SUTHERLAND, H. H., and D. M. C. MACEWAN: Organic complexes of vermiculite. Clay Minerals Bull. 4, 229 (1961).

SWOBODA, A. R., and G. W. KUNZE: Infrared study of pyridine adsorbed on montmorillonite surfaces. Clays and Clay Minerals 13, 277 (1964).

—— —— Reactivity of montmorillonite surfaces with weak organic bases. Soil Sci. Soc. Amer. Proc. 32, 806 (1968).

TAHOUN, S. A., and M.M. MORTLAND: Complexes of montmorillonite with primary, secondary and tertiary amides. I. Protonation of amides on the surface of montmorillonite. Soil Sci. 102, 248 (1966 a).

—— —— Complexes of montmorillonite with primary, secondary and tertiary amides. II. Coordination of amides on the surface of montmorillonite. Soil Sci. 102, 314 (1966 b).

TALBERT, R. E., and O. H. FLETCHALL: The adsorption of some s-triazines in soils. Weeds 13, 46 (1965).

TAYLOR, A. L.: Chemical treatment of the soil for nematode control. Adv. Agron. 3, 243 (1951).

TRICHELL, H. L., H. L. Morton, and M. G. MERKLE: Loss of herbicides in runoff water. Weeds 16, 447 (1968).

UPCHURCH, R. P.: The behavior of herbicides in soil. Proc. Brit. Weed Control Conf. 7, 1011 (1964).

—— Behavior of herbicides in soil. Residue Reviews 16, 45 (1966).

——, and W. C. PIERCE. The leaching of monuron from Lakeland sand soil. I. The effect of amount, intensity, and frequency of simulated rainfall. Weeds 5, 321 (1957).

—— —— The leaching of monuron from Lakeland sand soil. II. The effect of soil temperature, organic matter, soil moisture and amount of herbicides. Weeds 6, 24 (1958).

VAN DIJK, H.: Some physico-chemical aspects of the investigation of humus. In: The use of isotopes in soil organic matter studies. (Rept. FAO/IAEA Tech. Meeting, Brunswick-Volkenrode, Sept. 9-14, 1963). London: Pergamon (1966).

WALLING, C.: The acid strengths of surfaces. J. Amer. Chem. Soc. 72, 1164 (1950).

WARD, T. M., and K. HOLLY: The sorption of s-triazines by model nucleophilics as related to their partitioning between water and cyclohexane. J. Colloid Sci. 22, 221 (1966).

——, and R. P. UPCHURCH: Role of the amino group in adsorption mechanisms. J. Agr. Food Chem. 13, 334 (1965).

WEAR, J. I., and J. L. WHITE: Potassium fixation in clay minerals as related to crystal structure. Soil Sci. 71, 1 (1951).

WEAVER, J.: Reaction of certain plant growth regulators with ion exchangers. Botan. Gaz. 109, 72 (1947).

WEBER, J. B.: Molecular structure and pH effects on the adsorption of 13 s-triazine compounds on montmorillonite clay. Amer. Mineral. **51,** 1657 (1966).
——, and S. B. WEED: Adsorption and desorption of diquat, paraquat, and prometone by montmorillonite and kaolinitic clay minerals. Soil Sci. Soc. Amer. Proc. **32,** 485 (1968).
——, P. W. PERRY, and R. D. UPCHURCH: The influence of temperature and time on the adsorption of paraquat, diquat, 2,4-D, and prometone by clays, charcoal and an anion-exchange resin. Proc. Amer. Soc. Soil Sci. **29,** 678 (1965).
WEED, S. B., and J. B. WEBER: The effect of adsorbent charge on the competitive adsorption of divalent organic cations by layer-silicate minerals. Amer. Mineral. **53,** 478 (1968).
WEISS. A.: Mica-type layer silicates with alkylammonium ions. Clays and Clay Minerals 10, 191 (1963).
WHITE, J. L., A. JELLI, J. ANDRE, and J. J. FRIPIAT: Perturbation of OH groups in decationated Y-zeolites by physically-adsorbed gases. Trans. Faraday Soc. **63,** 461 (1967).
WOLF, D. E., R. S. JOHNSON, G. D. HILL, and R. W. VARNER. Herbicidal properties of neburon. Proc. N.C. Weed Control Conf. **15,** 7 (1958).
WOODFORD, E. K., and G. R. SAGAR: Herbicides and the soil. London: Blackwell (1960).
YARIV, S., J. D. RUSSELL, and V. C. FARMER: Infrared study of the adsorption of benzoic acid and nitrobenzene in montmorillonite. Israel J. Chem. **4.,** 201 (1966).
YUEN, Q. H., and H. W. HILTON: Soil adsorption of herbicides. The adsorption of monuron and diuron by Hawaiian sugarcane soils. J. Agr. Food Chem. **10,** 386 (1962).
ZETTLEMOYER, A. C.: Immersional wetting of solid surfaces. Ind. Eng. Chem. **57,** 27 (1965).

Mechanisms of adsorption of s-triazines
by clay colloids and factors
affecting plant availability

By

Jerome B. Weber[*]

Contents

I. Introduction

Many new organic agricultural chemicals have been developed which offer great potential in the production and preservation of food and fiber (Weber 1969). In the process of developing and utilizing these chemicals we must come to understand how they affect the target organisms and what becomes of them after they have done their job. The purpose of this review is to discuss[1] the behavior of one family of these organic compounds, the s-triazines, at the molecular level, in

[*] Departments of Crop Science and Soil Science, *North Carolina State University*, Raleigh, N.C.
[1] Many of the ideas and conclusions that are included herein are the result of communications with scientists of many disciplines, both in academic work and in industry, and I am grateful to them for their many contributions.

systems in which the chemicals are associated with clay minerals. Factors affecting the adsorption and release of s-triazines by clay colloids and the availability of the compounds to plant roots will also be included. The literature citations are limited to those studies in which s-triazines and clays were directly associated and situations in which other compounds behaved in a manner similar to the s-triazines.

The association of organic compounds and clay minerals has long been of interest to soil scientists. Many theses have been written concerning the benefits of organic materials on soil structure, permeability, water-holding capacity, etc. A wide variety of aliphatic and aromatic compounds have been isolated from soils. Even the compound cyanuric acid (2,4,6-trichloro-s-triazine), which is directly related to those to be discussed in this paper, has been isolated from an acid humus soil (SMOLIN and RAPOPORT 1959). Thus, some of the organic chemicals which man has made closely resemble those which may be found in nature. If we can develop an understanding of the behavior of the organic agricultural chemicals which are applied to the soil, we will at the same time come to better understand the soil system itself.

II. Structure and properties of clay minerals

The term *clay*, as used in this discussion, refers to the layered aluminosilicate minerals which are considered to be colloidal in size (less than about one to two μ in diameter). The clays occur in the soil in hydrated form. A large number of different types of clay minerals have been found in nature. Each clay mineral has a specific crystalline structure and the reader is referred to BAILEY (1966), GIESEKING (1949), and GRIM (1968) for a more detailed discussion of their structural differences.

All of the layered aluminosilicate clays have one common property: they are made up of sheets of tetrahedra of silicon oxide and sheets of octahedra of aluminum oxides and hydroxides. A silicon tetrahedron is formed when the small silica atom is surrounded by four oxygen atoms. An aluminum octahedron is formed when six oxygens or hydroxyls surround a large atom like aluminum. Some clays are made up of alternating sheets of silica tetrahedra and aluminum octahedra. The ratio of tetrahedral sheets to octahedral sheets is 1:1, and these are called the 1:1 clays.

In other clays, the unit layer is made up of two sheets of silica tetrahedra enclosing a sheet of aluminum octahedra. The ratio of tetrahedral sheets to octahedral sheets is 2:1, thus these are called the 2:1 clays. As unit layers are combined, two tetrahedral sheets lie next to each other. In many cases, these two identical sheets do not attract each other by chemical bonds, but only by relatively weak van der Waals forces, and can be readily separated by water molecules. Such clays therefore have the ability to swell on wetting and to shrink on drying. Montmorillonite is probably the best-known member of this type of clay.

Sometimes aluminum replaces silicon in the tetrahedra while iron, magnesium, manganese, and a few other cations of similar size may replace aluminum in the octahedra. This replacement of ions has been termed "isomorphous substitution." The valence of the replacing cations in many cases is lower than that of the original ions. This means that some of the negative valences of the oxygen atoms are not satisfied internally and this creates an overall net negative charge on the clay surface. This total negative charge has to be neutralized by cations; thus cations such as sodium, potassium, calcium, and hydrogen are attracted to the clay surfaces. These ions may be replaced by other cations in the soil environment and are known as "exchangeable cations," and the amount of exchangeable cations that a clay can retain is known as the clay's "cation exchange capacity" or CEC. Because the seats of the negative valences are at different distances from the surface, the energy with which the cations are held is different, and the CEC of a clay mineral may vary depending upon the type and concentration of replacing cation. Competitive-ion studies indicate that the charges are expressed as discrete adsorption sites rather than a smear of charge (WEED and WEBER 1968).

In addition to the exchange capacity resulting from isomorphic substitution, electric charges develop at the edges of clay crystals from broken bonds. These charges may be positive or negative and they also contribute to the exchange capacity of the clay mineral. Negative charges are thought to result from exposed hydroxyl groups which dissociate with changes in pH and hence are termed pH-dependent charges, as opposed to charges originating from isomorphic substitution which are considered to be independent of pH. For more detailed discussion of the interlayer forces and the origin of the charge in clay minerals, in addition to the previous references, see VAN OLPHEN (1954) and FRIPIAT (1965).

In the 1:1 clays exchangeable cations are adsorbed only on the exterior surfaces, but in 2:1 clays the cations may also be adsorbed on the interior surfaces between the tetrahedral sheets, along with the water molecules. The area between the silicate sheets is also known as the interlayer or basal spacing. For some of the 2:1 clays, potassium atoms are located between the neighboring tetrahedral sheets and hold the sheets tightly together with electrostatic bonds. Consequently these clays do not shrink or swell and are termed the "nonexpanding 2:1 clays." Expanding 2:1 clays have a much greater specific surface area and CEC than the nonexpanding 2:1 clays. Some of the characteristics of selected clay minerals are given in Table I.

The exchangeable ions on the clay surfaces are highest in the immediate vicinity of the surface and decrease with distance from the

Table I. *Characteristics of some common clay minerals*

Characteristics	Montmo- rillonite	Vermiculite	Illite	Kaolinite
Type of layering	2:1	2:1	2:1	1:1
Type of swelling	Expanding	Limited- expanding	Non- expanding	Non- expanding
CEC (meq./g.)	80–120	120–200	15–40	2–10
Specific surface (sq. m./g.)	700–750	500–700	75–125	25–50
C axis basal spacing, in ethylene glycol (A)	17	14	10	7.2

surface. The charged surface and the associated oppositely charged ions are known collectively as the "double layer." The distribution of the cations from the surface out into the bulk of the medium, which in most cases is aqueous solution, is considered to be diffuse or a more or less exponential distribution;[2] see WICKLANDER (1965) and MYSELS (1964) for a detailed discussion of the mathematical models of the diffuse double layer.

The adsorbed ions are hydrated to some extent in association with water molecules on the clay surfaces. Replaceability of these cations depends upon their valence and hydrated size. The water on the clay surface is less dense and more ordered than normal water and is said to have an "ice-like" structure. The reader is referred to BARSHAD (1950), HENDRICKS et al. (1940), KEENAN et al. (1951), and SHAINBERG and KEMPER (1966) for further discussions of the hydration of cations on clay surfaces and to LOW (1961), MARTIN (1962), MERING (1946), and ROSS (1966) for a more detailed explanation of the structural arrangement of water molecules on clay surfaces. Water structure at clay surfaces and in the bulk solution are discussed in section V.

Although clay minerals are in many ways like synthetic cation-exchange resins, they contain the element aluminum which makes them behave differently. The aluminum atom is of such size and character that it accommodates itself both to tetrahedral and octahedral coordination with oxygen. It may move from positions within the crystalline lattice to exchangeable positions and vice versa. It combines with water to form amphoteric hydroxides, which are relatively stable and extremely insoluble. The formation of these hydroxides has a direct effect on pH; this effect must be kept in mind when considering the acidity of clay systems. The relationships between hydrogen, aluminum,

---

[2] This theory states that the concentration of exchangeable ions is highest at the clay surface and decreases at first rapidly and then asymptotically into the bulk solution.

and clay minerals have been examined intensively by COLEMAN and CRAIG (1961), JACKSON (1960), LOW (1955), and MATHERS et al. (1954), and are discussed in detail by MARSHALL (1964) and BANIN and RAVIKOVITCH (1966).

III. Physical and chemical properties of s-triazines

The preparation and properties of the s-triazines and their derivatives have been reviewed by SMOLIN and RAPOPORT (1959) and ALLCOCK (1967). Only the chemical properties of the compounds and their behavior in aqueous solution will be presented here.

The s-triazine molecule displays few of the characteristics of an aromatic nucleus, because of the strong electronegativity of the nitrogen atoms in the ring. Much of the chemistry of the s-triazine compounds is simply the chemistry of the substituent groups on a ring which is not often involved in the reaction except for its effect on charge distribution. The chemistry of the triazine derivatives cannot be compared with that of the corresponding benzene derivatives. The hydroxyamino-s-triazines are amphoteric in nature and, as such, are readily soluble in strong inorganic acids and bases. Due to the weakly acidic and basic properties of the compounds, the salts are readily hydrolyzed in water.

Alkyl-amino-s-triazines, alkyl-diamino-s-triazines, phenyl-diamino-s-triazines, and melamine behave as weak bases in aqueous solutions. Melamine was titrated in aqueous solutions by a spectrophotometric technique by DIXON et al. (1947), who assumed that two principal species existed in solution, the molecular and the protonated forms; a basic dissociation constant of $K_b = 1.1 \times 10^{-9}$ was obtained.

HIRT and SCHMIDT (1958) concluded from ultraviolet spectral studies that the probable structure for the positive ion of melamine was as follows:

| Basic and neutral | Acidic | Very acid | [1] |

Infrared studies by BOITSOV and FINKEL'SHTEIN (1959) showed that the acid salts of oxy- and amino-derivatives of s-triazines resulted in the formation of NH_2^+ or NH_3^+ groups. The authors suggested that protonation occurred on the ring nitrogen atoms.

HIRT et al. (1961) showed that thio-s-triazines were uniformally weaker bases than their oxygen-containing analogs. They suggested that both derivatives had analogous structures and formed positive ions in acid solutions in a manner similar to melamine.

The variation in absorption spectra for s-triazines was conveniently rationalized by Boitsov *et al.* (1962) by assuming five possible distributions of double bonds in the s-triazine rings. The following structures are examples for melamine:

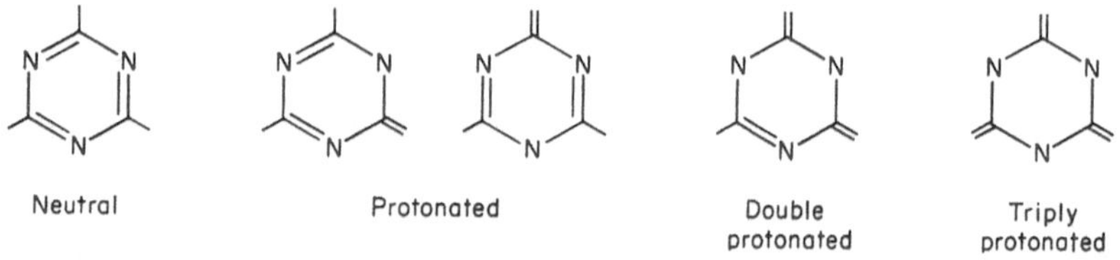

Neutral   Protonated   Double   Triply
                                        protonated   protonated

[2]

Nuclear magnetic resonance and infrared studies by Morimoto (1966 a and b) showed that 2-substituted-4,6-diamino-s-triazines (where $R = -CH_3, -C_2H_5, -OCH_3, -OC_2H_5, -OC_6H_5$) are protonated in the 1- and 5- positions. The molecules showed preferential protonation in the 5-position. The following structures were suggested:

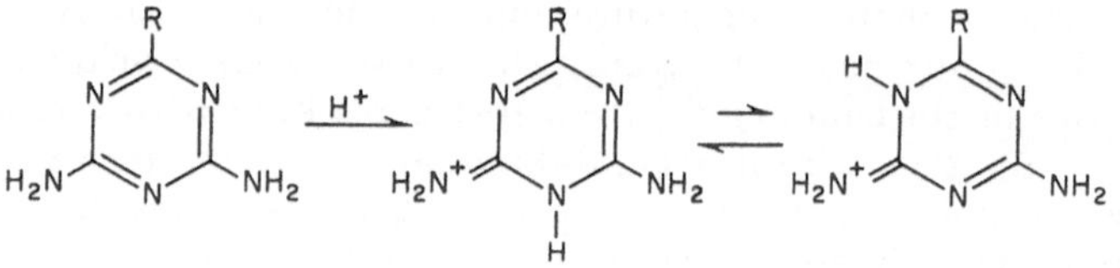

[3]

Spectrophotometrically determined ionization constants for 13 alkyl-amino-s-triazines and the relationships of molecular structure and basicity were reported by Weber (1967). Substitutions in the 2-position had the most significant effect on the basicity of the compounds. Hydroxy substituted compounds were more basic than methoxy-, methylthio-, and chloro-substituted compounds, respectively. The number of alkyl groups in the 4- and 6-positions also had a pronounced effect on basicity for methoxy substituted compounds; for instance, the greater the number of ethyl groups in place of hydrogen atoms, the more basic the compound.

Ward and Weber (1968) have shown that the aqueous solubility of the s-triazines is dependent upon the pH of the solution. Figure 1 was drawn from their data and clearly demonstrates that the compounds were much more soluble under acid conditions.

Recent studies by Ward and Weber (1969) show the influence of molecular structure on the ultraviolet spectra of 13 s-triazine compounds. Indications of protonation of ring nitrogen atoms in the 1- and 3-positions at low pH and tautomeric forms of the hydroxy-s-triazines are discussed.

Physical and chemical properties for a number of s-triazines have been reported by Gysin and Knüsli (1960) and Bailey and White

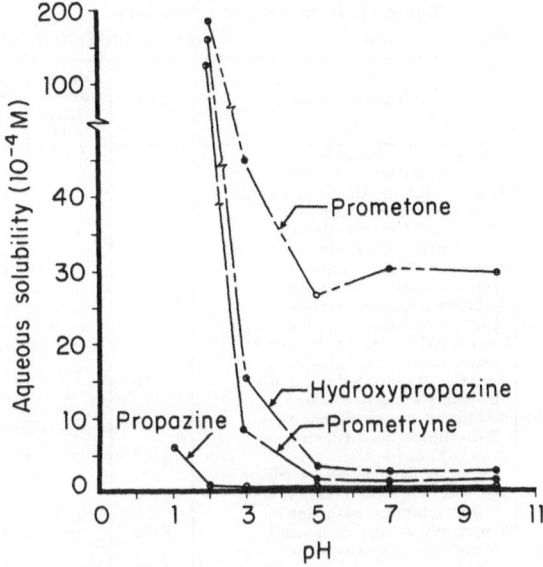

Fig. 1. Effect of pH on the aqueous solubility of s-triazines (WARD and WEBER 1968)

(1965). Some properties of s-triazines which are discussed in this paper are given in Table II (see also Reference Tables I and II in the Foreword).

IV. Aspects of adsorption from solution

The adsorption from aqueous solution of a solute by a solid is substantially more complex than the two-phase adsorption studies which have made the Freundlich and Langmuir equations popular. A comprehensive and realistic discussion on adsorption from solution has been prepared by KIPLING (1965). Other references which complement this treatise include the books of DAVIES and RIDEAL (1963) and EKWALL et al. (1965).

The important feature to recognize in studies concerning adsorption from solution is that the adsorption at the solid surface is a preferential or selective adsorption phenomenon. We assume that a solid in an aqueous system has water molecules adsorbed on its surfaces and that the water must be displaced if a solute is to be adsorbed. The system becomes even more complex if several different solute species are competing with water and with each other for sites on the solid surface. In a system such as the soil, which contains organic and inorganic ions and molecules in various stages of hydration in contact with inorganic and

Table II. *Properties of s-triazines*
(Weber 1967, Ward and Weber 1968, and manufacturer's data)

Common name[a]	Chemical name	pKa	Dipole moment[b] (Debye units)	Molar aqueous solubility (X10⁻⁴) pH 3	pH 7
Atrazine	2-chloro-4-ethylamino-6-isopropylamino-s-triazine	1.68	4.63	1.44	1.61
Ametryne	2-methylthio-4-ethylamino-6-isopropylamino-s-triazine	4.0	3.15	17.8	8.57
Atratone	2-methoxy-4-ethylamino-6-isopropylamino-s-triazine	4.20	—	90.4	76.0
Hydroxyatrazine	2-hydroxy-4-ethylamino-6-isopropylamino-s-triazine	—	—	11.5	0.30
Norazine	2-chloro-4-methylamino-6-isopropylamino-s-triazine	—	—	—	1.1
Desmetryne	2-methylthio-4-methylamino-6-isopropylamino-s-triazine	—	—	—	24
Noratone	2-methoxy-4-methylamino-6-isopropylamino-s-triazine	4.15	—	—	160
Ipazine	2-chloro-4-isopropylamino-6-diethylamino-s-triazine	1.85	—	1.15	1.13
Ipatryne	2-methylthio-4-isopropyl-amino-6-diethylamino-s-triazine	4.43	—	—	—
Ipatone	2-methoxy-4-isopropylamino-6-diethylamino-s-triazine	4.54	—	13.6	3.61
Hydroxyipazine	2-hydroxy-4-isopropylamino-6-diethylamino-s-triazine	5.32	—	41.2	30.1
Propazine	2-chloro-4, 6-bis (iso-propylamino)-s-triazine	1.85	4.52	0.21	0.20
Prometryne	2-methylthio-4, 6-bis (iso-propylamino)-s-triazine	4.05	3.54	8.53	1.67
Prometone	2-methoxy-4, 6-bis (iso-propylamino)-s-triazine	4.28	2.94	44.4	30.1
Hydroxypropazine	2-hydroxy-4, 6-bis (iso-propylamino)-s-triazine	5.20	—	15.4	1.96
Simazine	2-chloro-4, 6-bis (ethyl-amino)-s-triazine	1.65	—	0.29	0.25
Simetryne	2-methylthio-4, 6-bis (ethyl-amino)-s-triazine	—	—	31.7	20.8
Simetone	2-methoxy-4, 6-bis (ethyl-amino)-s-triazine	4.15	—	132	119
Hydroxysimazine	2-hydroxy-4, 6-bis (ethyl-amino)-s-triazine	—	—	9.60	0.22
Trietazine	2-chloro-4-ethylamino-6-diethylamino-s-triazine	1.88	4.77	1.22	1.25
Trietatryne	2-methylthio-4-ethylamino-6-diethylamino-s-triazine	—	—	0.74	0.09
Trietatone	2-methoxy-4-ethylamino-6-diethylamino-s-triazine	4.51	—	10.6	1.82
Hydroxytrietazine	2-hydroxy-4-ethylamino-6-diethylamino-s-triazine	—	—	11.1	1.60
Chlorazine	2-chloro-4, 6-bis(diethyl-amino)-s-triazine	1.74	4.43	0.92	0.86
Tetraetatone	2-methoxy-4, 6-bis(diethyl-amino)-s-triazine	4.76	—	—	—
GS-14260	2-methylthio-4-ethylamino-6-tert. butylamino-s-triazine	4.0	—	—	2.41
GS-16065	2-ethylthio-4-ethylamino-6-isopropylamino-s-triazine	—	—	—	2.41
GS-14253	2-methylthio-4-ethylamino-6-sec. butylamino-s-triazine	4.1	—	—	<40
SD-15418	2-(2-chloro-4-ethylamino-s-triazin-6-ylamino)-2-methylpropionitrile	1.0	—	—	3.65
Dyrene[c]	2,4-dichloro-6-o-chloro-anilino-s-triazine	—	—	—	<1
Benzoguanamine	2,4-diamino-6-phenyl-s-triazine	3.6	—	—	—

[a] In some cases, the names of the compounds were designated by the author. All compounds were obtained from the *Geigy Chemical Corporation*, with the exception of the herbicide SD-15418, which was obtained from the *Shell Development Company*, and the fungicide Dyrene, which was obtained from the *Chemagro Corporation*.

[b] Measured in dioxane at 20°C.

[c] Trade name.

organic colloidal surfaces, the picture of the adsorption phenomenon becomes extremely complex. It is for this reason that the use of theoretical adsorption equations generally do not adequately describe the results obtained from adsorption studies.

GILES *et al.* (1960) studied the adsorption from aqueous solution of a variety of solutes and developed a classification of adsorption isotherm types. The four basic types are given in Figure 2. The S-type

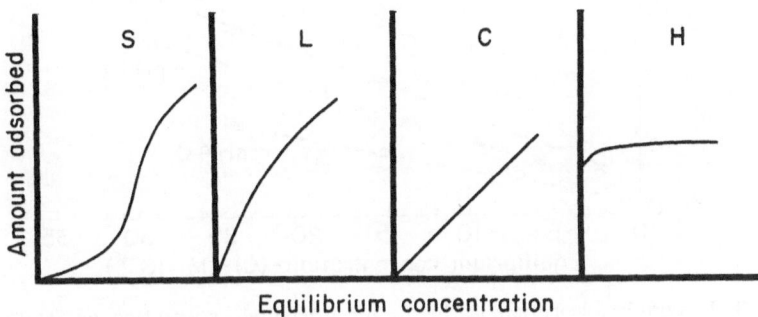

Fig. 2. Classification of adsorption isotherms (GILES *et al.* 1960)

isotherm is common when the solid has a high-affinity for the solvent. The L-type is the most common and occurs when the solid has a high-affinity for the solute. The constant partition or C-type isotherm is common when new sites become available as the solute is adsorbed from the solution. The H-type is rare and occurs only when the solute has a very high affinity for the solid.

The L-type of isotherm has been observed for the adsorption of *s*-triazines by Na-montmorillonite clay by WEBER (1966) and BAILEY *et al.* (1968). BAILEY found that the adsorption isotherms were well described by use of the Freundlich equation, but did not conform well to the Langmuir equation. Results of clay-prometone systems taken from WEBER (1966) were transformed by use of both equations and were found to be better described by the Langmuir equation than the Freundlich equation. The Langmuir adsorption isotherms for the clay-prometone systems are given in Figure 3. Note that the fit of the isotherms is usually poorest at the low concentration. The Freundlich equation is the analytical expression for the general parabola, and theoretically adsorption increases indefinitely with increasing concentration. The Langmuir equation is an expression of a rectangular hyperbola passing through the origin and tends to a definite limit for the adsorption of the solute on the surface. In the case of WEBER's data, it appears that adsorption was approaching a specific limit at each pH level. However, applicability of a Langmuir equation is not, in itself, an

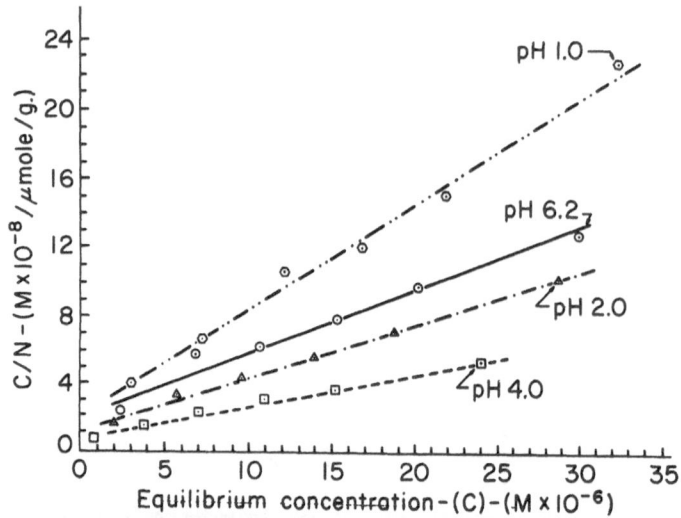

Fig. 3. Langmuir adsorption isotherms for prometone adsorption on Na-mont-morillonite at four pH levels (WEBER 1966)

indication that monolayer coverage has occurred or that any one particular adsorption mechanism was involved. In this case it just happened to describe the data. The pH dependence of the isotherm indicates that ionic bonding was involved and that the triazine molecules were complexing with hydrogen ions.

Because adsorption increased with a decrease in pH above values of pH 4 and decreased with increasing acidity below pH 4, it appeared that hydrogen ions were involved in two different ways. As the pH approached four from the basic side hydrogen complexed with the triazine molecules and adsorption increased; under more acid conditions, the hydrogen ions competed with triazine molecules for sites on the clay surface. The adsorption mechanisms which are believed to be involved will be discussed in detail later in this paper.

The H-type adsorption isotherm was observed for diquat [6,7-dihydrodipyrido (1,2-a:2',1'-c) pyrazidiinium dibromide] and paraquat (1,1'-dimethyl-4,4'-bipyridinium dichloride) on Na-montmorillonite and Na-kaolinite clays and for prometone on H-montmorillonite and on an Amberlite IR-120-H cation-exchange resin (WEBER et al. 1965 and 1968 b). Diquat and paraquat, being large organic cations, were adsorbed by the clay through a combination of coulombic and van der Waals forces. Adsorption of prometone by the hydrogen-saturated clay and resin was attributed to complexing of the triazine molecules with hydrogen atoms on the adsorbents.

V. Solvent effects

Adsorption of organic molecules on clay mineral surfaces is greatly affected by the kind and the amount of solvent involved. Because the reaction almost always occurs in aqueous solutions, the properties of water should be included in our discussion.

Liquid water has an open structure, rather than a close-packed one (NANCOLLAS 1966). This has been deduced from X-ray measurements on ice crystals. On melting, there is a breakdown of this structure and the molecules become more close packed with a resulting increase in density. The space occupied by an individual water molecule is largely determined by that of the oxygen atom. The two hydrogens take up practically no space (KOHNKE 1968). Water molecules, however, do not exist individually. The hydrogen in one water molecule serves to connect to the oxygen atom in another molecule, thus linking the molecules together. This hydrogen bonding results in the formation of a hexagonal lattice structure of many molecules held tightly together. Water can therefore be regarded as a giant polymer of hydrogen-bonded water molecules. It has been described as having a "cluster" or cage-type structure like the water in water clathrates, with monomeric water occupying the cavities within the structure (KOSOWER 1968). FRANK and WEN (1957) have proposed a new picture of water as consisting of "flickering clusters" of hydrogen-bonded molecules, in which the cooperative nature of cluster formation and relaxation is related to the partially covalent character which is postulated for the hydrogen bond.

Water is a good solvent in as much as it is a dipole and consequently can orient itself in such a direction that the negative pole will contact the positive pole of the compound, and the positive pole of the water will contact the negative pole of the compound. The dielectric constant of water is high because it takes much energy to displace the hydrogen bonds. In the process of dissolving, a compound must disrupt these hydrogen bonds. A compound is soluble when the attraction of its molecules for water molecules is stronger than the attraction of the molecules of the compound for each other. Organic molecules which are capable of hydrogen bonding have the strongest interaction with water. Both the hydrophobic and hydrophilic portion of the molecules must be considered active in influencing the water structure (AMIS 1966). Some ions and molecules are structure breakers, while others appear to promote water structures. Information about the effects of ions and molecules on the structure of water and the phenomenon of ion hydration is still only vaguely understood.

Water is a semi-strong base which dissociates as depicted in the equation:

$$H_2O \rightleftharpoons H^+ + OH^- \qquad [4]$$

AMIS (1966) points out that both hydrogen and hydroxyl ions are highly hydrated in solution and suggests that the species be referred to in the form H_3O^+ and $H_7O_4^-$, respectively. FUTRELL and TIERNAN (1968) have recently shown that an unsolvated proton cannot exist in solution. They propose that hydrogen exists in solution as the H_3O^+, $H_5O_2^+$, or $H_9O_4^+$ species, depending on the acidity of the system.

The interactions of water at clay surfaces has recently been reviewed by Low (1961). He reported that water adsorbed on clay surfaces is more ordered than that in bulk solution. Exchangeable ions on the clay surface were found to affect the heats of wetting and adsorption of water. In the presence of large or multivalent cations, there appeared to be little or no order in the adsorbed water. The specific volume, viscosity, and freezing resistance of adsorbed water was directly related to its structural development. Consequently the magnitudes of these properties decrease continuously with distance from the mineral surface. Nowhere was the structure of water so rigid that ions could not diffuse through it. Low stated that although adsorbed water has a precise molecular arrangement and is "ice-like," the structure is not that of ice.

Studies by MORTLAND et al. (1963) suggested that residual water in the interlayer surfaces of expanding clay minerals is dissociated to a greater degree than free water. Recent studies by HARTER and AHLRICHS (1967) indicate that the pH of water at the clay surface is lower than that in the suspension. When suspension pH was 6, 5, 4, and 3, it was estimated that the pH at the clay surface was approximately 4.5, 4.0, 3.2, and 2.5, respectively. Therefore, water on the surface of clay minerals is more acidic than in the bulk solution. It can be seen then that water affects the adsorption of organic compounds on clay surfaces in several ways, including a) hydration of cations and molecules in solution, b) adsorption on the clay surface directly with a resultant change in structure and acidity, c) reaction with hydrogen and hydroxyl ions to form highly hydrated species, d) reactions with aluminum ions to form various hydroxides, and e) interactions which influence the solubilities of the organic compounds.

VI. Clay-organic reactions

Reactions between different types of organic compounds and clay minerals have been investigated by a great many workers. A comprehensive review and discussion has been prepared by GRIM (1968). It has been found that organic cations and organic bases enter into cation-exchange reactions with clay minerals. In addition to the coulombic force, organic ions are held by van der Waals forces. Small ions are adsorbed only up to the CEC, whereas larger ions may be adsorbed in excess; and these excess molecules are not dissociated but are probably adsorbed by van der Waals forces and/or hydrogen bonds.

Adsorption of organic ions on the interlayer surfaces of montmorillonites causes a shift in the c-axis spacings as detected by X-ray diffraction measurements. Flat shaped ions lie with their flat surfaces parallel to the silicate sheets in the interlayer spacings. However, the ions may be adsorbed in various orientations; and in some cases more than one molecular layer may be adsorbed.

The water-adsorbing properties of montmorillonite are gradually reduced as the basal surfaces are coated with the organic ions. In general, the larger the organic ions, the greater the reduction in the water-adsorbing capacity.

Nonionic organic molecules of a polar character are also adsorbed by clay minerals. The polar water molecules are displaced from the clay surface, but the inorganic cations are not necessarily displaced. The organic dipolar molecules are generally adsorbed in the interlayer spacings of montmorillonite in positions as flat as possible. Dipolar organic molecule adsorption is generally attributed to hydrogen bonding of the $C=O \ldots$ cation, $C=O \ldots H_2O \ldots$ cation, $N-H \ldots O$, and $C-H \ldots O$ type. Nonpolar portions of organic molecules are thought to be adsorbed by van der Waals forces only.

Nonpolar molecules such as saturated hydrocarbons are not actively adsorbed on clay surfaces, and polar groups are necessary for organic compounds to be adsorbed directly by clay minerals.

GREENLAND (1965 a) recently reviewed the phenomena of clay-organic reactions and provides an excellent summary of the forces involved and the factors which affect adsorption, as reproduced in Tables III and IV. In addition to the factors listed in the tables, others have been observed by a number of investigators. TALIBUDEEN (1955) recognized that organic nitrogen bases reacted directly with a H-montmorillonite through a cation-exchange mechanism which was pH dependent. The reaction was also dependent on the size of the molecule and the number of proton-accepting groups on it. He suggested that

Table III. *Forces involved in the interactions between clays and organic compounds* (from GREENLAND 1965 a)

1. Coulombic attraction between:
   a. negatively charged surface and positively charged organic compound
   b. positively charged surface, or ion or oligo-ion[a] at surface, and organic anion
2. van der Waals forces, composed of:
   a. Polar
       (1) Charge—dipole interactions
       (2) Dipole—dipole interactions
           (including H-bonding)   between   a. surface and organic compounds
       (3) Charge—induced dipole interactions   b. adsorbed molecules of similar species
       (4) Dipole—induced dipole interactions   c. adsorbed molecules of dissimilar species

   b. Non-polar
       (1) Dispersion forces

[a] For example, positively charged hydroxy aluminum oligomers.

Table IV. *Summary of properties of components of the systems which influence the adsorption of organic compounds by clays* (from Greenland 1965 a)

Properties of surface	Properties of organic compound	Properties of medium
a. Extent (total surface area)	a. charge	Those which affect its behavior:
b. Accessibility (size and tortuosity of pores)	b. size	a. as solvent for organic compound
c. Chemical nature of atoms forming surface (O or OH)	c. polarity	b. as competitor with organic compounds for sites on clay surface
d. Charges at surface (spatial density and distribution, origin)	d. polarizability	
e. Exchangeable ions on surface	e. shape	
f. Configuration of surface	f. flexibility	

the basicity of the compounds was the most important property determining their adsorption. The more basic compounds were adsorbed in greater amounts than less basic ones.

McLaren *et al.* (1958) observed that the maximum adsorption of proteins on montmorillonite occurred in the region of the isoelectric point of the compounds. Less adsorption was noted at low pH and this was attributed to competition by H^+ ions for adsorption sites on the clay surfaces. This phenomenon was also noted by Armstrong and Chesters (1964), but they attributed the lesser adsorption at low pH to an increase in charge density of the protein as acidity increased, *i.e.*, the total number of positive charges carried/molecule increased and less protein was needed to satisfy the negative charge on the clay mineral. Decreased adsorption at high pH was attributed to the formation of protein anions which were repelled by the clay surface.

Weiss (1963) characterized the complexes of more than 8,000 different derivations of mica-type layer silicates. He found that the ion-exchange reaction of alkylammonium ions for inorganic cations on the clay surface was dependent upon the charge density of the silicate and the length of the *n*-alkyl chains of the organic molecules. Basal spacings of the clay minerals increased with increased length of the *n*-alkyl chains, but swelling with aromatic or branched-chain compounds gave mixed results. The orientation of these molecules depended on the nature of the molecule, its polar groups, and their size and shape. The spacings depended on the pretreatment of the samples, but analytically no differences were found, thus indicating that different arrangements of the alkyl chains are possible between the silicate sheets. With clays of low charge density, the chains were lying flat. With clays of higher charge density, the alkyl chains tended to be erect.

VII. Adsorption studies with s-triazines

General review articles concerning the reaction of pesticides and clay minerals have been prepared by BAILEY and WHITE (1964), GREENLAND (1965 a), and KUNZE (1966).

One of the most comprehensive reviews concerning the forces involved in the adsorption of organic herbicides by clay minerals was prepared by FRISSEL (1961). Some of these studies were summarized in a later publication (FRISSEL and BOLT 1962). FRISSEL studied the effects of suspension pH and electrolyte concentration on the adsorption of 14 organic compounds, including the chloro-s-triazines (simazine, chlorazine, and trietazine) on the clay minerals illite, montmorillonite, and kaolinite. He found that the adsorption process was pH-dependent and postulated that the triazines were adsorbed as neutral molecules in neutral and basic environments and as positively charged ions in acidic solutions. Montmorillonite adsorbed greater amounts of the triazines than illite or kaolinite, respectively. Chlorazine was adsorbed more strongly on Ca-illite at high salt concentrations ($> 0.2N$ calcium chloride) than at low salt concentrations and this was attributed to the salting-out effect. It was also observed that the adsorption of the triazines both at high salt concentrations and at low pH appeared to be fully reversible, and in 50 percent ethanol the adsorption of simazine was almost negligible. FRISSEL made calculations for the adsorption of trietazine on Na-montmorillonite based on the assumption that adsorption in the range pH 8 to 10 was due to van der Waals forces only and adsorption in the range pH 2.9 to 8.0 was due to a combination of van der Waals forces and coulombic forces. The calculated values in the pH range 5 to 9 were smaller than the experimental values and the greater adsorption which occurred was attributed to adsorption of the triazine molecule to SiOH-groups on the clay surface. In the pH range 2.9 to 4.0, the calculated adsorption was much too high and this was attributed to errors in the assumptions which did not take into account the limited solubility of the compound.

HARRIS and WARREN (1964) studied the adsorption from aqueous solution of several herbicides, including atrazine and simazine, by bentonite clay, muck soil, and exchange resins. Atrazine was totally adsorbed from solution by an Amberlite IR-120 cation-exchange resin. The exchangeable cation on the resin was not stated, but in view of the high adsorption it was undoubtedly the H-form. Adsorption of atrazine by bentonite was much greater at pH 4.1 than 8.2. The high adsorption at low pH was attributed to the association of the triazine molecule with protons on the clay surface as suggested by FRISSEL. Simazine was adsorbed in greater amounts than atrazine by bentonite at pH 8.5. Both compounds were adsorbed in greater amounts at 0° than at 50°C. The two compounds were also readily desorbed with deionized water from muck at pH 5.6 and from bentonite at pH 8.5.

The adsorption from aqueous solution (pH 7, saturated calcium hydroxide) of five s-triazines on 25 soil types, four clay minerals, and two peat soils was determined by TALBERT and FLETCHALL (1965). In almost all cases, adsorption decreased in the order: prometryne > prometone > simazine > atrazine > propazine. Adsorption was not related to the water solubility of the compounds. Correlation analyses indicated that adsorption of prometryne and prometone were more highly related to the clay content of the soils, whereas adsorption of the chloro-s-triazines was more highly related to organic matter content. Adsorption of simazine and atrazine by a Marshall silty clay loam (4.2 percent organic matter, 30 percent clay, pH 5.4) was well described by the Freundlich equation. Greater amounts were adsorbed at 0° than at 50°C. The herbicides were readily leached from the soil with deionized water and it was concluded that the adsorption mechanism involved was reversible. In adsorption studies with clay minerals, greater adsorption of atrazine and simazine occurred on montmorillonite than on illite or Putnam clays, respectively. Slightly greater adsorption of the two compounds occurred at pH 5 than at pH 7. No adsorption resulted on kaolinite clay.

The influence of temperature and time on the adsorption from aqueous systems (pH 6, $0.025M$ sodium phosphate buffer) of four herbicides, including prometone, on montmorillonite and kaolinite clays, charcoal, and Amberlite IR-411 anion-exchange resin in the chloride form was studied by WEBER et al. (1965). Adsorption of prometone by montmorillonite was nearly at equilibrium in one hour. The small amount of adsorption which occurred after one hour was attributed to slow diffusion of the triazine molecules into the interlayer spacings of the clay. Greater adsorption occurred at 10° than at 55°C. Adsorption was postulated to have occurred by H-bonding of the s-triazine molecules to oxygen in the silicate lattice of the clay and by association of the prometone molecules with protons in solution or at the clay surface with the resultant adsorption of the prometone cations by cation-exchange forces. Small amounts of prometone were adsorbed by the anion-exchange resins. The adsorption was temperature dependent and was attributed to weak physical adsorption forces. Prometone was not adsorbed by kaolinite clay in these systems. X-ray diffraction studies showed that prometone was adsorbed in the interlayer spacings of Na- and H-montmorillonite and that the $d(001)$ spacings increased with the amount adsorbed. Figure 4 was plotted with data from these studies. Basal spacings of 13.0 to 13.5 A correspond to adsorption of prometone in a flat position with the plane of the triazine ring parallel to the silicate sheets (Fig. 4). Spacings of 18.0 to 18.5 A could have resulted from various orientations of the s-triazine molecules between the clay plates. Molecular model measurements indicate that the s-triazine molecules adsorbed in a position perpendicular to the silicate sheets could result in spacings of 17.0 to 18.5 A. Adsorption of double

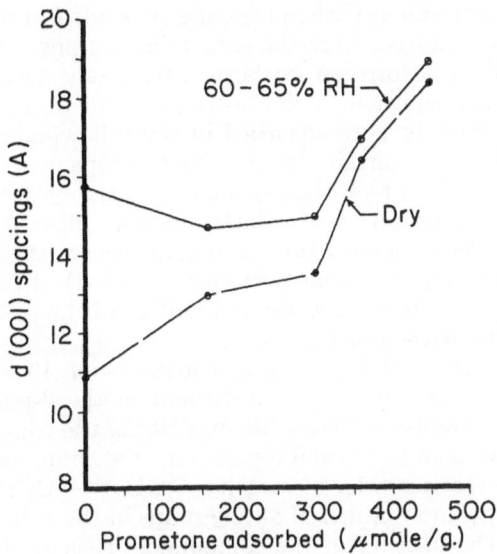

Fig. 4. Variation in basal spacings of Na-montmorillonite clay with the amount of prometone adsorbed (WEBER *et al.* 1965)

layers of prometone molecules between the clay plates could result in spacings of approximately 18.0 to 18.5 A. There is also the possibility of the s-triazine molecules stacking in the interlayer spacings at various angles.

Interlayer spacings of 17.6 A were reported by CRUZ *et al.* (1968) in studies on the adsorption of hydroxypropazine by H-montmorillonite. The authors suggested the presence of two layers of the hydroxypropazine cations between the montmorillonite sheets.

It was observed that prometone was adsorbed in large amounts at low pH and that prometone adsorbed by H-montmorillonite resulted in a hydrophobic organic-clay complex that floated on the surface of the aqueous suspension (WEBER *et al.* 1965). Additions of concentrated hydrochloric acid to the suspension caused the hydrophobic colloid to disperse and it appeared that hydrogen ions displaced prometone from the clay surface.

In adsorption studies with simazine and 18 soils in aqueous solution, NEARPASS (1965) found that the organic matter content and the titratable acidity of the soils were the factors most highly related to the amount of adsorption. Adsorption of simazine increased as pH decreased. In leaching studies, applications of lime increased the depth of penetration of simazine. In a later study, NEARPASS (1967) found that adsorption from aqueous solutions of atrazine and simazine by a Bayboro clay soil (6.8 percent organic matter, 55 percent clay, pH 4.95) was dependent upon the exchange acidity of the soil. Simazine was adsorbed in greater amounts than atrazine. The adsorption was similar when Ca, Mg, K, and Na were the saturating cations.

The Freundlich adsorption isotherm adequately describes adsorption from aqueous solution of five s-triazines by four soils (HARRIS 1966). The compounds were adsorbed in the following order: prometryne > prometone > simazine $\simeq$ atrazine $\simeq$ propazine. Later studies by HARRIS (1967), in which the movement of the herbicides through soil columns by subirrigation was determined by biosasay techniques, showed that the movement of the compounds was in the reverse order of their adsorption (prometone was not included). In other words, prometryne was adsorbed in greater amounts than chloro-triazines, and it also moved less with subirrigation.

Determinations of the ionization constants of 13 s-triazines by WEBER (1967) showed that the substituent in the 2-position on the molecule had the greatest effect on the basicity of the compounds. With similar substituents in the 4- and 6-positions, the compounds generally followed the decreasing order of basicity $-OH > -OCH_3 > -SCH_3 > -Cl$. The type and number of alkyl groups in the 4- and 6-positions also influenced the basicity of the compounds. Recent studies (WARD and WEBER 1968) show that molecular structure and pH also affects the water solubility of the s-triazines. Adsorption from aqueous solutions of the 13 triazines by Na-montmorillonite clay showed that adsorption was dependent upon the molecular structure of the compounds and the pH of the system (WEBER 1966). Maximum adsorption of each compound occurred in the vicinity of its pK_a. Adsorption as high as 700 μmole/g. suggested that the s-triazines were adsorbed as monovalent cations, but since adsorption was dependent on factors such as the amount of clay, pH, etc. it was concluded that competition from H_3O^+ and Na^+ ions confounded the picture. The more basic compounds were generally adsorbed in greater amounts than less basic compounds; but the key to the amount of adsorption was the molecular structure of the compounds. Decreased adsorption resulted as the 2-substituent was changed in the following order: $-SCH_3 > -OCH_3 > -OH > -Cl$. The type and number of alkyl groups in the 4- and 6-positions also influenced the amount of adsorption. For a series of ethyl substituent compounds, as the alkyl groups were replaced by H atoms, adsorption decreased in the order: $(C_2H_5)_4 > (C_2H_5)_3 > (C_2H_5)_2$. It appeared that the 2-substituent determined the primary adsorption mechanism and that changes in the alkyl groups in the 4- and 6-positions affected the basicity of the compounds and hence the amount of adsorption. The studies showed that Na^+ ions were displaced from montmorillonite surfaces as prometone was adsorbed, even at pH values as high as 8.2. Also, prometone could be displaced from the clay surfaces with a neutral salt solution and with $0.01N$ sodium hydroxide. Adsorption mechanisms were postulated and discussed, and these will be discussed and re-evaluated later in this paper.

McGLAMERY and SLIFE (1966) studied the adsorption from aqueous solution of atrazine on a Drummer silty clay loam soil [6.3 percent organic matter, 30 percent clay (predominantly vermiculite), pH 6] and on humic acid. Adsorption on the soil was greater at pH 3.9 than at 8.0 and at 0.5° than at 40°C. Adsorption on the humic acid was attributed to ionic bonding caused by protonation of the amino groups on the atrazine molecule at low pH. It was greater at low pH and at the higher temperature. The temperature affect on the humic acid was the opposite of that observed on the soil. Approximately 11 percent of the atrazine adsorbed on the humic acid was desorbed with deionized water and approximately 69 percent was desorbed when calcium carbonate was added to the system.

The adsorption from aqueous solution of 29 herbicides, including 19 s-triazines, by two soils [a sandy loam (12 percent organic carbon, 6.6 percent clay) and a chalky boulder clay (2.1 percent organic carbon, 39 percent clay)] has been correlated with their chromatographic movement (HANCE 1967). The author suggested that the chromatographic movement of the chemicals was an assessment of their hydrophilic-hydrophobic balance. Correlation coefficients for herbicide adsorption versus chromatographic movement were 0.90 and 0.85 for the loam and clay soils, respectively. It was concluded that the high correlations suggested that the basic mechanisms of adsorption involved forces of non-specific nature. HANCE proposed that the chromatographic technique might give a useful indicator of the soil adsorption behavior of nonionic compounds. However, most of the chemicals employed in the study were compounds that ionize appreciably in aqueous solutions, and it is unlikely that the chromatographic technique would be much simpler than the routine adsorption technique.

The adsorption from aqueous solutions of six s-triazines on Na- and H-montmorillonite was studied by BAILEY et al. (1968). Adsorption isotherms of the compounds on Na-montmorillonite were well described by the Freundlich equation but not by the Langmuir equation. The order of adsorbability on the Na-montmorillonite was simetone ≫ atratone > prometone > trietazine > propazine > atrazine. A direct relationship was found to exist between water solubility and adsorbability with the exception of atrazine. Adsorption of simetone on H-montmorillonite was found to occur in amounts in excess of the CEC, if the compound was assumed to be adsorbed as a divalent cation.

LAVY (1968) in adsorption studies of s-triazines on three soils [Sharpsburg silty loam (4.4 percent organic matter, 35 percent clay, pH 5.8), Keith sandy loam (2.9 percent organic matter, 13 percent clay, pH 6.2), and Anselmo sandy loam (1.7 percent organic matter, 10 percent clay, pH 7.0)] found that adsorption decreased in the order: simazine > propazine > atrazine. Adsorption was higher as acidity, organic matter, and clay content of the soils increased.

Adsorption-desorption studies by WEBER et al. (1968 b) showed that adsorption of prometone by charcoal in an aqueous system was reversible and was well described by the Freundlich equation. Prometone was adsorbed in small amounts by Amberlite IR-120-Na (cation) and IR-400-Cl (anion) exchange resins and was readily desorbed with deionized water. Adsorption was attributed to weak physical adsorption forces. Prometone was adsorbed by an Amberlite IR-120-H (cation) exchange resin in amounts approximating the CEC, if the compound is assumed to be adsorbed as a divalent cation. No significant change in pH occurred in the H-resin system indicating that prometone was associated with H atoms on the resin surface. Prometone was readily desorbed from the H-resins with $1M$ sodium chloride as hydroxypropazine. The high density of H^+ ions on the resin surface apparently caused the hydrolysis of the s-triazine molecules to the hydroxy form. Similar studies (WEBER and WEED 1968) with Na-saturated montmorillonite and kaolinite clays showed that adsorbed prometone was readily desorbed with deionized water and salt solutions. Less prometone was desorbed from Na-montmorillonite with $1M$ barium chloride than with deionized water. This was attributed to the "salting out" effect observed by FRISSEL (1961).

RUSSELL et al. (1968 b) and CRUZ et al. (1968) presented spectroscopic evidence on the interactions between chloro- and hydroxy-s-triazines and H- and NH_4-montmorillonites. The authors concluded that atrazine and propazine were hydrolyzed to the hydroxy forms on the clay surfaces and adsorbed as the protonated hydroxy species. The authors also suggested that the chloro-s-triazines are protonated on clay surfaces saturated with metallic cations by association with protons donated by the highly dissociated water on the clay surface. In recent infrared studies, SKIPPER et al.[3] suggested that H- and Al-montmorillonite were acidic enough to hydrolyze atrazine to hydroxyatrazine but that H-allophane was not acidic enough to accomplish this.

RUSSELL et al. (1968 a) found that 3-aminotriazole was strongly adsorbed by Ca-, Cu-, Ni-, and Al-montmorillonite. They postulated that the triazole molecule became protonated by the highly polarized water molecules directly coordinated to the cations on the clay surfaces and were adsorbed as organic cations.

Recent adsorption studies by WEBER et al. (1969 c) showed that the pH of the system and the molecular structure of the s-triazines influenced their adsorption on organic soil colloids in a matter analogous to their adsorption on clay minerals. Adsorption of the compounds was attributed to the complexing of the s-triazine molecules by ionizable H^+ ions on functional groups of the organic colloids and to adsorption of the protonated species by cation exchange. Triazine-organic matter complexes are discussed elsewhere in detail by HAYES (1969).

---

[3] SKIPPER, H. D., V. V. VOLK, M. M. MORELAND, and K. V. RAMAN: Unpublished results (1969).

Factors which influence the adsorption of s-triazines by clay colloids thus included a) molecular structure, basicity, and solubility of the compounds, b) type of clay mineral, c) acidity, d) type and concentration of ions in solution and on the clay surface, and e) temperature.

VIII. Mechanisms of adsorption of s-triazines by clay minerals

The effects of pH and molecular structures of s-triazines adsorbed on Na-montmorillonite were mentioned previously, and Figure 5 was taken from studies by WEBER (1966). Mechanisms of adsorption in that

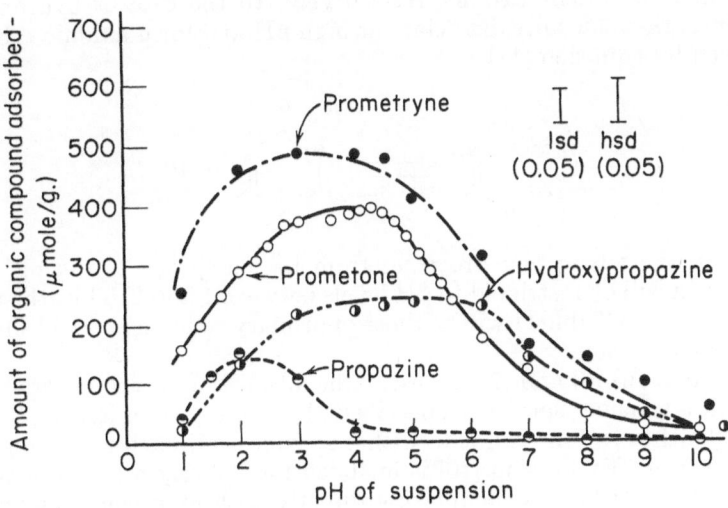

Fig. 5. Effect of pH on the adsorption of four related s-triazines [4,6-bis(isopropyl-amino) series] on Na-montmorillonite clay (WEBER 1966)

study were discussed by use of equilibrium equations. LAILACH et al. (1968) improved the symbolism and recently applied analogous equations to the adsorption of pyrimidines and purines by montmorillonite. I have further simplified the equations and will now use them to discuss the adsorption mechanisms operating between the s-triazines and montmorillonite for Figure 5. The following processes will be considered:

$$R + H^+ \rightleftharpoons RH^+ \qquad\qquad [5]$$
$$R + X\text{-mont.} \rightleftharpoons RX\text{-mont.} \qquad\qquad [6]$$
$$RH^+ + X\text{-mont.} \rightleftharpoons RH\text{-mont.} + X^+ \qquad\qquad [7]$$
$$H^+ + RH\text{-mont.} \rightleftharpoons H\text{-mont.} + RH^+ \qquad\qquad [8]$$
$$H^+ + X\text{-mont.} \rightleftharpoons H\text{-mont.} + X^+ \qquad\qquad [9]$$
$$R + H\text{-mont.} \rightleftharpoons RH\text{-mont.} \qquad\qquad [10]$$

where: R = triazine compound; RH^+ = triazine cation; X-mont. = montmorillonite with X = exchangeable cations Na, Li, $\frac{1}{2}$Ca, $\frac{1}{2}$Mg, etc.; H-mont. = hydrogen (aluminum) montmorillonite; and H^+ = hydrogen ion, as the hydrated H_3O^+ species.

Process [5] is governed by the basicity of the s-triazine compounds, indicated by the equilibrium constant K_a, and is given by:

$$pK_a = \log([RH^+]/[R]) + pH$$

The pK_a of the compound determines the pH at which maximum adsorption occurs. Each compound may have more than one pK_a because the molecules may be doubly or triply protonated under very acid conditions (see equations [1] through [3]). They would thus be divalent or trivalent cations, respectively. In the case of hydroxy-s-triazines, the molecules dissociate at high pH and form anionic species as given by equation [11]:

$$[11]$$

To keep the adsorption processes from becoming too confusing, the discussion will be restricted to pH levels between 1 and 8. Thus the pK_a values given in Table II will be those of primary concern and will suffice for this discussion.

Expressions [6] and [7] represent the adsorption of the neutral and cationic s-triazines species, respectively, by montmorillonite. In [7], [8], and [9], other cations compete with s-triazines cations for sites on the clay surfaces. Expression [10] is included for the case where the s-triazines are adsorbed by complexing directly with hydrogen ions on an H-montmorillonite.

Assume we begin with one of the s-triazines in aqueous solution with Na-montmorillonite at a neutral pH, say approximately pH 7 (see Fig. 5). At this pH, the principal mechanism of adsorption would be by process [6]. The organic molecules may be associated directly with cations X on the clay surface by replacing some or all of the water molecules or they may be directly in contact with the silicate surface. There is also the possibility that the s-triazine molecules may be associated with the cations X through H-bonding by way of the highly dissociated water molecules surrounding the cations, as discussed by FARMER and MORTLAND (1966) and RUSSELL et al. (1968 b).

As the pH is decreased, by the addition of acid, the s-triazine molecules become protonated as shown in expression [5] and these cations are adsorbed by the montmorillonite in exchange for cations X (X = Na in this case) by process [7]. This results in an increase in adsorption as shown in Figure 5. The addition of acid to the system also causes

some of the X cations on the clay to be replaced by hydrogen by process
[9] and the resulting H-montmorillonite may absorb neutral R molecules
by process [10]. In the vicinity of the pK_a (e.g., $pK_a = 4.28$ for prometone), adsorption reaches a maximum and then begins to decline with
a further decrease in pH. The decrease in adsorption is probably due to
competition from the H^+ ions according to expression [8]. It has also been
found (WEBER 1966) that the addition of other salts such as $0.1M$
barium chloride causes a large decrease in adsorption and this is because
expression [7] is driven to the left releasing the s-triazine cations to the
solution.

Water molecules participate in the adsorption mechanisms in
various ways (see the discussion on solvent effects). The organic and
inorganic molecules and ions are hydrated to various degrees. Montmorillonite clay is hydrated and the water at the clay surface is thought
to have a slightly different structure than that in the bulk solution (Low
1961). Water associated with the high-field cations (Mg^{2+}, Ca^{2+}) on the
clay surface appears to have a much-higher-than-normal degree of dissociation (MORTLAND 1963). The higher-than-normal state of dissociation has expressed itself as a lower effective pH in the interlayer volume
than in the bulk solution (HARTER and AHLRICHS 1967 and RUSSELL
et al. 1968 b).

Addition of acid to the system also causes some release of Al^{3+} ions
from the montmorillonite clay and the various aluminum hydroxide
species probably have some influence on the pH of the system. LAILACH
et al. (1968) found that the increase of exchangeable Al^{3+} due to hydrochloric acid treatment of montmorillonite for 24 hours was 0.3 me/100g.
at pH 4, and 7.3 me/100g. at pH 2. Therefore, for the results in Figure
5, in which the adsorption was completed within four hours, the
influence of Al^{3+} was probably very small.

The adsorption of prometone by several adsorbents known to
possess cation-exchange properties has recently been completed by the
author[4] and some of the data from these studies are presented in Figure
6. The curves are not comparable on a quantitative basis since differing
amounts of adsorbent were used, but qualitatively some comparisons
may be made. The Na-montmorillonite and vermiculite clays exhibited
similar characteristics and the adsorption mechanisms involved are
probably similar. Adsorption of prometone by the acid muck decreased
with pH at levels below the pK_a and approached zero adsorption at
pH 1. This adsorption has previously been postulated to occur through
complexing of the s-triazine molecules with ionizable H^+ ions on functional groups of the organic colloids and/or through adsorption of
triazine cations by cation-exchange forces (WEBER et al. 1969 c).
Adsorption of s-triazines by organic colloids are discussed in detail
elsewhere by HAYES (1969).

---

[4] WEBER, J. B.: Unpublished results (1969).

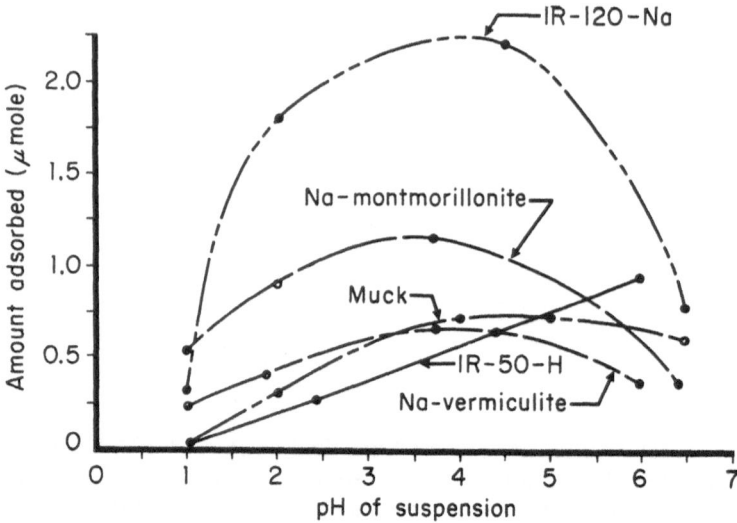

Fig. 6. Effect of pH on the adsorption of prometone on five adsorbents (40 ml. of 5 × 10⁻⁵M solutions used with 2.5, 10, 20, and 100 mg. of Na-montmorillonite, Amberlite IR-120-Na and IR-50-H resins Na-vermiculite, and muck, respectively) (see footnote 4)

Prometone adsorption on the weakly acidic Amberlite IR-50-H cation-exchange resin decreased linearly with a decrease in pH (Fig. 6). This is probably because the ionizable H^+ ions on the —COOH groups of the resin become undissociated at low pH and the s-triazines cations are not able to compete with the protons for sites on the exchange resin. The phenomenon is probably somewhat analogous to the muck system, but the differences in the shapes of the curves suggests that other factors or adsorption mechanisms are involved.

Adsorption of prometone by the strong acid Amberlite IR-120-Na cation-exchange resin increased with a decrease in pH in a manner similar to the clay minerals, but at pH levels below the pK_a both hydroxypropazine and prometone species were detected in the solution. The high surface charge density of the resin apparently caused hydrolysis of the prometone molecules, and as H^+ ions displaced the s-triazine molecules from the resin surface, both species were found in the bulk solutions. This acid catalysis of prometone hydrolysis was previously observed using Amberlite IR-120-H resin (WEBER et al. 1968 b). Hydrolysis of the s-triazines is discussed briefly later in this paper.

IX. Effects of molecular structure on adsorption

The effects of the molecular structure of s-triazines on their adsorption by montmorillonite clay was discussed previously (WEBER 1966).

Figure 5 showed the adsorption of four 4,6-bis(isopropylamino)-s-triazines by montmorillonite from equimolar solutions. By comparing adsorption of the compounds in the vicinity of their pK_a values, one observes that the amount of adsorption was dependent on the substituent in the 2-position and was in the order: —SCH_3 > —OCH_3 > —OH > —Cl. This corresponds to the series prometryne > prometone > hydroxypropazine > propazine, respectively. The same order was found for the series ipatryne > ipatone > hydroxyipazine > ipazine (WEBER 1966). Related studies recently completed by the author show that the relationship holds for the series ametryne > atratone > atrazine (Fig. 7). Unfortunately hydroxyatrazine was not included. The figures cannot be directly compared, however, since slightly different systems were involved. Greater adsorption of —SCH_3 substituted compounds than comparable —OCH_3 compounds has previously been attributed to electronegativity and steric differences (WEBER 1966). It could also be added that the lower solubility of the —SCH_3 compounds (see Table II and Fig. 1) indicates that these compounds do not hydrogen bond with water molecules as readily as —OCH_3 groups and might therefore be more inclined to leave the solution phase and adhere to the silicate surface.

The lower adsorption of the hydroxy compounds has previously been attributed to adsorption through partial positive charges at the 4- and 6-positions only, in comparison with partial positive charges at the 2-, 4-, and 6-positions of the methylthio and methoxy compounds (WEBER 1966). The low amount of adsorption of the chloro-s-triazines has previously been attributed primarily to their low basicity. Because of the complexity of the molecular structures (WEBER 1967) and solubility and solution pH effects (WARD and WEBER 1968), our understanding of the mechanisms is not completely clear.

Figure 7 shows the adsorption of SD-15418 on montmorillonite in comparison with atrazine. The compounds differ in structure in the 6-position only. The differing substituent groups were —C_3H_7 (iso) for atrazine and —$C(CH_3)_2CN$ for SD-15418. The substitution of —$C\equiv N$ for a hydrogen atom in the isopropyl groups resulted in less adsorption of the compound by the clay colloid.

It appears that the maximum adsorption of SD-15418 was not reached at the pH levels employed, but maximum adsorption has been found to occur in the vicinity of the pK_a. Spectrophotometric titrations of this compound show that its pK_a is approximately 1.0 (Table II). Since its basicity is lower than that of atrazine this may explain why it was adsorbed in much lesser amounts at pH levels greater than one.

The compound GS-16065, which differs from ametryne in the 2-position only, —SC_2H_5 versus —SCH_3, respectively, was included in the recent studies.[4] The GS-16065 was adsorbed in only slightly greater amounts than ametryne by Na-montmorillonite. Two other methylthio-s-triazines included in these studies[4] were GS-14253 and GS-14260.

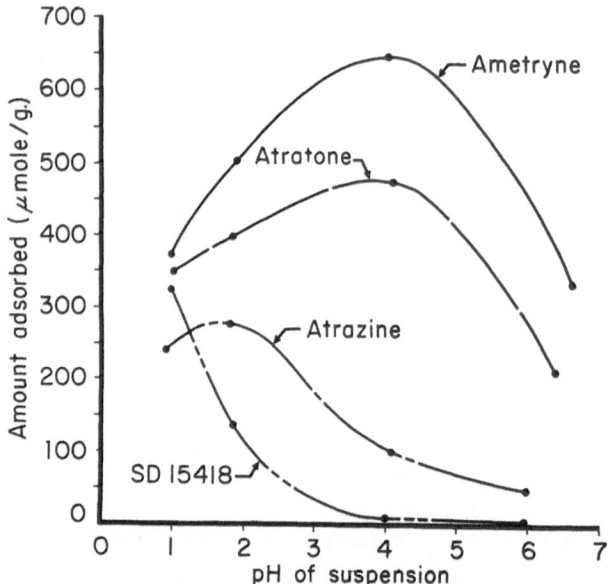

Fig. 7. Effect of pH and molecular structure on the adsorption of *s*-triazines on Na-montmorillonite (40 ml. of 5 × 10⁻⁵M solutions used with 2.5 mg. of clay) [WEBER, J. B.: Unpublished results (1969)]

The two compounds are similar to ametryne in all but the 6-position and differ as follows: ametryne, —C_3H_7 (iso); GS-14253, —C_4H_9 (*tert.*-butyl); and GS-14260 —C_4H_9 (*sec.*-butyl). The latter two compounds were adsorbed by Na-montmorillonite in similar amounts which were significantly greater than that observed for ametryne. This suggests that the slightly longer alkyl chain increased adsorption.

The fungicide, Dyrene, was included,[4] but only negligible adsorption occurred in the range pH 1.5 to 6.0. Since there are chloro-groups in the 2- and 4-positions, the compound is probably a very weak base and this may account for its low adsorption. The compound 2,4-diamino-6-phenyl-*s*-triazine was found[4] to be adsorbed by Na-montmorillonite in amounts similar to those of atratone. It has a pKₐ of approximately 3.6 (Table II). The amino groups in the 2- and 4-positions apparently impart enough basicity to the compound to allow it to be adsorbed by the same mechanisms as described for the *s*-triazine herbicides.

X. Hydrolysis of *s*-triazines

HORROBIN (1963) showed that hydrolysis of chloro-*s*-triazines occurred in steps, each involving the replacement of one chlorine atom. A slow hydrolysis occurred in the region of neutrality, but it increased with either increasing alkalinity or increasing acidity. The suggested

mechanism for alkaline and neutral hydrolysis and for buffer-ion catalysis was a direct attack by a base (OH⁻ ion, water, buffer anion) on an
electron-deficient carbon atom with replacement of chlorine. The suggested mechanism of acid catalysis was the uptake of a proton by the
triazine ring, followed by hydrolysis of the protonated molecule.

The degradation of chloro-s-triazine herbicides in soils has been
studied by HARRIS (1967). Approximately 30 to 50 percent was converted to the hydroxy form in eight weeks at 30°C. Formation of
hydroxy compounds occurred in much greater amounts in soil as compared with aqueous solutions. HARRIS suggested that the soil constituents catalyzed a nonbiological hydrolysis reaction.

HANCE (1967) measured the nonbiological decomposition of atrazine
and five other herbicides in aqueous solutions, in two soils, and in a
bentonite clay suspension. The half-life of atrazine was related to temperature; it ranged from 180 hours at 107° to 116 years at 20°C. in the
bentonite suspension. Atrazine half-life was approximately one-half
these values in the soils suspensions. In distilled water and in a pH 6
buffer solution, the half-life was also very long. The author concluded
that nonbiological chemical processes do not play an important part in
the loss of atrazine, from the soil, but all of the systems were at pH 6
and it has been shown that hydrolysis of the s-triazines is pH dependent.

ARMSTRONG et al. (1967) found that atrazine hydrolysis followed a
first-order kinetics reaction in sterilized soil and perfusion systems.
Atrazine hydrolysis in aqueous solution was found to be ten days at pH
2 and 12 and 100 days at pH 4 and 11, and 1,000 days at pH 6 and 10.
The addition of the sterilized soil caused an almost tenfold increase in
the rate of atrazine hydrolysis and the authors suggested that the
hydrogen ions around the soil particles created a zone of lower pH than
that measured in the bulk solution resulting in more rapid hydrolysis
than predicted from the pH of the system.

Infrared studies of s-triazine-montmorillonite complexes by RUSSELL
et al. (1968 b) were mentioned previously, but should be included here
also because of the catalytic effect of the H-clay surfaces on hydrolyz-
ing the s-triazine molecules. These investigators concluded that atrazine and propazine were hydrolyzed to the hydroxy forms on the
surfaces of H- and NH₄-montmorillonite.

THOMPSON[5] recently found no carbonyl band formation in infrared
studies with prometryne and H- and Na-montmorillonite. He concluded that hydroxypropazine was not formed at the clay surface.

WEBER et al. (1968 b) observed that prometone adsorbed on a strong
acid exchange resin was hydrolyzed to the hydroxy form and was
desorbed with 1M NaCl as hydroxypropazine. Recent studies[4] also
show that prometone adsorbed from acid solution by an Amberlite
IR-120-Na cation-exchange resin is also readily hydrolyzed to hydroxy-

---

[5] THOMPSON, R. P.: Unpublished results (1969).

propazine; in addition, propazine and prometryne are hydrolyzed by the strong acid exchange resins in a similar manner. These studies showed that propazine was readily hydrolyzed in aqueous solutions at pH 1, following first-order kinetics and with a half-life of 5.7 days at 25°C. Prometone and prometryne were hydrolyzed to hydroxypropazine in aqueous solutions at very low pH levels (> 1), and the hydrolysis appeared to be more rapid in systems in which small amounts of montmorillonite had been added. It is apparent that the hydrolysis of chloro-s-triazines in the vicinity of acidic colloid surfaces would be hydrolyzed more rapidly than where the compounds are in contact with neutral surfaces or are present in the bulk solutions. Methylthio- and methoxy-compounds were hydrolyzed to a much lesser extent than the chloro-s-triazines.

Because of its pH dependence, nonbiological hydrolysis of the s-triazines is probably an important process in relatively acid soils. Chloro-s-triazines are probably more readily hydrolyzed than the methoxy or methylthio compounds under these conditions. However, the relatively long persistence of atrazine and simazine in the Midwest suggests that hydrolysis, both biological and nonbiological, is not a very active process in soils which are calcareous or which are neutral in reaction.

XI. Biological availability of s-triazines

Several investigations have shown that the decomposition of organic compounds is greatly altered when they are adsorbed on clay surfaces. GREENLAND (1965 b) has recently reviewed this topic from the stand point of soil organic matter and clay complexes. He concluded that adsorption of organic compounds can increase the rate of bacterial activity, as well as depress it.

Adsorption by bentonite clay was found to interfere with the enzymatic hydrolysis of proteins in both acid and alkaline solutions (ENSMINGER and GIESEKING 1942, PINCK et al. 1954). Kaolinite had no significant effect on the protein decomposition. PINCK et al. (1961) found that basic antibiotics were strongly adsorbed by clay minerals. Microbial bioassays showed that the antibiotics were released from kaolinite but not from montmorillonite, illite, or vermiculite. Studies by ESTERMANN et al. (1959) showed that proteins adsorbed on kaolinite and montmorillonite were digested by enzymatic proteolysis, but proteins adsorbed on montmorillonite were degraded at a much slower rate than those adsorbed on kaolinite. A decrease in the basal spacings of montmorillonite was reported to have occurred following digestion of the protein-montmorillonite complex.

SCOTT and WEBER (1967) showed that the addition of montmorillonite and kaolinite clays to model soil systems greatly reduced the phytotoxicity of soil-applied paraquat and prometone. Paraquat adsorbed in the interlayer spacings of montmorillonite clay was unavailable to plant roots, but when the herbicide was adsorbed on the surfaces of kaolinite clay, it was available (WEBER and SCOTT 1966). Later studies by WEBER et al. (1969 a) showed that approximately five to ten percent of the paraquat adsorbed on montmorillonite was available to plants, providing the clay mineral surface was saturated with the herbicide. Paraquat adsorbed by the clay in amounts less than the CEC was not available to the plants. Adsorption of paraquat in the interlayer spacings of vermiculite clay reduced its phytotoxicity, but the herbicide was eventually available to the plants. Recent studies show that large, non-toxic, organic cations can displace paraquat from montmorillonite making it available for adsorption by the plants (WEBER et al. 1969 b). Another herbicide, diquat, was unavailable for microbial decomposition when it was adsorbed on montmorillonite (WEBER and COBLE 1968). Diquat adsorbed on kaolinite clay was readily available for microbial decomposition.

General knowledge concerning the behavior of herbicides in soils has been reviewed by a great many investigators such as AUDUS (1964), GOULD (1966), FREED et al. (1962), MARTIN (1963), UPCHURCH (1966), and BAILEY and WHITE (1964). Information pertaining more specifically to the s-triazines is in the works of HOLLY and ROBERTS (1963), HARRIS and SHEETS (1965), and SHEETS et al. (1962).

Many attempts have been made to correlate specific soil properties with herbicidal phytotoxicity, but in a large number of instances properties of the soils, such as CEC, surface area, exchangeable cations, etc., were confounded with the types of soil constituents, such as the contents of silt, clay minerals, or organic matter. In other instances, the herbicidal compounds were all lumped together without regard for their chemical properties. Compounds which ionize to anionic species in the soil solution would be expected to behave much differently than cationic and nonionic compounds. Aromatic and aliphatic compounds would also be expected to behave differently.

GYSIN and KNÜSLI (1960) reported that simazine did not leach readily in soils containing high amounts of organic matter. It was observed by BURNSIDE and BEHRENS (1961) that soils high in organic matter and/or clay content caused reduced simazine phytotoxicity as compared with soils low in these components. GROVER (1966) found that the addition of peat moss to soil reduced the phytotoxicity to oats of soil applied simazine.

DAY et al. (1968) also indicated that the soil constituent most highly related to the phytotoxicity of simazine applied to 65 soil types was organic matter. SCOTT and WEBER (1967) showed that the phytotoxicity of soil applied prometone was significantly reduced by the addition of montmorillonite and kaolinite clays and soil organic matter to model soil systems. The authors attributed the reduced activity of prometone to adsorption by the additives. Later studies showed that phytotoxicity of prometryne was also reduced by the additions of montmorillonite clay and soil organic matter to synthetic soils (WEBER *et al.* 1968 a). It was also shown that the activity of the herbicide in growth media containing the soil additives was much lower at pH 4.5 than at 6.5. The reduction in prometryne activity was attributed to the pH-dependent adsorption of the herbicide by the soil constituents.

Recent field studies by THOMPSON (1969)[5] show that incorporation of 20 tons/acre of Na-montmorillonite or peaty muck soil greatly influenced the phytotoxicity of prometryne. The muck reduced phytotoxicity to a greater extent than the clay.

It is apparent that adsorption of organic compounds by clay minerals greatly influences their decomposition by microbes and their availability to plants. In cases where the compounds were strongly adsorbed in the interlayer spacings of expanding clay minerals, utilization of the compounds was reduced greatly; in some cases it was reduced to zero. Adsorption of the compounds on exterior surfaces of clay minerals in some cases reduced their availability for microbial decomposition and in other cases decomposition was increased.

Adsorption of the s-triazines by clay minerals is dependent on the pH of the system, and many plant studies have shown that the s-triazines are generally less phytotoxic under acid conditions than under basic conditions. It is apparent that the reduction in herbicide concentrations resulting from adsorption by soil colloids results in less absorption by plant roots. Although the adsorption of the compounds by the soil colloids is in most cases reversible, the decrease in the rate of uptake as a result of dilution reduces herbicidal activity of the chemicals.

Acknowledgement

The author wishes to acknowledge the assistance of Miss E. Barefoot, Mrs. S. Storaasli, and Messrs. J. A. Best, R. F. Panza, and R. P. Thompson in the preparation of this paper.

Summary

The s-triazine herbicides are readily adsorbed by a variety of soil clay minerals including montmorillonite, illite, and kaolinite. Adsorption is dependent upon the acidity of aqueous systems and tends to increase with a decrease in pH, reaching a maximum in the vicinity of the ionization constants of the respective compounds. Adsorption of the compounds from neutral solution was attributed to hydrogen bonding and other nonionic forces. Adsorption at pH levels in which the compounds were in the cationic form was attributed to cation exchange and to complexing of the s-triazine molecules with hydrogen ions on the clay surfaces. Under highly acid conditions, hydronium ions competed with the s-triazine cations for sites on the clay surfaces. Other inorganic cations also competed for adsorption sites. Methylthio-s-triazines were adsorbed in greater amounts by montmorillonite clay than methoxy- and chloro-substituted compounds (*i.e.*, prometryne > prometone > propazine). The larger, more basic compounds, were adsorbed in greater amounts than the smaller, less basic, ones (e.g., tetraetatone > trietatone > simetone). Other factors which influenced adsorption of the s-triazines by clay minerals included the length of the alkyl chains of substituents in the 4- and 6-positions, steric effects, and solubilities of the compounds at different pH levels. The compounds were adsorbed in the interlayer spacings of expanding lattice clay minerals.

Adsorption by acidic clay and resin surfaces catalyzed hydrolysis of the s-triazines to the hydroxy forms. The chloro-s-triazines were hydrolyzed much more rapidly than methythio- and methoxy-substituted compounds.

The reduction in phytotoxicity of several soil applied s-triazine herbicides was related to the amount of organic matter and the types and amounts of clay minerals in the soils. The acidity of the soil also influenced the activity of the compounds. The compounds were more phytotoxic under neutral and basic conditions as compared with acidic conditions.

The s-triazine herbicides appear to exist in an equilibrium condition in soil systems, with a portion of the molecules being free in the solution phase and the rest being associated with the various soil colloids and other particles that comprise the system. The amount of triazine adsorbed at any particular moment depends upon many factors including the rate of removal of the triazine molecules from solution by plant roots, type and concentration of other ions and molecules in the system, constituency of the particles that make up the system, pH, etc. Changing one or more of these factors influences the amount of the herbicide adsorbed by the clay colloids and other soil particles and this in turn affects the amount of chemical available to the plants. For example, heavy fertilization or liming would be expected to have some influence upon the performance of the s-triazine herbicides under field conditions.

Résumé*

Adsorption des herbicides du groupe des triazines par les minéraux argileux

On a découvert que les herbicides du groupe des s-triazines étaient facilement adsorbés par une variété de minéraux argileux, y compris la montmorillonite, l'illite et la kaolinite. L'adsorption dépendait de

---

* Traduit par S. Dormal-van den Bruel.

l'acidité des systèmes aqueux et tendait à croître avec la diminution du pH, en atteignant un maximum dans la zone voisine des
constantes d'ionisation des composés respectifs. La liaison hydrogène
et d'autres forces non ioniques ont été attribuées à l'adsorption des
composés de solutions neutres. L'adsorption à des pH pour lesquels les
composés étaient présents sous la forme cationique a été attribuée à
l'échange de cations et à la formation de complexes entre les molécules de
s-triazines et les ions hydrogène sur les surfaces argileuses. Dans des conditions très acides, on a découvert une compétition entre les ions hydronium et les cations s-triazines pour l'emplacement sur les surfaces argileuses. D'autres cations inorganiques entrent également en compétition
en ce qui concerne l'emplacement de l'adsorption. Les méthylthio-
s-triazines ont été adsorbées en plus fortes quantités par les montmorillonites que les composés méthoxy- et chloro-substitués, respec-
tivement (c-à-d prometryne > prometone > propazine). Les molécules
plus grandes, plus fortement basiques ont été adsorbées en quantités
plus importantes que les molécules plus petites, moins basiques (c-à-d
tetraetatone > trietatone > simetone). D'autres facteurs influençant
l'adsorption des s-triazines par les minéraux argileux incluaient la
longueur des chaînes alkylées des substituants dans les positions 4 et
6, des effets stériques et les solubilités différées des composés à différents pH. On a découvert que les composés étaient adsorbés dans les
espaces entre les couches de minéraux argileux s'étendant en treillis.

On a trouvé que l'adsorption des s-triazines par les argiles acides et
les surfaces résineuses catalysait leur hydrolyse sous la forme hydroxy.
Les chloro-s-triazines étaient hydrolysées beaucoup plus rapidement
que les composés méthylthio et méthoxy substitués.

On a découvert que la diminution de la phytotoxicité de plusieurs
herbicides du groupe des s-triazines appliqués au sol était en relation
avec la quantité de matière organique, le type et l'importance des
minéraux argileux présents dans les sols. L'acidité du sol influençait
aussi l'activité de ces composés. Ceux-ci étaient plus phytotoxiques
dans des conditions neutres et basiques que dans des conditions acides.

Les herbicides du groupe des s-triazines semblent être dans un état
d'équilibre dans les sols, une fraction des molécules étant libre dans la
phase en solution et le reste étant associé aux différents colloïdes du sol
et aux particules contenues dans le système. La quantité de triazines
adsorbées à tout moment dépend de nombreux facteurs incluant la
vitesse de captage par les racines des plantes des molécules de triazine
en solution, le type et la concentration des autres ions et molécules
présents dans le système, la constitution des particules du système, le
pH, etc. Toute modification de l'un ou de plusieurs de ces facteurs
influence la quantité d'herbicide adsorbée par les colloïdes argileux et
d'autres particules du sol et ceci affecte en retour la quantité de produit
chimique disponible pour les plantes. On peut s'attendre, par exemple,
à ce qu'une forte fertilisation du sol ou un chaulage exerce une influence

sur les qualités des herbicides du groupe des s-triazines dans des conditions de plein champ.

Zusammenfassung*

Die Adsorption von Triazin-Unkrautvertilgungsmitteln in Ton-Mineralien

Es wurde gefunden, dass die s-Triazin-Unkrautvertilgungsmittel leicht von einer Reihe von Tonboden-Mineralien, einschliesslich Montmorillonit, Illit und Kaolinit, adsorbiert werden. Die Adsorption war vom Säuregehalt des wasserhaltigen Systems abhängig und neigte bei einer Abnahme des pH-Wertes zum Zunehmen. Sie erreichte ein Maximum in der Nähe der Ionisierungs-Konstanten der betreffenden Verbindung. Wasserstoffbindung und andere nicht-ionische Kräfte werden der Adsorption der Verbindungen durch neutrale Lösung zugeschrieben. Adsorption auf pH-Ebenen, auf denen sich die Verbindungen in kationischer Form befanden, wurde auf den Kationenaustausch und die Verflechtung der s-Triazin-Moleküle mit Wasserstoffionen an Tonoberflächen zurückgeführt. Man fand, dass unter sehr sauren Bedingungen Hydronium-Ionen mit den s-Triazin-Kationen um Plätze an den Tonoberflächen wetteiferten. Andere anorganische Kationen kämpften ebenfalls um die Adsorptionsplätze. Methylthio-s-Triazine wurden von Montmorillonit-Ton in grösseren Mengen adsorbiert als beispielsweise Methoxy- und durch Chlor ersetzte Verbindungen (z. B. Prometryn > Prometon > Propazin). Die grösseren basischeren Verbindungen wurden in grösseren Mengen adsorbiert als die kleineren weniger basischen (z. B. Tetraetaton > Trietaton > Simeton). Andere Faktoren, die die Adsorption der s-Triazine durch Ton-Mineralien beeinflussten, schlossen die Länge der Alkyl-Ketten der Substituenten in den 4 und 6 Positionen, sterische Wirkungen und die unterschiedliche Löslichkeit der Verbindungen in verschiedenen pH-Ebenen ein. Es wurde gefunden, dass die Verbindungen in den Räumen der Zwischenschichten der sich gitterartig ausbreitenden Ton-Mineralien adsorbiert werden.

Man stellte fest, dass die Adsorption der s-Triazine durch saure Ton- und Harzoberflächen ihre Hydrolyse zu den Hydroxy-Formen katalysiert. Die Chlor-s-Triazine wurden viel schneller hydrolysiert als die durch Methylthio- und Methoxy-ersetzten Verbindungen.

Man stellte fest, dass die Verminderung der Pflanzentoxizität von verschiedenen auf Boden angewandten Triazin-Unkrautvertilgungsmitteln in Beziehung zu der Menge organischer Substanz und den Arten und Mengen der Ton-Mineralien, die in den Böden anwesend waren, stand. Der Säuregehalt des Bodens beeinflusste ebenfalls die Wirksam-

---

* Übersetzt von M. Düsch.

keit der Verbindungen. Die Verbindungen waren—verglichen mit sauren Bedingungen—unter neutralen und basischen Bedingungen pflanzentoxischer.

Die s-Triazin-Unkrautvertilgungsmittel scheinen in einem Gleichgewichtszustand in Bodensystemen vorhanden zu sein, bei dem ein Teil der Moleküle frei in der Lösungsphase ist, während der Rest mit den verschiedenen Bodenkolloiden und -teilchen, die das System umfasst, verbunden ist. Die Triazin-Menge, die in jedem einzelnen Augenblick adsorbiert wird, hängt von vielen Faktoren, einschliesslich der Beseitigungsrate der Triazin-Moleküle aus der Lösung durch Pflanzenwurzeln, Art und Konzentration anderer Ionen und Moleküle im System, Wahlkreis der Teilchen, die das System aufbauen, pH, usw., ab. Das Ändern eines oder mehrerer dieser Faktoren beeinflusst die Menge des Unkrautvertilgungsmittels, die durch die Tonkolloide und andere Bodenteilchen adsorbiert wird; und dieses wiederum beeinflusst die Menge des Chemikals, das der Pflanze zur Verfügung steht. Zum Beispiel ist zu erwarten, dass starkes Düngen oder Kalken einigen Einfluss auf die Leistung der s-Triazin-Unkrautvertilgungsmittel unter Feldbedingungen haben würde.

References

ALLCOCK, H. R.: Heteroatom ring systems and polymers. New York: Academic Press (1967).

AMIS, E. S.: Solvent effects on reaction rates and mechanisms. New York: Academic Press (1966).

ARMSTRONG, D. E., and G. CHESTERS: Properties of protein-bentonite complexes as influenced by equilibration conditions. Soil Sci. **98**, 39 (1964).

—— ——, and R. F. HARRIS: Atrazine hydrolysis in soil. Soil Sci. Soc. Amer. Proc. **31**, 61 (1967).

AUDUS, L. J.: The physiology and biochemistry of herbicides. New York: Academic Press (1964).

BAILEY, G. W., and J. L. WHITE: Review of adsorption and desorption of organic pesticides by soil colloids, with implications concerning pesticide bioactivity. J. Agr. Food Chem. **12**, 324 (1964).

—— —— Herbicides—A compilation of their physical, chemical, and biological properties. Residue Reviews **10**, 97 (1965).

—— ——, and T. ROTHBERG: Adsorption of organic herbicides by montmorillonite. Role of pH and chemical character by adsorbate. Soil Sci. Soc. Amer. Proc. **32**, 222 (1968).

BAILEY, S. W.: The status of clay minerals. Clay and Clay Minerals, 14th Conf., p. 1 (1966).

BANIN, A., and S. RAVIKOVITCH: Kinetics of reactions in the conversion of Na- or Ca-saturated clays to H-Al clay. Clays and Clay Minerals, 14th Conf., p. 193 (1966).

BARSHAD, I.: The effect of the interlayer cations on the expansion of the mica type of crystal lattice. Amer. Min. **35**, 225 (1950).

BOITSOV, E. N., and A. I. FINKEL'SHTEIN: Optical investigations of the molecular structure of s-triazine derivatives. V. Infrared adsorption spectra of salts of oxy- and amino-derivatives of s-triazine. Optics and Spectroscopy **7**, 307 (1959).

—— ——, and V. A. PETUKOV: Dependence of the vacuum ultraviolet spectra of sym-triazine derivatives on their molecular structure. Optics and Spectroscopy **13**, 151 (1962).

BURNSIDE, O. C., and R. BEHRENS: Phytotoxicity of simazine. Weeds **9**, 145 (1961).

COLEMAN, N. T., and D. CRAIG: The spontaneous alteration of hydrogen clay. Soil Sci. **91**, 14 (1961).

CRUZ, M., J. L. WHITE, and J. D. RUSSELL: Montmorillonite-s-triazine interactions. Israel J. Chem. **6**, 315 (1968).

DAVIES, J. T., and E. K. RIDEAL: Interfacial phenomena. New York: Academic Press (1963).

DAY, B. E., L. S. JORDAN, and V. A. JOLLIFFE: The influence of soil characteristics on the adsorption and phytotoxicity of simazine. Weed Sci. **16**, 209 (1968).

DIXON, J. K., N. T. WOODBERRY, and G. W. COSTA: The dissociation constants of melamine and certain of its compounds. J. Amer. Chem. Soc. **69**, 599 (1947).

EKWALL, P., K. GROTH, and V. RUNNSTROM-REIO: Surface chemistry. New York: Academic Press (1965).

ENSMINGER, L. E., and J. E. GIESEKING: Resistance of clay-adsorbed proteins to proteolytic hydrolysis. Soil Sci. **53**, 205 (1942).

ESTERMANN, E. F., G. H. PETERSON, and A. D. McLAREN: Digestion of clay-protein, lignin-protein, and silica-protein complexes by enzymes and bacteria. Soil Sci. Soc. Amer. Proc. **23**, 31 (1959).

FARMER, V. C., and M. M. MORTLAND: An infrared study of the coordination of pyridine and water to exchangeable cations in montmorillonite and saponite. J. Chem. Soc. (A), p. 344 (1966).

FRANK, H. S., and W. WEN: Structural aspects of ion-solvent interaction in aqueous solutions. A suggested picture of water structure. Disc. Faraday Soc. **24**, 133 (1957).

FREED, V. H., J. VERNETTI, and M. MONTGOMERY: The soil behavior of herbicides as influenced by their physical properties. Proc. W. Weed Control Conf. **19**, 21 (1962).

FRIPIAT, J. J.: Surface chemistry and soil science. In E. D. Hallsworth and D. V. Crawford (ed.): Experimental pedology, pp. 3-13. London: Butterworths (1965).

FRISSEL, M. J.: The adsorption of some organic compounds, especially herbicides, on clay minerals. Versl. Landbouwk. Onderz. N.R. 67.3, Wageningen, 54 pp. (1961).

——, and G. H. BOLT: Interaction between certain ionizable organic compounds (herbicides) and clay minerals. Soil Sci. **94**, 284 (1962).

FUTRELL, J. H., and T. O. TIERNAN: Ion-molecule reactions. Science **162**, 415 (1968).

GIESEKING, J. E.: The clay minerals in soils. Adv. Agron. **1**, 159 (1949).

GILES, C. H., T. H. MACEWAN, S. N. NAKHWA, and D. SMITH: Studies in adsorption. Part XI. A system of classification of solution adsorption isotherms, and its use in diagnosis of adsorption mechanisms and in measurement of specific surface areas of solids. J. Chem. Soc., p. 3973 (1960).

GOULD, R. F.: Organic pesticides in the environment. Adv. Chem. Series **60**, 309 (1966).

GREENLAND, D. J.: Interactions between clays and organic compounds in soils. Part I. Mechanisms of interaction between clays and defined organic compounds. Soils and Fert. **28**, 415 (1965 a).

—— Interaction between clays and organic compounds in soils. Part II. Adsorption of soil organic compounds and its effect on soil properties. Soils and Fert. **28**, 521 (1965 b).

GRIM, R. E.: Clay Mineralogy. New York: McGraw-Hill (1968).

GROVER, R.: Influence of organic matter, texture, and available water on the toxicity of simazine in soil. Weeds **14**, 148 (1966).

GYSIN, H., and E. KNÜSLI: Chemistry and herbicidal properties of triazine derivatives. Adv. Pest Control Research 3, 289 (1960).
HANCE, R. J.: Relationship between partition data and the adsorption of some herbicides by soils. Nature 214, 630 (1967).
―― Decomposition of herbicides in the soil by non-biological chemical processes. J. Sci. Food Agr. 18, 544 (1967).
HARRIS, C. I.: Adsorption, movement, and phytotoxicity of monuron and s-triazine herbicides in soil. Weeds 14, 6 (1966).
―― Fate of 2-chloro-s-triazines in soil. J. Agr. Food Chem. 15, 157 (1967).
―― Movement of herbicides in soil. Weeds 15, 214 (1967).
――, and T. J. SHEETS: Influence of soil properties on adsorption and phytotoxicity of CIPC, diuron, and simazine. Weeds 13, 215 (1965).
――, and G. F. WARREN: Adsorption and desorption of herbicides by soil. Weeds 12, 120 (1964).
HARTER, R. D., and J. L. AHLRICHS: Determination of clay surface acidity by infrared spectroscopy. Soil Sci. Soc. Amer. Proc. 31, 30 (1967).
HAYES, M. H. B.: The role of soil organic matter in the adsorption of triazine herbicides. Residue Reviews, this volume (1969).
HENDRICKS, S. B., R. A. NELSON, and L. T. ALEXANDER: Hyrdation mechanism of the clay mineral montmorillonite saturated with various cations. J. Amer. Chem. Soc. 62, 1457 (1940).
HIRT, R. C., and R. G. SCHMITT: Ultraviolet adsorption spectra of derivatives of symmetric triazine—II. Oxo-triazines and their acyclic analogs. Spectrochimica Acta 12, 127 (1958).
―― ――, H. L. STRAUSS, and J. G. KOREN: Spectrophotometrically determined ionization constants of derivatives of symmetric triazines. J. Chem. Eng. Data 6, 610 (1961).
HOLLY, K., and H. A. ROBERTS: Persistence of phytotoxic residues of triazine herbicides in soil. Weed Research 3, 1 (1963).
HORROBIN, S.: The hydrolysis of some chloro-1,3,5-triazines. Mechanism-structure and reactivity. J. Chem. Soc., p. 4130 (1963).
JACKSON, M. L.: Structural role of hydronium in layer silicates during soil genesis. Trans. 7th Internat. Congress Soil Sci., p. 445 (1960).
KEENAN, A. G., R. W. MOONEY, and L. A. WOOD: The relation between exchangeable ions and water adsorption on kaolinite. J. Phys. Chem. 55, 1462 (1951).
KIPLING, J. J.: Adsorption from solutions of non-electrolytes. New York: Academic Press (1965).
KOHNKE, H.: Soil physics. New York: McGraw-Hill (1968).
KOSOWER, E. M.: Physical organic chemistry. New York: Wiley (1968).
KUNZE, G. W.: Pesticides and clay minerals. In: Pesticides and their effects on soils and water. A.S.A. Special Publ. No. 8, Madison, Wis., pp. 49–70 (1966).
LAILACH, G. E., T. D. THOMPSON, and G. W. BRINDLEY: Adsorption of pyrimidines, purines, and nucleoses by Lidi-, Na-, Mg-, and Ca-montmorillonite. Clays and Clay Minerals, 16th Conf., p. 285 (1968).
LAVY, T. L.: Micromovement mechanisms of s-triazines in soil. Soil Sci. Soc. Amer. Proc. 32, 377 (1968).
LOW, P. F.: The role of aluminum in the titration of bentonite. Soil Sci. Soc. Amer. Proc. 19, 135 (1955).
―― Physical chemistry of clay-water interaction. Adv. Agron. 13, 269 (1961).
·―― Influence of adsorbed water on exchangeable ion movement. Clays and Clay Minerals, 9th Conf., p. 219 (1962).
MARSHALL, C. E.: The physical chemistry and mineralogy of soils. New York: Wiley (1964).
MARTIN, J. P.: Influence of pesticide residues on soil microbiological and chemical properties. Residue Reviews 4, 96 (1963).

MARTIN, R. T.: Adsorbed water on clay. A review. Clays and Clay Minerals, 9th Conf., p. 28 (1962).

MATHERS, A. C., S. B. WEED, and N. T. COLEMAN: The effect of acid and heat treatment on montmorillonoids. Clays and Clay Minerals, 2d Conf., p. 403 (1954).

McGLAMERY, M. D., and F. W. SLIFE: The adsorption and desorption of atrazine as affected by pH, temperature, and concentration. Weeds, 14, 237 (1966).

McLAREN, A. D., G. H. PETERSON, and I. BARSHAD: The adsorption and reactions of enzymes and proteins on clay minerals. IV. Kaolinite and montmorillonite. Soil Sci. Soc. Amer. Proc. 22, 239 (1958).

MERING, J.: On the hydration of montmorillonite. Trans. Faraday Soc. 42B, 205 (1946).

MORIMOTO, G.: Dissociation constants of 6-substituted-diamino-1,3,5-triazines. J. Chem. Soc. Japan 87, 790 (1966 a).

—— Infrared adsorption of 6-substituted-2,4-diamino-1,3,5-triazines. J. Chem. Soc. Japan 87, 797 (1966 b).

MORTLAND, M. M., J. J. FRIPIAT, J. CHAUSSIDON, and J. UYTTERHOEVEN: Interaction between ammonia and the expanding lattices of montmorillonite and vermiculite. J. Phys. Chem. 67, 248 (1963).

MYSELS, K. J.: Introduction to colloid chemistry. New York: Interscience (1964).

NANCOLLAS, G. H.: Interactions in electrolyte solutions. New York: Elsevier (1966).

NEARPASS, D. C.: Effects of soil acidity on the adsorption, penetration, and persistence of simazine. Weeds 13, 341 (1965).

—— Effect of the predominating cation on the adsorption of simazine and atrazine by Bayboro clay soil. Soil Sci. 103, 177 (1967).

PINCK, L. A., R. S. DYAL, and F. E. ALLISON: Protein-montmorillonite complexes, their preparation and the effects of soil micro-organisms on their decomposition. Soil Sci. 78, 109 (1954).

——, D. A. SOULIDES, and F. E. ALLISON: Antibiotics in soils. II. Extent and mechanism of release. Soil Sci. 91, 94 (1961).

ROSS, M.: The torbenite minerals as model compounds for the hydrous layer silicates. Clays and Clay Minerals, 13th Conf., p. 65 (1966).

RUSSELL, J. D., M. I. CRUZ, and J. L. WHITE: The adsorption of 3-aminotriazole by montmorillonite. J. Agr. Food Chem. 16, 21 (1968 a).

—— —— ——, G. W. BAILEY, W. R. PAYNE, JR., J. D. POPE, JR., and J. I. TEASLEY: Mode of chemical degradation of s-triazines by montmorillonite. Science 160, 3140 (1968 b).

SCOTT, D. C., and J. B. WEBER: Herbicide phytotoxicity as influenced by adsorption. Soil Sci. 104, 151 (1967).

SHAINBERG, I., and W. D. KEMPER: Conductance of adsorbed alkali cations in aqueous and alcoholic bentonite pastes. Soil Sci. Soc. Amer. Proc. 30, 700 (1966).

SHEETS, T. J., A. S. CRAFTS, and H. R. DREVER: Influence of soil properties on the phytotoxicities of the s-triazine herbicides. J. Agr. Food Chem. 10, 458 (1962).

SMOLIN, E. M., and L. RAPOPORT: s-Triazines and derivatives. New York: Interscience (1959).

TALBERT, R. E., and O. H. FLETCHALL: The adsorption of some s-triazines in soils. Weeds 13, 46 (1965).

TALIBUDEEN, O.: Complex formation between montmorillonoid clays and aminoacids and proteins. Trans. Faraday Soc. 51, 582 (1955).

UPCHURCH, R. P.: Behavior of herbicides in soil. Residue Reviews 16, 46 (1966).

VAN OLPHEN, H.: Interlayer forces in bentonite. Clays and Clay Minerals, 2d Conf. p. 418 (1954).

Ward, T. M., and J. B. Weber: Aqueous solubility of alkylamino-s-triazines as a function of pH and molecular structure. J. Agr. Food Chem. **16,** 959 (1968).
—— —— Ultraviolet adsorption spectra of alkyamino-s-triazines. Spectrochimica Acta **25A,** 1167 (1969).
Weber, J. B.: Molecular structure and pH effects on the adsorption of 13 s-triazine compounds on montmorillonite clay. Amer. Mineralog. **51,** 1657 (1966).
—— Spectrophotometrically determined ionization constants of 13 alkylamino-s-triazines and the relationships of molecular structure and basicity. Spectrochimica Acta **23A,** 458 (1967).
—— Revolution in farm chemicals. Crops and Soils. **21**(5), 11 (1969).
——, and H. D. Coble: Microbial decomposition of diquat adsorbed on montmorillonite and kaolinite clays. J. Agr. Food Chem. **16,** 475 (1968).
——, R. C. Meek, and S. B. Weed: The effect of CEC on the retention of diquat, and paraquat by three-layer type clay minerals. II. Plant availability of paraquat. Soil Sci. Soc. Amer. Proc. **33,** 382 (1969 a).
——, P. W. Perry, and K. Ibaraki: Effect of pH on the phytotoxicity of prometryne applied to synthetic soil media. Weed Sci. **16,** 134 (1968 a).
—— ——, and R. P. Upchurch: The influence of temperature and time on the adsorption of paraquat, diquat, 2,4-D, and prometone by clays, charcoal, and an anion-exchange resin. Soil Sci. Soc. Amer. Proc. **29,** 678 (1965).
——, and D. C. Scott: Availability of a cationic herbicide adsorbed on clay minerals to cucumber seedlings. Science **152,** 1400 (1966).
——, and S. B. Weed: Adsorption and desorption of diquat, paraquat, and prometone by montmorillonitic and kaolinitic clay minerals. Soil Sci. Soc. Amer. Proc. **32,** 584 (1968).
——, T. M. Ward, and S. B. Weed: Adsorption and desorption of diquat, paraquat, prometone, and 2,4-D by charcoal and exchange resins. Soil Sci. Soc. Amer. Proc. **32,** 197 (1968 b).
——, S. B. Weed, and J. A. Best: Displacement of diquat, a cationic herbicide, from clay and its phytotoxicity. J. Agr. Food Chem. **17,** 1075 (1969 b).
—— ——, and T. M. Ward: Adsorption of s-triazines on soil organic matter. Weed Sci. **17,** 417 (1969 c).
Weed, S. B., and J. B. Weber: The effect of adsorbent charge on the competitive adsorption of divalent organic cations by layer-silicate minerals. Amer. Mineralog. **53,** 478 (1968).
Weiss, A.: Mica-type layer silicates with alkylammonium ions. Clays and Clay Minerals, 10th Conf., p. 191 (1963).
Wicklander, L.: Cation and anion exchange phenomena. In F. E. Bear (ed.): Chemistry of the soil, pp. 163–205. New York: Reinhold (1965).

Adsorption of triazine herbicides on
soil organic matter, including a short
review on soil organic matter chemistry

By

M. H. B. HAYES*

Contents

* Department of Chemistry, P.O. Box 363, *University of Birmingham*, Birmingham 15, England.

132   M. H. B. HAYES

I. Introduction

The earlier conclusions which indicated that soil organic matter
could reduce the phytotoxicity of s-triazines and other preemergent
herbicides were drawn from the analysis of data by simple correlation
analysis and by multiple regression analysis. These techniques also
indicated that the cation-exchange capacities and, in some instances,
the clay contents of soils could also be inversely correlated with pre-
emergent herbicide phytotoxicity. BAILEY and WHITE (1964) have
pointed out, however, that these and other soil properties which can be
highly or significantly correlated with the lowering of herbicide
phytotoxicity can also be highly or significantly inter-correlated.

There are, according to LAMBERT et al. (1965), five major categories
of factors which affect the fate of soil applied chemicals which can be
summarized as: (1) the type of soil (content of clay, silt, sand, organic
matter etc.), (2) the type of chemical (its stability, solubility, physical
properties etc.), (3) the climatic conditions (rainfall, temperature, sunlight, etc.), (4) the soil biological populations, and (5) the method of
application of the chemical (as a granule, wettable powder, etc.). The
same authors considered the soil organic matter component to be the
most important factor for the adsorption of all essentially neutral
organic chemicals onto soil and they showed, by use of the data of
SHERBOURNE and FREED (1954), that the sorption coefficient, Kp, (g.
of compound adsorbed/g. of organic matter divided by the concentration of compound in solution) for the adsorption of monuron [3-(p-
chlorophenyl)-1,1-dimethylurea] was reasonably constant for a wide
range of soils. By use of their slotted-tube soil profile technique, in
which the soil is regarded as being essentially a chromatographic
column, they further showed that Kp values did not change for soils
which ranged from zero to 45 percent organic matter contents.

The evidence which has accumulated to demonstrate how the
phytotoxicity of s-triazine herbicides is lowered by soil organic matter
and certain other organic compounds will be reviewed in this communication. Some of this evidence was obtained by inference from studies
on adsorption in the laboratory, some from studies on plant growth in
the greenhouse, and some from field studies. It must be emphasized,
however, that laboratory determined adsorption data, and even green-

house studies can give erroneous indications of the performances of herbicides in the same soils under field conditions where the effects of climatic factors such as rainfall, temperature, and light intensity (BURNSIDE and BEHRENS 1961, MUZIK and MAULDIN 1964, BUCHHOLTZ 1965, UPCHURCH et al. 1966, HANCE et al. 1968), and even the effects of fertilizer applications (ADAMS 1965) can influence herbicide phytotoxicity.

II. Effects of soil organic matter on triazine adsorption and phytotoxicity

Clay as well as soil organic matter (the two most important soil separates which influence soil fertility) are thought to effect the phytotoxicity of soil applied herbicides. The adsorption of triazine herbicides on clay minerals is thoroughly reviewed elsewhere in this volume (WEBER 1969). The author intends to review the involvement of clay in triazine phytotoxicity and adsorption only in cases where organic matter as well as clay were linked in soil-triazine interactions.

a) Relative importance of clay and organic matter in soil-triazine interactions; the special case of simazine

Simazine, because of its early impact on the soil-applied herbicide market, has been extensively researched. It would appear, however, that its behaviour in soils cannot always be regarded as the norm for all triazine herbicides. Early work by ALBERS and HOMBURG (1959) indicated that the phytotoxicity of this herbicide was lowered in soils with high contents of clay and of organic matter. Later, correlations by BURNSIDE and BEHRENS (1961) and by NEARPASS (1965) suggested that both clay and organic matter were involved in adsorbing simazine onto soils, and NEARPASS found that the type of clay present and the exchangeable acidity were more highly correlated with the adsorption effect than was the total clay content of the soil. That a special relationship exists in the case of the adsorption of simazine onto montmorillonite, even in the presence of organic matter, appears to be evident from the work of HARRIS and SHEETS (1965). They determined the amounts of CIPC [isopropyl N-(3-chlorophenyl)carbamate], diuron [3-(3,4-dichlorophenyl)-1,1-dimethylurea] and simazine, applied to 32 different soils, required to reduce the fresh weight of oat seedlings by 50 percent (the ED_{50} value, defined as the equivalent dose for 50 percent response), and they correlated these values with the clay and organic matter contents of the soils and with the amounts of the herbicides adsorbed by the soils. In general CIPC was the most and simazine the least strongly adsorbed of the three herbicides, but it was noted that the relative extents of adsorption of the three herbicides were not the same on all soils. The soils which adsorbed more simazine than CIPC and diuron had high clay relative to organic matter contents, and montmorillonite was the predominating clay in soils where simazine was adsorbed most. The ED_{50} values for CIPC and diuron correlated with the soil organic matter contents but those for simazine did not.

The work of ADAMS (1966) is of especial interest when trying to resolve the relative significances of clay and organic matter in the adsorption and phytotoxicity of simazine. He noted from bioassay studies that, in general, the soil organic matter content had the greatest effect in reducing the phytotoxicity of the herbicide. The release of simazine, adsorbed on soils and clays, during Soxhlet extraction in 95 percent ethanol was a first-order reaction but the slopes of the straight lines (from the plot of simazine recovered *versus* time of extraction) were different for each soil or clay sample. It became evident to ADAMS, when he related the K_{rc} value (the release constant value, or the slope of the line from the first-order plot for the release of simazine from a particular soil or clay sample) to the content of clay, organic matter, and to the cation-exchange capacity in each case, that the release of simazine was inversely related to the clay content and cation-exchange capacity of each sample, but increased as the content of soil organic matter increased. Because organic matter and clays tended to have opposing influences on the K_{rc} values it was presumed that an ethanol-soluble, simazine-organic matter complex was formed which prevented the herbicide from getting into contact with the clay.

The use of ethanol in herbicide-soil adsorption studies may not, however, give a realistic indication of performance under field conditions, and for that reason the work of HARRIS and WARREN (1964) may have a special relevance. In their work simazine was more strongly adsorbed than monuron by montmorillonite, but both compounds had similar affinities for an organic soil. Desorption (in water) of the herbicides took place more readily from the clay than from the organic soil. These results cannot be critically compared with those by ADAMS (1966) since solvents, techniques, and possibly important aspects of the chemistry of organic matter and mineral separates were different in the separate experiments. It should be noted that whereas simazine was adsorbed more strongly than monuron by montmorillonite in the experiments of HARRIS and WARREN, and in those by HILTON and YUEN (1943) (who worked with a wide range of soils), diuron tended to be more strongly adsorbed than simazine in the experiments of HARRIS and SHEETS (1965).

The results of WALKER and CRAWFORD (1968) indicate that the strong montmorillonite-simazine interactions quoted are not typical of chlorotriazine-clay interactions in general. They established a good correlation between carbon contents of 36 English soils (containing between 0.1 and 27.1 percent carbon) and the adsorption of atrazine, propazine, prometone, and prometryne by the soils. Correlation analyses between adsorption and soil properties showed that there was little difference, where adsorption was concerned, between the types of organic matter in soils which had high organic matter ($>$ five percent carbon) contents. The clay content was seen to make some contribution to the adsorption of the herbicides when the soil carbon content was less than five percent. These soils, (containing $<$ five percent carbon) when treated with hydrogen peroxide to reduce their carbon contents to between 20 and 50 percent of their original values and then saturated with calcium, invariably adsorbed less chlorotriazine compounds. The reduced adsorption was especially noticeable in the case of atrazine adsorbed onto the soil with the largest surface area and highest clay content. On the other hand, adsorption of prometone and prometryne was significantly increased after treatment of soils with high clay contents and large surface areas with hydrogen peroxide. The composition of the clay separates in the different soils has not been given. It will be seen later that the predominance or otherwise of montmorillonite in soils which contain as much as four percent organic matter can influence the relative extents of adsorption of different triazine herbicides.

b) *Evidence which indicates that soil organic matter is the most active soil component in reducing the phytotoxicity of triazines*

The work of BURSCHEL (1961) implicated soil organic matter in the lowering of the herbicidal activity of simazine in soil. The positive role of organic matter in this regard was later more definitely established by UPCHURCH and MASON (1962). They included simazine among 12 herbicides used in bioassay tests and showed that the amount of this compound (and others also) required to produce a GR_{50} (the amount of herbicide required to reduce plant growth by 50 percent) value was highly and positively correlated with soil organic matter content. The amount of simazine required to produce a GR_{50} was five times greater in a soil which contained 20 percent than in one which contained four percent organic matter. SHEETS *et al.* (1962), by means of simple correlation analysis, showed that the ED_{50} value for simazine was positively correlated with soil organic matter and cation-exchange capacity and negatively correlated with pH. It is interesting to note in these experiments that triazine ED_{50} values were less on a Madera silty clay soil than in a Yolo silt loam even though the silty clay had the higher content (3.03 *versus* 2.44 percent organic matter) of organic matter and clay.

DAY and JORDAN (1968) surveyed the phytotoxicity and the adsorption of simazine in 65 soils and showed that adsorption increased as the amounts of herbicide added to the soils were increased. Their data indicated that correlations between the amounts of simazine required to produce GR_{50} values and the clay contents and pH values

of the soils were not significant, but that correlations between these
values and contents of organic matter, cation-exchange capacities, and
equilibrium concentrations of herbicide in solution were significant.

GROVER (1966) added increments of peat moss to a Regina heavy
clay soil (52 percent clay, 2.89 percent organic matter) in order to
increase the soil organic matter contents to 5.49 and to 7.68 percent
and showed that the ED_{50} value of simazine for oats increased as the
organic matter content was increased. The choice of the type of organic
matter added might have some significance in this instance since
TALBERT and FLETCHALL (1965) have shown that a peat moss (with
the pH adjusted to 7.0) adsorbed more simazine and atrazine than
did a Wisconsin peat at the same pH value. Also there were no significant differences in the amounts of the herbicides adsorbed when the
pH values of the Wisconsin peat were adjusted to five or seven.

By a bioassay technique using cucumber seedling test plants
CHAPMAN (1966) clearly showed that the phytotoxicity of atrazine
was lowered as the amount of Everglades muck added to sand was
increased. The addition of Wyoming bentonite to sand-muck mixtures
had little influence on the phytotoxicity and Chapman concluded that
the varying clay contents in soils probably had little effect on the
herbicidal performance of atrazine.

c) Relative affinities of different triazines for soil organic matter

It is not surprising, because of the extent of their application, that
the behaviour of simazine and atrazine in soils has been more thoroughly researched than that of other triazine herbicides. There are thus
only a limited amount of data available which compare the affinities
of several triazines for soil organic matter. The first significant comparison was made in several Missouri soils by TALBERT and FLETCHALL
(1965). In the case of their Marshall silty clay loam, which had the
highest content of organic matter (4.2 percent) among the soils tested,
absorption of five triazine herbicides decreased in the order: prometryne
> prometone > simazine > atrazine > propazine. Prometryne and
prometone were bound significantly more than the other three herbicides. The work of HARRIS (1966), with soils which contained 1.9, 3.3,
and 4.3 percent organic matter, corroborates the order of adsorption
found by TALBERT and FLETCHALL for the same herbicides. Only in
the case of one soil, which contained 4.4 percent organic matter, was
propazine not the least strongly adsorbed of these herbicides. This
soil had a clay content of 38 percent and a cation exchange capacity
of 7.6 meq./100 g. The soil which contained 4.3 percent organic matter
had values of 48 and 12.5 for clay content and cation-exchange capacity, respectively.

The soil organic matter contents of about four percent quoted in
the foregoing examples can be regarded as similar to the content in the

soil which contained 2.7 percent carbon in the experiments of WALKER and CRAWFORD (1968). Adsorption in this instance decreased in the order: prometryne > propazine > atrazine > prometone. WALKER (1968), in discussing his results for the several English soils studied, found that the above order of adsorption was obeyed when organic matter surfaces were principally involved in adsorption. When mineral surfaces were involved (after soils had been treated with hydrogen peroxide) the order was: prometryne > prometone > propazine > atrazine. When explaining the discrepancy which occurred between his results and those by TALBERT and FLETCHALL (1965) he noted that only six of the 25 Missouri soils studied were non-montmorillonitic and that the order for adsorption on these six soils (for the four herbicides which WALKER had used in his experiments) was prometryne > atrazine > prometone > propazine. It seems therefore that even in the case of soils which contain as much as four percent organic matter montmorillonite can exert an important influence on the mode and extent of adsorption of triazine herbicides.

It can be deduced from the graphs and data presented by HAYES et al. (1968) that the extent by which four triazine herbicides were adsorbed onto a Fenland muck soil (79 percent organic matter and ten percent of the cation-exchange sites saturated with hydrogen ions) decreased in the order: prometryne > propazine > atrazine. Prometone was not included in the experiments. The data for the adsorption of propazine is extrapolated from the results given for its adsorption onto barium saturated soil. It can be seen that the order of the extent of adsorption for prometryne, propazine, and atrazine agrees with that by WALKER (1968) for adsorption on an organic matter surface. The extent of adsorption order found by HANCE (1969) for triazines agrees also with that of WALKER (1969).

III. Effects of soil organic matter on movement of triazines in soils

The reversibility of adsorption should determine the movement of herbicides in soils which contain adequate moisture. HARTLEY (1964) has pointed out the unlikelihood of completely irreversible adsorption taking place in soils. Accordingly the extent to which a herbicide could penetrate into the soil in phytotoxic concentrations should depend on the partition of the chemical between the soil surface and the water phase. When the nature of the soil surface remains the same the partition between this surface and water should also stay the same and hence, according to HARTLEY, the conditions for transport downwards into the soil should follow the theory for partition chromatography. However, the theoretical plate theory for chromatography appears to be too simple for application in a heterogeneous system such as soil in the field though the work of LAMBERT et al. (1965) indicates that the principles of chromatography can have applications for their "slotted tube test" technique for the study of herbicide movement in soil.

The sand-Everglades muck column technique used by CHAPMAN (1966) in order to study the movement of atrazine can be regarded as analogous to a series of column-chromatography experiments. He applied atrazine in increasing concentrations to separate 26-inch long columns consisting of varying proportions of sand and Everglades muck. Columns with low contents of organic matter were irrigated daily with one inch applications of water and those with the higher organic matter contents were irrigated twice daily with the same amounts of water each time. Columns were allowed to drain for 72 hours following the final applications of water on the fifth day and soil samples, taken at successive two inch depths, were assayed for herbicidal phytotoxicity. The depths to soil where lethal concentrations of atrazine could be detected became progressively shorter as the concentrations of organic matter in the columns increased, and phytotoxic concentrations were not detected below two inches when the columns contained 16 percent muck (even though these columns had been irrigated with a total of ten inches of water over the five-day period). THOMPSON (1965), in a frontal analysis chromatography experiment, passed solutions of simazine and atrazine for prolonged periods through short columns of Fenland muck and failed to detect the herbicides in the eluates. These experiments indicate that the partition of simazine and atrazine between water and muck strongly favours the muck soil phase. It should be realised, however, that muck soils and soil organic matter in general are heterogeneous chemical entities and that variations can be expected to occur from sample to sample in the mechanisms and extents of sample-triazine interactions.

Adsorption, and not solubility, regulates the extent of movement into or from soils. The comparative affinities of five triazine herbicides (HARRIS 1966) for four soils with organic matter contents of 1.9 to 4.4 percent has been referred to already. Prometryne (water solubility 48 p.p.m.) and prometone (750 p.p.m.) which were adsorbed most in these studies moved least in the soils. Propazine (8.6 p.p.m.), the least adsorbed, moved most and movements of simazine (five p.p.m.) and atrazine (33 p.p.m.) were intermediate. Later HARRIS (1967), by use of column and bioassay techniques, compared the movement of atrazine, prometryne, propazine, and simazine with that of monuron in a Haggerston silty clay loam (4.3 percent organic matter, 30 percent clay) and in a Lakeland sandy loam (3.3 percent organic matter, 10 percent clay). Prometryne moved least but to the same extent in both soils. The relative mobilities of the other triazines were similar in each soil but movement in the silty clay was less than in the sandy loam.

RODGERS (1968) has further illustrated the lack of correlation between solubility and leachability of triazines by showing that the leachability (as determined by bioassay techniques at depths of 22 and 36 inches, respectively, in the field and in soil columns) decreased in the order: atratone (solubility in water 1,654 p.p.m.) > propazine > atrazine > simazine > ipazine (40 p.p.m.) > ametryne (193 p.p.m.) > prometryne. The fact that soil type can alter the order of leachability can be seen from earlier experiments by RODGERS (1962). The order in a sandy soil was atrazine > simazine-atratone > ipazine and little leaching of any of these herbicides could be detected at depths greater than one inch in a muck soil.

The work of ARMSTRONG et al. (1967) indicates that the hydrolysis of atrazine to hydroxyatrazine is catalyzed by adsorption onto organic matter. They showed also that hydroxyatrazine is more strongly adsorbed than atrazine. If this relationship should hold true for all hydrolyzed triazines the ratios of hydrolyzed to non-hydrolyzed products should decrease with depth into any soil profile containing sufficient organic matter to react with the herbicides.

IV. Adsorption of triazines on to soil organic matter fractions[1]

The study of the adsorption of triazines, and soil-applied herbicides in general, by the different fractions of soil organic matter has received little attention. It is realised that it is the manner in which a herbicide reacts with the total soil organic matter fraction which is important (from the herbicide-organic matter interaction point of view) in any particular soil. However, since soil organic matter can vary in its proximate analysis and in its degree of saturation with different cations from soil to soil, it is possible that its reaction with herbicides can vary accordingly. It would appear to be worthwhile, therefore, to fractionate organic matter and to study the interaction of herbicides with different fractionated preparations. The relative proportions of these fractions present, and the extent to which any herbicide might react with each fraction or preparation might be of value in predicting herbicide behaviour in a particular instance. The author has no conclusive evidence, at this stage, to indicate that the behaviour of a herbicide towards the different organic matter fractions will predict its behaviour towards the composite material.

Triazines have been shown not be adsorbed by soil polysaccharides and by the brown fulvic acid materials extracted from a muck soil with dilute sulfuric acid (HAYES et al., 1968). The acid extracted material did, however, cause some degradation of atrazine (THOMPSON 1965).

---

[1] The reader is referred to the section "The chemistry of soil organic matter," for definitions, and a summary of the chemistry of fulvic acids, humic acids, and humin, the classical fractions of soil organic matter.

McGlamery and Slife (1966) showed that humic acid adsorbed more atrazine at pH 2.5 than at pH 7.0 (the pH was adjusted to 7.0 by addition of calcium carbonate). The work of Hayes et al. (1968) shows that atrazine, propazine, norazine, desmetryne, and prometryne were adsorbed strongly (where tested) onto hydrogen ion (H^+)-, calcium ion (Ca^{++})-, and barium ion (Ba^{++})-saturated humic acid and humin preparations. Adsorption onto the H^+ was invariably greater than onto the Ca^{++}- or Ba^{++}-saturated preparations. The amounts of herbicides adsorbed onto soil fractions saturated with a particular cation where greater than those adsorbed onto the parent organic soil saturated with the same cation. More detailed comparative studies on the adsorption of norazine and desmetryne onto H^+-saturated humic acid and humin preparations indicated that more norazine than desmetryne was adsorbed onto the humin, and that the reverse was true for the humic acid preparation. In similar studies with Ca^{++}-saturated preparations norazine was the more strongly adsorbed by the humic acid and humin materials.

Sullivan and Felbeck (1968) reacted the 95 percent ethanol extracts of humic acids from a muck soil, a latosol, and a silty clay loam by refluxing for 60 to 90 minutes, or by intermittently shaking for ten days at 0°C with solutions of atrazine, prometryne, simazine and trietazine. Unreacted herbicide was removed by liquid-liquid extraction for 24 to 48 hours with n-hexane or by distilled chloroform in a separatory funnel. It was possible to study the extent of reaction of the herbicides from nitrogen analysis data and by means of infrared analysis. The latter indicated that carboxyl, phenolic hydroxyl, and quinone groups were involved in the herbicide adsorption processes and the analytical data indicated that adsorption (or reaction) was greater at the higher than at the lower temperature.

V. Adsorption of triazines on to non-humic materials

a) Adsorption on to ion-exchange resins

Soil organic matter in general, and humic acid in particular, has many properties similar to cation-exchange resins. Thus it is not surprising that resins have been used as models for soil organic matter in adsorption studies.

Sheets (1959) first noted that simazine was adsorbed onto cation-exchange resins and that these, when incorporated in soil, could protect sensitive plants but that anion-exchange resins could not. Later Weber et al. (1965) showed that prometone was adsorbed onto Amberlite IRA-411 anion-exchange resin. They indicated, since the compound was less adsorbed at 55° than at 10°C., that physical adsorption was involved. The more recent studies by Weber et al. (1968 b) have shown that prometone adsorbed in small amount onto Amberlite

IR-120 (Na$^+$ form) cation-exchange resin and onto Amberlite IR-400 (Cl$^-$ form) anion-exchange resin, but was readily desorbed by water in each case. The herbicide could not be desorbed with water after adsorption onto Amberlite IR-120 (H$^+$ form) but hydroxypropazine was released when the resin was subsequently treated with $1M$ salt solution. The work of ARMSTRONG et al. (1968) has shown that whereas atrazine was strongly adsorbed onto carboxylic- and phenolic-resins, the carboxylic-resins alone catalyzed the hydrolysis to hydroxyatrazine.

TURNER and ADAMS (1968) compared the adsorption of atrazine and atratone onto Dowex-2 anion-exchange (OH$^-$, Cl$^-$, and H$_2$PO$_4^-$ forms) and onto Dowex-50W cation-exchange (H$^+$, Na$^+$, K$^+$, NH$_4^+$, Ca^{2+}, Fe^{3+}, and Al^{3+} forms) resins. Atrazine was 92 percent adsorbed onto Dowex-2 (OH$^-$ form) but atratone, regardless of the resin saturating anion, was relatively weakly adsorbed. Atratone was adsorbed significantly more onto Dowex-50W (H$^+$ form) than onto the resin when saturated with the other cations listed. Atratone was strongly adsorbed when this resin was saturated with H$^+$, Fe^{3+}, or Al^{3+} ions. These data indicate that the predominant cation involved in saturating the cation-exchange capacity of soil organic matter can influence the adsorption and phytotoxicity of the triazines.

b) Adsorption on to charcoal

Charcoal cannot be regarded as a component of soil organic matter. It can, however, accumulate in soils where woody plants or, in some instances, crop residues are burned. Since the material is insoluble in the solvents used to extract soil organic matter it can be mistakenly classified as part of the humin materials. It has been shown to have a high adsorptive capacity for triazine herbicides and can be used to protect sensitive crop roots against the herbicides (see SHEETS 1959, SHEETS and HARRIS 1965, GAST 1962, AHRENS 1964, ROBINSON 1965, WEBER et al., 1965 and 1968).

c) Adsorption on to decomposing residues in soil

WALKER and CRAWFORD (1968) have shown in separate experiments that straw and lucerne, immediately on addition to a soil with a low organic matter content, adsorbed atrazine, propazine, prometone, and prometryne. The amount adsorbed onto the decomposing straw did not significantly change during 81 weeks of incubation in soil. Lucerne adsorbed more triazine than did the straw initially but the amounts adsorbed onto both plant materials were the same, subsequent to about 27 weeks of incubation. Sucrose additions to the soil did not affect herbicide adsorption. The authors point out that the lignin components of the plant materials might have been involved in the adsorption process since HANCE (1965) showed that diuron was strongly adsorbed to non-polar surfaces such as lignin but was only weakly taken up by cellulose and chitin.

d) Adsorption on to some synthetic and natural polymers

It is generally agreed that triazines do not adsorb onto cellulose but that they do adsorb onto cellulose acetates (Ward and Holly 1966, Armstrong and Chesters 1968), and onto nylon (Ward and Holly 1966). The significance of studies with such model systems will be discussed in the section on mechanisms of triazine adsorption.

VI. Effects of temperature on triazine adsorption on to organic matter materials

Adsorption processes are exothermic, and when attractive forces between adsorbent and adsorbate are weak (e.g., van der Waals forces) adsorption should decrease with even moderate increases in temperature. Only a limited amount of work has been done on the effects of temperature on the adsorption of triazines onto soil organic matter materials. Early studies by Harris (1964) indicated that adsorption of atrazine onto an organic soil was the same at 50° as at 0°C. That humic acid materials behave differently is indicated in the work of McGlamery and Slife (1966). They showed that adsorption of atrazine onto pH 2.5 (H$^+$-humic acid) and pH 7.0 calcium carbonate-neutralized (Ca^{++}-humic acid) humic acid increased as the temperature was raised from 0.5° to 40°C. Adsorption, however, was ten times greater at the lower pH value. Stronger adsorption forces became involved in binding the atrazine to the H$^+$-humic acid as the temperature was raised since recovery of the herbicide with deionized water tended to decrease with increasing temperatures. Recovery from the Ca^{++}-humic acid was significantly greater and was relatively unaffected by the temperature range studied.

Hayes et al. (1968) have shown that more atrazine was adsorbed onto H$^+$-humic acid at 20° than at 70°C. during the first 24 hours of solution-humic acid contact. The amounts adsorbed after 72 hours of contact was the same at both temperatures. The same trend was evident in the case of adsorption of the herbicide onto Ca^{++}-humic acid and Ca^{++}-humin (Ca^{++} saturated with calcium acetate). However, adsorption onto the Ca^{++}-preparations was less than that onto the H$^+$-humic acid, and Ca^{++}-humin adsorbed more atrazine than Ca^{++}-humic acid at 20°C. even though adsorption onto the two Ca^{++}-preparations was the same at 70°C. In similar experiments desmetryne was more strongly adsorbed than norazine onto Ca^{++}-humic acid at 20°C. Both compounds were adsorbed to the same extent, but less extensively (than at 20°C.), during the first eight hours of contact with the adsorbent at 70°C. Thereafter the norazine was the more strongly adsorbed during the subsequent 64 hours of the experiment.

Temperature studies are very important when trying to understand adsorption mechanisms. The author will attempt to explain the above results in terms of the mechanisms which might be involved in triazine organic-matter interactions. This, however, will require an insight into the chemistry of soil organic matter and of its fractions.

VII. The chemistry of soil organic matter

a) Definitions and scope of this review

Soil organic matter may be broadly classified into non-humic and humic substances. The former consist of the unaltered remains of plants, animals, and microorganisms. The latter, derived from altered or transformed components of plants, animals, and microorganisms, have some resistance to further degradation and can be intimately associated with the mineral components of the soil (MORTENSEN and HIMES 1964, KONONOVA 1966).

The term 'humic substances', as used in this review, will refer to all chemically or biochemically altered organic compounds found in soil. These are composed of humic acids (organic materials precipitated on acidification to pH one of aqueous alkali or organic solvent soil extracts), fulvic acids (organic compounds from soil not precipitated on acidification of soil extracts), and humin (soil organic materials insoluble in aqueous acid or alkali or organic solvents). An overlap can occur in this classification system in the case of non-humic substances and humin. The inclusion of organic solvents in the foregoing classification system causes the definitions to deviate from the classical humic acid, fulvic acid, and humin terms which are based on extraction with aqueous alkali.

Though considerable world-wide interest has been, and is, directed towards the study of soil organic matter chemistry it is still the least understood field of soil chemistry. It is probably true to say that the mechanisms of herbicide-soil organic matter reactions will not be adequately appreciated until considerable progress has been made in the study of the composition and shape of the humic acid molecule in particular.

It is proposed to confine this section primarily to a review of the features of soil organic matter chemistry which can be of most help in the study of herbicide-organic matter interactions. For that reason emphasis will be given to techniques for the extraction, fractionation, and purification of fractions of soil organic matter, to a résumé of methods for estimating the amounts of organic matter in soils, and to the techniques which can be used for functional group analyses in humic substances. Reference will be made to current theories on the nature and composition of humic acids.

Little emphasis, apart from the techniques used in their study, will be placed on the much researched Podzol B (B_H) humic substances. This material, deposited in the B horizon from A horizon-solubilized organic matter, appears to be appreciably different to that in the plow layer where herbicide-organic matter interactions take place.

For more complete reviews on the chemistry of soil organic matter the reader is referred to the contributions by Dubach and Mehta (1963), Mortensen and Himes (1964), Felbeck (1965 a), to several contributors in *Soil Biochemistry* (McLaren and Peterson 1967), and to the future contributions in *Soil Biochemistry*, Volume II (McLaren and Skujins 1970) which is being prepared for publication.

b) Extraction, fractionation, and purification of fractions of humic substances

1. Extraction.—According to Dubach and Mehta (1963) the criteria for satisfactory preparative extraction of organic matter are (1) universal applicability of methods to obtain comparable fractions from all soils, (2) isolation of unaltered materials, (3) selective extraction of humic substances free from contamination, and (4) to insure representation of fractions of the whole molecular weight range. A suitable solvent or solvent system which might satisfy the above criteria should have five qualities (Whitehead and Tinsley 1964): (1) the ability to immobilize cations, (2) the ability to break hydrogen bonds and to form new solvent-organic matter bonds, (3) be of small molecular size in order to penetrate organic matter bonds, (4) have a high polarity and dielectric constant in order to disperse charged organic matter polymers, and (5) be of such a nature as not to alter organic matter significantly. As yet no solvent system has been found which will satisfy the criteria listed earlier.

The more successful techniques for the extraction of soil organic matter involve the prewashing of the soil with dilute acid (1N hydrochloric acid) to remove cations and then with water to remove the acid and soluble salts. Small amounts of fulvic acid materials are lost during this process. Fats, waxes, and resins present can lower the extracting efficiency of aqueous solvents, and may be removed by continuous extraction with ether or with ethanol-benzene mixtures.

Acid washed soils (H^+-soils) are usually extracted with neutral salt solutions, aprotic solvents, or dilute (0.5N) sodium hydroxide. Neutral salt solutions (Bremner 1954) of pyrophosphate, oxalate, nitrate, and fluoride (neutralized with sodium, lithium, potassium, or ammonium ions) are capable of forming soluble complexes or, in some instances, insoluble precipitates with polyvalent cations which insolubilize humic substances. These mild solvents, and neutral solutions of the disodium salt of ethylenediaminetetraacetic acid (EDTA) (Dubach et al. 1964), as well as organic compounds capable of complexing metal ions, such as cupferron, 8-hydroxyquinoline, and acetylacetone (MARTIN and REEVE 1955, 1957a and b) have been successfully used for the extraction of Podzol B_H organic matter. They do not, however, extract sufficient quantities, and hence sufficiently representative materials from the plow-layer organic matter.

SWIFT (1968) found that the ability of four aprotic solvents to extract organic matter from a H^+-muck soil increased in the order: N,N-dimethylformamide (DMF) > sulfolane > dimethylsulfoxide (DMSO) > pyridine. He also provided evidence from exhaustive, extraction experiments to indicate that each appropriate solvent extracted the organic materials solvated by the less efficient solvents in the series, and went on to suggest that the sequential extraction approach might provide an improved technique for the fractionation of organic matter. Exhaustive extraction experiments also suggested that ethylenediamine (EDA, which extracted 63 percent of the organic matter as compared to 60, 36, and 18 percent by $0.5N$ sodium hydroxide, pyridine, and DMF, respectively) might be an efficient solvent for H^+-muck. However, WARDLE (1968) has shown that these EDA extracted samples contained about ten times more free radicals than were present in the parent H^+-soil. Furthermore when the extracts with neutral sodium pyrophosphate solution and the aprotic solvents listed were dissolved in $2.5M$ EDA and then fractionated into humic acid and fulvic acid fractions, an increase (especially significant in the case of the humic acids) in the free radical contents of the fractions was observed. The EDA extracted humic acid material had high nitrogen contents (probably the result of Schiff base formation) even after prolonged washing with and dialysis against dilute acid.

Extracts from the H^+-muck with water, DMF, pyridine (under aerobic conditions), and sulfolane had free radical contents similar to that in the parent material (WARDLE 1968). Those extracted with pyridine (under anaerobic conditions) and DMSO had twice, and those with sodium hydroxide (under aerobic and anaerobic conditions) had three times the free radical content of the H^+-muck. Despite this disadvantage, the author considers that sodium hydroxide is still the best solvent available. Its principal advantage is in the amount of material which it extracts. Known disadvantages are that it increases oxygen uptake by the organic matter (BREMNER 1950) during extraction, that it lowers the molecular weights of the humic substances in contact with it (DUBACH et al. 1964, MEHTA et al. 1963), and increases the free radical contents (STEELINK 1966, ATHERTON et al. 1967, WARDLE 1968). Attempts which have been made to overcome some of these disadvantages include extraction in an atmosphere of nitrogen (SCHNITZER et al. 1958) or in an atmosphere of nitrogen in the presence of stannous chloride (CHOUDRI and STEVENSON 1957).

2. **Fractionation and purification of organic matter extracts.**—The humic acid materials are readily isolated by acidifying the extract.

Silica can be removed by redissolving the precipitate in base, adjusting to pH 8.5 and centrifuging (WILDUNG et al. 1965), and the process of precipitation, redissolving, and centrifugation can be repeated as often as necessary. The acid is removed by washing with and/or dialysis against distilled water. It is well known that some humic acid materials are solubilized in water after removal of the free acid. During recent observations the author noted that about 25 percent of humic acid (extracted in 0.5N sodium hydroxide from a H^+-Fenland muck) was solubilized during the course of ten water washings and centrifugations following the removal of hydrochloric acid from the humic acid precipitate. The freeze-dried washings readily redissolved in water and contained about 70 percent carbohydrate material which had been, presumably, coprecipitated with the humic acid.

A degree of fractionation of humic acids may be obtained by isolating materials precipitated at acidity increments between pH 3.0 and 1.0, by availing of the different solubilities in different solvents, by precipitation with salt, polyvalent cations, or detergents, by use of electrophoresis techniques (GILLET and PIRGHAYE 1960, JOHNSTON 1961, WALDRON and MORTENSEN 1961, MACKENZIE and DAWSON 1962, DAVIES et al. 1964), and by means of gel filtration techniques (FERRARI and DELL'AGNOLA 1963, MEHTA et al. 1963, POSNER 1963, LADD and BUTLER 1966, LINDQVIST 1967, SWIFT 1968). None of these techniques has, however, given highly satisfactory fractionations of humic acids.

The techniques used for isolating the fulvic acid components depend on the extraction solvents used. Dialysis, followed by freeze-drying or rotary evaporation below 40°C., is the simplest process but low molecular weight materials are lost in the process. Pressure filtration through membranes impermeable to molecules of molecular weight greater than 500 (e.g., Diaflow UM-1 membrane filter, Amicon Corp., Lexington, Mass.) should overcome the disadvantages inherent in the dialysis technique.

Combinations of gel filtration and ion-exchange chromatography techniques have been successfully applied (BARKER et al. 1967, HAYES et al. 1968, SWINCER et al. 1968, SWIFT 1968) to separate soil polysaccharide materials from brown components of fulvic acids.

c) Analysis of soil organic matter functional groups

Work in this field has been confined largely to oxygen-containing functional groups. There is no general agreement about the amounts and even the existence of the same functional groups in all similar fractions of soil organic matter. This might result from the lack in consistency in the soil types from which the materials were extracted (assuming for example as is often suggested, that no two humic acids are alike), and in the methods for extraction and purification adopted.

1. Total acidity.—The acidity in humic substances is thought to arise from the carboxyl, phenolic, enolic, and hydroquinone functional groups. The acidities of these functional groups, especially in poly-basic molecules, are known to overlap and the contribution of each to the total acidity cannot therefore be determined by pK-dependent methods such as titration (BECKWITH 1959, MARTIN and REEVE 1958). Various aqueous and non-aqueous titrimetric techniques have, however, been used to determine the total acidity in humic substances. In their nonaqueous titration experiments WRIGHT and SCHNITZER (1960) found that EDA was a better medium than pyridine or DMF for dissolving or suspending various podzol extracts during titration against sodium aminoethoxide. The work of VAN DIJK (1960) suggests that high-frequency titration is more reliable than aqueous and non-aqueous potentiometric and conductometric titrations for the deter-mination of the total acidity in humic acid, but even this method, he indicated, might be determining only the more strongly acidic func-tional groups.

The establishment of equilibrium during titration is essential for the accurate determination of total acidity. For that reason discon-tinuous titration techniques have their champions in the organic matter field. By use of discontinuous potentiometric titrations over long periods of time POMMER and BREGER (1960) showed that the acidity of their humic acid fell markedly over a 52-day period and they noted an increase in the carbonyl stretch over the same period. They pointed out that humic acid changes in composition when allowed to stand in solution. Thus its equivalent weight will have a precise mean-ing only when the method of humic acid extraction is stated in detail, and when the time in, and the temperature of, solution for titration is known, and the times between addition of titrant and measurement of pH are recorded.

Reaction with diborane can be used as a measure of the acidity of humic substances (MARTIN *et al.* 1963, DUBACH *et al.* 1964). In this technique the progress of the reaction can be followed by measuring the hydrogen released and by observing the disappearance of the car-bonyl stretch frequency in the infrared spectrum. The method is, how-ever, limited in application to humic substances soluble in inert solvent (e.g., tetrahydrofuran) in which the reactions are carried out.

For practical purposes the cation-exchange capacity (CEC) of humic substances, which can be determined by relatively elementary procedures, should be regarded as an acceptable measure of acidity for herbicide-soil organic matter interaction studies. The barium hydroxide method, as developed by BROOKS and STERNHELL (1957) for measuring acidity in brown coals, has been successfully used by WRIGHT and SCHNITZER (1960) to determine the CEC of materials extracted with sodium hydroxide from the A_O and B_H Podzol horizons. Readers are referred to the work of BROADBENT and coworkers (see

RANDHAWA and BROADBENT 1965 a and b and references therein) on the determination of the CEC of humic substances.

The studies of WILDUNG et al. (1965) are especially relevant to our interests. They determined the CEC values of lignins and humic acids by saturating the colloids with barium (Ba^{++}) cations, then replacing these with NH_4^+ ions (from an ammonium acetate solution), and measuring the barium released. They showed that considerable hydrolysis (replacement of Ba^{++} by H^+ ions) errors were introduced into the determinations when excess Ba^{++} ions were removed from the saturating medium with distilled water instead of with dilute barium salt solutions. Errors were greatest at pH 2.5, the lowest medium pH used. They also showed that the CEC measurements increased as the pH of the buffering media used were raised from 2.5 to 8.0. Consideration of these studies is especially important when choosing techniques for use in determining the effects of the exchange capacity saturating cations in soil organic matter-herbicide interactions.

2. Carboxyl groups.—BROADBENT and BRADFORD (1952) estimated the carboxyl content of humic substances by methylating one portion with diazomethane (to react with most functional groups containing acidic hydrogen) and another portion with dimethylsulfate (DMS, to react with acidic hydrogens other than those in carboxyl groups). The carboxyl content was estimated from the reduction of the CEC in DMS-treated samples and also from the release of methyl groups upon saponification of the diazomethane treated samples.

Other methods include the measurement of acetic acid liberated on reacting the organic matter with calcium acetate, decarboxylation [which can give low carboxyl content values (WRIGHT and SCHNITZER 1960)] by heating the sample directly or in a suitable solvent, and the iodometric method (VAN KREVELEN and SCHUYER 1957, WRIGHT and SCHNITZER 1960).

WAGNER and STEVENSON (1965) concluded, from chemical and infrared spectroscopy studies, that one-third of the carboxyl groups in their humic acid materials occurred sufficiently close together to form cyclic anhydrides.

3. Hydroxyl groups.—It has been suggested that the hydroxyl groups in humic substances are present as phenolic, alcoholic, enolic, hydroquinone, and pyrone forms (MARTIN et al. 1963, FELBECK 1965, STEELINK 1966).

The total hydroxyl content of samples has been measured (WRIGHT and SCHNITZER 1960, MARTIN et al. 1963) by acetylation at 90°C. with acetic anhydride and pyridine, then determining the amount of unreacted acetic anhydride (found by MARTIN et al. to give unreliable results), or, alternatively, saponifying the ester and measuring the acetic acid released. Phenolic hydroxyls have been determined by WRIGHT and SCHNITZER (1959) by refluxing organic materials with an excess of alcoholic potassium hydroxide, filtering, washing to remove the alkali, suspending in 85 percent alcohol, then saturating with carbon dioxide, filtering and washing, and finally titrating against standard acid to determine potassium carbonate. The amount of potassium released by carbon dioxide saturation is equivalent to the phenolic hydroxyl group content.

WAGNER and STEVENSON (1965), by their combination of chemical and infrared analysis techniques, estimated that the hydroxyl content of their humic acid was two-thirds phenolic. Reaction with diazomethane suggested the presence of two "types" of phenolic hydroxyl but spectroscopic evidence for this was inconclusive.

4. Carbonyl groups.—Standard reagents for carbonyl groups such as sodium borohydride (MARTIN et al. 1963, DUBACH et al. 1964) and hydroxylamine (WRIGHT and SCHNITZER 1960) gave unsatisfactory results when applied to humic acid analysis. SCHNITZER and SKINNER (1965) found evidence from infrared, ultraviolet, and visible spectra for the formation of 2,4-dinitrophenylhydrazone, phenylhydrazone, semicarbazone, and oxime derivatives of a Podzol B_H extract. On the basis of results from reductive acetylation followed by infrared analysis of the product, and from reductometric titrations, they concluded that their samples did not contain measureable amounts of quinone groups and that ketonic carbonyl groups had reacted with the reagents. Later (SCHNITZER and SKINNER 1966), they estimated the carbonyl content by refluxing their sample with 2,4-dinitrophenylhydrazine and polarographically measuring the amount of unreacted reagent.

d) Physical methods for the characterization of humic substances

1. Molecular weight measurements.—It is important when comparing the molecular weight values for humic substances quoted in the literature to take into account the origins of these materials, the methods used for their extraction, fractionation and purification, as well as the procedures used for the estimation of the molecular weight. As in the cases of other naturally occurring polymers the values quoted for the molecular weight of humic substances has tended to increase over the years.

Techniques for humic acid molecular weight measurements have included osmometry (WRIGHT et al. 1968), which gave higher number average molecular weight values than freezing-point depression (SCHNITZER and DESJARDINS 1962), viscometric and ultracentrifugation techniques (PIRET· et al. 1960), light scattering (ORLOV and GORSHKOVA 1965), and gel filtration. MEHTA et al. (1963), by use of Sephadex gels, have estimated molecular weight values of 5,000 to 100,000 for their materials. It has not been possible, however, rigorously to calibrate these or other gels for humic substances.

2. Spectroscopic studies.—Humic substances strongly absorb through the visible and ultraviolet wavelengths, but the spectra are relatively featureless and show mainly a steady increase in absorption with decreasing wavelength (DUBACH and MEHTA 1963, ZIECHMAN 1964, KLEIST and MUCKE 1966). The relative absorption of light in the visible range, the $E_4:E_6$ ratio (KUMADA 1959, DURODA 1966) can be used to characterize humic materials and can be, according to KONONOVA (1966), related to the degree of condensation of the aromatic nuclei.

Fluorescence in humic acids is observed at low solution concentrations (FREUDENBERG 1959, WALDRON and MORTENSEN 1962, SWIFT 1968), and PAULI (1966) has monitored the humification of straw particles by use of fluorescence-microscopy and fluorimetry.

The infrared spectroscopy technique has been used for characterization and differentiation between humic acid materials (SCHNITZER et al. 1959, JOHNSTON 1961, TURNER and SCHNITZER 1962, POSNER et al. 1968), to follow the course of reactions [e.g., oxidation or reduction of a humic material (SCHNITZER et al. 1959, FARMER and MORRISON 1960, ORLOV et al. 1962, OADES and TOWNSEND 1963, MENDEZ and STEVENSON 1963, STEVENSON and MENDEZ 1967)], and to study the reaction of triazines with humic acids (SULLIVAN and FELBECK 1968).

KUMADA and AIZAWA (1958) detected infrared absorption bands which indicated the presence of hydrogen-bonded hydroxyl, aromatic, and alifatic C-H, carboxyl, carbonyl, and carbon-carbon double-bond structures. SCHNITZER et al. (1959) noted that humic materials extracted with different solvents gave similar infrared spectra but found no direct evidence for carbonyl groups.

Great care should be taken when preparing humic materials for infrared spectroscopy. THENG et al. (1966) have shown that spurious bands are obtained in the -OH stretch and in the carbonyl absorption regions when samples are not rigorously dried.

Nuclear magnetic resonance (N.M.R.) spectroscopy has been applied by BARTON and SCHNITZER (1963) for the analysis of a methylated low molecular weight organic extract from a Podzol B_H horizon. They reported the absence of aromatic and of olefinic protons, the presence of methyl and methylene signals and the expected O-methyl signal. This technique may prove to be of value in studies on herbicides-organic matter reaction mechanisms.

Electron spin resonance (E.S.R.) has been applied by REX (1960) to detect the presence of stable free radicals in wood, lignins, and humic acid. Since then the study has been extended by STEELINK and co-workers (STEELINK et al. 1960, STEELINK and TOLLIN 1962 and 1967, STEELINK 1964), by ATHERTON et al. (1967), and by WARDLE (1968).

STEELINK and TOLLIN (1962) have estimated a free-radical content in the order of 10^{18} spins/g. for humic acid, and their analysis of spectra suggested the presence of two stable coexistent free radicals. STEELINK (1964) suggested that these species were an *ortho-* or *para-*semiquinone anion coexistent with quinhydrone or quinone species.

Later STEELINK (1966) demonstrated a marked increase in radical concentration after conversion of humic acid into its sodium salt and interpreted this as proof that the quinhydrone species exists in the macro-molecule and could provide a means of weak interpolymer cross-linking.

The types of spectra obtained by ATHERTON et al. (1967) from analyses on a wide range of soils and lake bed deposits were mainly determined by the pH of the parent soil rather than by its age or type of plant cover.

e) The structure of humic substances

1. Principles involved in degradation studies.—It is necessary, in order to elucidate the structure of any naturally occurring organic polymer by the techniques up till now available, to degrade the polymer at some stage into units that can be completely characterized. An ideal degradation reaction should result in the fragmentation of the polymer into monomer, dimer, and oligomer units by the selective cleavage of certain types of linkages. The conditions used should be sufficiently drastic to bring about the necessary cleavages, yet sufficiently mild to prevent the complete destruction of the molecule or lead to the production of artefacts. The percentage yield of products should be such that it can be safely assumed that these originated from the humic materials and not from contaminants.

2. Degradation studies on fulvic acids.—No polymeric materials which satisfy all of the criteria of molecular homogeneity have, as yet, been isolated from soil fulvic acid extracts though several pure low molecular-weight organic compounds have been isolated and identified. Polysaccharides, homogeneous from the point of view of molecular weight and charge density (FINCH et al. 1966 and 1968, BARKER et al. 1967), have been isolated, but the numbers of sugars present in the hydrolysates suggested that these isolates were mixtures. It should be worthwhile to apply the classical techniques used in polysaccharide structural work and the newer techniques such as enzyme induction (BARKER et al. 1967) only when pure compounds are isolated. The brown components in fulvic acid materials are often regarded as humic acid precursors or even humic acid fragmentation products, and can be subjected to similar degradation techniques as used for humic acid structural studies.

3. Degradation studies on humic acids.—The techniques for humic acid degradation have included hydrolysis with acid and alkali and fusion with alkali (COULSON et al. 1959, COFFIN and DELONG 1960, STEELINK 1960, GREEN and STEELINK 1962, JAKAB et al. 1961 and 1963, CHESHIRE et al. 1968, SWIFT 1968), oxidation with compounds such as hydrogen peroxide (MEHTA et al. 1963), alkaline permanganate (SCHNITZER and DESJARDINS 1964, HANSEN and SCHNITZER 1966,

Cheshire et al. 1967), nitric acid (Hayashi and Nagai 1961, Hansen and Schnitzer 1967), reduction with sodium amalgam (Burges et al. 1964, Mendez and Stevenson 1966, Stevenson and Mendez 1967, Martin et al. 1967), zinc distillation (Cheshire et al. 1967 and 1968), and high temperature and pressure hydrogenation using kaolin or Raney nickel catalysts (Felbeck 1965 b).

Hydrolysis with acid and alkali at temperature up to 120°C. has released phenolic products which could result from lignin degradation. However, m-hydroxy- and 3,5-dihydroxybenzoic acids and resorcinol (products not generally isolated from the degradation of lignin) were detected when a humic acid from a Swiss Podzol was hydrolyzed at 170° and at 250°C. with 5N sodium hydroxide. About 30 phenolic compounds were detected among the products of this reaction (Jakab et al. 1961 and 1963). Treatment of humic acids with alkali in the presence of copper sulfate released 20 percent of the products as ether-soluble products which contained vanillin, vanillic acid, p-hydroxy-benzaldehyde, p-hydroxybenzoic acid, and 3,5-dihydroxybenzoic acid (Steelink et al. 1960, Green and Steelink 1962). From the point of view of yields of ether-soluble products, fusion of humic acids with solid potassium hydroxide (40 to 70 percent yield, Coffin and Delong 1960), and digestion at 250°C. with a saturated aqueous solution of sodium sulfide (60 percent yield, Swift 1968) appear to be the best of the alkaline treatment techniques.

Nitric acid oxidation, which effectively degrades lignins, has released 30 to 60 percent of humic acid as ether-soluble products (Hayaski and Nagai 1961). Nitrophenols, nitrobenzoic acid, and hydroxy benzoic acid have been identified among these ether-soluble products. Hansen and Schnitzer (1967) have identified similar compounds among the 2N and 7.5N nitric acid oxidation products of a Podzol B_H humic acid.

Products (1-naphthalene, anthraquinone, fluorene, and xanthone) which suggested that humic acid contained a polycyclic aromatic core were isolated (Cheshire et al. 1967) from the oil (0.5 percent yield) obtained from decarboxylated alkaline permanganate oxidation products of an acid-hydrolyzed peat humic acid.

The yields of ether-soluble products from humic acid reduction have in general been disappointing, but several phenolic compounds have been isolated from the sodium amalgam reduction processes (Mendez and Stevenson 1967, Martin et al. 1967). Perhaps the most interesting application of the technique has been that by Martin et al. (1967) who have shown, by means of two-dimensional thin-layer chromatography, the presence of 14 phenolic compounds (including hydroxytoluenes) in the 15 to 20 percent yield of ether-soluble materials in the degradation products of a humic acid material synthesized by Epicoccum nigrum.

The products isolated in low yield from zinc-dust distillation at

550°C. of acid-boiled humic acid (CHESHIRE *et al.* 1967) included naphthalene, anthracene, benzfluorene, pyrene, and several other polycyclic aromatic structures. In a later study by the same group (CHESHIRE *et al.* 1968) it was shown that the fused aromatic structures could have arisen from dihydroxybenzoic acids. However, since the yield of complex hydrocarbons from the zinc-dust distillation at 400°C. of dihydroxybenzoic acids was low the authors concluded that it was unlikely that the majority of the complex hydrocarbons found in the treatment of the humic acid at 400°C. (in lower yield than at 550°C.) could have arisen as secondary products.

FELBECK (1965 b) isolated a series of normal alkanes with chain lengths of up to C_{25} or C_{26} units from the distillation of oils which resulted from hydrogenation (Raney nickel catalyst) at 300°C. and 6,000 pounds/sq. inch pressure of the non-hydrolyzable fraction of a muck soil.

SWIFT (1968) has pointed out that hydrogenation reactions might be of value in future humic acid structural studies. The reaction conditions used in the past have been drastic but these can be varied (temperature, pressure, type of catalyst) and, in principle, adjusted to cleave any type of chemical bond.

f) Postulated structures for humic acid

Soil chemists' ideas on the origins and chemistry of humic acids were governed for a long time by the ligno-protein theory of WAKSMAN and IYER (1933). A number of workers have shown that only a small proportion of the nitrogen in humic acids can be accounted for as protein nitrogen. Also differential thermal analysis thermograms (HAYES 1960) for an oxidized-lignin complex with casein resembled those for lignin and not those for a humic acid.

The leading modern theories suggest that amino acids, formed by proteolysis or microbial synthesis, react with polyphenols from the degradation of lignin to form humic acids (LAATSCH *et al.* 1952, FLAIG 1960, KONONOVA 1966). Amino acids could condense through their amino groups (leaving the carboxyl group free) with quinones (SWABY and LADD 1962, FLAIG *et al.* 1963). Cross-linking could take place through the reaction of diamino amino acid and cysteine residues with adjacent quinones. Polyphenols need not necessarily be supplied from lignin since they can be synthesized by microorganisms (FLAIG *et al.* 1963, KANG and FELBECK 1965, HAIDER and MARTIN 1967). It is most unlikely that any systematic or reproducible structure would be built up from the combination at random of units of these types. Oxidized phenols when reacted with simple amino acids, mixtures of amino acids, peptides, or with proteins, at pH 8.0 and 45°C., produced $0.5N$ sodium hydroxide-soluble materials which were precipitated by acid at pH 1.0 (LADD and BUTLER 1966). The percentage of amino acid-nitrogen released after acid hydrolysis indicated that products from the reactions with proteins and peptides were closer in composition to natural humic acids than were those from the reactions with amino acids.

The occurrence of peptide bonds in humic acids is probable (HOBSON and PAGE 1932, SHARPENSEEL and KRAUSE 1962), and LADD and BRISBANE (1967) have shown that pronase hydrolyzed almost 40 percent of the amino acids released from humic acid by acid hydrolysis.

Synthetic polymers which resemble humic acids have been synthesized by heating solutions of glucose and glycine (MAILLARD 1916) and methylglyoxal and glycine (ENDERS 1948). HAYES (1960) has shown that when appropriate ratios of reactants are chosen and the polymerisation process controlled these synthetic humic acids can have charge density and other physico-chemical properties similar to humic acid isolated from a muck soil.

FELBECK (1965 a and b), on the basis of his organic matter hydrogenation studies, has proposed a humic acid structure in which 4-pyrones act as a central structural unit for phenolic and amino acid side chains.

HAWORTH and his co-workers (CHESHIRE et al. 1967) suggest that humic acid possesses a chemically resistant polycyclic aromatic core to which polysaccharides, proteins, and phenolic acids could be linked and also linked to each other. Such a molecule would have the ill-defined structure typical of humic substances, and the central core, which would contain the stable semiquinone radicals, should be resistant to microbial attack.

VIII. Preparation of organic matter fractions for triazine-organic matter adsorption studies

The views expressed here are merely the author's impressions of the most suitable and simple techniques available at this time for the preparation of organic matter fractions for herbicide-organic matter studies. It is hoped that better techniques will become available which can be applied by the majority of workers in the field. Only when uniform materials become available, and used widely, will it be possible critically to compare results from laboratory to laboratory.

For the most complete extraction of organic matter soils should be hydrogen-ion saturated, preferably with dilute ($1N$) hydrochloric acid. Several subsequent extractions with dilute ($0.5N$) sodium hydroxide are necessary in order to remove all of the humic and fulvic acid materials. However, the author's experience indicates that the materials extracted in the later extractions contain higher proportions of materials of lower charge density which probably approach the structures of humin materials. It is suggested that the humic acids, therefore, be precipitated from the first two extractions with base.

It is suggested that the humic acid materials be recovered from the acid precipitate (at pH 1.0) of the alkaline extract. If a fractionation on the basis of charge density differences is desired the precipitates at various pH values between 3.0 and 1.0 can be recovered by centrifugation. Free acids can be removed by water washing. For more representative true humic acid materials the precipitate should be washed several times subsequent to the removal of free acids. The recovered washings (supernatants from high speed centrifugation) will be found to contain high concentrations of polysaccharides and amino-acid containing materials. The sodium salt of the residue can be fractionated on Sephadex G-100 columns (LADD and BUTLER 1966). Temperatures greater than 50°C. should be avoided since humic materials are known to decarboxylate above this temperature (DEUEL and DUBACH 1958).

Silicates are likely to be coextracted from mineral soils. These can be precipitated prior to humic acid precipitation by adjusting the pH of the alkaline extract to 8.0 (WILDUNG et al. 1965) and centrifuging. For more complete removal of these materials the humic acid precipitates can be redissolved in base, the pH adjusted to 8.0, and the centrifugation process repeated. Alternatively the HCl-HF treatment of MORTENSEN and SCHWENDINGER (1963) can be applied to the humic acid preparations.

If fractionation on the basis of solubility differences is desired exhaustive extraction with each solvent in the aprotic solvent series DMF, sulfolane, DMSO, and pyridine can be used. In such cases it is necessary to dialyze, following humic acid precipitation and washing, in order to remove the solvents adequately.

Extraction of humin from mineral soils is difficult, particularly since it is suspected that this material can be intimately associated with the soil inorganic colloid content (KONONOVA 1966). The author has noted that dry humin materials from an organic soil were soluble in concentrated sulfuric acid. He suggests that this component of soil organic matter could be extracted from mineral soils (previously exhaustively extracted with sodium hydroxide) by drying the soils, then extracting with concentrated sulfuric acid, and precipitating the humin from this extract by the addition of water. Water addition should be slow, with the temperature carefully controlled by use of an ice bath.

In the case of organic soils humin materials can be separated from the inorganic components (in the absence of appreciable contents of fine clays) by centrifugation. Partially decomposed plant materials can be removed by passing an aqueous suspension of the organic material through fine sieves.

Fulvic acids can be recovered from the supernatants (after precipitation of humic acids in the alkaline soil extracts) by dialysis to remove salt, and then freeze-drying the non-dialysable products. These products can, as indicated earlier, be fractionated by gel-filtration techniques. Much of the brown materials lost during dialysis is soluble in alcohol. These can be recovered by concentrating the dialyzate and extracting the concentrate with ethanol.

IX. Estimation of the soil organic matter content

The reader is referred to the procedures outlined by JACKSON (1958).

Occasionally workers in the herbicide field refer to the organic carbon content of soils. This would appear to be more reliable than values for organic matter contents since the latter are often obtained by use of a conversion factor from the carbon contents obtained from dry combustion. Conversion factors are merely an approximation. Different organic matter components contain varying amounts of organic matter and hence conversion factors should vary from soil to soil (depending on the relative distributions of the different components in the soil organic materials). It is essential to remove carbonates by pre-washing the soil with acid prior to using the dry combustion method. Pre-washing with HCl-HF mixtures also degrades silicates. The pre-washing technique, however, results in the loss of soluble organic materials (JACKSON 1958).

CAROLAN (1948) has provided a quick colorimetric method, based on dichromate-sulfuric acid oxidation, for the estimation of soil organic matter. Other 'wet combustion' methods (in which organic matter is oxidized by hydrogen peroxide or potassium dichromate and the unused reagent titrimetrically or colorimetrically determined) which can give more reliable results have been described by EVANS (1959), BREMNER and JENKINSON (1960 a and b) and ENWEZOR and CORNFIELD (1965).

X. Mechanisms of adsorption of triazines on to soil organic matter

a) Techniques for adsorption studies

The reader is referred to the text *Adsorption from Solutions of Non-electrolytes* (KIPLING 1965).

Herbicidal inactivation of triazines by organic matter is usually regarded as representative of adsorption at the liquid-solid interface. The slurry technique (which involves the bringing together of known amounts of solid and herbicide solution, allowing equilibrium to become established, measuring the decrease in solution concentration of herbicide and expressing results in terms of the amounts adsorbed/unit weight of adsorbent), though widely used in the study of adsorption of herbicides onto soils, can have a number of shortcomings in herbicide-soil organic matter interaction studies. It has been pointed out that soil organic matter is a mixture of polymeric components, some components of which are peptizable or soluble in water. The slurry technique cannot be satisfactorily applied to a study of binding

by such components which cannot be sedimented by centrifugation. In their studies on the adsorption of triazines onto hydrogen-ion
saturated humic acid preparations HAYES *et al.* (1968) have enclosed
the adsorbate suspension in dialysis tubing, immersed the assembly in
a solution of the appropriate triazine, and allowed equilibrium to take
place through the dialysis tubing. Adsorption by the total material
was calculated from the decrease in concentration of the external
solution.

Adsorption onto non-peptizable and water insoluble components
can also be studied by the frontal analysis chromatography technique.
CLAESSON (1946) has described the applicability of this technique to
other systems. Because of the high adsorptivity of organic matter for
triazines it is suggested that short columns be used and that the
adsorbent be diluted with an inert carrier before packing the columns.

The more recent developments of the column chromatography
technique for binding studies involve also the principles of gel filtration.
In principle when a small molecule is bound to a soluble macromolecule
the bound complex should pass through the column as a unit, provided
the attractive forces of the small molecule for the macromolecule are
greater than those of either component for the gel. The theoretical
developments of this technique were made by GILBERT and JENKINS
(1959) and NICHOL and OGSTON (1965), and a useful practical application, as developed by HUMMEL and DREYER (1962), is currently being
used in our laboratories by BURNS and GRICE (1969) to study the binding of bipyridylium and triazine herbicides by sodium humate solutions. We are attempting to determine whether or not each of the
sodium humate fractions binds the herbicides to the same extent.

The principle and application of the technique for the study of
other binding systems has been described succinctly by KELLETT (1967).
Let us assume, for simplicity, that we are dealing with a humic macromolecule which is eluted in the void volume of Sephadex G-100. The
column is packed in a herbicide (preferably radioactive) solution of
known concentration, and the humic material, dissolved in a solution
containing the same concentration of herbicide, is applied to the top
of the column, and eluted with the same herbicide solution. Should
the humic material bind or adsorb herbicide then the concentration
of free herbicide in the mixture would be lowered below the background
concentration. Equilibrium would be reestablished with the occluded
herbicide in the column during passage of the polymer-herbicide complex through the column. Thus the concentration of herbicide associated with the polymer solution would be greater than that in the
background solution, and this should be represented as a peak corresponding to the column void volume. For mass to be conserved this
excess should then be matched by a corresponding deficiency, and a
trough should be formed to correspond to the elution volume of the
herbicide. Herbicides are relatively small molecules and their passage

through gels such as Sephadex G-100 is retarded by the gel filtration process. The area of the trough should provide a direct measure of the number of moles of herbicide bound by the humic materials. FAIRCLOUGH and FRUTON (1966) have used the technique for the evaluation of equilibrium constants, from SCATCHARD (1949) plots, when studying the binding of tryptophan and its derivatives to bovine serum albumin. The system described deals with a macromolecule and herbicide system eluted as a single component from Sephadex G-100. The technique can also be applied to a study of binding in macromolecular systems which can be fractionated on gels.

b) Possible mechanisms for the adsorption of triazines on to organic matter

The reader is referred to the chapter "Sorption" (HAYES and THOMPSON 1970) in the provisionally entitled *Organic Chemicals in the Soil Environment* being prepared for publication.

The mechanisms most likely to be involved in the adsorption of triazines onto soil organic matter are (1) van der Waals forces, (2) hydrophobic bonding, (3) hydrogen bonding, (4) ion exchange, (5) charge transfer forces, (6) ligand exchange, and (7) chemisorption.

Some ideas on the mechanisms involved can be obtained from adsorption isotherm studies at different temperatures. Physical adsorption, which includes van der Waals forces and hydrophobic bonding, involve weak attractive forces which can be disrupted with even moderate temperature increases. The van der Waals bonds have energies (one to two kcal/mole) only slightly greater than that of kinetic motion (0.6 kcal/mole) at 25°C. (WATSON 1965).

The increase in molecular vibrational energy in going from a temperature T1 to a higher temperature T2 is given by (BARROW 1962):

$$\Delta E = R(T2 - T1)(3n - 6)$$

where ΔE is the increase in vibrational energy, R is the gas constant (1.987 cal/deg/mole), and n is the number of atoms in the molecule. The value of ΔE for atrazine, on increasing the temperature from 20 to 70°C., is 7.8 kcal/mole, which is certainly greater than the energy involved in its van der Waals and hydrophobic bonding to organic matter. This value is also greater than the energy involved in binding the atrazine to the organic matter through one, or perhaps even two, hydrogen bonds (VICKERSTAFF 1954, WATSON 1965). Since the distinction between ionic bonds and hydrogen bonds is sometimes arbitrary and the energies of ionic bonds fall within the range for hydrogen bonds (three to seven kcal/mole), ionically bound atrazine might also be desorbed at 70°C. Also the possibility of ion-exchange to free acid groups is less at 70° than at 20°C. because of the possible decrease in the ionization constants of the carboxyl groups (MacINNES

1961) in the organic matter, and the amine groups (ALBERT and SERJEANT 1962) in the triazines.

Increased adsorption at higher temperatures should indicate chemisorption or the formation of covalent bonds. Thus it can be seen that carefully controlled adsorption studies at different temperatures can give indications of the mechanisms of adsorption involved.

Ligand exchange (HILL et al. 1965) and various types of charge exchange or charge transfer (PULLMAN 1965, BRIEGLEB 1961, ANDREWS and KEEFER 1964), particularly $n - \pi$ and $\pi - \pi$ interactions between the triazine ring and the postulated aromatic residues in the organic matter molecule should be less affected by temperature changes.

THOMPSON (1968) has suggested that ligand exchange adsorption might occur at sites where transition metal ions are incompletely chelated by the acidic groups or organic nitrogen functions, and where the water of solvation acts as a ligand. The ligand water could be displaced by the secondary amine groups of the triazine. He suggested that this mechanism might be tested by attempting to displace the adsorbed triazine with diamine ligands containing similar inter-amine nitrogen distances as those present in the triazine herbicides.

c) Suggested mechanisms for the adsorption of triazines on to organic matter

It is necessary, in order to attempt an explanation for the mechanisms suggested for the adsorption of triazines onto organic matter, to construct a hypothetical model of the systems involved in the adsorption process. The simplest models which we can construct for humic acids, based on the earlier review of the chemistry of soil organic matter, should contain phenolic and possibly quinone compounds, aromatic carboxylic acids, at least some aliphatic compounds which release amino acids on hydrolysis, and possibly also fused aromatic structures. It can be assumed that this material is highly cross-linked, highly branched, and sterically complex. It might be difficult, because of this steric complexity, for organic molecules to penetrate to all of the active adsorption sites in the polymer. Though the material has a higher charge density/unit weight than clays, the spacing of the charged groups need not be as close or as symmetrically arranged as they are in clays. Furthermore the shape of the molecule can be predicted to change (like all organic polyelectrolytes capable of rotation around carbon-carbon single bonds) as the chemical environment is changed.

When a humic acid material is saturated with Group IIa metal cations numerous inter- and intramolecular bridges can be formed. This should result in an overall contraction of the polyelectrolyte surface area. When saturated with monovalent cations, the same material, because of charge repulsion between the organic anions, can be expected to be highly expanded and thus to approach its maximum poly-electrolyte volume. When hydrogen ion saturated a certain amount of inter- and intramolecular hydrogen bonding can be expected to result in some contraction in the polymer.

The cation-exchange capacity of humin can be expected to be as low as 20 percent of that of humic acid (THOMPSON 1968). The material is more hydrophobic, and might contain a relatively high proportion of fused aromatic structures as compared with humic acid.

The brown fulvic acid materials have higher charge densities and lower molecular weight values than humic acids (WARDLE 1968), but otherwise it is possible that the composition of the two materials is somewhat similar.

The form in which the triazine molecules are present in the soil organic matter environment should also be considered. The pKa values of chlorotriazine, methylthiotriazine, and methoxytriazine herbicides are within the ranges 1.65 to 1.88, 3.05 and 4.51, and 4.15 to 4.51, respectively. The differences in the values between the different groups can be attributed to the activating mesomeric effects of the methylthio- and methoxy-groups and the deactivating inductive effects of the chlorine atom. It is necessary, in order to bring about appreci- able protonation of the different triazines, to have the pH of the medium approach that of the pKa value of a particular triazine. How- ever, if soil organic matter can be regarded as colloidally analogous to biopolymers, it can be assumed that the pH at the surface of the hydro- gen ion saturated organic matter will be 0.5 to two units lower than that of the liquid environment (McLAREN 1960). Because of the difficulty (as the result of steric hindrance, etc.) of completely saturat- ing organic matter with cations it is possible that localized areas of low pH could exist on a supposedly metal cation saturated humic polymer.

Such information as has been outlined is partly consistent with the theory that ion-exchange mechanisms can be involved in binding triazines to soil organic matter. The interesting results of McGLAMERY and SLIFE (1966), for the adsorption of atrazine onto H^+- and Ca^{++}- saturated humic acid materials, show that adsorption onto H^+-humic acid was ten times greater than that onto the Ca^{++}-material. They cautiously stated that "this pH effect may be due to increased ionic bonding caused by protonation of the amino groups on the atrazine molecule at the low pH. Humic acids have functional groups such as phenolic hydroxyl, alcoholic hydroxyl and carboxyl groups which, if dissociated, can participate in exchange reactions."

The question as to whether or not the humic acid materials are sufficiently dissociated at the pH required to protonate atrazine is of especial relevance. WARDLE (1968) found that humic acid materials, somewhat analogous to those used in triazine adsorption studies by HAYES et al. (1968), had average pKa values of approximately 5.5. This caused HAYES et al. (1968) to postulate that ion exchange could account for only a small proportion of the atrazine and norazine bound by their humic acid preparation of pH 3.8. Their argument was based on the fact that, even allowing for up to two pH units difference between the acidity in the bulk solutions and that at the polymer surface, protonation of the trazines could be expected to be low. Furthermore, they argued, the percent ionization of humic acid at pH 3.8, as obtained by the equation (ALBERT and SERJEANT 1962):

$$\% \text{ ionized} = \frac{100}{1 + \text{antilog}_{10} \, (\text{pKa} - \text{pH})}$$

should be only about two percent. Then they drew on literature references which showed that cationic counterions are only weakly associated with the polymer at low degrees of ionization of polymeric acids. These arguments would militate against the involvement of ion-exchange reactions between atrazine and the H^+-humic acid from Leonardite, pH 2.5, used by McGLAMERY and SLIFE (1966), even though the atrazine could well be protonated in this medium. However, the author recognizes that ion-exchange could have been involved in the case of the two works cited. No pKa data are available for the Leonardite preparation but pKa values between about four and 5.5 are cited in the literature for humic acid materials. It can be assumed that the Leonardite pKa value was somewhere in that range. In any case only average values are listed for the pKa of humic acid materials. These values can change between locations within humic acid polymers, depending on the nature of the component acidic groups, and on their proximity to each other.

The proposals of WEBER et al. (1968) for the mechanisms of binding of prometryne to soil organic matter are of relevance to the contexts cited. They postulated that increased adsorption at the lower pH values parallelled the increase in the protonated form of the prometryne. The prometryne could be protonated through association with carboxyl groups on the organic matter and this protonated triazine could then be held to the carboxylate anions by coulombic forces. They further inferred that hydrogen bonding could take place between carbonyl groups on the organic matter and the amino groups of the herbicide, that interactions could take place between the triazine and hydroxyl or phenolic groups on the organic matter, and that complexes could be formed between the triazine molecule and electron pair donor groups on the organic matter. They favoured, however, the ionic mechanism because of the pH dependence of the adsorption.

SULLIVAN and FELBECK (1968) indicated that hydrogen bonding and ion-exchange mechanisms could both be involved in the binding of chlorotriazines to their humic acid preparations. After careful examination of their infrared data they suggested that ion exchange could take place between a protonated secondary amine group on the triazine and a carboxylate anion on the humic acid, and that hydrogen bonding could take place between the second (non-protonated) secondary amine group and carbonyl or quinone groups on the polymer. The more recent evidence which indicates that protonation takes place in the triazine ring would leave, on the basis of the foregoing argument, both triazine secondary amine groups free to participate in hydrogen bonding reactions.

The present author is of the opinion that the participation of hydrogen-bonding mechanisms in triazine-organic matter interactions have not been sufficiently stressed. The results of McGLAMERY and SLIFE (1966) and HAYES et al. (1968) have clearly shown that the adsorption of atrazine onto Ca^{++}-humic acids is very much less than its adsorption onto the H^+-materials. Ion exchange can be expected to play an insignificant role in adsorption onto Ca^{++}-preparations. The possibility still exists that hydrogen bonding could take place in these preparations between (stereochemistry permitting) carbonyl and quinone groups on the organic matter and the hydrogens in the secondary amine groups on the triazine. The oxygen of the methoxy groups in the 2-position of triazines could also act as a strong donor group to hydrogens on alcoholic and phenolic groups on the humic acid. The chlorine and the sulphur of the methylthio-groups could, in a similar manner, act as feeble electron donor units. This postulates a "three point attachment" between the triazine and the organic materials. The extent to which a triazine could participate in such attachments would depend on the availability of the appropriate sites on the Ca^{++}-saturated preparations. It has been pointed out that humic acid, when calcium saturated, should approach the minimum polyelectrolyte volume. This could be expected to hinder the approach of the herbicides to the active sites. In the case of the H^+-saturated preparations, however, the molecule could be expected to have a more expanded structure and more sites available for the above type interactions to take place. More important still is the fact that carboxyl groups would be made available for participation in the hydrogen bonding reactions.

HAYES (1960) has estimated that a humic acid from a muck soil contained approximately one readily dissociable acid group/nine carbon atoms. It is likely that most of these groups in humic acids are carboxylic and, where aromatic and alicyclic structures are concerned, it is plausible to imagine these in close proximity. The hydrogen-saturated humic acid material, therefore, should provide a very much better hydrogen bonding environment than Ca^{++}-humic acid. Should triazine ring protonation occur and the possibility of ion exchange be real, then, depending on the steric arrangement of possible interacting functional groups, the possibility of a "four point attachment" should exist.

McGLAMERY and SLIFE (1966) have not considered the possibility of atrazine hydrolysis and the adsorption of the hydroxyatrazine formed onto their Leonardite H^+-humic acid. ARMSTRONG et al. (1967) have indicated that atrazine hydrolysis is catalyzed by adsorption onto organic matter. The adsorption, they thought, could take place through interactions between the ring or side chain nitrogens of the atrazine and the weak acid groups on the organic matter. Such a bonding process would further decrease the electron deficiency around carbon atom number two of the triazine and thereby facilitate solvolysis to hydroxytriazine. The involvement of carboxyl groups in the organic matter in the hydrolysis process can be inferred from later work by ARMSTRONG and CHESTERS (1968), who showed that though carboxylic acid and phenolic resins adsorbed atrazine, the carboxylic acids alone were involved in its hydrolysis.

The radioisotope techniques used to determine the C^{14} ring-labelled atrazine in solution in the slurry experiments of McGLAMERY and SLIFE (1966) could not distinguish between the hydroxy- and the chlorotriazine. By combining the techniques of polarographic analysis (HAYES et al. 1967) and scintillation counting, HAYES et al. (1968) have shown that most of the atrazine in solution in a H^+-humic acid environment was presumably hydrolyzed at 70°C. in three days. The same herbicide in contact with a dilute humic acid material in solution (pH 3.8) was 56 percent hydrolyzed at 70°C. in the same time. A significant amount of hydrolysis could have, therefore, taken place in the pH 2.5 humic acid suspensions of McGLAMERY and SLIFE even during two hours of contact, particularly at the higher temperatures (up to 40°C.). ARMSTRONG et al. (1967) have indicated that hydroxyatrazine is more strongly adsorbed than atrazine to organic matter. This is readily acceptable. On theoretical grounds hydroxytriazines should be more strongly adsorbed than all of their methoxy- and chlorotriazine analogs. The hydroxyl group itself is more capable of involvement in hydrogen bonding than the other groups mentioned. Also, since the hydroxytriazines can be predicted to exist in equilibrium with their substituted amide forms, four points of hydrogen bond attachment are capable of existence.

The extent to which all of the possible points of attachment for hydrogen bonding between triazines and soil organic matter can be involved will depend, of course, on the stereochemistry of the organic matter to which they are adsorbed. This governs the availability of multiple donor or acceptor systems on the organic matter.

The increased adsorption as temperature is increased would appear to rule out the involvement of physical adsorption and increase the possibilities of chemisorption in the experiments of McGLAMERY and SLIFE (1966). However, the experiments of HAYES et al. (1968) have shown that less atrazine and appreciably less "total atrazine" (atrazine plus hydroxyatrazine) was adsorbed onto H^+-humic acid during the initial stages of contact at 70°C. than at 20°C. The amount of "total atrazine" (mainly hydroxyatrazine) in solution at 70°C. after 72 hours was approximately the same as the amount of atrazine in solution at 20°C.

The same temperature effects were evident for the adsorption of atrazine onto Ca^{++}-humic acid and Ca^{++}-humin preparations. The effects of temperature were more pronounced in the case of the humin materials. These data suggest that weak physical bonds were involved as an initial mechanism of adsorption. It is suggested that the herbicide held by such attachments readily desorbed and diffused to other more strongly binding sites. The energy difference of 7.8 kcal/mole in going from 20°C. to 70°C. did not allow van der Waals or hydrophobic bonding to occur.

Little emphasis has been placed on charge transfer forces as a mechanism for the adsorption of triazines onto organic matter. In seeking an explanation for the fact that more norazine than desmetryne was adsorbed onto H^+-humin while more desmetryne than norazine was adsorbed onto H^+-humic acid, HAYES et al. (1968) reasoned, without experimental evidence, as follows. If, they assumed, humin, the more hydrophobic of the two colloidal materials, contained aromatic structures it might act as a donor in charge transfer reactions with norazine whose aromatic nucleus, relative to that of desmetryne, can be regarded as deactivated. On the other hand the desmetryne might act as a donor to carboxyl group deactivating aromatic structures in humic acids. However, it should be pointed out here that phenolic groups in H^+-humic acids could act as donor groups in charge transfer reactions.

This author considers that charge transfer or chemisorption reactions are particularly important when trying to explain the high adsorption of methylthiotriazines on to organic matter. The differences in adsorption to organic sites between methylthio and methoxytriazine compounds must lie, all other factors considered, in the chemistry of the sulfur and oxygen atoms. The polarizability of the sulfur relative to that of oxygen can make the methylthio-group a soft base and cause it to react with soft acids such as humic acids. On the other hand involvement of the d orbitals of sulfur would mean that methylthio-groups could accept electrons and result in a chemisorption process. The d-orbital involvement could also result in binding to transition metals shown by WARDLE (1968) to be present in humic acids. This theory can fit into the ligand-exchange mechanism postulated in section X b).

Chlorotriazines are more readily bound to organic matter sites than are methoxytriazines (WALKER 1968, WALKER and CRAWFORD 1968). Hydrogen bonding and ion-exchange mechanisms would suggest the opposite effect. This author suggests, without adequate evidence, that chlorotriazines might be hydrolyzed on the organic matter, and the hydroxytriazines could be expected to be more tightly bound, on the basis of the ion-exchange and hydrogen bonding mechanisms discussed already, than the methoxytriazines.

Adsorption studies using appropriately selected model compounds of known structures might be of value when investigating the mechanisms of adsorption of triazines onto soil organic matter and its fractions. WARD and HOLLY (1966) used powdered preparations of cellulose, cellulose triacetate, and nylon as adsorbents and found that the cellulose preparation did not adsorb any of the triazine solutions tested. Cellulose triacetate had a slightly higher affinity for the triazines than did the nylon preparation. This caused the authors to suggest that the carbonyl group, common to both adsorbents (and not the imino group on nylon) was involved in the adsorption process. The affinity of triazines, which differed only in the 2-position substituents, for the adsorbents decreased in the order methylthio- > chloro- > methoxy- > hydroxytriazine. The adsorptive order for any particular 2-substituent parallelled an increasing complexity of the alkyl substituents at the 4,6-positions. It is of interest to note that the affinity order methylthio-, chloro-, methoxytriazine corresponds to that found by WALKER (1968) for adsorption of triazines onto soil organic matter surfaces.

WARD and HOLLY (1966) did not consider that the amino hydrogen on monoalkyl substituted triazines was involved in the adsorption process. They drew on evidence from infrared spectra data which did not indicate hydrogen bonding between amino hydrogens of triazines and the carbonyl oxygen of acetone. Furthermore they concluded that amino hydrogen atoms of monoalkyl substituted triazines are sterically blocked in isopropyl amino compounds. Through a series of reasoning processes they concluded that the substituent atom adjacent to the 2-position, that is oxygen, chlorine, or sulfur, presented the electropositive adsorptive site on the triazine to act as the electrophile to the carbonyl oxygen atoms in the adsorbents. This author believes that there are many arguments against this suggestion.

Cellulose triacetate and nylon are aliphatic and lack, as the authors realized, the complexity of functional groups and the steriochemistry of humic substances. Such compounds can give only limited information on what might happen in soils. Resins containing phenolic and carboxyl groups should prove of greater value. It is suggested that a study of complex formation with donor and acceptor molecules in charge transfer reactions might give more information, at this stage, however.

Summary

Evidence which demonstrates the importance of soil organic matter in adsorbing triazines, in reducing their phototoxicity, and in affecting their movement in soils, is reviewed and discussed. Some evidence is presented which indicates that montmorillonite clay could have a stronger binding effect than soil organic matter on simazine. Reference is made to observations that this particular clay separate can alter the relative affinities of methoxy- and chlorotriazines for soils which contain four to five percent organic matter.

Methods for extraction, fractionation, purification of fractions, and for determination of acidity and functional groups in soil organic matter are outlined. The techniques used in, and the results obtained from degradation studies on humic acids as well as some theories on their composition and structure, and on the composition of humin materials, are presented. The importance of finding suitable techniques for handling organic matter which can be universally used in herbicide-soil organic matter interaction studies is stressed.

Classical and newly emerging techniques for the study of binding, and their applicability to the field of herbicide-organic matter studies, are outlined. The composition, and the postulated effects of the ion-exchange saturating hydrogen and metallic cations on shapes, sizes, and stereochemistry of humic acids in particular are emphasized in a discussion on the mechanisms of triazine adsorption onto organic matter. Evidence which indicates that van der Waals forces, ion exchange, and hydrogen bonding can be involved in the mechanisms of adsorption is reviewed and the possible involvement of ligand exchange reactions, charge transfer processes, and chemisorption is discussed.

Résumé*

Adsorption des herbicides du groupe des triazines sur les matières organiques du sol et examen de la chimie des matières organiques du sol

On examine les faits qui ont permis de mettre en évidence le rôle important des matières organiques du sol dans la réduction de la phytotoxicité des triazines et de leur déplacement dans les sols. On discute l'observation selon laquelle dans certaines circonstances (et en particulier dans le cas de la simazine) les argiles peuvent être plus importantes que les matières organiques du sol dans la réduction de la phytotoxicité.

En vue d'apprécier les mécanismes impliqués dans les actions réciproques des triazines et des matières organiques du sol, on examine la chimie, les structures, les dimensions et les formes des différents composants des matières organiques du sol et la manière selon laquelle ces composants sont associés entre eux et avec les constituants inorganiques du sol. Il y a lieu de souligner cependant que notre connaissance de la composition des matières organiques du sol se laisse distancer par celle des autres composants du sol.

On examine les techniques appliquées à l'extraction, au fractionnement, à la purification, à la détermination des groupes fonctionnels et de l'acidité, et à la dégradation des substances humiques. On attire l'attention sur les renseignements que ces procédés peuvent fournir pour faciliter la compréhension de l'adsorption et de l'inactivation des

---

* Traduit par S. DORMAL-VAN DEN BRUEL.

triazines par les matières organiques du sol. Cependant, on insiste sur l'avantage de disposer de techniques acceptables (pour caractériser la matière organique du sol) qui puissent être adoptées par tous les chercheurs dans le domaine des herbicides.

On souligne l'importance des techniques classiques et récentes pour l'étude des liaisons et leur applicabilité dans le domaine des recherches sur les herbicides en relation avec les matières organiques.

Il semble probable que, dans des conditions acides, un échange d'ions et une liaison hydrogène entre les colloïdes organiques du sol (qui peuvent être considérés comme des polyélectrolytes) et les triazines puissent avoir lieu. Un échange d'ions devient moins probable lorsque le pH du milieu est voisin de la neutralité. L'implication probable des forces de van der Waal et l'importance plausible du transfert des charges, la sorption chimique et les mécanismes d'échange par liaison sont discutés.

Il apparaît de plus en plus probant que l'hydrolyse des groupes labiles en position deux des triazines herbicides est catalysée par les matières organiques du sol, en particulier lorsqu'il s'agit d'une forme saturée en ions hydrogène.

Zusammenfassung*

Die Adsorption von Triazin-Unkrautvertilgunsmittel durch organische Stoffe des Bodens und ein Überblick über die Chemie der organischen Stoffe des Bodens

Beweismaterial, das zur Annahme der wichtigen Rolle, die die organischen Stoffe des Bodens in der Verminderung der Pflanzentoxizität von Triazinen und in der Abnahme ihrer Tätigkeit in Böden geführt hat, wird besprochen. Die Beobachtung, dass unter gewissen Umständen (besonders im Falle von Simazin) Ton für die Verminderung der Pflanzentoxizität wichtiger sein kann als der Gehalt an organischer Bodensubstanz, wird diskutiert.

Um den Mechanismus, der in die organische Bodensubstanz-Triazin-Wechselwirkung verwickelt ist, richtig einzuschätzen, werden die Chemie, Strukturen, Grösse und Form der verschiedenen Bestandteile der organischen Bodensubstanz und die Art in der diese Bestandteile miteinander und mit den anorganischen Bodenbestandteilen verbunden sind, besprochen. Es muss jedoch hervorgehoben werden, dass unsere Kenntnisse über die Zusammensetzung der organischen Bodensubstanz hinter denjenigen der anderen Bodenbestandteile zurückbleiben.

Methoden, die für die Extraktion, Fraktionierung, Reinigung, Bestimmung der funktionellen Gruppen und des Säuregehaltes und die Degradierung der humischen Substanzen entwickelt worden sind,

---

* Übersetzt von M. DÜSCH.

werden besprochen. Es wird Nachdruck auf Kenntnisse gelegt, die sich von diesen Verfahren herleiten und die zu unserem Verständnis der Adsorption und Inaktivierung der Triazine durch organische Bodenbestandteile beitragen können. Es ist jedoch sehr wünschenswert, annehmbare Methoden (zur Charakterisierung von organischen Bodenbestandteilen) zu finden, die von allen auf dem Gebiet der Unkrautvertilgungsmittel Arbeitenden übernommen werden können.

Klassische und neu hervorgegangene Methoden für die bindende Arbeit und ihre Anwendungsmöglichkeit auf dem Gebiet der Studien der Unkrautvertilgungsmittel-organische Substanz, werden umrissen.

Unter sauren Bedingungen scheint es möglich zu sein, dass Ionenaustausch und Wasserstoffbindung zwischen organischen Bodenkolloiden (die als Polyelektrolyten betrachtet werden können) und Triazinen stattfinden können. Ionenaustausch wird weniger wahrscheinlich, wenn sich die pH-Werte der Medien der Neutralität nähern. Die wahrscheinliche Verwicklung der van der Waal'schen Kräfte und die mögliche Wichtigkeit von Ladungsverlagerung, Chemisorption und Austauschmechanismen von Liganden wird diskutiert.

Wachsende Beweismittel deuten an, dass die Hydrolyse labiler Gruppen in der Zweier-Position der Triazin-Unkrautvertilgungsmittel durch organische Bodensubstanz katalysiert wird, besonders wenn sie in der wasserstoffionengesättigten Form vorliegt.

References

ADAMS, R. S.: Phosphorus fertilization and the phytotoxicity of simazine. Weeds
    13, 113 (1965).
—— Soxhlet extraction of simazine from soils. Proc. Soil Sci. Soc. Amer. 30, 689
    (1966).
AHRENS, J. F.: Detoxification of simazine treated soil with activated charcoal.
    Weed Abstr. 13, 335 (1964).
ALBERS, E., and K. HOMBURG: De Inactivering en Penetratie van Simazin in de
    Grond. Meded. Land-Hogesch. Gent. 24, 893 (1959).
ALBERT, A., and E. P. SERJEANT: Ionization constants of acids and bases, p. 14.
    London: Methuen (1962).
ANDREWS, L. J., and R. M. KEEFER: Molecular complexes in organic chemistry.
    San Francisco: Holden-Day (1964).
ARMSTRONG, D. E., and G. CHESTERS: Adsorption catalysed chemical hydrolysis
    of Atrazine. Environ. Sci. Technol. 2, 683 (1968).
—— ——, and R. F. HARRIS: Atrazine hydrolysis in soil. Proc. Soil Sci. Soc. Amer.
    31, 61 (1967).
ATHERTON, N. M., P. A. CRANWELL, A. J. FLOYD, and R. D. HAWORTH: Humic
    acid. I: E.S.R. spectra of humic acids. Tetrahedron 23, 1653 (1967).
BAILEY, G. W., and J. L. WHITE: Review of adsorption and desorption of organic
    pesticides by soil colloids, with implications concerning pesticide bioactivity.
    J. Agr. Food Chem. 12, 324 (1964).
BARKER, S. A., M. H. B. HAYES, R. G. SIMMONDS, and M. STACEY: Studies on soil
    polysaccharides. I. Carbohydrate Research 5, 13 (1967).
BARROW, G. M.: Introduction to molecular spectroscopy, chap. 6. New York:
    McGraw Hill (1962).
BARTON, D. H. R., and M. SCHNITZER: New experimental approach to the humic
    acid problem. Nature 198, 217 (1963).

BECKWITH, R. S.: Titration curves of soil organic matter. Nature **184**, 745 (1959).

BREMNER, J. M.: Oxidation of soil organic matter in the presence of alkali. J. Soil Sci. **1**, 198 (1950).

—— A review of recent work on soil organic matter, II. J. Soil Sci. **5**, 214 (1954).

——, and D. S. JENKINSON: Determination of organic carbon in soil. I. Oxidation by dichromate of organic matter in soil and plant materials. J. Soil Sci. **11**, 394 (1660a).

—— —— Determination of organic carbon in soil. II. Effect of carbonized materials. J. Soil Sci. **11**, 403 (1960 b).

BRIEGLEB, G.: Elektronen—Donator—Acceptor—Komplexe. Berlin: Springer-Verlag (1961).

BROADBENT, F. E., and G. R. BRADFORD: Cation-exchange groupings in the soil organic fractions. Soil Sci. **74**, 447 (1952).

BROOKS, J. D., and S. STERNHELL: Chemistry of brown coals. I. Oxygen-containing functional groups in Victorian brown coals. Aust. J. Applied Sci. **8**, 206 (1957).

BUCHHOLTZ, K. P.: Factors influencing oat injury from triazine residues in soil. Weeds **13**, 362 (1965)

BURGES, A., H. M. HURST, and B. WALKDEN: The phenolic constituents of humic acid and their relation to the lignin of the plant cover. Geochim. Cosmochim. Acta. **28**, 1547 (1964).

BURNS, I. G., and R. E. GRICE: Personal communication (1969).

BURNSIDE, O. C., and R. BEHRENS: Phytotoxicity of simazine. Weeds **9**, 145 (1961).

BURSCHEL, P.: Untersuchungen über das Verhalten von Simazin im Boden. Weed Res. **1**, 131 (1961).

CAROLAN, R.: Modification of Graham's method for determining soil organic matter by colorimetric analysis. Soil Sci. **66**, 241 (1948).

CHAPMAN, L. J.: Influence of an organic soil in mixtures with sand on the phytotoxicity of atrazine. Ph.D. Diss., Univ. of Florida (1966).

CHESHIRE, M. V., P. A. CRANWELL, and R. D. HAWORTH: Humic acid-III. Tetrahedron **24**, 5155 (1968).

—— ——, C. P. FALSHAW, A. J. FLOYD, and R. D. HAWORTH. Humic acid II. Structure of humic acids. Tetrahedron **23**, 1669 (1967).

CHOUDRI, M. B., and F. J. STEVENSON: Chemical and physico-chemical properties of soil humic colloids. III: Extraction of organic matter from soils. Proc. Soil Sci. Soc. Amer. **21**, 508 (1957).

CLAESSON, S.: Studies on adsoption and adsorption analysis with special reference to homologous series. Arkiv. Kemi. Min. Geol. **23A**, P133 (1946).

COFFIN, D. E., and W. A. DELONG: Extraction and characterization of organic matter of a Podzol B. horizon. Trans. 7th Internat. Congress Soil Sci., Madison, Wis. **3**, 91 (1960).

COULSON, C. B., R. I. DAVIES, and E. J. A. KHAN: Humic acid investigations. III. The chemical properties of certain humic acid preparations. Soil Sci. **88**, 191 (1959).

DAVIES, R. I., C. B. COULSON, and D. A. LEWIS: Polyphenols in plant, humus, and soil. IV. Factors leading to increase in biosynthesis of polyphenols in leaves and their relation to mull and mor formation. J. Soil Sci. **15**, 310 (1964).

DAY, B. E., and L. S. JORDAN: The influence of soil characteristics on the adsorption and phytotoxicity of simazine. Weed Sci. **16**, 209 (1968).

DEUEL, H., and P. DUBACH: Decarboxylierung der organischen Substanz des Bodens III. Extraktion und Fraktionierung decarboxylierbarer Humusstoffe. Helv. Chim. Acta **41**, 1310 (1958).

DIJK, H. VAN: Electrometric tibrations of humic acids. Scien. Proc. Roy. Dublin Soc. **AI**, 163 (1960).

DUBACH, P., and N. C. MEHTA: The chemistry of soil humic substances. Soils and Fert. **26**, 293 (1963).

—— ——. T. JAKAB, F. MARTIN, and N. ROULET: Chemical investigations on soil humic substances. Geochim. Cosmochim. Acta **28**, 1567 (1964).

Enwezor, W. O., and A. H. Cornfield: Determination of total carbon in soils by wet combustion. J. Sci. Food Agr. **16,** 277 (1965).

Evans, L. T.: The use of chelating reagents and alkaline solutions in soil organic-matter extractions. J. Soil Sci. **10,** 110 (1959).

Fairclough, G. F., Jr., and J. S. Fruton: Peptide-protein interaction as studied by gel filtration. Biochem. **5,** 673 (1966).

Felbeck, G. T.: Structural chemistry of soil humic substances. Adv. Agron. **17,** 327 (1965 a).

Felbeck, G. T.: Studies on the high pressure hydrogenolysis of the organic matter from a muck soil. Proc. Soil Sci. Soc. Amer. **29,** 48 (1965 b).

Ferrari, G., and G. Dell' Agnola: Fractionation of the organic matter of soil by gel filtration through Sephadex. Soil sci. **96,** 418 (1963).

Finch, P., M. H. B. Hayes, and M. Stacey: Studies on soil polysaccharides and on their interaction with clay preparations. Trans. Comm. II and IV, Internat. Soc. Soil Sci., Aberdeen, Scotland, p. 19 (1966).

Flaig, W., J. C. Salfeld, and K. Haider: Zwischenstufen bei der Bildung von natürlichen Huminsäuren und synthetischen Vergleichsubstanzen. Landw. Forsch. **16,** 85 (1963).

Gast, A.: Beiträge zur Kenntnis des Verhaltens von Triazinen im Boden. Meded. Landbhogesch. Gent **27,** 1252 (1962).

Gilbert, G. A., and R. C. L. Jenkins: Sedimentation and electrophoresis of inter-acting substances. II. Asymptotic boundary shape for two substances inter-acting reversibly. Proc. Roy. Soc. **A253,** 420 (1959).

Gillet, A. C., and A. Pirghaye: Chromotography and separation of raw an-thranilic acid. Scien. Proc. Roy. Dublin Soc. **AI,** 133 (1960).

Greene, G., and C. Steelink: Structure of soil humic acid. II. Copper oxidation products. J. Org. Chem. **27,** 170 (1962).

Grover, R.: Influence of organic matter, texture and available water on the toxicity of simazine in soil. Weeds **14,** 148 (1966).

Haider, K., and J. P. Martin: Synthesis and transformation of phenolic com-pounds by *Epicoccum nigrum* in relation to humic acid formation. Proc. Soil Sci. Soc. Amer. **31,** 766 (1967).

Hance, R. J.: Observations on the relationship between the adsorption of diuron and the naure of the adsorbent. Weed Research **5,** 108 (1965).

—— Personal communication to A. Walker (1969).

——, S. D. Hocombe, and J. Holroyd.: The phytotoxicity of some herbicides in field and pot experiments in relation to soil properties. Weed Research **8,** 136 (1968).

Hansen, E. H., and M. Schnitzer: Oxidative degradation of Danish illuvial organic matter. Trans. Commision II and IV, Internat. Soc. Soil Sci. Aber-deen, p. 87 (1966).

—— —— Nitric acid oxidation of Danish illuvial organic matter. Proc. Soil Sci. Soc. Amer. **31,** 79 (1967).

Harris, C. I.: Adsorption, movement, and phytotoxicity of monuron and *s*-triazine herbicides in soil. Weeds **14,** 6(1966).

—— Movement of herbicides in soil. Weeds **15,** 214 (1967).

——, and T. J. Sheets: Influence of soil properties on adsorption and phytotoxic-ity of CIPC, diuron, and simazine. Weeds **13,** 215 (1965).

——, and G. F. Warren: Adsorption and desorption of herbicides by soil. Weeds **12,** 120 (1964).

Hartley, G. S.: Herbicide behavior in the soil. I. Physical factors and action through soils. In L. J. Audus (ed.): The physiology and biochemistry of herbicides, III. New York: Academic Press (1964).

Hayashi, T., and T. Nagai: Components of soil humic acids. VIII. Oxidation decomposition with alkaline potassium permangonate and nitric acid. Soil and Plant Food **6,** 170 (1961).

HAYES, M. H. B.: Subsidence and humification in peats. Ph.D. Thesis, Ohio State Univ. (1960).
——, and J. M. THOMPSON: Sorption. In C. A. I. Goring (ed.): Organic chemicals in the soil environment. New York: Dekker. (1970).
——, M. STACEY, and J. STANDLEY: Studies on the humification of plant tissue. Trans. IXth Int. Congress Soil Sci. **3**, 247 (1968).
—— ——, J. M. THOMPSON: Polarographic analysis of s-triazine herbicides. Chem. of Ind., p. 1222 (1967).
—— —— —— Adsorption of s-triazine herbicides by soil organic-matter preparations. In: Isotopes and radiation in soil organic-matter studies, p. 75. Internat. Atomic Energy Agency, Vienna (1968).
HELFFERICH, F.: Ion exchange, p. 166. New York: McGraw-Hill (1962.)
HILL, A. G., R. SEDGELEY, and H. F. WALTON: Separation of amines by ligand exchange. III. A comparison of different cation exchangers. Anal. Chem. Acta **33**, 84 (1965).
HILTON, H. W., and QUAN H. YUEN: Adsorption of several preemergence herbicides by Hawaiian sugar cane soils. J. Agr. Food Chem. **11**, 230 (1963).
HUMMEL, J. P., and W. J. DREYER: Measurement of protein-binding phenomena by gel filtration. Bichim. Biophys. Acta. **63**, 530 (1962).
JACKSON, M. L.: Organic matter determination for soils, chapt. 9. In: Soil chemical analysis. London: Constable (1958).
JAKAB, T., P. DUBACH, N. C. MEHTA, and H. DEUEL: Abbau von Huminstoffen. I. Hydrolyse mit Wasser und Mineralsäuren. Z. Pflernähr. Düng. Bodenk. **96**, 213 (1961).
—— —— —— Abbau von Huminstoffen. III. Abbau mit Alkali. Z. Pflernähr. Düng Bodenk. **102**, 8 (1963).
JOHNSTON, H. H.: Soil organic matter. II. Origin and chemical structure of soil humic acids. Proc. Soil Soc. Amer. **25**, 32 (1961).
KANG, E. S., and G. T. FELBECK, JR.: A comparison of the alkaline extract of tissues of *Aspergillus niger* with humic acids from three soils. Soil Sci. **99**, 175 (1965).
KELLETT, G. L.: The study of interacting protein systems by molecular sieve chromatography. Lab. Practice **16**, 857 (1967).
KIPLING, J. J.: Adsorption from solutions of non-electrolytes. New York: Academic Press (1965).
KLEIST, H., and D. MUCKE: Optische Untersuchungen an Huminsäuren. Albrecht-heer-Arch. **10**, 471 (1966).
KONONOVA, M. M.: Soil organic matter. London: Pergamon (1966).
KREVELEN, D. W. VAN, and J. SCHUYER: Coal science. Amsterdam and New York: Elsevier (1957).
KUMADA, K., and K. AIZAWA: Infrared spectra of humic acids. Soil and Plant Food **3**, 152 (1958).
—— Absorption spectra of humic acids. Soil and Plant Food **1**, 29 (1959).
KURODA, N.: Organic matter in core samples of 906 metre borings at Ianakomoto-machin, Minato-ku, Osaka. Pedologist **10**, 95 (1966).
LADD, J. N., and J. H. A. BUTLER: Comparison of some properties of soil humic acids and synthetic phenolic polymers incorporating amino derivatives. Austral. J. Soil Research **4**, 41 (1966).
LAMBERT, S. M., P. E. PORTER, and R. H. SCHIEFERSTEIN: Movement and sorption of chemicals applied to the soil. Weeds **13**, 185 (1965).
LINDQVIST, I.: Adsorption effects in gel filtration of humic acid. Acta Chem. Scand. **21**, 2504 (1967).
McGLAMERY, M. D., and F. W. SLIFE: The adsorption and desorption of atrazine as affected by pH, temperature, and concentration. Weeds **14**, 237 (1966).
MACINNES, D. A.: The principles of electrochemistry, p. 205. New York: Dover (1961).

MacKENZIE, A. F., and J. E. DAWSON: A study of soil horizons using electrophoretic techniques. J. Soil Sci. **13**, 160 (1962).

McLAREN, A. D.: Enzyme action in structurally restricted systems. Enzymologia **21**, 356 (1960).

——, and G. H. PETERSON: Soil biochemistry, 1 ed. New York: Dekker (1967).

——, and J. J. SKUJINS: Soil Biochemistry, vol. II, 1 ed. New York: Dekker. (1970).

MARTIN, A. E., and REEVE, R.: The extraction of organic matter from Podzolic B horizons with organic reagents. Chem. & Ind., p. 356 (1955).

—— —— Chemical studies on podzolic illuvial horizons. I. The extraction of organic matter by chelating agents. J. Soil Sci. **8**, 268 (1957 a).

—— —— Chemical studies on podzolic illuvial horizons. II. The use of acetylacetone as extractant of translocated organic matter. J. Soil Sci. **8**. 279 (1957 b).

—— —— Chemical studies of podzolic illuvial horizons. III. Titration curves of organic matter suspensions. J. Soil Sci. **9**, 89 (1958).

MARTIN, F., P. DUBACH, N. C. MEHTA, and H. DEVEL: Determination of the functional groups of humic substances. Z. Pflernähr. Düng. Bodenk. **103**, 27 (1963).

MARTIN, J. P., S. J. RICHARDS, and K. HAIDER: Properties and decomposition and binding action in soil of "humic acid" synthesized by *Epicoccum nigrum*. Proc. Soil. Sci. Soc. Amer. **31**, 657 (1967).

MEHTA, N. C., P. DUBACH, and H. DEUEL: Untersuchungen über die Molekulargewichtsverteilung von Huminstoffen durch Gelfiltration an Sephadex. Z. Pflernähr. Düng. Bodenk. **102**, 128 (1963).

MENDEZ, J., and F. J. STEVENSON: Reductive cleavage of humic acids with sodium amalgam. Soil Sci. **102**, 85 (1966).

MORTENSEN, J. L., and F. L. HIMES: Soil organic matter. In: F. E. Bear (ed.): New York: Reinhold (1964).

——, and SCHWENDINGER: Electrophoretic and spectroscopic characterization of high molecular weight components of soil organic matter. Geochim. Cosmochim. Acta **27**, 201 (1963).

MUZIK, T. J., and W. G. MAULDIN: Influence of environment on the response of plants to herbicides. Weeds **12**, 142 (1964).

NEARPASS, D. C.: Effects of soil acidity on the adsorption and persistence of simazine. Weeds **13**, 341 (1965).

NICHOL, L. W., and A. G. OGSTON: A generalized approach to the description of interaction boundaries in migrating systems. Proc. Roy. Soc. **B163**, 343 (1965).

OADES, J. M., and W. N. TOWNSEND: The influence of iron on the stability of soil organic matter during peroxidation. J. Soil Sci. **14**, 134 (1963).

ORLOV, D. S., and E. I. GORSHKOVA: Size and shape of humic acid particles from chermozem and soddy podzolic soils. Dokl. Vyssh. Shk. Biol. Nanki **1**, 207 (1968).

RANDHAWA, N. S., and F. E. BROADBENT: Soil organic matter-metal complexes: 5. Reactions of zinc with model compounds and humic acid. Soil Sci. **99**, 295 (1965 a).

—— —— Soil organic matter-metal complexes: 6. Stability constants of zinc-humic acid complexes at different pH values. Soil Sci. **99**, 362 (1965 b).

PAULI, F. W.: Physicochemical methods in modern humus research. Jena Rev. **11**, 36 (1966).

REX, R. W.: Electron paramagnetic resonance studies of stable free radicals in lignins and humic acids. Nature (London) **188**, 1185 (1960).

PIRET, E. L., R. G. WHITE, H. C. WALTHER, JR., and A. J. MADDEN: Some physicochemical properties of peat humic acids. Scien. Proc. Roy. Dublin Soc. **AI**, 69 (1960).

POMMER, A. M., and I. A. BREGER: Equivalent weight of humic acid from peat. Geochim. Cosmochim. Acta **20**, 45 (1960).

POSNER, A. M., B. K. G. THENG, and J. R. H. WAKE: The extraction of soil organic matter in relation to the humification. Trans. Internat. Congress Soil Sci. Adelaide, Australia 3, 153 (1968).

ROBINSON, D. W.: The use of sorbents and simazine on newly planted strawberries. Weed Research 5, 43 (1965).

RODGERS, E. G.: Leaching of four triazines in three soils as influenced by varying frequencies and rates of simulated rainfall. Proc. S. Weed Conf. 15, 268 (1962).
—— Leaching of seven s-triazines. Weed Sci. 16, 117 (1968).

SCATCHARD, G.: The attraction of proteins to small molecules and ions. Ann. N.Y. Acad. Sci. 51, 660 (1949).

SCHNITZER, M., and J. G. DESJARDINS: Molecular and equivalent weights of the organic matter of a podzol. Proc. Soil. Sci. Soc. Amer. 26, 362 (1962).
—— —— Further investigations on the alkaline permangonate oxidation of organic matter extracted from a Podzol Bₕ horizon. Can. J. Soil Sci. 44, 272 (1964).

——, D. A. SHEARER, and J. R. WRIGHT: A study in the infrared of high-molecular weight organic matter extracted by various reagents from a podzolic B horizon. Soil. Sci. 87, 252 (1959).

——, and S. I. M. SKINNER: The carbonyl group in a soil organic matter preparation. Proc. Soil Sci. Soc. Amer. 29, 400 (1965).
—— —— Polarographic method for the determination of carboxyl groups in soil humic compounds. Soil Sci. 101, 120 (1966).

——, J. R. WRIGHT, and J. G. DESJARDINS: A comparison of the effectiveness of various extractants for organic matter from two horizons of a podzol profile. J. Can. Soil Sci. 38, 49 (1958).

SHEETS, T. J.: The comparative toxicities of monuron and simazine in soil. Weeds 7, 189 (1959).
——, A. S. CRAFTS, and H. R. DREVER: Soil effects on herbicides. Influence of soil properties on the phytotoxicities of the s-triazine herbicides. J. Agr. Food chem. 10, 458 (1962).

STEELINK, C.: Free radical studies of lignin, lignin degradation products, and soil humic acid. Geochim. Cosmochim. Acta 28, 1615 (1964).
—— Electron paramagnetic resonance studies of humic acid and related model compounds. Adv. Chem. Series 55, 80 (1966).

——, and G. TOLLIN: Stable free radicals in soil humic acids. Biochim. Biophys. Acta. 59, 25 (1962).
—— —— Free radicals in soil. In A. D. McLaren and G. H. Peterson (ed.): New York: Dekker (1967).

——, J. W. BERRY, A. HO, and H. E. NORDBY: Alkaline degradation products of soil humic acid. Proc. Roy. Dublin Soc. AI, 59 (1960).

STEVENSON, F. J., and M. MENDEZ: Reductive cleavage products of soil humic acids. Soil Sci. 103, 383 (1967).

SULLIVAN, J. D., JR., and G. T. FELBECK, JR.: The interaction of s-triazine herbicides with humic acids from three different soils. Soil Sci. 106, 42 (1968).

SWABY, R. J., and J. N. LADD: Chemical nature, microbiol resistance, and origin of soil humus. Trans. Internat. Soc. Soil Sci. Comm. IV and V, New Zealand, p. 197 (1962).

SWIFT, R. S.: Physico-chemical studies on soil organic matter. Ph. D. Thesis, Univ. of Birmingham, Eng. (1968).

SWINCER, G. D., J. M. OADES, and D. J. GREENLAND: Studies on soil polysaccharides. I. The isolation of polysaccharides from soil. Austral. J. Soil. Research 6, 211 (1968).

TALBERT, R. E., and O. H. FLETCHALL: The adsorption of some s-triazines in soils. Weeds 13, 46 (1965).

THENG, B. K. G., J. R. H. WAKE, and A. M. POSNER: The infrared spectrum of humic acid. Soil Sci. 102, 70 (1966).

THOMPSON, J. M.: Some studies of the adsorption of s-triazine herbicides on soil organic matter. M. Sc. Thesis, Univ. of Birmingham (1695).

TURNER, M. A., and R. S. ADAMS, JR.: Adsorption of atrazine and atratone by anion and cation exchange resins. Proc. Soil Sci. Soc. Amer. **32**, 62 (1968).

TURNER, R. C., and M. SCHNITZER: Thermogravimetry of the organic matter of a podzol. Soil Sci. **93**, 225 (1962).

UPCHURCH, R. P., and D. D. MASON: The influence of soil organic matter on the phytotoxicity of herbicides. Weeds **10**, 9 (1962).

——, F. L. SELMAN, D. D. MASON, and E. J. KAMPRATH: The correlation of herbicidal activity with soil and climatic factors. Weeds **14**, 42 (1966).

VICKERSTAFF, T.: The physical chemistry of dyeing, chapt. 6 and 7. Edinburgh: Oliver and Boyd (1954).

WAGNER, G. H., and F. J. STEVENSON: Structural arrangement of functional groups in soil humic acid as revealed by infrared analysis. Proc. Soil Sci. Soc. Amer. **29**, 43 (1965).

WAKSMAN, S. A., and K. R. N. IYER: Contributions to our knowledge of the chemical nature and origin of humus: IV. Fixation of proteins by lignins and formation of complexes resistant to microbiol decomposition. Soil Sci. **36**, 69 (1933).

WALKER, A.: Personal communication (1969).

—— Physico-chemical aspects of the behaviour of triazine herbicides in soils. Ph.D. Thesis, Univ. of Nottingham, Eng. (1968).

——, and D. V. CRAWFORD: The role of organic matter in adsorption of the triazine herbicides by soils. In: Isotopes and radiation in soil organic matter studies, p. 91. Internat. Atomic Energy Agency, Vienna (1968).

WALDRON, A. C., and J. L. MORTENSEN: Soil nitrogen complexes. II. Electrophoretic separation of organic components. Proc. Soil Sci. Soc. Amer. **25**, 29 (1961).

—— —— Soil nitrogen complexes. III. Distribution and identification of nitrogenous constituents in electrophoretic separates. Soil Sci. **93**, 286 (1962).

WARD, T. M., and K. HOLLY: The sorption of s-triazines by model nucleophiles as related to their partitioning between water and cyclohexane. J. Colloid and Interface Sci. **22**, 221 (1966).

WARDLE, R. E.: The extraction and characterisation of soil organic matter. M.Sc. Thesis, Univ. of Birmingham, Eng. (1968).

WATSON, J. D.: Molecular biology of the gene, chap. 4. New York and Amsterdam: Benjamin (1965).

WEBER, J. B.: Adsorption of triazine herbicides on clay minerals. Residue Reviews **32**, 93 (1970).

——, P. W. PERRY, and E. IBARAKI: Effect of pH on the phytotoxicity of prometryne applied to synthetic soil media. Weed Sci. **16**, 134 (1968 a).

—— ——, and R. P. UPCHURCH: The influence of temperature and time on the adsorption of paraquat, diquat, 2,4-D, and prometone by clays, charcoal, and an anion-exchange resin. Proc. Soil Sci. Soc. Amer. **29**, 678 (1965).

——, T. M. WARD, and S. B. WEED: Adsorption and desorption of diquat, paraquat, prometone, and 2,4-D by charcoal and exchange resins. Proc. Soil Sci. Soc. Amer. **32**, 197 (1968 b).

WHITEHEAD, D. C., and J. TINSLEY: Extraction of soil organic matter with dimethylformamide. Soil Sci. **97**, 34 (1964).

WILDUNG, R. E., G. CHESTERS, and S. O. THOMPSON: Cation-exchange capacity of plant lignins and soil humic colloids as affected by pH and hydrolysis error. Proc. Soil Sci. Soc. Amer. **29**, 688 (1965).

WRIGHT, J. R., M. SCHNITZER, and R. LEVICK: Some characteristics of the organic matter extracted by dilute inorganic acids from a podzolic B horizon. Can. J. Soil Sci. **38**, 14 (1958).

ZIECHMAN, W.: Spectroscopic investigations of lignin humic substances and peat. Geochim. Cosmochim. Acta. **28**, 1555 (1964).

Movement of s-triazine herbicides in soils

By

Charles S. Helling[*]

Contents

I. Theory

The movement of an organic pesticide[1] placed onto or within the soil may influence both its effectiveness and its potential as a contaminant in adjacent soil, water, or air. Movement may occur while in solution or adsorbed on migrating soil particles, or by transfer as vapors. Volatilization, the latter, is discussed in another part of this volume (Jordan et al. 1970). The two general processes of non-gaseous move-

---

[*] Crops Research Division, Agricultural Research Service, *U.S. Department of Agriculture*, Beltsville, Maryland.
  [1] Pesticide chemicals mentioned in this text and their common or trade names are given in Table I.

ment, mass transfer and diffusion, will be considered within the soil profile; movement at the soil surface will be discussed separately.

a) Sub-surface movement

Pesticides entering the soil are moved, or leached, predominantly in a vertical direction in the zone of aeration above the water table (LeGrand 1966). Lateral movement occurs when pesticides reach a zone of water saturation, and near the boundary of relatively dry and wetted (by rainfall or irrigation) soils inasmuch as some water movement occurs laterally into the drier soil. Movement of water and pesticide may occur upward as well as downward, as drying of the soil surface takes place. Sub-irrigation is a practical example of this. The natural wetting-drying cycles in the soil may tend toward the uniform distribution of moderately mobile pesticides (Schuldt et al. 1957). Recognition of this may explain in part why a chemical which appears rather mobile in laboratory tests will, in the field, remain within the upper several feet of the soil profile.

Hartley (1960 and 1964) has aptly described the movement of herbicides in soils. Any transport over a considerable distance must occur by mass transfer, i.e., flow or convection. Figure 1, adapted in part from Hartley (1964), illustrates the distribution of atrazine and

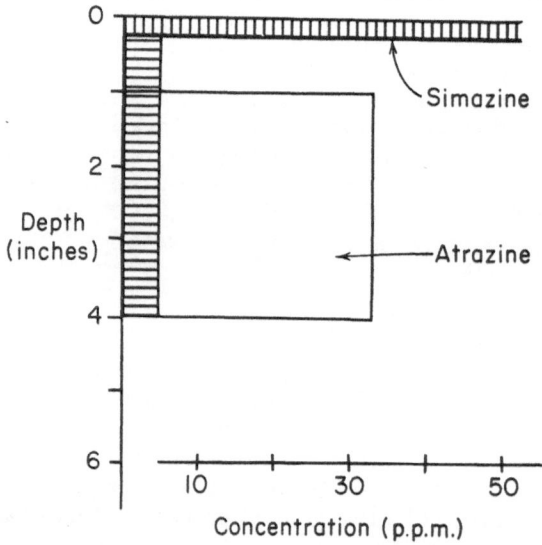

Fig. 1. Theoretical distribution of atrazine and simazine by vertical flow, considering only the effect of solubility; parameters: one cu. inch/sq. inch water, 25 percent (v/v) field capacity, five lb./acre application rate (adapted from Hartley 1964)

simazine in a soil if solubility were the only consideration in the pesticide transfer. Because atrazine (33 p.p.m.) is more soluble in water than simazine (5 p.p.m.), only 0.75 inch rain would be necessary to totally dissolve five lb./acre. It is thus shown moving downward as a three-inch band. Simazine would, in this overly simple model, also appear at the leading edge of the water front (at four inches depth), but would require five inches of rain to be completely dissolved.

A number of factors invalidate the "solubility" model for movement. First, the bands are depicted as having sharp boundary concentrations. Even in the best chromatographic systems some blurring of the sample front occurs due to hydrodynamic dispersion and diffusion. The former is caused by water percolating downward through the centers of soil pores more rapidly than along the sides. Furthermore, pores of various sizes are present; water velocity therefore varies, leading to a velocity distribution and thus mixing. In general, the more heterogeneous the system, the greater the extent of dispersion. Unsaturated soils have many cul-de-sacs partially filled with entrapped air, and a considerable degree of spreading may be expected.

A second factor influencing the distribution pattern of pesticides in soils is diffusion. According to Fick's laws of diffusion:

$$J = -D \frac{\partial C}{\partial x} \quad \text{and} \quad \frac{\partial C}{\partial t} = D \left(\frac{\partial^2 C}{\partial x^2} \right)$$

the rate of movement (J) of a chemical over distance x and time t is directly proportional to its concentration, C, and a diffusion coefficient, D. Diffusion coefficients of pesticides are 1 to 3 $\times$ 10^4 times greater in air than in water. According to GORING (1967), chemicals with water/air ratios (i.e., the ratio of weights of the chemical, at equilibrium, in equal volumes of water and air) under 1 $\times$ 10^4 should diffuse primarily through air. Those with ratios over 3 $\times$ 10^4 diffuse principally through water. Five s-triazine herbicides were assigned ratios of 18 $\times$ 10^6 (prometone, atrazine) to 69 $\times$ 10^6 (simazine), indicating that most diffusive movement of triazines within the soil occurs in aqueous solution. Diffusive movement in saturated soils, of course, can only occur in solution. A theory for diffusion in saturated soils was recently developed and tested with 2,4-D (LINDSTROM et al. 1968). The overall effect of longitudinal diffusion is to spread the concentration curve.

Interaction with the soil itself represents a third, highly significant factor which mediates the movement of herbicides in soils. Adsorption, discussed in greater detail in section III(a), retards the movement of triazines such that little or no chemical occurs at the water front (at four inches in Fig. 1). The rate of chemical and biological transformations, usually insignificant in short-term laboratory studies, may retard the amount and depth of chemical leached in the natural environment.

The use of chromatographic theory has been applied to characterize movement of both organic and inorganic compounds in soils. Frissel and Poelstra (1967 a) evaluated existing chromatographic theories and compared (1967 b) calculations of ion distribution patterns with those obtained experimentally for [45]Ca, [85]Sr, and [90]Sr. They suggested that for herbicides in soils, the following assumptions are often valid: the material under consideration is adsorbed by the adsorbent phase according to a linear adsorption isotherm, the exchange process is thermodynamically irreversible, and the longitudinal diffusion coefficient and other dispersion effects are significant and must be included. They classify theories as rate theories and concentration (or plate) theories. The former emphasizes the kinetic process, *i.e.*, the rate of adsorption-desorption, flow behavior, or diffusion. Plate theories consider the column divided into a number of segments or plates, individually at equilibrium between stationary and mobile phases.

Movement of herbicides has recently been described via chromatographic theory. Lindstrom *et al.* (1967) developed a mathematical model for vertical movement of chemicals in saturated soil. They assume adsorption follows the Freundlich isotherm. Their rate theory also includes contributions of the diffusion coefficient, percolation velocity of the water, average particle size in the soil, and the fractional number of sorbing sites. Lambert *et al.* (1965) applied a fairly simple plate theory to assess movement of a neutral herbicide, SD 9515-B, on soil columns. The depth to maximum concentration was calculated by measuring the distribution coefficient (between soil organic matter and water), void volume, water and organic matter contents, and volume of water added.

b) Surface movement

Loss of a pesticide at the soil surface, apart from volatilization, occurs by water or wind erosion. Hydrogeologic aspects of movement across the soil surface have been discussed by LeGrand (1966). Topographic relief, permeability, and precipitation are the chief factors governing accepted *versus* rejected recharge. Intense precipitation occurring on soils of low permeability (e.g., high clay content) usually causes some runoff. Thus certain pesticides may appear briefly in streams after a storm, depending on the extent of previous dissipation and the proximity of the stream. Steep topography accentuates runoff loss, as soil erosion is more severe and water infiltration is reduced, increasing movement directly on soil particles.

Some of the factors influencing surface movement of pesticides can be deduced from the following observations. Residues of picloram, a herbicide of moderate mobility and persistence, appeared periodically (after rainstorms) up to 11 months in the stream below the Arizona watershed to which it was applied (DAVIS *et al.* 1967). Atrazine runoff from fallow Cecil soil, 6.5 percent slope, was estimated to be $\leq$0.1 lb/ acre during normal field conditions (WHITE *et al.* 1967). Greater losses occurred when intense rain was applied immediately after herbicide application. Initial runoff loss of dicamba, picloram, and 2,4,5-T was greater on sod than fallow plots, but after four months losses were reduced to less than one percent of those first observed (TRICHELL *et al.* 1968). Formulation of 2,4-D as amine salts substantially retarded runoff loss as compared to the less soluble isooctyl and butyl ether esters (ANONYMOUS 1968 b, International Pest Control 1967). UPCHURCH (1966) noted the paucity of information regarding lateral movement from soil surfaces. He observed, however, that field plots receiving moderate rates of diuron or simazine maintain rather clear boundaries of weed control, suggesting that they have been made resistant to runoff movement by some vertical leaching.

Movement of pesticides, especially herbicides, from the soil surface by wind erosion has received little attention. One instance of long-range transport was reported by COHEN and PINKERTON (1966). Dust originating from a storm in the Southern High Plains region of Texas and deposited in Cincinnati, Ohio, carried a significant amount of adsorbed pesticides. The seven compounds identified and their concentrations in p.p.m. were: DDT (0.6), chlordane (0.5), DDE (0.2), Ronnel (0.2), heptachlor epoxide (0.04), 2,4,5-T (0.40), and dieldrin (0.003). High winds occurring immediately after surface application sharply reduced the performance of five pre-emergence herbicides on Willacy fine sandy loam (MENGES 1964). Wind-blown soil collected on the leeward side of experimental plots contained considerable pesticide. MENGES cited unpublished research indicating that winds could redistribute treated soil to increase crop injury and decrease weed control. Surface transport (excluding vapor loss) of pesticides by wind is most likely under the following conditions: high soil erodibility, particularly related to fine sands, smooth soil surfaces, and strong winds, and surface-application of the pesticide on a dry soil with no subsequent rainfall or irrigation before winds occur.

II. Methods

The techniques used to evaluate movement of the *s*-triazine herbicides in soil have also been used for virtually all types of soil-applied pesticides. They are broadly classified as field observations and laboratory (or greenhouse) studies. A more significant distinction is that the former is subjected to interwoven complexities of the total environment; the latter, usually by design, seeks to isolate pesticide mobility from effects of microbial degradation, plant uptake, variations in moisture and temperature, etc.

180

a) Field observations

Eventually, research data acquired in the laboratory are usually tested under field conditions. The mobility of pesticides is, in nearly all cases, assessed by analysis for the compound at increasing depths. This is undertaken after intervals of weeks for highly mobile compounds, to years for relatively immobile and persistent substances. Often, data from the latter category accrues from residue analyses—the primary objective being a persistence study. Field data of s-triazine mobility during natural conditions usually represents a span of six months to one year.

Sampling technique is important for proper evaluation of pesticide mobility data. By analysis of several soil samples per treatment, one can recognize anomalous behavior. For example, deep movement might be caused by physical transport of surface soil into animal burrows or into the deep cracks formed when soils high in montmorillonitic clay become dried. Periodic multiple sampling of a small plot could inadvertently find a point where a soil core had previously been removed.

Errors in sampling technique are less likely in short-term field experiments. For the triazines, the latter are represented by artificial irrigation of plots in studies of leaching and/or runoff. In addition to assessing the effect of specific methods and amounts of water application, these experiments circumvent long range changes in temperature, variable periods of precipitation, and microbial degradation. At the same time, movement has been measured in a natural soil profile.

A less apparent type of error may occur from restricted drainage, e.g., from a hardpan. NASH and WOOLSON (1968) suggested that gravel and tile drainage under some *U.S. Department of Agriculture* plots at Beltsville restricted the leaching of organochlorine insecticides.

Limited mobility research has been conducted using lysimeters besides the sampling of long- and short-term field plots. Although use of a lysimeter permits precise monitoring of water and pesticide movement (as a special case of effluent analysis) in a large mass of soil maintained outdoors, few investigators have employed this device. They remain comparatively inaccessible, apparently. DDT movement through a forest floor and into the underlying gravelly loam was analyzed with the aid of a lysimeter, however (RIEKERK and GESSEL 1968). It consisted of a 6.13 sq. dm. porous plate against which a 0.1 atmosphere tension was maintained, simulating that of the underlying soil at field capacity. The authors claim a higher sensitivity for the measurement of DDT movement as compared to the usual residue analyses.

b) Soil columns

Nearly all leaching studies in the laboratory have utilized a vertical column containing soil and applied pesticide. Water percolates downward or upward and pesticide movement is somehow measured. Commonly, the column has been pre-sectioned. After leaching, the sections are removed by slicing the soil with a spatula and each segment is analyzed to obtain a profile of pesticide movement. Alternately, soil is extruded directly from a straight glass column, then analyzed (GEISSBÜHLER et al. 1963). Examples of conventional columns, typical and bizarre, include stacked aluminum rings, each three-inch (i.d.) × one-inch (HARRIS 1966 and 1967); sectioned monel metal pipe (UPCHURCH and PIERCE 1957); stovepipe sections, with soil added to approximate the natural horizons (RODGERS 1968); stacked tin cans (SHEETS 1958); perforated Teflon (KAY and ELRICK 1967); rectangular plastic pipe (HOROWITZ 1968); 7.5-cm. (i.d.) × 20-cm. polyethylene bags (KAZARINA 1965); Cellophane cylinders (OGLE and WARREN 1954); glass, consisting of 1.7-inch (i.d.) × 3-inch sections (GRAY and WEIRICH 1968); four-inch (i.d.) × six-inch waxed paper columns (FRIESEN 1965); and paper carton cylinders (KOZLOWSKI and KUNTZ 1963). An unusual modification was the split-cylinder, a 17.5-cm. metal column which was longitudinally halved (GAST 1959). The slotted column of LAMBERT et al. (1965) allowed bioassay of movement in situ as a continuous gradient. They cut a 0.5-inch × 12-inch longitudinal slot into an 18.5-inch glass column, and covered the slot with tape before filling the column with soil. After leaching, the tape was removed and the exposed soil sown with an indicator plant (for herbicide testing) or spore suspension (for soil fungicides). Growth of the test organism gave a living profile of pesticide movement. The technique of slotted columns was also used by SHAHIED and ANDREWS (1966) to study leaching of prometryne, fluometuron, linuron, and trifluralin. Although most workers use relatively homogeneous soil, POELSTRA and FRISSEL (1965 and 1967) acquired intact field soil cores, 12 cm. (i.d.) × 100 cm., for some leaching studies.

When leaching is conducted as a saturated system, i.e., the total soil pore volume is filled with water, care is taken to exclude air within the column. This has been achieved by prior evacuation (FRISSEL and POELSTRA 1967 b, POELSTRA and FRISSEL 1965), slow saturation with water from bottom to top (SWOBODA and THOMAS 1968), and incorporation of small vent holes along the top side of a horizontally-placed column (DAVIDSON and SANTELMANN 1968). Water flow is often uncontrolled in leaching experiments, being dependent on the infiltration capability of the soil column. Simple constant-head devices are commonly used. SWOBODA and THOMAS (1968) controlled the leaching rate by mixing 50 g. of soil with 12.5 g. of Whatman cellulose powder, after first ascertaining that the cellulose did not adsorb parathion. FRISSEL and POELSTRA (1967 b) maintained a constant water head but varied flow rate by attaching different lengths of 0.07 mm. (i.d.) capillary tubing to the column exit. A constant-volume pump is often employed in more elaborate apparatus.

The pesticide in the columns described above was usually analyzed as that remaining in various sections of the column. In other cases, especially with mobile pesticides or when using relatively non-absorbent media such as glass beads, direct analysis is made of the column effluent (DAVIDSON and SANTELMANN 1968, ELRICK and MACLEAN 1966, ESHEL and WARREN 1967, RODGERS 1968). A simple example permitting this was the use of a Büchner funnel as the soil column (SMITH et al. 1957). The technique of miscible displacement, first used on soils for inorganic ion movement, has now been employed in studies of pesticides such as lindane (KAY and ELRICK 1967), 2,4-D (ELRICK and MACLEAN 1966), fluometuron, diuron (DAVIDSON and SANTEL- MANN 1968), and atrazine (GREEN et al.1968 b, SNELLING et al. 1969). Plots of the reduced concentration, C/C_0 (C_0 is initial concentra- tion of solute, C is that measured in the effluent), versus effective pore volume, Qt/V_0 (Q is the liquid volume flux, t is time, V_0 is the volume in a column occupied by fluid), are useful in evaluating dis- persion, diffusion, and adsorption occurring during flow through porous media. If a herbicide solution is used to displace water in a column, the resultant concentration profile is termed a breakthrough curve; the opposite technique yields an elution curve. Determination of elution profiles has additional merit in gross estimation of the amount of water required to leach pesticides to a specific soil depth. Effluent analysis probably affords the most precise concentration distribution profile for pesticides of sufficient mobility to be leached from soil columns. POELSTRA and FRISSEL (1965 and 1967) have developed a "soil column scanner," i.e., a single-channel γ-spectometer which employs a heavily shielded γ-detector to scan the column at variable speed, for research on the vertical transport of ions in soils. This apparatus is noteworthy in that it gives a dynamic picture of the progressing distribution pattern without disrupting the column. The use of such a column scanner would be a boon to pesticide mobility investigations, but the low energy and/or activity of the isotopes commonly used—^{14}C, ^{36}Cl, ^{35}S—may preclude this application. Other isotopes, ^{131}I and ^{32}P, have potential despite their short half-lives (8.04 and 14.3 days, respectively).

SWOBODA and THOMAS (1968) estimated movement of parathion in the field from column data, based on the equation

$$R = \frac{H\rho L}{d\theta}$$

where R is the amount of rainfall needed to move half the chemical to depth L, H is the height of water used to displace parathion to depth d in the column, and ρ and θ are the bulk densities of the soil in the field and columns, respectively. Movement may also be related to the distribution coefficient, K_d, by the equation

$$R = L\left(K_d\rho + \frac{V_v}{V}\right)$$

where V_v/V, the porosity, is the ratio of void volume to total column volume. The main assumption in converting laboratory to field data is that behavior of the chemical is analogous in both systems. Variations in profile characteristics, flow rate, amount of soil water, surface evaporation, etc. may restrict the comparison.

A special case of soil column technique is the study of pesticide diffusion. LOPEZ-GONZALEZ and VALENZUELA CALAHORRO (1968 a and b) determined the diffusion rates of DDT and DDE in H$^+$-bentonite as a function of temperature. They employed a lucite column consisting of 100 individual plates, 40 mm. (i.d.) $\times$ 2.5 mm., and extracted each five mm. interval at the conclusion of the experiment to assess movement. LINDSTROM *et al.* (1968) allowed ^{14}C 2,4-D to diffuse into wet soil from an overhead reservoir during a three-day period; the column was then frozen and sliced into two-mm. segments for analysis. MASSINI (1961) inserted a ^{14}C dichlobenil-impregnated filter paper disc in the middle of dry soil columns, then determined radioactivity distribution after 30 days. Diffusion of monuron and neburon into hydrosoil was simulated using 55-gallon barrels containing six inches of water over nine inches of sediment (FRANK 1966 a). After spraying the herbicides onto the water surface, the soil was sampled at one- to 128-week intervals by removal and analysis of previously embedded plastic columns.

c) *Soil thin-layer chromatography*

Until recently, methods for investigating mobility of non-volatile pesticides within soils were based on the use of field analyses or soil columns. In 1968, HELLING and TURNER introduced soil thin-layer chromatography (or soil TLC) as an alternate procedure. It is analogous to conventional TLC, with substitution of soil as the adsorbent phase.

In their initial report, HELLING and TURNER used three soils, Lakeland sandy loam, Chillum silt loam, and Hagerstown silty clay loam. Medium sand (>250 μ) was removed by dry-sieving from Chillum and Hagerstown soils, and coarse sand (>500 μ), from Lakeland. Aqueous slurries were prepared and 500-μ (silt loam, silty clay loam) or 750-μ (sandy loam) thick layers spread using conventional TLC apparatus. After drying, six or seven radioactive pesticides were applied near the base of a 20 $\times$ 20 cm. plate and developed ten cm.

with water by ascending chromatography. Pesticide movement was
visualized by autoradiography.

The relative mobilities of various herbicides appear in Figures 2

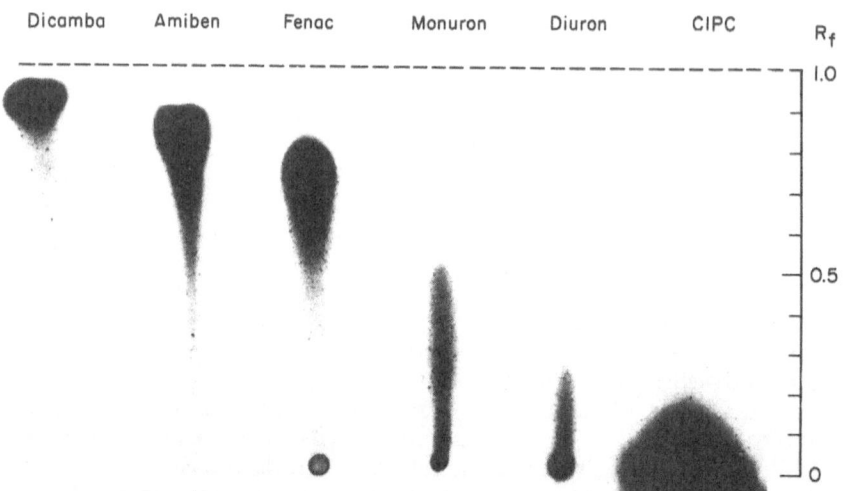

Fig. 2. Movement of six herbicides on a Hagerstown silty clay loam plate (10 cm.
× 500 μ) by soil TLC (Helling and Turner 1968) (reproduced with
permission)

(Helling and Turner 1968) and 3 (Helling 1969). The near absence
of tailing in the highly mobile dicamba indicates that its adsorption
isotherm is nearly linear. Compounds of lesser mobility exhibit
increased tailing. Movement was expressed by the conventional R_f
designation, although this referred to the front of pesticide movement
rather than its maximum concentration. The soil TLC data are most
appropriately compared to other mobility data which indicate the
depth to which a pesticide may be leached. The ranking of pesticides in
order of mobility is in good agreement with general trends previously
reported. Although the patterns of movement of atrazine, propazine,
and prometryne (Fig. 3) are similar, the depths of leaching are clearly
different.

Absolute movement on soil TLC plates cannot be transposed
directly to field or soil column experiments. Soil structure in the TLC
system was considerably more homogeneous than that in most other
systems. For this reason, band spreading may be somewhat less than
field or column regimes, by analogy with conventional chromatography.
Flow rates are also higher than those occurring naturally: infiltration
into Hagerstown silty clay loam was equivalent to rainfall of ca. 1.2

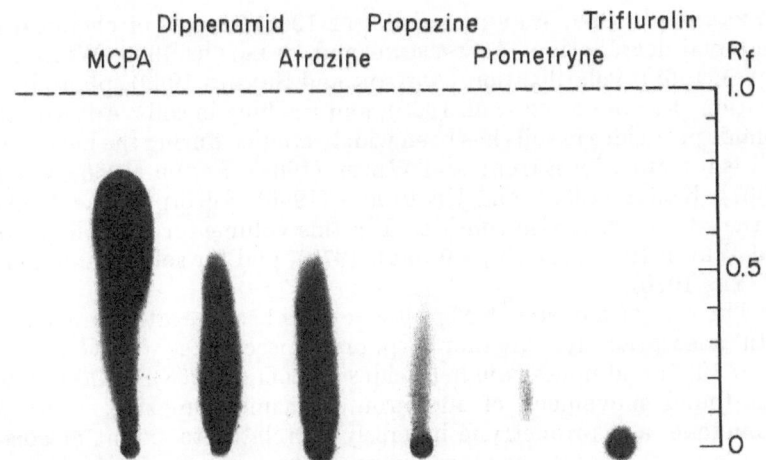

Fig. 3. Movement of six herbicides, including three triazines, on a Hagerstown silty clay loam plate (10 cm. × 250 μ) by soil TLC

cm./hr. (HELLING 1969). High flow rates are usually associated with increased mobility,[2] as later correlations (HELLING 1968) bore out.

UPCHURCH (1966) noted the expense and difficulty involved in conducting leaching studies. HELLING and TURNER (1968) indicated that soil TLC offered a rapid, simple, and inexpensive procedure, suggesting it could be used for establishing a general mobility classification of pesticides. HANCE (1967) suggested another, slightly related method for use in screening and evaluation of pesticides. He compared movement of 29 herbicides, including 19 alkylamino-s-triazines, on a normal partition TLC system with the log μmolar solute adsorbed/g. of soil. Movement of the herbicides, all non-ionic at pH 7, was negatively correlated with increasing adsorption. Extrapolations of partition data would seem best made only for soils whose adsorption characteristics were well known.

III. Parameters influencing movement

a) Adsorption

Adsorption, *i.e.*, attraction or repulsion of a solute at a surface, is probably the most significant factor affecting the overall behavior of s-triazine herbicides (and most pesticides) in soils. The rate and extent to which a compound is adsorbed and subsequently desorbed may govern its biological toxicity (COGGINS and CRAFTS 1959, SUND 1964,

---

[2] The parameter of flow rate is discussed in section IIIc).

WEBER et al. 1968, WEBER and SCOTT 1966), degree of chemical and microbial degradation (ARMSTRONG and CHESTERS 1968, WEBER and COBLE 1968), volatilization (ASHTON and SHEETS 1959), photodecomposition (FUNDERBURK et al. 1966), and leaching in soils. Adsorption of organic pesticides in soils has been widely studied during the past decade and is reviewed by BAILEY and WHITE (1964), FREED (1966), GORING (1967), KUNZE (1966), and UPCHURCH (1966). Adsorption as it relates to the s-triazines is also considered in this volume for the soil (BAILEY and WHITE 1970), for clay (WEBER 1970), and for soil organic matter (HAYES 1970).

The rate of movement of pesticides has been negatively correlated with adsorption for organophosphorus insecticides (MCCARTY and KING 1966) and non-acidic herbicides (HELLING 1968). HARRIS (1966) also found movement of monuron, atrazine, simazine, propazine, prometone, and prometryne inversely correlated to extent of adsorption on four soils. One or more soil properties such as clay and organic matter contents, and cation-exchange capacity are often correlated positively with adsorption of specific pesticides (BAILEY and WHITE 1964). These properties may, in turn, be negatively correlated with leaching.

The order of decreasing mobility of seven s-triazines (RODGERS 1968) seems to coincide with increasing relative adsorption (ANONYMOUS 1968 d, BAILEY et al. 1968, HILTON and YUEN 1963, NEARPASS 1967, TALBERT and FLETCHALL 1965). The mobility of amitrole was in reverse order to the adsorption capacity of quartz sand and three soils (DAY et al. 1961). FREED (1966) reviewed the behavior of nine herbicides from a qualitative standpoint: "slight" adsorption corresponded to "ready" leaching, and "strong" adsorption, to "resistant" or "moderately resistant" leaching.

Soil organic matter is the primary soil parameter responsible for adsorption (and possibly absorption) of most pesticides. GORING (1967) cites 40 references affirming this. For example, the adsorption of ten ureas from slurries of 11 soils was related only to organic matter content (HANCE 1965). Adsorption of simazine by 18 acid soils was highly correlated with organic matter and titratable acidity, and somewhat less well with clay content (NEARPASS 1965). Similarly, organic matter was highly correlated with retention against leaching of mevinphos (GETZIN and CHAPMAN 1959) and NPA (OGLE and WARREN 1954). BAILEY and WHITE (1964) cite many references in which less leaching of pesticides occurred in organic soils and heavy textured soils than in lighter textured soils. Thus, the depth of penetration of propazine was inversely proportional to organic matter content (SKOB 1963). Higher organic matter (1.44 versus 0.87 percent) greatly increased retention of monuron in Lakeland sand (UPCHURCH and PIERCE 1958) and dalapon in Iowa soils (HOLSTUN and LOOMIS 1956). Likewise, picloram mobility had been greatest after 15 months in soils of lowest organic matter content (HERR et al. 1966); Atrazine retention, as measured by the difference in volume of solution required for atrazine and chloride (nonadsorbed) breakthrough in saturated columns, was highly correlated with soil organic matter content, surface area, and cation exchange capacity (SNELLING et al. 1969).

Clays adsorb pesticides, but the interaction is sensitive to many factors. Clays maintaining a high cation-exchange capacity (e.g., montmorillonite and vermiculite) strongly adsorb the cationic pesticides such as diquat and paraquat; anionic compounds (dicamba, 2,4-D, etc.) are much more weakly adsorbed. Reduction of pH generally increases adsorption of both neutral and anionic compounds. Since the montmorillonite charge arises primarily from isomorphous substitution, the pH effect must be predominantly on the pesticide itself. This increased adsorption of various s-triazines by montmorillonite is generally considered a result of protonation of the somewhat basic molecule (BAILEY and WHITE 1964, BAILEY et al. 1968, FRISSEL 1961, FRISSEL and BOLT 1962, HARRIS and WARREN 1964, TALBERT and FLETCHALL 1965). The type of clay also affects pesticide adsorption. DSMA is more strongly adsorbed on kaolinite than on montmorillonite or vermiculite (DICKENS and HILTBOLD 1967). In general, however, decreasing adsorption of s-triazine and other neutral compounds occurs on montmorillonite > illite > kaolinite (BAILEY and WHITE 1964, FRISSEL and BOLT 1962, TALBERT and FLETCHALL 1965, WEBER and WEED 1968). Desorption of atrazine, simazine, and prometryne from the Ca^{++} forms of these clays is enhanced in the presence of electrolytes (MALQUORI et al. 1967).

Although adsorption onto soil organic matter seems both greater in magnitude and more generally applicable to pesticides than adsorption onto clay, the role of clay is probably important because most soils contain much more clay than organic matter. This reasoning is analogous to comparison of the contributions of organic matter and clay to cation exchange capacity of 60 soils: although the CEC of clay at pH 7 was only ⅓ that of organic matter, it contributed 1.7 times more to the total CEC (HELLING et al. 1964). The surface area of montmorillonite and vermiculite are comparable to soil organic matter; illite, chlorite, and kaolinite are much less (BAILEY and WHITE 1964).

The effect of clay on pesticide leaching must be construed from the preceding model adsorption studies and others. Typical soil observations compare soils in which both clay and organic matter increase, usually without multiple regression analyses to isolate the governing parameters. Thus, simazine had greater mobility in a sandy soil than in a clay soil (BURSCHEL 1961, STROUBE 1961). Atrazine was comparatively mobile in a fine sandy loam, less so in Drummer clay loam (McGLAMERY 1965). Movement of simazine was inhibited in a high clay soil (24 percent clay, 4.1 to 4.2 percent organic matter) but exhibited moderate leaching in Schweizerhalle soil (10.5 percent clay, 4.6 to 4.9 percent organic matter), indicating clay was influential in restricting simazine mobility (GAST 1959).

The role of soil pH was inferred during the preceding discussion of clays and adsorption. Thus, BAILEY et al. (1968) recently found that adsorption of 22 herbicides and related compounds on montmorillonite always was greater at acidic than neutral pH. Atrazine adsorption onto Drummer clay loam increased markedly as pH decreased below six; desorption increased as pH and temperature increased (McGLAMERY and SLIFE 1966). Adsorption on humic acid (from leonardite) was ten-fold greater at pH 2.5 than at 7.0. Since the process was irreversible, atrazine would thus be rendered immobile to leaching in this model system. The mobility of simazine was greater in limed than unlimed soil, and NEARPASS (1965) suggested that decreased adsorption occurred at increased pH. Titratable acidity, but not soil pH, was highly correlated with simazine adsorption. Trifluralin also leached more in limed than in unlimed soils (ANONYMOUS 1968 a). Dicamba was readily leached in soils except at pH levels more acidic than 4.2 (ANONYMOUS 1968 f). Adsorption of the urea herbicides may be unrelated to pH, however (HANCE 1965, HILTON and YUEN 1966, OBIEN et al. 1966).

The nature of the adsorbate also influences adsorption and movement of pesticides (BAILEY and WHITE 1964, LAMBERT 1968). WEBER (1966) compared adsorption of 13 s-triazines onto montmorillonite. Maximum adsorption occurred near the pK_a of each compound. Substituted triazines, in order of adsorption, were—SCH_3 > —OCH_3 > —OH > —Cl. Dialkylamino-s-triazines were adsorbed more readily than monoalkylamino derivatives. Adsorption of five ^{14}C-labeled s-triazine herbicides by 25 Missouri soils was in the order prometryne > prometone > simazine > atrazine > propazine (TALBERT and FLETCHALL 1965). Correlation coefficients revealed that parameters important in the adsorption of prometryne and prometone apparently differed in part from those governing chloro-s-triazine behavior. In another study (ANONYMOUS 1968 d), the specific adsorption on sand and silt loam soils was ametryne > atrazine.

b) Solubility

Water solubility as a parameter influencing pesticide movement in soils remains imprecisely defined. In theory it may limit the ability of a chemical to diffuse into the flowing water in soil. HARTLEY (1960) calculated that 30 hours is necessary to dissolve simazine in a stagnant medium, assuming spheres ten μ in diameter and a diffusion coefficient, D, of 10^{-6} sq. cm./sec. It appears that increased solubility may correlate with decreased adsorption (and therefore, increased mobility) for some individual classes of compounds. Thus, LEOPOLD et al. (1960)

found a strong inverse correlation of solubility and adsorption on carbon of 17 chlorophenoxyacetic acid derivatives. BAILEY *et al.* (1968),
however, concluded that the magnitude of adsorption onto montmorillonite was directly related to and governed by the degree of water
solubility, when comparing members within a chemical family. Included, in decreasing order of solubility, were six s-triazines: simetone,
atratone, prometone, atrazine, trietazine, and propazine. Adsorption
on Na-montmorillonite also decreases in this order. The leaching of
seven triazines in sandy loam corresponded rather poorly with solubility; RODGERS (1968) concluded that measurement of relative
adsorption was a better predictor. Mobility of six thiocarbamate
herbicides in soils was directly correlated with solubility (GRAY and
WEIRICH 1968, KOREN *et al.* 1967). Similarly, leaching of amiben
derivatives seemed to correlate with solubility (BAKER *et al.* 1966).
Urea herbicide adsorption may be unrelated to solubility (HANCE
1965).

c) Flow rate and amount

Application of additional water to an immobile pesticide will not
induce perceptible leaching. Contrariwise, the amount of water added
to a pesticide exhibiting some movement will influence the depth to
which it travels. Greater amounts of water increased leaching for
bromacil (HOROWITZ 1968), DNBP (DAVIS *et al.* 1954), dicamba
(FRIESEN 1965), and diphenamid (DUBEY and FREEMAN 1965); also
for the ureas diuron (BAYER 1967, HOROWITZ 1968), fluometuron,
neburon (HOROWITZ 1968), linuron (DUBEY and FREEMAN 1965), and
monuron (SHERBURNE *et al.* 1956), and the s-triazines atrazine, simazine (RODGERS 1962, TERENT'EVA 1962), ametryne, GS-14260, and
prometryne (HOROWITZ 1968). The same *relative* movement of five
thiocarbamate herbicides occurred with four as with eight inches of
water (GRAY and WEIERICH 1968). In the field, depth of penetration of
atrazine and simazine during the growing season was unexpectedly
similar for a silt loam and sandy loam, presumably because less precipitation occurred on the sandy loam (TERENT'EVA 1962). SHERBURNE
et al. (1956) calculated the depth monuron would leach in soil columns,
as a function of applied water:

$$y = xe^{-c/x}$$

where y is the depth of maximum concentration, x is inches of "rainfall," and c is a constant dependent on the soil and its initial moisture
content. The equation was derived empirically.

A generalized concentration profile as a function of the quantity of
added water may appear as in Figure 4 (GORING 1967). As the chemical
is leached deeper, the resultant curve becomes broad and frequently

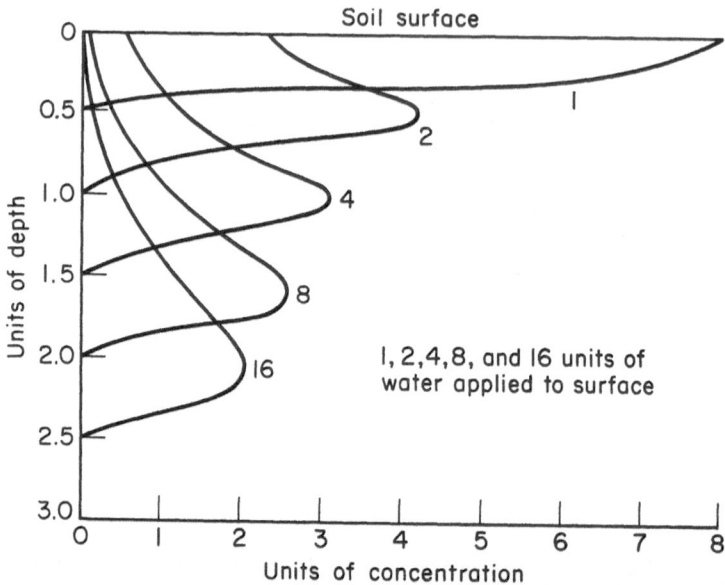

Fig. 4. Relation between amounts of water applied and concentration of surface-
applied pesticide at various depths (GORING 1967) (reproduced with
permission)

skewed toward the surface. Even chloride ion shows such tailing if
significant adsorption can occur (THOMAS 1963).

HARTLEY (1964), in his discussion of the movement of herbicides in
soils, emphasized the importance of the rate at which water enters the
soil. Most leaching experiments, he claimed, are carried out using an
unrealistically high flow rate. If one assumes that percolating water has
time only to reach 50 percent equilibration with each theoretical plate
in a soil column, the distribution of a pesticide with depth may appear
as shown in Figure 5 (HARTLEY 1964), curves "b" and "d". Initial
pesticide application was at the surface. It is clear that at early stages
of leaching (curve "b"), more pesticide remains near the surface than
might have been expected. Band spreading should also depend in part
on flow rate. By analogy with chromatographic columns, vertical
diffusion would be an important contributor at extremely low flow
rates, whereas turbulence and the decreased influence of lateral diffusion would affect spreading at high rates.

Some laboratory observations do not agree entirely with Figure 5.
Many pesticides on both soil columns (MUNNECKE 1961, YOUNGSON
et al. 1967) and soil TLC (HELLING 1968) were more mobile at higher
rates of percolation. Perhaps this indicates flow through large channels
with less exposure to adsorptive surfaces, or gross movement of micro-

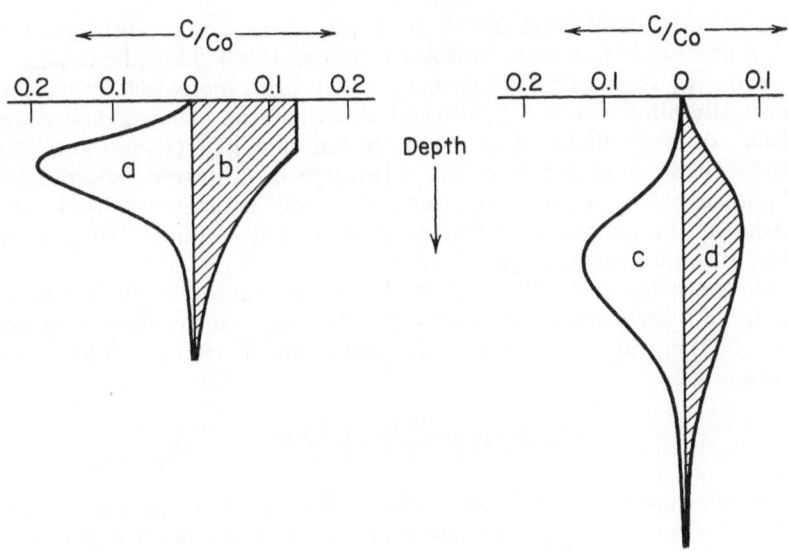

Fig. 5. Theoretical distribution of a pesticide by vertical flow in a multiplate
column; curves "a" and "c" show total equilibrium, "b" and "d" 50 percent equilibration/plate between static and dynamic water; the volume of
water passed is four-fold that of a single plate in "a" and "b", ten-fold that
in "c" and "d" (adapted from HARTLEY 1964)

crystals of the pesticide. On the other hand, slow leaching of dicamba
somewhat increased its mobility (FRIESEN 1965). HARRIS (1964)
attributed greater mobility of dicamba and diphenamid in subirrigated
columns to the slower flow rate, allowing capillary movement through
small pores and more desorption. Elution of atrazine from saturated
columns of a latosolic soil was virtually identical at flow velocities of
0.6, 3.5, and 7.0 cm./hr. (GREEN et al. 1968 b).

The technique of adding water becomes more complicated when
comparing frequency of addition and original soil moisture content.
Movement of dicamba and diphenamid was reportedly greater on
Hagerstown silty clay loam when water was applied in 0.25-inch rather
than one-inch increments (HARRIS 1964). Higher mobilities of simazine,
atrazine, atratone, and ipazine on Lakeland and Greenville fine sands
also coincided with more frequent additions of water (RODGERS 1962).
On the other hand, UPCHURCH and PIERCE (1957) found frequency of
artificial rainfall important only when longer drying periods reduced
the net downflow. Pesticides in irrigation water (YOUNGSON et al. 1967)
or paraquat (TAYLOR and AMLING 1966), monuron, fenuron, and
various acid herbicide derivatives (WIESE and DAVIS 1962) applied
directly to the soil were distributed to greater depths if the soil was
initially moist rather than dry. However, neburon leached less in a

sandy loam at field capacity than at wilting capacity (GEISSBÜHLER
et al. 1963), as did monuron (SHERBURNE et al. 1956). This phenomenon,
according to HARTLEY (1960), may result from increased by-passing
of partially filled channels in the soil. Another study showed soil water
content to have little effect on monuron mobility (UPCHURCH and
PIERCE 1958). Atrazine was eluted through unsaturated soil columns
(57 percent water) somewhat more slowly and with greater peak concentration than when corresponding saturated columns (79 percent
water) were used (GREEN et al. 1968 b).

To overcome low solubility as a limiting factor in the diffusive
transfer of a herbicide from a static to a flowing element, GEISSBÜHLER
et al. (1963) calculated the rate of dissolution of chloroxuron. The Fick's
equation was:

$$m = D \cdot q \frac{C_0 - C_1}{\delta} t$$

where m = mass of herbicide applied, D = diffusion coefficient (ca.
4×10^{-6} cm.^{2}/sec.), q = surface area of herbicide (as 5 μ spheres),
C_0 = solubility of herbicide, C_1 = final concentration of dissolved
herbicide, δ = thickness of diffusion layer (10^{-2} cm.), and t = time.
The time required was 41 hours, and GEISSBÜHLER et al. allowed 45
hours for leaching, ensuring complete equilibrium dissolution. Impre-
cise assignments of D, δ, and q may hamper application of the equation
in most experiments, however.

d) Formulation

The formulation in which a particular pesticide is applied in many
cases influences its mobility. The only reference found directly relating
formulation to s-triazine mobility was the work of GAST (1962). He
noted that the mobility of simazine was increased by incorporation in
granulated formulation with a strongly dissociating carrier, $(NH_4)_2SO_4$,
as compared to lime. Cation-rich carriers [H_2SO_4, $Al_2(SO_4)_3$, $FeCl_3$],
water-insoluble organic acids (SMITH et al. 1957), or volatile amine
soaps (DOWLING and SOBELSKI 1968) had the opposite effect on NPA's
mobility. This increased residual weed control by maintaining a high
herbicide concentration at the soil surface under wet conditions and
minimized injury to deeply seeded crops. Mobility of EPTC was
reduced by formulating it in paraffinic hydrocarbons of greater than 18
carbon atoms (LIU and ILNICKI 1965). Solution formulations of some
nonvolatile fungicides were more mobile than suspensions (MUNNECKE
1961). Most published research compares acidic herbicides with their
chemical derivatives: the free acid or its salt is more mobile than its
ester or amide, as shown for amiben (Amchem Products 1965,
COOKE 1966, McLANE et al. 1967), 2,4-D (ALDRICH and WILLARD 1952,
SMITH and ENNIS 1953, WIESE and DAVIS 1962), 2,4,5-T, silvex

(WIESE and DAVIS 1962), and PCP (TSUNODA 1965). Loss by runoff was greater, however, for 2,4-D esters than for amine salts (ANONYMOUS 1968 b).

Mobility of diuron in soil columns was increased by formulating with one percent solutions of 23 surfactants (BAYER 1967). In a few cases, increasing the surfactant concentration from one percent to ten percent further increased the depth of leaching. Lauryl pyridinium chloride and aniline in methanol both increased the desorption of atrazine from an acid soil (McGLAMERY 1965). This may suggest a practical approach to increasing mobility of the triazines in soils.

e) Rate of pesticide application

HARTLEY (1964) described the pattern of leaching of simazine expected if two rates were each leached into a soil having no adsorptive capacity and a 20 percent pore volume. At one lb./acre, one inch of rain could dissolve all the simazine, producing a zone in the upper five inches saturated with the herbicide. Further rain (in this simple model) would move the simazine as a five-inch band. At ten lb./acre, ten inches would completely dissolve and give a 50-inch band of simazine. Monuron at ten lb./acre would require only 0.2 inch of rain, because of greater solubility; additional precipitation would move a one-inch band. Obviously adsorption greatly reduces this movement and boundaries become diffuse.

Experimentally, some similarity to HARTLEY's (1964) model was observed by HELLING and TURNER (1968) on soil TLC plates. Although the frontal movement of various herbicides remained constant over application rate changes of ten- to 400-fold, streaking greatly increased at high rates. As expected, this was most obvious for weakly adsorbed pesticides, e.g., fenac, and in a soil of low adsorption capacity, Lakeland sandy loam.

Higher application rates of a number of herbicides reportedly increase their observed mobilities. The triazines ametryne and prometryne (HOROWITZ 1968, RODGERS and WILCOX 1965), atrazine (IVEY and ANDREWS 1965), GS-14260 (HOROWITZ 1968), and simazine (DAWSON et al. 1968, IVEY and ANDREWS 1965, SHEETS 1959) all exhibited this effect, as did amiben, CIPC, 2,4-D (ESHEL and WARREN 1967), DCPA (IVEY and ANDREWS 1965), diphenamid (DUBEY and FREEMAN 1965), diuron (DAWSON et al. 1968, IVEY and ANDREWS 1965), linuron (DUBEY and FREEMAN 1965), monuron (DAWSON et al. 1968, SHEETS 1959), and trifluralin (ESHEL and WARREN 1967). Little simazine, applied at 1.5 kg./ha., appeared below ten cm. after one year; at 25 kg./ha., simazine concentration was greater in the ten-to-20 cm. zone than nearer the surface of the loamy sand (ZURAWSKI and PLOSZYNSKI 1968). Degradation was a complicating factor in assessing mobility of simazine.

f) Rate of degradation

Mobility data derived from short-term laboratory experiments do not include the factor of pesticide degradation. The probable effect in the field is a reduction in the amount of pesticide reaching deeper soil, assuming normal rates and amounts of water infiltration. The problem has been considered (KING 1966, McCARTY and KING 1966), and theoretical curves were prepared of concentration with depth as a function of half-life (HARTLEY 1964). The half-life of simazine is ca. 40 days at pH 2 (GYSIN and KNÜSLI 1960). Although very acidic, this pH may occur at the surface of montmorillonite clay in acid soils (BAILEY et al. 1968). In the field, chemical degradation may combine with strong adsorption to greatly inhibit simazine leaching. Contrariwise, liming and intermittant drying reduced simazine decomposition, with higher mobility observed on limed than unlimed soils (NEARPASS 1965). Atrazine may be equally affected, since its half-life in Alabama field soils was ca. 30 days; although significant amounts were lost by leaching, decomposition was the primary mode (HILTBOLD 1967).

In apparent confirmation of degradation as a factor in triazine movement, SPIRIDONOV and YAKOVLEV (1967) observed penetration of atrazine, simazine, and prometryne to ca. 70 cm. without significant decomposition, when the soil temperature was $<10°$C. At 20°C., the triazines penetrated only to 20 to 30 cm. and persisted only three to five months even at doses of 20 to 30 kg./ha. Considering only the increased solubility and decreased adsorption (HARRIS and WARREN 1964, TALBERT and FLETCHALL 1965, WEBER et al. 1965) at the higher temperature, the triazines might have been expected to leach further at 20° than at 10°C. GREEN et al. (1968 b) found a substantial enrichment of hydroxyatrazine in the tail portion of the atrazine elution curve after an 18-day experiment. Their results suggest not only a significant rate of degradation in the pH 4.6 soil, but also a mobility order of atrazine $>$ hydroxyatrazine.

IV. *s*-Triazine mobility—a survey

In the following section, a comprehensive survey of *s*-triazine herbicide mobility is presented. It is intended to be complete through December 1968. The author is aware of only two prior compilations devoted to the triazines and containing brief sections on mobility (ERCEGOVICH 1964, HARRIS et al. 1968).

Evaluation of the movement of a specific triazine in soils is perhaps best accomplished by comparison with the behavior of other chemicals. Larger studies of this type include the examination of 28 herbicides by HARRIS (1967). He found the relative mobility of triazines and other compounds in columns of Hagerstown silty clay loam and Lakeland sandy loam to be fenac $\gg$ diphenamid $>$ monuron $=$ atrazine $=$

simazine > propazine ≥ EPTC > diuron > prometryne ≥ CIPC > trifluralin. Thus, monuron may be an index to the leaching behavior of atrazine and simazine. HARRIS' method of presenting data, the weighting of normalized growth reduction in each column segment, does not necessarily indicate which pesticide will penetrate farthest. This would require knowledge of the depth versus concentration curves.

HELLING and TURNER (1968) compared mobilities of 16 herbicides on soil TLC plates of Chillum silt loam and the same Hagerstown and Lakeland soils as HARRIS (1967) used. Selected compounds leached in the order fenac ≫ monuron > atrazine ≥ simazine > propazine > diuron > prometryne > CIPC > trifluralin. The order of mobility compares very favorably with HARRIS' column experiments and seems to give somewhat greater selectivity, as adjudged by other observations, i.e., monuron > simazine (DAWSON et al. 1968, GAST 1959) and atrazine > simazine (ASHTON 1961, BURNSIDE et al. 1963, KOZLOWSKI and KUNTZ 1963, MONTGOMERY and FREED 1959, RODGERS 1962, STROUBE 1962, TERENT'EVA 1962). The soil TLC method measures the frontal movement, or R_f of each pesticide. The mobilities of all compounds increased in the order Chillum silt loam < Hagerstown silty clay loam < Lakeland sandy loam, assumed to reflect their decreasing organic matter contents. Later correlations using 13 pesticides on 12 soils confirmed the importance of organic matter and clay in atrazine leaching, clay in simazine leaching, and rate of water movement for movement of both compounds (HELLING 1968).

An earlier study by HARRIS (1966) using Lakeland, Chillum, and Hagerstown soils plus Wehadkee silt loam indicated the mobility order monuron = atrazine = simazine > propazine > prometone > prometryne. RODGERS (1968) established the order atratone > propazine > atrazine > simazine > ipazine > ametryne > prometryne in greenhouse and field experiments with Lakeland fine sand.

HARRIS et al. (1968) concluded that the s-triazines were rather immobile in soil. Many references tend to confirm this for atrazine (BAUMAN and WILLIAMS 1966, BIRK and ROADHOUSE 1964), simazine (BIRK and ROADHOUSE 1964, BOUCHET 1967, BURNSIDE et al. 1961, BURSCHEL 1961, CLAY and McKONE 1968, DAWSON et al. 1968, DEWEY 1960, KOZACZENKO 1965, KOZLOWSKI and KUNTZ 1963, ROADHOUSE and BIRK 1961, SHEETS 1959, STROUBE and BONDARENKO 1960), and prometryne (ANDREEVAFETVADZHIEVA 1967, HARRIS 1966, RODGERS 1968, RODGERS and WILCOX 1965). Such a generalization must be tempered by acknowledging that leaching does usually occur or increase in sandy soils and that mobility differences do exist among individual triazines. Furthermore, some reports emphasize location of the maximum herbicide concentration, whereas others indicate the maximum depth of leaching. The difference may be quite significant. DAWSON et al. (1968) applied simazine at three lb./acre annually for six years on a Warden silt loam, under low rainfall conditions (six to 12 inches/yr.). One year after the final application, simazine was found in the four-to-eight inch depth. Monuron and diuron, being somewhat more mobile and/or persistent, were recovered at eight to 12 inches in the vineyard soil after applications at 7.2 lb./acre rates. However, *most* pesticide was found at zero to four inches. ROADHOUSE and BIRK (1961) found negligible movement of simazine below four inches, one or two years after field application of 0.5 to 20 lb./acre. Most remained in the upper one-inch of soil. AELBERS and HOMBURG (1959) also found little simazine movement below 15 cm. in six months in a variety of soils. Atrazine exhibited some movement on Plainfield sand, but simazine and propazine were immobile (KOZLOWSKI and KUNTZ 1963). Water leaching through columns of Lakeland fine sandy loam up to 36 inches in length contained sufficient atrazine (applied at two lb./acre) to cause phytotoxicity (RODGERS 1966). MONTGOMERY and FREED (1959) found the maximum concentration of ^{14}C-atrazine to be at seven to eight inches, after leaching a Chehalis sandy loam column with 12 inches of water during three days. Simazine, on the other hand, largely remained at the zero-to-one inch depth. Most simazine and monuron remained in the upper three-inch section of Hesperia sandy loam columns after two inches of added water, but were present only in the seven- to 20-inch sections after ten inches of water (SHEETS 1959). A higher rate of simazine (20 lb./acre *versus* one lb./acre) increased its movement. Much less movement occurred on Yolo clay loam.

Many other reports indicate mobility of triazines increases in the soil order cl < l < sil < sl or when organic matter decreases (ANONYMOUS 1968 c, ASHTON 1961, BURSCHELL 1961, CHAMBERLAIN 1963, GAST 1959, HELLING and TURNER 1968, IVEY and ANDREWS 1965, KOZACZENKO 1965, McGLAMERY 1965, RODGERS 1962, SHAHIED and ANDREWS 1966, SNOB 1963, TERENT'EVA 1962). For example, GAST (1959) compared the mobility of simazine on 17.5-cm. columns of four Swiss soils, after adding 20 cm. of water. Movement was nil in Witzwill (27 to 30 percent organic matter) and limited to seven cm. in Wintessingen (24 percent clay). Moderate leaching occurred in Schweizerhalle (4.6 to 4.9 percent organic matter, 11 percent clay), *i.e.*, traces to all depths but fronting at ca. 12 cm. Extensive movement, to 17.5 cm., resulted in Bex sand. More than 20 cm. of $0.05N$ calcium acetate was required to reduce atrazine concentration to half that maintained at equilibrium in 15-cm. soil columns; total removal would have required leaching with more than one meter of solution (SNELLING *et al.* 1969). Atrazine appeared to be retained much more tenaciously by a muck than by six mineral soils.

Movement of triazines below the plow layer has been shown in Nebraska field soils (BURNSIDE *et al.* 1963). In Rosebud loam plots, higher concentrations of atrazine and simazine were found below 12

inches than in the upper six inches, 16 months after application.
STROUBE and BONDARENKO (1960) reported movement of simazine
into the three-to-six inch depth of Brookston silty clay loam five
months after application of eight lb./acre. After 12 months, the simazine residue was 1.5 lb./acre in zero to three inches, 1.25 lb./acre in
three to six inches, and 0.1 lb./acre in six to nine inches. Rainfall was 28
inches during this period at the Ohio location. Within six months after
field application of atrazine in neighboring Indiana, the herbicide had
leached to two to four inches, though most remained at zero to two
inches (BAUMAN and WILLIAMS 1966). Laboratory and field studies in
Wyoming showed that atrazine and ametryne moved to three inches
after three months (eight inches of rain), and atrazine to five inches
after nine months (13.5 inches of rain) in a sandy loam soil (CHAMBER-
LAIN 1963). The maximum depth of leaching of atrazine and simazine
in a light Chestnut soil was ten cm., when applied at six and four kg./
ha., respectively (STEPANOVA 1965). At lower rates, two to four kg./ha.,
TERENT'EVA (1962) observed penetration of atrazine to 17 to 20 cm. in
a sandy loam and 15 to 17 cm. in a dermo-podzolic silt loam during a
growing season. Simazine moved to 15 to 17 cm. in the sandy loam
and to 15 cm. in the silt loam. Less precipitation on the sandy loam
apparently reduced the impact of soil differences. Under conditions of
high precipitation and low soil temperature (<10°C.), atrazine,
simazine, and prometryne penetrated to ca. 70 cm. (SPIRIDONOV and
YAKOVLEV 1967). At 20°C., monuron, diuron, and the triazines penetrated only to 20 to 30 cm., which the authors suggest indicated faster
degradation. Mobility in red soils was somewhat greater than in mea-
dow-swampy soils.

ASHTON (1961) found the relative order of lateral mobility to be
atratone > atrazine > simazine, following simulated furrow irrigation
after band application of the herbicide, RODGERS (1962) modified the
series: atrazine > simazine = atratone > ipazine. Ipazine did not
leach below two inches in fine sand columns. Mobility of triazines in a
heavily leached chernozem smolnitsa soil was simazine > atrazine >
prometryne (ANDREEVA-FETVADZHIEVA 1967). The maximum depth of
leaching was 10 to 12 cm. for simazine, and four to six cm. for prometryne. Atrazine and simazine mobility in several soils was greater than
that of diuron (IVEY and ANDREWS 1965); atrazine was less mobile than
dicamba (ANONYMOUS 1968 c); and simazine was much less mobile
than DNC or KOCN (KOZACZENKO 1965).

Ametryne was slightly more mobile than prometryne in field and
column studies (RODGERS and WILCOX 1965). It moved to three inches
in a Lakeland fine sand plot, after application at eight lb./acre.
Prometryne was nearly immobile. These two mercapto-s-triazines and
a third, GS-14260, were also compared by HOROWITZ (1968) on columns
of Newe Ya'ar clay, a brown alluvial soil from Israel. His order of
mobility was bromacil > fluometuron > ametryne > diuron = prome-

tryne $\geq$ GS-14260 $\gg$ neburon. High rates of water addition may have induced somewhat greater leaching than otherwise expected. Prometryne appeared in the six to nine inch layer of a sand after only three months (ANONYMOUS 1968 e), but had begun to dissipate after two months.

The order of mobility (and persistence) of prometone and other herbicides in three Puerto Rican clay and clay loam soils was fenac > prometone > picloram > diuron after one year (DOWLER et al. 1968). All compounds were found at 36 to 48 inches after three months.

Propazine, a chloro-s-triazine herbicide, had medium (HARRIS 1966, 1967, HELLING and TURNER 1968, RODGERS 1968) to low (KOZLOWSKI and KUNTZ 1963) mobility relative to other triazines. Leaching was reported in Lakeland fine sandy loam (RODGERS 1968), but was only slight in Plainfield sand (KOZLOWSKI and KUNTZ 1963). Propazine at lower rates (1.0 to 1.5 kg./ha.) penetrated to ten cm. in an alluvial soil (3.1 percent organic matter) during the growing season (POPOV 1966). At a higher rate (2.0 kg./ha.), penetration was to ten to 20 cm., but 60 percent was retained in the zero-to-five cm. zone. In another Russian publication, SKOB (1963) indicated the depth of maximum concentration of propazine in soil columns. The depth (six to nine cm. in sandy loam and dermo medium podzolic soils, nine to 12 cm. in clayey silts, and zero to three cm. in chernozems) was inversely proportional to organic matter content.

A variety of new derivatives of the s-triazines are undergoing testing. Information on the soil mobility of most of these compounds is unavailable. Movement of aminoalkylnitrile derivatives on columns of *Shell Development Company's* standard soil, Ripperdan sand (1.0 percent organic matter), is expressed as the depth eluted by two inches of water: SD 15417, 6.5 inches > SD 15418, 4.5 inches = atrazine, 4.5 inches > SD 15419, 3.5 inches (HUGHES et al. 1967). There seems to be no correlation with solubility in this order. In another report (CHAPMAN et al. 1968), movement of SD 15418 (WL 19805) was compared to that of atrazine using both the slotted tube technique (LAMBERT et al. 1965) and thin-layer soil plates. SD 15418 leached to 12.0 and 13.7 cm. by column and plate methods, respectively; atrazine was nearly identical, moving to 12.5 and 13.1 cm. No other experimental details were offered. The azido-s-triazine C-7019 had low soil mobility and persisted 11 to 16 weeks when applied at 2.5 kg./ha. (GREEN et al. 1968 a). On soil TLC plates, prometryne, C-7019, and C-8250 were similar in mobility (HELLING 1969).

Lateral movement at the soil surface has been the subject of several investigations. Atrazine runoff on Cecil soil (6.5 percent slope) after pesticide application of three lb./acre was studied by WHITE et al. (1967) using simulated rainfall. Losses of 7.3 percent occurred after a ten-year frequency storm, i.e., 2.5 inches of rain in one hour. A common storm, 0.5 inch in one hour, caused a two percent loss. UPCHURCH et al.

(1968) applied simazine and other herbicides at soil sterilization rates (40 lb./acre initially, 20 lb./acre annually thereafter) in North Carolina. The soils, Norfolk sandy loam, Cecil clay loam, and Cecil sandy loam, averaged ten percent in slope and were subject to high rainfall. The relative order of lateral movement was isocil > diuron > simazine > monuron. This is the reverse order one would expect for vertical movement of the latter three.

V. Implications of s-triazine movement

The understanding of movement of s-triazines, an important facet of their behavior in soils, is significant when considering their effectiveness and their potential for environmental contamination. The influence of soil itself was emphasized in the preceding discussion and probably accounted for much of the difference in reported mobilities.

Utilization of this knowledge aids prediction of the optimum type and position of herbicide placement for a specific soil, climate, and use situation. Similarly the specific herbicide of choice may depend on its mobility. For example, LINSCOTT and HAGIN (1969) placed three triazine herbicides in a subsurface band about 2.5 inches from each row of seeded alfalfa. They speculated that atrazine, and especially, G-36393 were sufficiently mobile to reach the rooting zone of the legume and significantly reduce its yield. Simazine was less mobile and afforded good weed control without contacting the alfalfa. Thus, LINSCOTT and HAGIN (1969) recommended against the subsurface placement of G-36393 when establishing alfalfa. The greater mobility of atrazine than ametryne was considered to cause more toxicity by atrazine to bananas in Hawaiian soils (BARBA and ROMANOWSKI 1969). Selectivity thus seemed attributable to differential position of the banana roots in relation to herbicide placement.

The effects of pesticide movement may be related to persistence in soil. An investigation of widespread sugar beet injury in Ontario, Canada, revealed atrazine present (analyzed by bioassay) in both zero-to-three inch and three-to-six inch depths of sandy loam soils from many fields treated with atrazine the previous year (FRANK 1966 b). Atrazine largely remained in the zero-to-three inch zone in clays and clay loams. Injury was often pronounced in low areas of a field and this was ascribed to surface movement of atrazine. The carryover, however, may also have stemmed from reduced decomposition in the wetter, cooler soil of the low areas.

An unanswered question is whether movement of a triazine into the soil profile increases its persistence. HARRIS et al. (1969) recently surveyed persistence of atrazine and fenac placed three, nine, and 15 inches deep in soils from 12 locations throughout the United States and Puerto Rico. In general, persistence increased with depth, lower temperature, and higher organic matter content. More research is needed

firmly to establish these trends, however. Persistence appears to favor
deep leaching of inherently mobile pesticides.

The possible routes toward environmental contamination by
pesticides include vertical leaching to ground water, surface movement
in solution or suspension, volatilization, and uptake by living organisms.
The low to moderate mobility of s-triazines reduces the possibility of
contamination by the first route, vertical leaching. Surface movement
is somewhat more likely, especially for compounds of low mobility
applied to slopes and subject to intense rainfall after application. Its
effect will likely be confined within a field, however. In either case, persistence, discussed in this volume by SHEETS (1970), is a major consideration as well.

Two mechanisms of pesticide movement in soils have not previously
been described in this report. Both are speculative but perhaps worthy
of examination. PHILLIPS (1968) suggested that deep penetration of
2,3,6-TBA may have occurred, in part, from translocation downward
and exudation through field bindweed roots. He cited, as a precedent,
the report of LINDER et al. (1964) that 2,3,6-TBA, dicamba, and several
other growth-regulating compounds were released unchanged from
bean roots. Picloram and dicamba were also exuded into soil from bean
(FOY and HURTT 1967) and Canada thistle (VANDEN BORN and
CHANG 1967) roots. GRAHAM and BUCHHOLTZ (1968) observed leakage
of atrazine from soybean roots into the surrounding water medium.
However, it is not known whether the triazines could be absorbed by
roots near the surface, then translocated via deeper roots and released
farther into the soil, either before or after the plant dies.

A second, totally different mode of pesticide movement may occur
while adsorbed to a migrating soil particle. In conjunction with disposal
of radionuclides in soil, CARLILE (1968) recently demonstrated movement of radioactive particulate material through soil columns. Migration was greatly reduced by addition of a non-ionic polymer and
aluminum sulfate to the leaching solution. Parameters of composition
and volume of the leaching water causing particle movement were
similar to those for irrigation water used in western United States. If
pesticides are used, mass movement of a pesticide-particulate complex
may occur in the young, relatively coarse-textured soils now being
irrigated in the West. An "immobile" pesticide may be moved far into
the soil profile. Investigations into this aspect of pesticide mobility
would seem to be appropriate.

Summary

Movement of s-triazine herbicides in and on soils was considered
from the aspects of theory, methodology and parameters influencing
movement, with applicability to most pesticides. A survey of s-triazine

Table I. *Chemical designations of pesticides mentioned in text*

Common name	Chemical name
Ametryne	2-ethylamino-4-isopropylamino-6-methylthio-*s*-triazine
Amiben	3-amino-2,5-dichlorobenzoic acid
Amitrole	3-amino-1,2,4-triazole
Atratone	2-ethylamino-4-isopropylamino-6-methoxy-*s*-triazine
Atrazine	2-chloro-4-ethylamino-6-isopropylamino-*s*-triazine
Bromacil	5-bromo-3-*sec*-butyl-6-methyluracil
C-7019	2-azido-4-isopropylamino-6-methylmercapto-*s*-triazine
C-8250	2-azido-4-*sec*-butylamino-6-methylmercapto-*s*-triazine
Chlordane	1,2,4,5,6,7,8,8-octachloro-2,3,3a,4,7,7a-hexahydro-4,7-methanoindene
Chlorpropham (CIPC)	isopropyl *N*-(3-chlorophenyl)carbamate
Chloroxuron	3-[*p*-(*p*-chlorophenoxy)phenyl]-1,1-dimethylurea
2,4-D	2,4-dichlorophenoxyacetic acid
DCPA	dimethyl tetrachloroterephthalate
p,p'-DDE	1,1-dichloro-2,2-bis(*p*-chlorophenyl) ethene
p,p'-DDT	1,1,1-trichloro-2,2-bis(*p*-chlorophenyl) ethane
Dicamba	3,6-dichloro-2-methoxybenzoic acid
Dichlobenil	2,6-dichlorobenzonitrile
Dieldrin	1,2,3,4,10,10-hexachloro-6,7-epoxy-1,4,4a,5,6,7,8,8a-octahydro-1,4-*endo*,*exo*-5,8-dimethanonaphthalene
Dinoseb (DNBP)	2-*sec*-butyl-4,6-dinitrophenol
Diphenamid	*N*,*N*-dimethyl-2,2-diphenylacetamid
Diquat	6,7-dihydrodipyrido[1,2-a:2′,1′-*c*]pyrazidinium salt
Diuron	3-(3,4-dichlorophenyl)-1,1-dimethylurea
DNOC (DNC)	4,6-dinitro-*o*-cresol
DSMA	disodium methanearsonate
EPTC	*S*-ethyl dipropylthiocarbamate
Fenac	(2,3,6-trichlorophenyl)acetic acid
Fenuron	1,1-dimethyl-3-phenylurea
Fluometuron	1,1-dimethyl-3-(α,α,α-trifluoro-*m*-tolyl)urea
G-36393	2-isopropylamino-4-(3-methoxypropyl)amino-6-methylthio-*s*-triazine
GS-14260	2-*tert*-butylamino-4-ethylamino-6-methylthio-*s*-triazine
Heptachlor epoxide	1,4,5,6,7,8,8-heptachloro-2,3-epoxy-2,3,3a,7a-tetrahydro-4,7-methanoindene
Ipazine	2-chloro-4-diethylamino-6-isopropylamino-*s*-triazine
Isocil	5-bromo-3-isopropyl-6-methyluracil
KOCN	potassium cyanate
Lindane	γ-1,2,3,4,5,6-hexachlorocyclohexane
Linuron	3-(3,4-dichlorophenyl)-1-methoxy-1-methylurea
Mevinphos (Phosdrin®)	methyl 3-hydroxy-α-crotonate dimethyl phosphate
Monuron	3-(*p*-chlorophenyl)-1,1-dimethylurea
Naptalam (NPA)	*N*-1-naphthylphthalamic acid
Neburon	1-butyl-3-(3,4-dichlorophenyl)-1-methylurea
Paraquat	1,1′-dimethyl-4,4′-bipyridinium salt
Parathion	*O*,*O*-diethyl *O*-*p*-nitrophenyl phosphorothioate
PCP	pentachlorophenol
Picloram	4-amino-3,5,6-trichloropicolinic acid
Prometone	2,4-bis(isopropylamino)-6-methoxy-*s*-triazine
Prometryne	2,4-bis(isopropylamino)-6-methylthio-*s*-triazine

Table I. (Continued)

Common name	Chemical name
Propazine	2-chloro-4,6-bis(isopropylamino)-*s*-triazine
Ronnel	*O,O*-dimethyl *O*-2,4,5-trichlorophenyl phosphorothioate
SD 9515-B	4-chloro-2-nitrophenyl methyl sulfone
SD 15417	2-(4-chloro-6-methylamino-*s*-triazin-2-ylamino)-2-methylpropionitrile
SD 15418	2-(4-chloro-6-ethylamino-*s*-triazin-2-ylamino)-2-methylpropionitrile
SD 15419	2-(4-chloro-6-ethylamino-*s*-triazin-2-ylamino)-2-methylbutyronitrile
Silvex	2-(2,4,5-trichlorophenoxy)propionic acid
Simazine	2-chloro-4,6-bis(ethylamino)-*s*-triazine
Simetone	2,4-bis(ethylamino)-6-methoxy-*s*-triazine
2,4,5-T	2,4,5-trichlorophenoxyacetic acid
2,3,6-TBA	2,3,6-trichlorobenzoic acid
Trietazine	2-chloro-4-diethylamino-6-ethylamino-*s*-triazine
Trifluralin	α,α,α-trifluoro-2,6-dinitro-*N,N*-dipropyl-*p*-toluidine
WL 19805	see SD 15418

mobility literature suggested that triazines have low to moderate mobility, dependent on the soil adsorptive capacity and the triazine adsorptive strength. Surface (runoff and wind erosion) and sub-surface movement were discussed, with emphasis on vertical leaching. Methods for evaluating pesticide mobility include field observations, soil column techniques, and soil thin-layer chromatography. Parameters generally affecting mobility include adsorption, solubility, flow rate and amount, rates of pesticide application and degradation, and formulation.

Résumé*

Déplacement des herbicides du groupe des *s*-triazines dans les sols

On a pris en considération le déplacement des herbicides du groupe des *s*-triazines sur et dans les sols selon les aspects théorique, méthodologique et des paramètres influençant les déplacements, avec leurs applications possibles à la plupart des pesticides. Un examen de la littérature sur le déplacement des *s*-triazines a suggéré que les triazines ont une mobilité faible à modérée, selon la capacité d'adsorption du sol et la force d'adsorption des triazines. On a discuté le déplacement en surface (érosion par écoulement et par le vent) et sous la surface, en soulignant le lessivage vertical. Les méthodes permettant d'évaluer le déplacement des pesticides comprennent des observations de plein champ, des techniques utilisant des colonnes de sols et la chromatographie des sols surcouche mince. Les paramètres affectant générale-

---

* Traduit par S. Dormal-van den Bruel.

ment le déplacement comprennent l'adsorption, la solubilité, le débit et la quantité, les doses d'application du pesticide, la dégradation et la formulation.

Zusammenfassung*

Die Tätigkeit der s-Triazin-Unkrautvertilgungsmittel in Böden

Die Tätigkeit der s-Triazin-Unkrautvertilgungsmittel in und auf Böden wurde im Lichte von Theorie, Methodologie und Parameter beeinflussende Tätigkeit, mit Anwendung auf die meisten Pestizide, betrachtet. Ein Ueberblick über die Literatur der Triazin-Beweglichkeit deutete an, dass die Triazine eine niedrige bis mässige Beweglichkeit, bedingt durch die Adsorptionsfähigkeit des Bodens und die Adsorptionsstärke der Triazine, haben. Die Tätigkeit an der Oberfläche (Abfluss und Winderosion) und unter der Oberfläche wurden diskutiert und die Betonung auf das vertikale Durchsickern gelegt. Methoden zur Auswertung der Pestizid-Beweglichkeit schliessen Feldbeobachtungen, Bodensäulen-Techniken und Boden-Dünnschicht-Chromatographie ein. Parameter, die im allgemeinen die Beweglichkeit beeinflussen, schliessen Adsorption, Löslichkeit, Flussverhältnis und -Menge, Verhältnis der Pestizid-Anwendung und -Degradierung und Formulierung ein.

References

AELBERS, F., and K. HOMBURG: The breakdown and penetration of simazine in the soil. 11th Internat. Symposium Crop Protection, Ghent **82E**, 14 (1959).

ALDRICH, R. J., and C. J. WILLARD: Factors affecting the pre-emergence use of 2,4-D in corn. Weeds **1**, 338 (1952).

Amchem Products Inc.: Amiben-Dinoben derivatives. Tech. Service Data Sheet E-170, p. 3 (1965).

ANDREEVA-FETVADZHIEVA, N.: Penetration of simazine, atrazine, and prometryne and their detoxication in various soil layers. Rastenievud. Nauki. **4**(6), 85 (1967); through Chem. Abstr. **68**, 86351d (1968).

ANONYMOUS: Behavior of trifluralin in soil. 1967 Progress Rept. Pesticides and Related Activities, *U.S. Department of Agriculture* and Cooperators (Georgia Agr. Expt. Sta.), p. 181 (1968 a).

——— Loss of 2,4-D in washoff from cultivated fallow land. 1967 Progress Rept. Pesticides and Related Activities, *U.S. Department of Agriculture* and Cooperators (ARS, USDA, Watkinsville, Georgia), p. 182 (1968 b).

——— Mobility of herbicides in soil differ depending on the solubility and soil type. 1967 Progress Rept. Pesticides and Related Activities, *U.S. Department of Agriculture* and Cooperators (Nebraska Agr. Expt. Sta.), p. 181 (1968 c).

——— Movement of triazines in soils. 1967 Progress Rept. Pesticides and Related Activities, *U.S. Department of Agriculture* and Cooperators (Utah Agr. Expt. Sta.), p. 183 (1968 d).

* Übersetzt von M. DÜSCH.

—— Organic matter content of soil influenced breakdown of prometryne. 1967 Progress Rept. Pesticides and Related Activities, *U.S. Department of Agriculture* and Cooperators (Oklahoma Agr. Expt. Sta.), p. 182 (1968 e).

—— Progress Rept. Crops Research Div., Agr. Research Service, *U.S. Department of Agriculture* (Univ. Illinois), p. 188 (1968 f).

ARMSTRONG, D. E., and G. CHESTERS: Adsorption-catalyzed chemical hydrolysis of atrazine. Environ. Sci. Technol. **2**, 683 (1968).

ASHTON, F. M.: Movement of herbicides in soil with simulated furrow irrigation. Weeds **9**, 612 (1961).

——, and T. J. SHEETS: The relationship of soil adsorption of EPTC to oats and injury in various soil types. Weeds **7**, 88 (1959).

BAILEY, G. W., and J. L. WHITE: Review of adsorption and desorption of organic pesticides by soil colloids, with implications concerning pesticide bioactivity. J. Agr. Food Chem. **12**, 324 (1964).

—— —— Factors influencing the adsorption, desorption, and movement of pesticides in soil. Residue Reviews, This volume (1970).

—— ——, and T. ROTHBERG: Adsorption of organic herbicides by montmorillonite: Role of pH and chemical character of adsorbate. Soil Sci. Soc. Amer. Proc. **32**, 222 (1968).

BAKER, H. R., R. E. TALBERT, and R. E. FRANS: The leaching characteristics and field evaluation of various derivatives of amiben for weed control in soybeans. Proc. S. Weed Conf. **19**, 117 (1966).

BARBA, R. C., and R. R. ROMANOWSKI, JR.: Differential phytotoxicity of atrazine and ametryne to bananas. Weed Research **9**, 114 (1969)."

BAUMAN, T. T., and J. L. WILLIAMS: Atrazine activity and resulting residue in Indiana soils. Weed Soc. Amer. Abstr., p. 3 (1966).

BAYER, D. E.: Effect of surfactants on leaching of substituted urea herbicides in soil. Weeds **15**, 249 (1967).

BIRK, L. A., and F. E. B. ROADHOUSE: Penetration of and persistence in soil of the herbicide atrazine. Can. J. Plant Sci. **44**, 21 (1964).

BOUCHET, F.: Etude de l'influence de la nature du sol sur l'action herbicide de le simazine. Weed Research **7**, 102 (1967).

BURNSIDE, O. C., C. R. FENSTER, and G. A. WICKS: Dissipation and leaching of monuron, simazine, and atrazine in Nebraska soils. Weeds **11**, 209 (1963).

——, E. L. SCHMIDT, and R. BEHRENS: Dissipation of simazine from the soil. Weeds **9**, 477 (1961).

BURSCHEL, P.: Untersuchungen über das Verhalten von Simazin im Boden Weed Research **1**, 131 (1961).

CARLILE, B. L.: Use of synthetic polymers in soil waste disposal systems. Amer. Soc. Agron. Abstr., p. 132 (1968).

CHAMBERLAIN, E. W.: The movement and persistence of two triazine compounds as affected by moisture and soil types. W. Weed Control Conf. Research Progress Rept., p. 90 (1963).

CHAPMAN, T., D. JORDAN, D. H. PAYNE, W. J. HUGHES, and R. H. SCHIEFERSTEIN: WL 19805—A new triazine herbicide. Proc. Brit. Weed Control Conf. **9**, 1018 (1968).

CLAY, D. V., and C. E. McKONE: The persistence of chlorthiamid, lenacil, and simazine in uncropped soil. Proc. Brit. Weed Control Conf. **9**, 933 (1968).

COGGINS, C. W., Jr., and A. S. CRAFTS: Substituted urea herbicides: Their electrophoretic behavior and the influence of clay colloid in nutrient solution on their phytotoxicity. Weeds **7**, 349 (1959).

COHEN, J. M., and C. PINKERTON: Widespread translocation of pesticides by air transport and rain-out. Adv. Chem. Series **60**, 163 (1966).

COOKE, A. R.: Controlled studies on the interaction of rainfall and preemergence herbicide activity. Meded. Rijksfac. Landbouwwetensch., Gent **31**, 1165 (1966); through Chem. Abstr. **69**, 58520k (1968).

DAVIDSON, J. M., and P. W. SANTELMANN: Displacement of fluometuron and diuron through saturated glass beads and soil. Weed Sci. **16**, 544 (1968).

DAVIS, E. A., P. A. INGEBO, and C. P. PASE: Effect of a watershed treatment with picloram on water quality. Forestry Research Highlights, Ann. Rept. Rocky Mountain Forest and Range Expt. Sta., *U.S. Department of Agriculture*, p. 22 (1967).

DAVIS, F. L., F. L. SELMAN, and D. E. DAVIS: Some factors affecting the behavior of dinitro herbicides in soils. Proc. S. Weed Conf. **7**, 205 (1954).

DAWSON, J. H., V. F. BRUNS, and W. J. CLORE: Residual monuron, diuron, and simazine in a vineyard soil. Weed Sci. **16**, 63 (1968).

DAY, B. E., L. S. JORDAN, and R. T. HENDRIXSON: The decomposition of amitrole in California soils. Weeds **9**, 443 (1961).

DEWEY, O. R.: Further experimental evidence on the fate of simazine in the soil. Proc. Brit. Weed Control Conf. **5**, 91 (1960).

DICKENS, R., and A. E. HILTBOLD: Movement and persistence of methanearsonates in soil. Weeds **15**, 299 (1967).

DOWLER, C. C., W. FORESTIER, and F. H. TSCHIRLEY: Effect and persistence of herbicides applied to soil in Puerto Rican forests. Weed Sci. **16**, 45 (1968).

DOWLING, R. J., and A. SOBELESKI: Controlling weeds in soil. U.S. Pat. 3,393,065 (to *Uniroyal, Inc.*), July 16, 1968; through Chem. Abstr. **69**, 58532r (1968).

DUBEY, H. D., and J. F. FREEMAN: Leaching of linuron and diphenamid in soils. Weeds **13**, 360 (1965).

ELRICK, D. E., and A. H. MACLEAN: Movement, adsorption and degradation of 2,4-dichlorophenoxyacetic acid in soil. Nature **212**, 102 (1966).

ERCEGOVICH, C. D.: What happens to the triazines in soil. Presented Wash. State Weed Control Conf., Yakima, Wash. (1964). Nat. Agr. Library, 79, ER 2 (1964).

ESHEL, Y., and G. F. WARREN: A simplified method for determining phytotoxicity, leaching, and adsorption of herbicides in soils. Weeds **15**, 115 (1967).

FOY, C. L., and W. HURTT: Further studies on root exudation of exogenous growth regulators in *Phaseolus vulgaris* L. Weed Soc. Amer. Abstr., p. 40 (1967).

FRANK, P. A.: Persistence and distribution of monuron and neburon in an aquatic environment. Weeds **14**, 219 (1966 a).

FRANK, R.: Atrazine carryover in production of sugar beets in Southwestern Ontario. Weeds **14**, 82 (1966 b).

FREED, V. H.: Chemistry of herbicides. In: Pesticides and their effects on soils and water, p. 25. Madison, Wis.: Amer. Soc. Agron. Special Publ. **8** (1966).

FRIESEN, H. A.: The movement and persistence of dicamba in soil. Weeds **13**, 30 (1965).

FRISSEL, M. J.: The adsorption of some organic compounds, especially herbicides, on clay minerals. Verslag. Landbouwk. Onderzoek. **67.3**, 54 (1961).

——, and G. H. BOLT: Interactions between certain ionizable organic compounds (herbicides) and clay minerals. Soil Sci. **94**, 284 (1962).

——, and P. POELSTRA: Chromatographic transport through soils. I. Theoretical evaluations. Plant Soil **26**, 285 (1967 a).

—— —— Chromatographic transport through soils. II. Column experiments with Sr- and Ca-isotopes. Plant Soil **27**, 20 (1967 b).

FUNDERBURK, H. H., JR., N. S. NEGI, and J. M. LAWRENCE: Photochemical decomposition of diquat and paraquat. Weeds **14**, 240 (1966).

GAST, A.: Unpublished data, *J. R. Geigy S. A.*, Basel, Switzerland, (1959), as reported by H. GYSIN and E. KNÜSLI: Chemistry and herbicidal properties of triazine derivatives. Adv. Pest Control Research **3**, 289 (1960).

—— Contributions to the knowledge of the behavior of triazine herbicides in soil. Meded. Landb.-Hogesch. Gent **27**, 1252 (1962); through Weed Abstr. **12**, 593 (1963).

GEISSBÜHLER, H., C. HASELBACH, and H. AEBI: The fate of N'-(4-chloro-phenoxy-phenyl)-N,N-dimethylurea (C-1983) in soils and plants. I. Adsorption and leaching in different soils. Weed Research **3**, 140 (1963).

GETZIN, L. W., and R. K. CHAPMAN: Effect of soils upon the uptake of systemic insecticides by plants. J. Econ. Entomol. **52**, 1160 (1959).

GORING, C. A. I.: Physical aspects of soil in relation to the action of soil fungicides. Ann. Review Phytopathol. **5**, 285 (1967).

GRAHAM, J. C., and K. P. BUCHHOLTZ: Alteration of transpiration and dry matter with atrazine. Weed Sci. **16**, 389 (1968).

GRAY, R. A., and A. J. WEIRICH: Leaching of five thiocarbamate herbicides in soils. Weed Sci. **16**, 77 (1968).

GREEN, D. H., L. EBNER, and J. SCHULER: New selective triazine [herbicide] for Brassica cultures. C. R. Journees Etud. Herbic., Conf. COLUMA 4th **1**, 1 (1967). Publ. (1968 a); through Chem. Abstr. **70**, 86469q (1969).

——, V. K. YAMANE, and S. R. OBIEN: Transport of atrazine in a latosolic soil in relation to adsorption, degradation, and soil water variables. Trans. Internat. Congr. Soil Sci. 9th (Adelaide) **1**, 195 (1968 b).

GYSIN, H., and E. KNÜSLI: Chemistry and herbicidal properties of triazine derivatives. Adv. Pest Control Research **3**, 289 (1960).

HANCE, R. J.: The adsorption of urea and some of its derivatives by a variety of soils. Weed Research **5**, 98 (1965).

—— Relationship between partition data and the adsorption of some herbicides by soils. Nature **214**, 630 (1967).

HARRIS, C. I.: Movement of dicamba and diphenamid in soils. Weeds **12**, 112 (1964).

—— Adsorption, movement, and phytotoxicity of monuron and *s*-triazine herbicides in soil. Weeds **14**, 6 (1966).

—— Movement of herbicides in soils. Weeds **15**, 214 (1967).

——, D. D. KAUFMAN, T. J. SHEETS, R. G. NASH, and P. C. KEARNEY: Behavior and fate of *s*-triazines in soils. Adv. Pest Control Research **8**, 1 (1968).

——, and G. F. WARREN: Adsorption and desorption of herbicides by soil. Weeds **12**, 120 (1964).

——, E. A. WOOLSON, and B. E. HUMMER: Dissipation of herbicides at three soil depths. Weed Sci. **17**, 27 (1969).

HARTLEY, G. S.: Physico-chemical aspects of the availability of herbicides in soils. In E. K. WOODFORD and G. R. SAGAR (eds.): Herbicides and the soil, p. 63. Oxford: Blackwell (1960).

—— Herbicide behavior in the soil. I. Physical factors and action through the soil. In L. J. AUDUS (ed.): The physiology and biochemistry of herbicides, p. 111. New York: Academic Press (1964).

HAYES, M. H. B.: Adsorption of triazine herbicides on soil organic matter, including a short review on soil organic matter chemistry. Residue Reviews, This volume (1970).

HELLING, C. S.: Pesticide mobility investigations using soil thin-layer chromatography. Amer. Soc. Agron. Abstr., p. 89 (1968).

—— Unpublished data (1969).

——, G. CHESTERS, and R. B. COREY: Contribution of organic matter and clay to soil cation-exchange capacity as affected by the pH of the saturating solution. Soil Sci. Soc. Amer. Proc. **28**, 517 (1964).

——, and B. C. TURNER: Pesticide mobility: Determination by soil thin-layer chromatography. Science **162**, 562 (1968).

HERR, D. E., E. W. STROUBE, and D. A. RAY: The movement and persistence of picloram in soil. Weeds **14**, 248 (1966).

HILTBOLD, A. E.: Leaching and inactivation of atrazine and diuron in the field. Amer. Soc. Agron. Abstr., p. 90 (1967).

HILTON, H. W., and Q. H. YUEN: Adsorption of several pre-emergence herbicides by Hawaiian sugar cane soils. J. Agr. Food Chem. **11**, 230 (1963).

————— Adsorption and leaching of herbicides in Hawaiian sugarcane soils. J. Agr. Food Chem. **14**, 86 (1966).

HOLSTUN, J. T., and W. E. LOOMIS: Leaching and decomposition of 2,2-dichloropropionic acid in several Iowa soils. Weeds **4**, 205 (1965).

HOROWITZ, M.: Investigations on the influence of climatic and edaphic factors on the activity and persistence of newer soil-applied herbicides. Ann. Rept. Project A10-CR-68, Grant FG-Is-248, Agr. Research Service, *U.S. Department of Agriculture* (1968).

HUGHES, W. J., T. CHAPMAN, D. JORDAN, and R. H. SCHIEFERSTEIN: SD-15418—A new corn herbicide. Proc. N.C. Weed Control Conf., p. 27 (1967).

International Pest Control: Amines resist wash-off better. Internat. Pest Control **9**, 6 (1967); through Weed Abstr. **16**, 2384 (1967).

IVEY, M. J., and H. ANDREWS: Leaching of simazine, atrazine, diuron and DCPA in soil columns. Proc. S. Weed Conf. **18**, 670 (1965).

JORDAN, L. S., W. FARMER, J. GOODIN, and B. E. DAY: Nonbiological detoxication of the *s*-triazine herbicides. Residue Reviews, this volume (1970).

KAY, B. D., and D. E. ELRICK: Adsorption and movement of lindane in soils. Soil Sci. **104**, 314 (1967).

KAZARINA, E. M.: The movement of the herbicide monuron in the soil in relation to various factors. Trudy vses. nauchno-issled. Inst. Zashch. Rast. **24**, 36 (1965); through Weed Abstr. **16**, 1909 (1967).

KING, P. H.: The movement of pesticides through soils. Ph.D. Thesis, Stanford Univ. (1966). Diss. Abstr. **27**, 3549B (1967).

KOREN, E., C. L. FOY, and F. M. ASHTON: Adsorption, leaching and lateral diffusion of four thiolcarbamate herbicides in soils. Weed Soc. Amer. Abstr., p. 72 (1967).

KOZACZENKO, H.: Factors affecting the efficiency of herbicides. Biul. Warzyw. **8**, 31 (1965); through Weed Abstr. **15**, 1780 (1966).

KOZLOWSKI, T. T., and J. E. KUNTZ: Effects of simazine, atrazine, propazine, and Eptam on growth and development of pine seedlings. Soil Sci. **95**, 164 (1963).

KUNZE, G. W.: Pesticides and clay minerals. In: Pesticides and their effects on soils and water, p. 49. Madison, Wis.: Amer. Soc. Agron. Special Publ. **8** (1966).

LAMBERT, S. M.: Omega (Ω), a useful index of soil sorption equilibria. J. Agr. Food Chem. **16**, 340 (1968).

————, P. E. PORTER, and R. H. SCHIEFERSTEIN: Movement and sorption of chemicals applied to the soil. Weeds **13**, 185 (1965).

LEGRAND, H. E.: Movement of pesticides in the soil. In: Pesticides and their effects on soils and water, p. 71. Madison, Wis.: Amer. Soc. Agron. Special Publ. **8**, (1966).

LEOPOLD, A. C., P. VAN SCHAIK, and M. NEAL: Molecular structure and herbicide adsorption. Weeds **8**, 48 (1960).

LINDER, P. J., J. W. MITCHELL, and G. D. FREEMAN: Persistence and translocation of exogenous regulating compounds that exude from roots. J. Agr. Food Chem. **12**, 437 (1964).

LINDSTROM, F. T., L. BOERSMA, and H. GARDINER: 2,4-D diffusion in saturated soils: A mathematical theory. Soil Sci. **106**, 107 (1968).

————, R. HAQUE, V. H. FREED, and L. BOERSMA: Theory on the movement of some herbicides in soils. Linear diffusion and convection of chemicals in soils. Environ. Sci. Technol. **1**, 561 (1967).

LINSCOTT, D. L., and R. D. HAGIN: Precision placement of herbicides for weed control in seedling alfalfa. Weed Sci. **17**, 46 (1969).

LIU, L. C., and R. D. ILNICKI: The movement of several formulations of EPTC in soil. Proc. N.E. Weed Control Conf. **19**, 355 (1965).

LOPEZ-GONZALEZ, J. DE D., and C. VALENZUELA CALAHORRO: Diffusion of DDT in homoionic acid bentonite. An. Quim. **64**, 139 (1968 a); through Chem. Abstr. **68**, 94847k (1968).

———— Diffusion of DDT in homoionic acid bentonite. An. Quim. **64,** 359
    (1968 b); through Chem. Abstr. **69,** 43018d (1968).
MALQUORI, A., P. FUSI, and G. STACCIOLI: Soil-herbicide interactions. II. Effect of
    electrolytes on absorption and release of symmetrical triazines by clay min-
    erals. Chim. Ind. (Milano) **49,** 279 (1967); through Chem. Abstr. **67,** 63188w
    (1968).
MASSINI, P.: Movement of 2,6-dichlorobenzonitrile in soils and plants in relation to
    its physical properties. Weed Research **1,** 142 (1961).
McCARTY, P. L., and H. P. KING: The movement of pesticides in soils. Purdue
    Univ. Eng. Bull., Ext. Ser. No. **121,** 156 (1966); through Chem. Abstr. **67,**
    81422x (1967).
McGLAMERY, M. D.: Studies of atrazine-soil relationships and methods of deter-
    mining atrazine residues in soils. Ph.D. Thesis, Univ. Illinois (1965). Diss.
    Abstr. **26,** 2405 (1965).
——, and F. W. SLIFE: The adsorption and desorption of atrazine in soil as affected
    by pH, temperature, and concentration. Weeds **14,** 237 (1966).
McLANE, S. R., M. D. PARKINS, and A. R. COOKE: Physical and biological attri-
    butes of several amiben derivatives. Weed Soc. Amer. Abstr., p. 69 (1967).
MENGES, R. M.: Influence of wind on performance of preemergence herbicides.
    Weeds **12,** 236 (1964).
MONTGOMERY, M., and V. H. FREED: A comparison of the leaching behavior of
    simazine and atrazine in Chehalis sandy loam. W. Weed Control Conf.
    Research Progress Rept., p. 79 (1959).
MUNNECKE, D. E.: Movement of nonvolatile, diffusible fungicides through soil
    columns. Phytopathol. **51,** 593 (1961).
NASH, R. G., and E. A. WOOLSON: Distribution of chlorinated insecticides in cul-
    tivated soil. Soil Sci. Soc. Amer. Proc. **32,** 525 (1968).
NEARPASS, D. C.: Effects of soil acidity on the adsorption, penetration, and per-
    sistence of simazine. Weeds **13,** 341 (1965).
———— Effect of the predominating cation on the adsorption of simazine and atrazine
    by Bayboro clay soil. Soil Sci. **103,** 177 (1967).
OBIEN, S. R., R. H. SUEHISA, and O. R. YOUNGE: The effects of soil factors on the
    phytotoxicity of neburon to oats. Weeds **14,** 105 (1966).
OGLE, R. E., and G. F. WARREN: Fate and activity of herbicides in soils. Weeds **3,**
    257 (1954).
PHILLIPS, W. M.: Persistence and movement of 2,3,6-TBA in soil. Weed Sci. **16,**
    144 (1968).
POELSTRA, P., and M. J. FRISSEL: Methods used for investigation into the move-
    ment of ions in soils by means of radioactive tracers. Proc. Internat. Atomic
    Energy Agency, Isotopes and radiation in soil-plant nutrition studies (Vienna),
    p. 55 (1965).
———— ———— Migration of water and·ions in undisturbed soil columns and its descrip-
    tion by simulation models. Proc. Internat. Atomic Energy Agency, Isotope
    and radiation techniques in soil physics and irrigation studies (Vienna), p. 203
    (1967).
POPOV, N. T.: The behavior of propazine in the soil. Vest. sel'.-khoz. Nauki, Mosk.
    11, (12), 134 (1966); through Weed Abstr. **16,** 1916 (1967).
RIEKERK, H., and S. P. Gessel: The movement of DDT in forest soil solutions.
    Soil Sci. Soc. Amer. Proc. **32,** 595 (1968).
ROADHOUSE, F. E. B., and L. A. BIRK: Penetration of and persistence in soil of the
    herbicide 2-chloro-4,6-bis(ethylamino)-s-triazine (simazine). Can. J. Plant
    Sci. **41,** 252 (1961).
RODGERS, E. G.: Leaching of four triazines in three soils as influenced by varying
    frequencies and rates of simulated rainfall. Proc. S. Weed Conf. **15,** 268 (1962).
——— Leaching characteristics of certain herbicides in selected soils. Rept. Fla. Agr.
    Expt. Sta., p. 56 (1966); through Weed Abstr. **16,** 2394 (1967).
———— Leaching of seven s-triazines. Weed Sci. **16,** 117 (1968).

——, and M. Wilcox: Leaching characteristics of certain herbicides in selected soils. Rept. Fla. Agr. Expt. Sta., p. 63 (1965); through Weed Abstr. **15**, 1781 (1966).

Schuldt, P. H., H. P. Burchfield, and H. Bluestone: Stability and movement studies on the new experimental nematocide 3,4-dichlorotetrahydrothiophene-1,1-dioxide in soil. Phytopathol. **47**, 534 (1957).

Shahied, S., and H. Andrews: Leaching of trifluralin, linuron, prometryne, and Cotoran in soil columns. Proc. S. Weed Conf. **19**, 522 (1966).

Sheets, T. J.: The comparative toxicities of four phenylurea herbicides in several soil types. Weeds **6**, 413 (1958).

—— The comparative toxicities of monuron and simazin in soil. Weeds **7**, 189 (1959).

—— Persistence of triazine herbicides in soils. Residue Reviews, This volume (1970).

Sherburne, H. R., V. H. Freed, and S. C. Fang: The use of C^{14} carbonyl labeled 3(*p*-chlorophenyl)-1,1-dimethyl urea in a leaching study. Weeds **4**, 50 (1956).

Skob, V. A.: Study of propazine displacement in various soils by the soil-column method. Dokl. s-kh. Akad. Timiryazeva **84**, 316 (1963): through Soils Fert. **28**, 501 (1965).

Smith, A. E., A. W. Feldman, and G. M. Stone: Mobility of *N*-1-naphthylphthalamic acid (Alanap-1) in soil. J. Agr. Food Chem. **5**, 745 (1957).

Smith, R. J., Jr., and W. B. Ennis, Jr.: Studies on the downward movement of 2,4-D and 3-chloro-IPC in soils. Proc. S. Weed Conf. **6**, 63 (1953).

Snelling, K. E., J. A. Hobbs, and W. L. Powers: Effects of surface area, exchange capacity, and organic matter content on miscible displacement of atrazine in soils. Agron. J. **61**, 875 (1969).

Spiridonov, Y. Y., and A. I. Yakovlev: Infiltration and detoxication rate of soil herbicides under moist subtropical conditions. Khim. Sel. Khoz. **5**, 431 (1967); through Chem. Abstr. **67**, 89981z (1967).

Stepanova, Z. A.: Depth of penetration of simazine and atrazine in light Chestnut soils. Sb. Nauch.-Issled. Rab. Aspir. Molodykh. Uch., Vses. Nauch.-Issled. Inst. Agrolesomelior. No. **49**, 138 (1965); through Chem. Abstr. **68**, 2175p (1968).

Stroube, E. W.: The movement and persistence of simazine and atrazine in soil, and some related studies. Ph.D. Thesis, Ohio State Univ. (1961). Diss. Abstr. **22**, 3339 (1962).

——, and D. P. Bondarenko: Persistence and distribution of simazine applied in the field. Proc. N.C. Weed Control Conf. **17**, 40 (1960).

Sund, K. A.: An evaluation of atrazine, simazine, monuron and diuron on ten Hawaiian sugar cane plantations. Weeds **12**, 215 (1964).

Swoboda, A. R., and G. W. Thomas: Movement of parathion in soil columns. J. Agr. Food Chem. **16**, 923 (1968).

Talbert, R. E., and O. H. Fletchall: The adsorption of some *s*-triazines in soils. Weeds **13**, 46 (1965).

Taylor, T. D., and H. J. Amling: Penetration and persistence of 1,1'-dimethyl-4,4'-bipyridilium salt in soil. Weed Soc. Amer. Abstr., p. 54 (1966).

Terent'eva, M. I.: Movement of simazine and atrazine in soil. Trudy Vsesoyuz. Nauch.-Issled. Inst. Udobr. Agropochvoved. **39**, 241 (1962); through Biol. Abstr. **45**, 52740 (1964).

Thomas, G. W.: Kinetics of chloride desorption from soils. J. Agr. Food Chem. **11**, 201 (1963).

Trichell, D. W., H. L. Morton, and M. G. Merkle: Loss of herbicides in runoff water. Weed Sci. **16**, 447 (1968).

Tsunoda, H.: Pentachlorophenol (PCP) derivatives as weedkillers. 1. Herbicidal effect of PCP derivatives. 2. Movement and decomposition of PCP derivatives in soil. J. Sci. Soil Manure (Tokyo) **36**, 195 (1965); through Soils Fert. **29**, 2774 (1966).

Upchurch, R. P.: Behavior of herbicides in soil. Residue Reviews 16, 46 (1966).
——, J. A. Keaton, and F. L. Selman: Soil sterilization properties of monuron, diuron, simazine, and isocil. Weed Sci. 16, 358 (1968).
——, and W. C. Pierce: The leaching of monuron from Lakeland sand soil. Part I. The effect of amount, intensity, and frequency of simulated rainfall. Weeds 5, 321 (1957).
—— —— The leaching of monuron from Lakeland sand soil. Part II. The effect of soil temperature, organic matter, soil moisture, and amount of herbicide. Weeds 6, 24 (1958).
Vanden Born, W. H., and F. Y. Chang: Translocation and persistence of dicamba and picloram in Canada thistle. Weed Soc. Amer. Abstr., p. 40 (1967).
Weber, J. B.: Molecular structure and pH effects on the adsorption of 13 s-triazine compounds on montmorillonite clay. Amer. Mineral. 51, 1657 (1966).
—— Mechanisms of adsorption of s-triazines by clay colloids and factors affecting plant availability. Residue Reviews, This volume (1970).
——, and H. D. Coble: Microbial decomposition of diquat adsorbed on montmorillonite and kaolinite clays. J. Agr. Food Chem. 16, 475 (1968).
——, P. W. Perry, and K. Ibaraki: Effect of pH on the phytotoxicity of prometryne applied to synthetic soil media. Weed Sci. 16, 134 (1968).
—— ——, and R. P. Upchurch: The influence of temperature and time on the adsorption of paraquat, diquat, 2,4-D, and prometone by clays, charcoal and an anion-exchange resin. Soil Sci. Soc. Amer. Proc. 29, 678 (1965).
——, and D. C. Scott: Availability of a cationic herbicide adsorbed on clay minerals to cucumber seedlings. Science 152, 1400 (1966).
——, and S. B. Weed: Adsorption and desorption of diquat, paraquat, and prometone by montmorillonitic and kaolinic clay minerals. Soil Sci. Soc. Amer. Proc. 32, 485 (1968).
White, A. W., A. P. Barnett, B. G. Wright, and J. H. Holladay: Atrazine losses from fallow land caused by runoff and erosion. Environ. Sci. Technol. 1, 740 (1967).
Wiese, A. F., and R. G. Davis: Movement of herbicides in soil columns. Proc. S. Weed Conf. 15, 87 (1962).
Youngson, C. R., C. A. I. Goring, and R. L. Noveroske: Laboratory and greenhouse studies on the application of Fumazone in water to soil for control of nematodes. Down to Earth 23(1), 27 (1967).
Zurawski, H., and M. Ploszynski: Investigations on the disappearance of simazine from light soil. Proc. Brit. Weed Control Conf. 9, 115 (1968).

Influence of triazine herbicides on
soil microorganisms*

By

P. KAISER,** J. J. POCHON,** and R. CASSINI***

Contents

I. Introduction

On reviewing the literature published during the last 20 years concerning the action of chemical substances used as herbicides, fungicides, insecticides, nematocides, etc., certain facts become evident.

First of all, there is a growing number of such active substances; there is an unquestionable effectiveness of those which have been selected after close examination and are commercially available; there

---

* Translated by Demetrios M. Yermanos; translation approved by authors.
** Soil Microbiology Section, *Pasteur Institute*, Paris.
*** Mycology Section, INRA, Dijon.

is an efficiency tested for specific uses for which these compounds are recommended; and there is the lack of concern, at times, shown by industry regarding immediate, secondary, or belated effects, other than the specific effect expected (while meeting all the requirements set forth by legislation).

Furthermore, as a result of the above situation, there is serious concern, especially by ecologists, veterinarians, and public health personnel, regarding the breakdown of the biological equilibrium with its vast and often irreversible consequences, possible harmful effects on human and animal health, and the biological deterioration of soils with its unfavorable consequences on their "conservation."

Therefore, we felt that it would be of interest to prepare an analytic and then synthetic review of the state of our knowledge on the influence of triazines on the bacterial and fungal microflora of the soil. The major role played by these compounds on the soil biology and their effects on the metabolism of its organic substances, on the movements of its mineral elements, and on the nutrition and well-being of plants are well known.

Upon reviewing the literature on this particular topic, which though limited is of capital importance, we realized the difficulties of our undertaking due to gaps and imperfections in experimentation, particularly due to the absence of any standardization of methods used. The comparison of results and sometimes their interpretation is, therefore, often difficult and misleading.

When the available data refer to the action upon soil bacteria or fungi or phytopathogenic substances, experiments have been conducted either in the soil itself, which is often not well defined, and therefore in the presence of populations and of a considerable number of physical and chemical factors which are not controllable, or on pure cultures in the laboratory where conditions are well defined but remote from the natural environment.

In the first part of this report we will review the field experiments and in the second those in pure cultures; then we will attempt a synthesis, summarizing and discussing the information obtained.

II. Effects of triazines on microbial activity in the soil

a) *Overall microbial activity*

The release of carbon dioxide from a soil provides a good idea regarding the overall activity of microorganisms. It has been measured many times in soils treated with simazine.[1]

GUILLEMAT *et al.* (1960) conducted a laboratory test with a garden soil in which they incorporated simazine at 0.1/1,000, which corre-

---

[1] See Foreword to this volume for chemical designations of triazines mentioned in text.

sponds to 300 kg./ha. This high dosage does not cause any direct depression in the overall biological activity of the soil: mg. of carbon dioxide released in 10 days was 259 without simazine and 264 with simazine (Table I).

Table I. *Determination of overall activity in the check soil and in soil treated with simazine* (GUILLEMAT et al. 1960)

Hours	Mg. of CO_2 released/20 g. of soil samples			Daily average x	$(x - \bar{x})^2$
	1	2	3		
Check soil					
1	7.7	13.2	12.4	11.1	222.
2	15.	35.	21.	23.7	5.3
3	15.4	23.6	34.	24.3	2.9
4	18.2	39.5	22.5	26.7	0.5
5	18.1	27.	32.	25.7	00.9
6	23.	49.	39.	37.	121.
7	20.8	17.6	22.	20.1	34.8
8	21.5	36.5	45.5	34.5	72.3
9	13.	38.	31.	27.3	1.7
10	25.7	34.	28.	29.2	10.2
				259.6 $\bar{x} = 26$	470.
Soil treated with simazine					
1	20.2	24.	18.	20.7	32.5
2	19.8	18.3	19.8	19.3	50.4
3	13.	26.4	24.2	21.3	26.
4	20.7	40.	26.9	29.2	7.8
5	35.9	26.2	28.2	30.1	13.7
6	40.	24.6	23.3	29.3	8.4
7	19.8	29.9	26.4	25.2	1.4
8	21.5	35.2	20.9	25.5	0.8
9	30.8	28.3	37.2	32.1	32.5
10	30.	21.3	43.3	31.5	26.
				264.2 $\bar{x}' = 26.4$	199.5

BURNSIDE et al. (1961 and 1963), ENO (1962), and BAUER (1967) arrived at similar results. Dosages from zero to 8,100 p.p.m. of triazine are without effect. CHANDRA et al. (1960) showed, by contrast, that the release of carbon dioxide is inhibited by treatments of from five to 100 p.p.m. in simazine. Nine soils from Oregon were compared: the depressing effect lasted 28 days, following which biological activity returned to normal levels. The same holds true for monuron, diuron [3-(3,4-dichlorophenyl)-1,1-dimethylurea], and 2,4-D[(2,4-dichlorophenoxy)-acetic acid]. In only one case the depressive effect lasted 55 days. The authors emphasized the importance of soil type in this process. The weakest depression was noticed in the cases where the organic matter content was the highest. Dallyn, cited by Chandra et al. (1960), believed also that the effect of herbicides on microorganisms depends on soil type. However, the degree of depression is not always parallel to the level of humus and clay in the soil. Teuteberg (1967) observed a slight depression in the respiration of a soil treated with simazine. Bartha et al. (1967) studied the effect of numerous herbicides on soil respiration and nitrification. Atrazine and simazine are among the least harmful; atrazine at first causes a slight stimulation but later a slight depression of carbon dioxide release.

b) Algae

Triazines specifically inhibit photosynthesis (Hill reaction). One must expect a considerable repression of these phototrophic organisms in soils treated with triazines. The analyses conducted in this area of research are not too many.

Kaiser and Reber (1966) counted the algae in the maize root zone cultured in liquid medium with a strong dose of simazine. The number of algae/ml. of medium reduced to 1,500 (check 250,000). They noticed that some survived; doubtless those were the more resistant to this herbicide. According to Kiss (1967), certain blue-green, golden, and chloro algae are especially sensitive to herbicides and, therefore, are good indicators of the level of herbicides in the soil.

Atkins and Tchan (1967) used a strain of *Chlorella* to determine the quantity of atrazine contained by some soils. With this technique, the authors were able to detect levels of less than 0.5 p.p.m. of a triazine in solution. The phytotoxicity of atrazine varied according to soils. With the same quantity of atrazine, the soils containing a high level of organic matter inhibited algal growth much less than those containing low levels. In soils rich in organic matter, atrazine appeared to be adsorbed and inactivated.

c) Fungi

In a basic study Guillemat et al. (1960) analyzed the number and the distribution of fungi species in a number of soils with treatments of from six to 300 kg./ha. of simazine. Steinbrenner et al. (1960) conducted a similar experiment with maize using treatments of simazine ranging from 1.5 to 50 kg./ha. Eno (1962) compared the effects of simazine and atrazine.

Keller (1961), De Vries (1962), Ghinea (1964), Koltcheva and Markova (1964), Klyuchnikov et al. (1964), and Babak and Presman (1965) experimented with simazine, atrazine, or prometryne on different soil types.

Cassini and Cassini (1966), continuing the experiments of Guillemat et al. (1960), prepared a biological and physicochemical balance sheet of 10 years of weed control with simazine in a monoculture of maize. At the same time, KAISER (unpublished data) studied the effect of simazine on the root zone of maize grown in nutrient cultures or in soil.

The results of these various experiments demonstrate that triazines have highly variable effects on the soil mycoflora. They have no action at times (four authors), they stimulate the mycoflora or certain species (six authors), or inhibit fungi (eight authors).

It is important to continue these counts for long periods of time; inhibition or stimulation may appear 10 to 15 months after treatment. Furthermore, inhibitions are followed a few months later by stimulations. All authors found a return to normalcy after a more or less long period of time; the situation in regard to fungi returns to the original one as it was prior to treatments.

The more noticeable changes occur when strong dosages of herbicides are applied to poor soils (sandy acid soils, nutrient cultures), that is, in environments unfavorable for microbial growth.

ARPAD and JANOS (1958–1960), KAUFMANN (1964), MILLER and AHRENS (1964), GOGUADZE (1967), CHOPRA and CURL (1968), ROD-RIQUEZ-KABANA et al. (1968), TEUTEBERG (1968), and CASSINI and CASSINI (1969) paid particular attention to phytopathogenic fungi and their antagonists.

UHLI's (1966) work is the only one related to the effects of simazine on mycorhyzas. During four years, plantings of young pines were treated with dosages of five, 7.5, and 12.5 kg. of simazine. Analyses of the level of mycorhization show that the treatment did not have any influence on the ectotrophic development. Rather, it produced a slight stimulation of mycorhiza development in roots having strong mycorhization and a slight inhibition in roots having weak mycorhization.

The review of these various reports shows that the action of triazines on phytopathogenic fungi is diverse. Four authors mention a direct inhibitory action of triazines, while two others consider it indirect. Triazines produce growth stimulations in germs which are antagonistic to phytopathogens. Therefore, in some cases they behave like specific fungicides. Two authors do not report any sensitivity of phytopathogens to these herbicides, and one author notes a weak stimulation of a fusarium with a weak dosage of atrazine.

d) Actinomycetes and bacteria

Numerous authors report little or no change in the total microflora (number of germs) following a treatment with triazines with ordinary dosages (two to six kg./ha.) and even with strong dosages.

In their basic research, GUILLEMAT et al. (1960) analyzed four soils treated with six kg./ha. of simazine, three weeks after application of the treatment.

Tables and curves of biological activity show that the treatment had practically no action on the soil bacterial microflora. Treatments of 30, 60, 150, and 300 kg./ha. did not influence the quantity of bacteria significantly.

Several other researchers arrived at similar results: RANKOV et al. (1962), PETZOLDT (1962), ENO (1962), ALIEV (1962), AMANTAEV et al. (1963), SMIRNOVA and TRETJAKOV (1965), PANTOS and GYURKO (1962), TODOROVIC and CRBIC (1965), PORSCHEVA et al. (1966), and KAMPF (1968).

FINK et al. (1968) observed, however, certain changes of species and an augmentation of bacterial colonies following a combination treatment of simazine and nitrogenous fertilizer.

STEINBRENNER et al. (1960), as well as DE VRIES (1962 and 1963), observed, but only after several months, a reduction in bacteria and an increase in actinomycetes due to simazine.

NEPOMILUEV (1966) also observed stimulatory effects following a moderate inhibition.

A complete analysis of the microflora prompted GHINEA (1964) to conclude that treatment of various soils with aminotriazines has as a consequence the modification of certain microbiological characteristics of the soils; these modifications are reflected in the composition of the microflora as well as in the rhythm of certain biochemical processes. They are offset to a large degree 30 days after treatment, but certain ones persist throughout the entire vegetative period of maize.

Other authors examined the action of triazine on soil microorganisms. According to PORSCHNEVA et al. (1966) small dosages of atrazine stimulate the microflora in contrast to strong dosages which inhibit it. During a three year period KOLCHEVA and MARKOVA (1964) analyzed soil from a vineyard treated with simazine. According to these authors, one should alternate treatments of herbicides with tillage to avoid negative effects on the microflora.

Inhibitory effects were equally noticed by KLYUCHNIKOV et al. (1964), SOSNOVSKAYA and PASCHENKO (1967), and TEUTEBERG (1967). The latter pointed out two indirect actions in particular: the absence of weeds on the chemically treated plots decreases the amount of organic matter in the soil and the lack of tillage is less favorable to microorganisms. These two factors themselves can bring about a decrease of microorganisms. To remedy this effect, small quantities of organic matter should be added to the soil to stimulate the microflora and thus to balance the action of herbicides. KAMPF (1968) arrived at the same conclusions.

In summary, the behavior of actinomycetes and soil bacteria as far as triazines are concerned is very variable. All authors counted microorganisms after treatment with herbicides during a more or less long period of time and for different kinds of soils.

According to 17 authors, the triazines have little or no influence on the quantity and quality of microorganisms. Five authors note a stimulation of microflora and 12 authors a decrease in the number of microorganisms. Two other authors observed a compensation of this depression upon addition of organic matter to the soil. Sometimes the decrease in number was noticeable only 15 months after the treatment; certain decreases lasted for long periods (10 months) but the return to normalcy took place always after a more or less long delay. For actinomycetes and proteolytic bacteria, two authors report a depression followed by a stimulation.

A more thorough analysis of different species of soil microorganisms reveals that certain species are resistant to the herbicides and replace others less resistant: cocci bacteria are favored over rod shaped ones and the nonsporulating types are more resistant than the sporulating types. These changes in flora often last for several months. They are derived from indirect actions. The quantity and quality of herbicides used is of major importance. The influence of the environment is overpowering; the inhibitions are more evident in poor environments, and even more so when the dosage of herbicides is important.

In conclusion, we can say that the influence of triazines on soil microbes is dependent on the sensitivity of the species, the environment, and the quality and quantity of the herbicides.

III. Influence of triazines on some functional groups in the soil

a) Nitrogen fixation

Experiments on this subject are often contradictory; some indicate favorable, others unfavorable action.

GUILLEMAT et al. (1960) reported an increase in the aerobic N-fixing bacteria with strong dosages of simazine; the anaerobic Clostridium decreased.

GHINEA (1964) arrived at similar conclusions with agronomic dosages of atrazine and simazine; the nitrogen fixed in the soil increases by 30 to 50 percent following the treatments (Fig. 1). SMIRNOVA and TRETJAKOV (1965) noticed an azotobacter enrichment in the root zone of maize treated with simazine and atrazine. SZABO (1964) showed that a 0.3 percent solution of prometryne stimulated Azotobacter in nutrient cultures and inhibited Rhizobium. STEINBRENNER (1960), in a field experiment, obtained results similar to those of GUILLEMAT et al. (1960): in certain cases (50 kg./ha. first period after treatment) the number of Azotobacter increased by 33 percent, but it returned to the normal level in a few weeks. In synthetic media enriched with simazine, the number of Azotobacter colonies varied little; it diminished only with the higher dosages of herbicides (1.6 g./l.). Thus Azotobacter appear to be barely sensitive to this compound.

According to RANKOV et al. (1966), the number of Azotobacter and

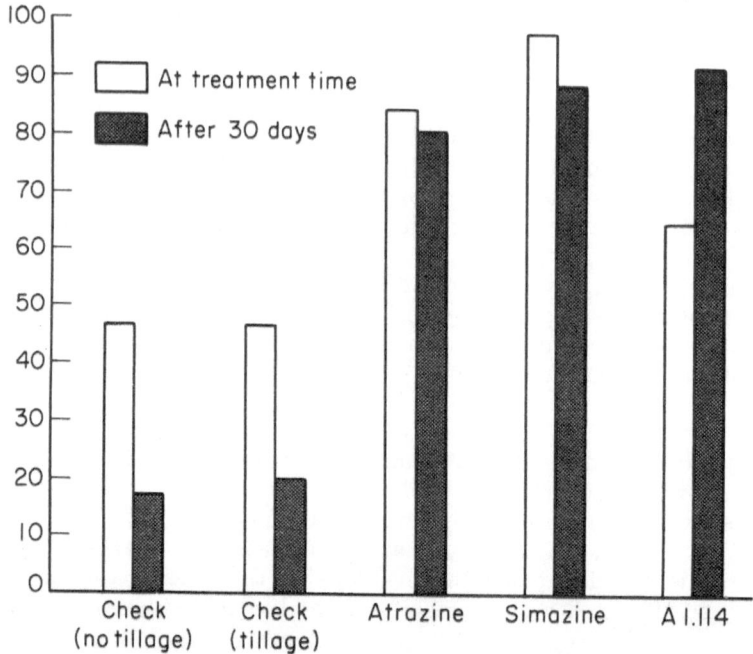

Fig. 1. Balance sheet of nitrogen fixed in the soil (Ghinea 1964)

Clostridium pasteurianum remained unchanged after a soil treatment with simazine and prometryne. Babak and Presman (1965) did not report any influence of prometryne on the growth of *Azotobacter chroococcum*.

Some reports mention a reduction in the numbers of *Azotobacter* and *Clostridium* after soil treatments (Koltcheva and Markova 1964, Smirnova 1963, Dubovska *et al.* 1965).

Triazines occasionally appear to be favorable to *Rhizobium*. Nepomiluev *et al.* (1967) showed that chlorazine at two to four kg./ ha. promoted the formation of nodules indirectly by stimulating the growth of beans. Rankov *et al.* (1966) also reported an increase in bean nodules. Avrov (1966) studied the action of different herbicides, particularly those of symmetric triazines, on *Rhizobium* of peas maintained in a suspension of sterile soil. None of the herbicides proved to be toxic to these bacteria. In another experiment, application of three kg./ha. of simazine during three years in a field of maize decreased nodulation of peas planted the following year. The residual activity of atrazine was greater than that of simazine; this herbicide increased the weight of lupines planted on podzolic soil when the seeds had been inoculated with *Rhizobium*. Kaszubiak (1966) did not report any toxicity of triazine toward *Rhizobium* although the substituted ureas were toxic.

In conclusion, *Azotobactor* is slightly sensitive to triazines (no action or stimulating action, seven authors and negative action, three authors).

The capacity for fixation of soils can increase greatly after triazine treatments. *Clostridium pasteurianum* is more sensitive than *Azotobacter*. Few experiments relate to the action of triazines on *Rhizobium;* these bacteria possess, apparently, little sensitivity and the action on nodulation takes place often indirectly. It is primarily legumes which are inhibited or stimulated by the triazines.

b) Nitrification

TsvETKOVA (1966) stated that the level of nitrates in a podzolic soil, rich in humus (two percent), increased ten times after an ordinary treatment with simazine and atrazine; the levels of K_2O and of P_2O_5 were equally raised.

TULABAEV and AZIMBEGOV (1967) arrived at similar conclusions: six days after treatment of a field of maize with simazine and atrazine, the quantity of NO_3 passed from 31 to 39 percent and from 31 to 48 percent. According to ZAVARZIN and BELJAEVA (1966) simazine, atrazine, and trietazine increased the level of NO_3 and of P_2O_5 in the soil. Similar observations were made by AMANTAEV et al. (1963) and SHAZKINA and SHARKOW (1966).

The data of TARLAPAN and ZAKHARYA (1966) point in the same direction; after a three-year treatment of a calcareous Tchernosem soil with simazine, its level of nitrogen was raised from 29 to 45 percent, that of phosphorus from 58 to 80 percent and that of potassium from four to 14 percent, and nitrification was stimulated.

None of the authors reported any influence of triazines on nitrification or on the number of nitrifying bacteria in the soil (POCHON et al. 1960, ALIEV 1962, PORSCHNEVA et al. 1966). The experiments of STEINBRENNER et al. (1960) proved that the number of nitrifying bacteria of a treated soil remains more or less equal to that of the check.

CASELEY and LUCKWILL (1964), in a study of percolation in the soil, did not observe any action of atrazine or simazine on nitrification.

BARTHA et al. (1967) showed that simazine has no influence on nitrification in lemon fields but that atrazine caused an inhibition, especially in the beginning of the treatment. SZABO (1964) found that prometryne applied to the soil delayed nitrification and growth of microorganisms considerably. PORSCHNEVA et al. (1966) found a decrease only when the dosage of atrazine was high (20 kg./ha.).

In the experiments of GHINEA (1964), the soil capacity for nitrification was reduced after triazine treatments, but 30 days later it became normal. This observation corresponded to those made by HULEA et al. (1961).

ENO (1962) found that a decrease of nitrification in sandy soils occurred only with very high dosages of simazine or atrazine, higher than 1,000 p.p.m.

FARMER et al. (1965) studied the influence of simazine on nitrification of a soil percolated with a solution of ammonium sulfate and on nitrification in pure strains of Nitrobacter and Nitrosomas. In percolated soil, nitrification was inhibited starting with dosages from six p.p.pm. of simazine: the nitrites accumulated in the percolating solution. Similar effects had been found with pure cultures of Nitrobacter: the utilization of nitrites was delayed with six p.p.m.; complete utilization took place in 70 days without herbicide and in 110 days with six p.p.m. of simazine. The activity of Nitrobacter was, therefore, slowed down but not that of Nitrosomonas. The growth of Nitrobacter was slowed down in agitated cultures; that of Nitrosomonas was not. The respiration of Nitrobacter agilis was not inhibited with 100 p.p.m. of simazine (FARMER and BENOIT 1965).

HAUCK and STEPHENSON (1964) showed that the amine groups of nonherbicide triazines could be nitrified after separation of the molecule. This nitrification was slow and varied inversely with the number of amine groups in the triazine nucleus. However, these nondegrading substances inhibited nitrification of ammonia in a percolated soil. Similar data have already been reported by CLARK et al. (1959).

In conclusion, the results obtained are contradictory: six authors observed an increase in the level of NO_3 in the soil, six observed a decrease, and one reported an accumulation of nitrites.

An analysis of the experiments explains, in part, the contradictory results observed in the field. These contradictions stem from, no doubt, the differences in environments; Nitrobacter, a sensitive species to herbicides, are inhibited mainly in poor, sandy soils, and with strong dosages of herbicides; soils richer in organic matter should, therefore, protect them; the stimulations observed have not been explained yet. On the other hand, the experiments of REID (1960) have shown that Nitrobacter became progressively adapted to repeated treatments of herbicides.

Some researchers proved that the amine groups of triazines can be weakly nitrified after degradation of the nucleus of triazines. The microorganisms responsible for this nitrification have not yet been isolated.

c) Cellulolysis

The works of GHINEA (1964) already cited showed that the cellulolytic bacteria were inhibited right after the treatment but that the normal microflora was re-established a little later. Other data reveal a negative influence of herbicides on cellulolysis. KOLTCHEVA and MARKOV (1964) and STEPANOVA (1967) found that simazine, propazine, and atrazine (three kg./ha.), when first applied to the soil, inhibited the degradation of cellulose and the formation of amino acids; later (three to four months) they produced a stimulation of cellulolytes and a secretion of amino acids.

SOSNOVSKAYA and PASCHENKO (1967) attributed the inhibition of cellulolysis by simazine in maize fields to an indirect action: the absence of weed roots that the cellulolytes decompose after the death of these plants.

A stimulation effect was reported by SPIRIDONOV and JAKOVLEV (1968): atrazine introduced at the rate of 10 to 20 kg./ha. in a subtropical soil increased the activity of cellulolytic organisms by 30 percent; with simazine, at the rate of five to 10 kg./ha., the rate of increase reached 11 to 27 percent. With 20 kg./ha. of simazine the cellulolytes were inhibited during the first three months, later their activity increased by 30 percent when simazine had partially disappeared from the soil due to leaching and degradation. This increase in the cellulolytic activity could be attributed to the activation of aerobic cellulolytic microbes.

In conclusion, four authors observed a decrease of cellulolytic activity in the presence of simazine. Some time after the application, cellulolysis returned to normalcy. Another author reported a stimulation after any inhibition. The fungi exhibited a smaller sensitivity than the bacteria. In general, cellulolytic microorganisms behave just like the other soil microbes.

IV. Action of triazines on soil microorganisms in pure culture and artificial media

a) Fungi

Numerous authors have compared the growth of various fungi in culture media with or without increasing dosages of triazine. It proved to be difficult to compare growth in the laboratory (artificial environment) and in the soil.

GUILLEMAT et al. (1960) cultivated 31 strains of fungi isolated from the soil in a normal Czapek medium containing 1.5 percent of simazine. The growth was measured in mm. of diameter of the colony. At these rates all the strains were severely inhibited, half of them did not exhibit any growth, and the other half produced colonies with very small diameter.

STEINBRENNER et al. (1960) inoculated malt media containing increasing concentrations of simazine with dilutions-suspensions of soil. With 65 p.p.m. of herbicide the number of colonies formed was reduced by 15 percent, with 650 p.p.m. by 50 percent, and with 1,625 p.p.m. by more than 90 percent.

The study of the action of herbicides on mycelium growth (measured in mm.) showed that fungi react in two ways. The genera

Zygorrhynchus, Rhizopus, Alternaria, Trichoderma, and *Stemphyllium* were inhibited at all concentrations of simazine (65 to 1,625 p.p.m.) while the *Fusarium* and *Actinomycetes* were at times stimulated; for one strain of *Fusarium* the authors noticed a stimulation of +86 with 325 p.p.m. and of 66 percent with 1,625 p.p.m.

ENO (1962) estimated the weight of mycelium formed in liquid media without agitation for different concentrations of triazine up to 8,000 p.p.m. and for different fungi. *Aspergillus niger* was stimulated with dosages of simazine higher than 32 p.p.m. and up to 4,000 p.p.m.

Atrazine, by contrast, has no influence on the growth of *A. niger.* It is necessary to use herbicide dosages of 8,000 p.p.m. to bring about growth inhibition. Two other strains of *Aspergillus* react entirely differently. *Aspergillus* strain No. 1, to be stimulated, requires simazine concentrations of at least 1,000 p.p.m.; atrazine, however, exerts a fungistatic activity on this fungus starting with 32 p.p.m.; with 256 p.p.m. growth is nil. *Aspergillus* strain No. 2 is inhibited by simazine concentrations exceeding 1,000 p.p.m. Two species of *Trichoderma viride* were also included in the test. The first reacted like *Aspergillus* No. 1; the second was inhibited by atrazine and simazine.

PANTOS (1964) studied the effect of one triazine on strains of *Aspergillus fumigatus,* of *Chaetonium globosum,* and of *C. elatum* in pure cultures. The weak concentrations of herbicides used in agronomic treatments did not influence the growth of these strains. The mycorhizial fungi such as *Boletus edulis, Suillus grevillei,* and *Leccinum aurantiacum* also remain insensitive to weak concentrations, even up to 100 p.p.m. of herbicide. Fungi of the genera *Penicillium, Aspergillus, Chaetonium,* and *Pacilomyces* are resistant to dosages of one part/1,000 of atrazine and simazine.

VORDERBERG (1961) cultivated aseptically several strains of fungi in Petri dishes in various media and observed their growth in the presence of eight different herbicides. The results indicated that the action of herbicides (stimulation or inhibition) depend largely on the nutritive medium.

ARPAD and GEZA (1958 and 1960) observed an inhibition of growth in *Aspergillus niger, Fusarium oxysporum, F. culmorum,* and *Trichothecium roseum.*

The different species of mycorhizial fungi also have variable reactions to triazines (UHLIG 1966). A concentration of 6,000 p.p.m. of simazine stimulates the growth of *Scleroderma vulgare* but the same concentration inhibits the growth of *Tricholoma pessundatum;* a dosage of 100 p.p.m. is sufficient to bring about inhibition of *Hebeloma crustiliniforma.*

According to experiments by the same author, *Schleroderma* is capable of degrading the herbicide; the stimulation, therefore, with this strong dosage would be due to a utilization of this compound as a source of carbon or nitrogen.

Sikka *et al.* (1965) determined the effect of several concentrations of atrazine on mycelian growth of the following four common soil fungi cultivated in liquid medium: *Trichoderma, Fusarium, Penicillium,* and *Geotrichum.* The herbicide stimulates the growth of the four fungi at weak concentrations of one to 10 p.p.m. With 10 p.p.m. the weight of mycelium is increased by almost 100 percent. Thus, this stimulation does not depend upon the utilization of the herbicide as a nutritional source. The herbicide must increase the consumption of nitrates or glucides as is the case with plants (Ries *et al.* 1967).

The Funderbunk school (Curl and Funderbunk 1965) analyzed these stimulation or inhibition phenomena on phytopathogenic germs. *Sclerotium rolfsii,* a fungus attacking the crown of plants, is cultivated in the presence of other soil microorganisms (added by a soil suspension) in an agitated liquid medium. The addition of atrazine then brings about a rapid decrease in the phytopathogenic population. With 20 p.p.m. of herbicide, the antagonists of *S. rolfsii* are stimulated to the maximum and their zones of inhibition toward the pathogen are broadened. The same school (Rodriguez-Kabana *et al.* 1967) is proceeding with research on *S. rolfsii* and *Trichoderma viride* and is determining the influence of the dosage of atrazine in each case.

Millikan and Fields (1964) observed a decrease in the growth of *Fusarium culmonum* of 71 percent with 10 p.p.m. of simazine added to the culture medium of Fries. This same concentration decreased the growth of *Trichoderma* and of *Rhizoctonia* by 92 and 93 percent, respectively.

Kaiser (unpublished data) cultivated 32 strains of fungi of the maize root zone in an environment rich in glutamate and glycerol, to which 10 mg. of simazine was added/30 ml. of medium. Under these conditions, five strains were slightly inhibited by it.

In conclusion, the behavior of fungi in pure cultures and in artificial media to which triazine was added varies greatly depending on the dosage, the quality of herbicide, and the species studied. Eight authors noted clear-cut growth inhibition. Five authors utilized very strong dosages, three others medium or weak dosages. One author did not find any action of the herbicide on growth. Four authors found growth stimulation with strong doses of herbicides; without doubt, fungi utilize the herbicide as a source of carbon or nitrogen. The kind of herbicide is of importance, also: working with the same fungi, one author observed inhibition with atrazine and stimulation with simazine.

There are, therefore, sensitive (inhibited) and insensitive (non-inhibited or stimulated) strains. One wonders if the insensitive strains are not exactly those which have the ability to modify the structure of the molecule in one way or another. Very small dosages of triazine (one to 10 p.p.m.) accelerate the growth of certain fungi to the same degree (two authors); consequently, the herbicide must increase the consumption of nitrates as is the case with plants.

b) Actinomycetes and bacteria

To understand the contradictory results obtained in the soil, it is necessary to study the behavior of pure cultures of microorganisms in the presence of triazine.

The influence of the culture medium on this behavior was first shown by Voderberg (1961). *Bacillus mycoides*, cultivated at two simazine concentrations, was inhibited in potato broth agar, while it was stimulated in a culture of peptone-glucose agar. The same remarks apply to a *Micrococcus*, to two *Steptomyces* and, to a certain extent, to *Azotobacter chroococcum*.

In the presence of other herbicides: [(4-chloro-*o*-tolyl)oxy]-acetic acid (MCPA), 3,5-dinitro-*o*-cresol (DNOC), and, especially, isopropyl-*N*-phenylcarbamate (IPC), the inhibitions are stronger.

The influence of the environment is, therefore, important but also that of the herbicide and its concentration. The different species do not respond to identical herbicides in the same manner.

This area of research was broadened by Balicka *et al.* (1964). The growth of several strains of bacteria, in pure cultures, was measured with or without atrazine. Certain very sensitive strains had their growth inhibited by one p.p.m. of atrazine in all media; other less sensitive strains did not suffer any unfavorable effects on their growth except in synthetic mineral media and were not inhibited by herbicide concentrations of one to ten p.p.m. Organic environments (potato broth and soil extracts) partially neutralize the toxic action. Under these conditions, some strains remain insensitive to higher concentrations and it is possible to distinguish four groups:

1. Growth inhibition with strong dosage (1,000 p.p.m.); no inhibition with weak dosage (one to 10 p.p.m.).
2. Inhibition with strong dosage, stimulation with weak dosage.
3. Inhibition at all concentrations.
4. Insensitive to all concentrations.

Amino acid secretion is influenced in all microorganisms; it is either inhibited or stimulated depending on the dosage of herbicide or the kind of microorganism.

Caseley and Luckwill (1964) observed that in pure culture *Nitrobacter* and *Nitrosomonas* were inhibited with all herbicides used. *Nitrobacter* are more sensitive than *Nitrosomonas*.

Simazine is the less toxic herbicide because of its low solubility; commercial herbicides (with wetting agents) have a higher toxicity than pure herbicides.

Kaiser (unpublished data) studied the behavior of pure strains of actinomycetes and bacteria isolated from the root zone of maize, with various dosages of simazine. The stimulatory or inhibitory effects depend on the dosage of herbicide and on the strain cultivated. The stimulation of nitrate-reductase could, as in fungi and higher plants, explain the positive effects. Furthermore, the increased excretion of organic acids could be due to a disorder in the functioning of pyridines-nucleotides.

In conclusion, the behavior of bacteria and actinomycetes in pure culture clarifies the results obtained in soil. The microorganisms are divided in sensitive and insensitive strains to the herbicides. The first are more or less inhibited according to the kind and the quantity of herbicides and the environment.

The presence of organic matter enhances the sensitivity of organisms. The metabolism of germs is transformed, at times, upon contact with the herbicide; the secretion of amino acids and organic acids is modified in quantity and quality.

c) Selection of a microflora which decomposes the triazines

Many authors have shown that repeated applications of triazines, on the soil or on artificial media inoculated with soil, caused the appearance of a different microflora which decomposed the herbicide. According to KAUFMAN et al. (1965), the microbial population of a mineral medium containing simazine and inoculated with a soil is different than that cultivated without simazine. They isolated primarily fungi, Streptomyces and Arthrobacters. BRYANT (1963) studied in a thesis the decomposition of 19 herbicides; he carried out seven successive transfers of the microflora on mineral media containing herbicides. All the triazines used were members of the 6-isopropylamine series and were differentiated by substitution in the second and fourth position (in the atrazine series); it is the substitution in the fourth position which determines the dominant microflora:

1. With the ethylamine group, selection of Arthrobacter (simazine).
2. With the diethylamine group, selection of Pseudomonas (ipazine).
3. With the group isopropylamine, selection of two genera at the same time (propazine).

DUKE (1964), following the same principle, isolated microorganisms capable of decomposing atrazine, atratone, and ametryne. The release of carbon dioxide is stronger in soils treated with triazines and inoculated with microorganisms decomposing these compounds than in soils without herbicides, inoculated with the same microorganisms.

CHARPENTIER and POCHON (1962) noted the appearance of a bacterial flora decomposing simazine: the genera Empedobacter, Achromobacter, Microbacterium, and Bacterium were selected.

Reid (1960) isolated *Corynebacterium* and *Streptomycetes*. After eight transfers (each period lasted four days) the population in the medium increased from 5 × 10² to 5 × 10⁶ bacteria.

Swietochowski *et al.* (1965) stated that the decomposition of a simazine becomes more and more rapid in fields which receive herbicide applications every year.

He concludes that there is a progressive enrichment in microorganisms which metabolize the herbicides; Scudder (1963) has the same opinions; Baur (1967) contradicts these data and concludes that there is no enrichment in catabolytic microorganisms after repeated applications of herbicides. The decomposition of triazines would be a passive type, not zymogenous. It should be added that the duration of experiments conducted by Baur was short: three to four weeks. Most authors who studied the decomposition of triazines point out the slowness of the decomposition (several weeks).

V. Behavior of viruses

Cole *et al.* (1968) investigated the action of various herbicides and pesticides on the growth of maize infected with the maize mosaic virus. The virus disease, alone, decreased the total fresh weight of the plant and the weight of grain. The latter was increased with atrazine and propazine, considering infected and non-infected parts together. The combined action of atrazine treatments (six p.p.m.) and the virus inoculation increased the fresh weight of leaves (atrazine, alone, has a small influence; the virus, alone, reduces maize growth). The other triazines, alone, diminish the quantity of maize. Atrazine or ametryne do not raise the grain yield to a higher level than that obtained when maize is inoculated with the virus. These interesting results deserve further investigation.

VI. Conclusions

The triazine herbicides exert a very variable influence on the quantity, quality, and activity of microorganisms in the soil.

The results regarding the count of fungi, actinomycetes, and bacteria after the application of herbicides are contradictory. The authors who found no influence and those who noted an inhibition of microorganisms are about equal in number, others, less numerous, found a stimulation of the soil microflora. The inhibitions and stimulations persist occasionally for long times (10 to 15 months), which is to be expected with triazines; inhibition is followed by stimulation, but all authors mention a return to normalcy after a more or less long delay.

Similarly, the various microbial functional groups react in a variable way: *Azotobacters* and *Rhizobia* appear to be relatively insensitive to triazines. In pure culture, *Nitrobacters* are inhibited but *Nitrosomonas* remains insensitive; in the soil, nitrification may be stimulated or retarded. The cellulolytes behave the same way as other soil microorganisms. Some phytopathogenic fungi are inhibited directly or indirectly by triazines. The algae are very sensitive, but few reports relate to this subject. No research was conducted on protozoa.

A more detailed analysis of these reports explains in part these results. The influence of soil type appears to be predominant: microbial activity is less disturbed by herbicides in soils which are rich in organic matter and clay. Addition of organic matter offsets the inhibitions observed in poor soils. Next in order of importance is the sensitivity of the species: certain resistant ones replace others less resistant. Different reports show that a microflora may appear which decomposes the herbicide. Changes in the microflora come from indirect actions also: no tillage, no weeds.'

The behavior of microorganisms in pure culture and in artificial media clarifies the results obtained in the soil, which are often contradictory. In the same way as plants, the microorganisms are classified into sensitive and insensitive species to triazine. The resistance mechanism to triazine is still unknown. In part, it would be due to the ability of the organism to degrade or modify the herbicide. The sensitive species are more or less inhibited depending on the quality and quantity of herbicide and the surrounding environment. The presence of organic matter attenuates or suppresses the sensitivity of organisms to herbicides. The metabolism of microbes is sometimes transformed in the presence of triazines. With strong dosages, the excretion of organic acids increases in higher plants, fungi, and, sometimes, in bacteria. The excretion of aminoacids can also be modified in bacteria. A detailed analysis of these phenomena could explain the mode of action of triazines. With low rates of triazines, an increase in cell weight is observed, from time to time, in fungi and in bacteria. Without doubt, it appears to be due to a stimulation of nitrate-reductase, as in chlorophyl bearing plants.

The increase in weight, with strong dosages of herbicide, observed in fungi, results from the utilization of the herbicide as a source of carbon or of nitrogen.

In conclusion, microorganisms, like higher plants, respond to triazines in a variable way depending on their sensitivity or their ability to break down the herbicide and the surrounding environment.

Summary

The effect of triazine herbicides on soil microorganisms has been related by numerous authors after trials conducted both in fields and in laboratory.

In field trials, the effect of triazine herbicides on the count of populations of fungi, actinomycetes, and soil bacteria is examined, as well as the reaction of various microbial functional groups. The results are often contradictory probably because of the different types of soil. The microbial activity is less disturbed by herbicides in soils rich in organic matter and clay. Changes in microflora also come from direct action (specific microflora degrading the herbicide may appear) or indirect action (absence of tillage or absence of certain weeds), but there is eventually a return to normal.

Trials conducted in laboratory in a pure culture and in artificial media show that microorganisms are sensitive to triazines and also develop a mechanism of triazine resistance both very variable according to the species.

The more or less important reasons causing this sensitivity are often debated.

Some observations pertaining to viruses are related.

In conclusion, microorganisms like the higher plants respond in very different ways to triazine herbicides.

Résumé*

Influence des triazines herbicides sur les microorganismes du sol

L'action des triazines herbicides sur les microorganismes du sol a été étudiée par de nombreux auteurs, dans des essais aux champs et en laboratoire.

Dans les essais aux champs, l'influence des triazines sur l'importance des populations de champignons, actinomycètes et bactéries du sol, est examinée, de même que la réaction de quelques groupements fonctionnels microbiens. Les résultats sont contradictoires, sans doute à cause des différences de nature du sol. La présence de matière organique ou d'argile réduit l'effet des triazines. Des changements de microflore peuvent se produire sous l'effet direct (une microflore spécifique dégradant l'herbicide peut apparaître) ou indirect des herbicides (absence de labour et disparition des mauvaises herbes), mais il y a généralement retour à la normale aprés un dèlai plus ou moins long.

Les essais conduits en laboratoire, sur cultures pures et en milieu artificiel, montrent que les microorganismes présentent une sensibilité ou une résistance aux triazines très variable suivant les espèces. Les causes de cette sensibilité plus ou moins grande sont discutées. Quelques observations sont rapportées concernant les virus. En conclusion, les microorganismes semblent réagir de façon très variée aux triazines herbicides, comme le font les végétaux supérieurs.

---

* Traduit par les auteurs.

Zusammenfassung*

Der Einfluss der Triazine als Unkrautvertilgungsmittel auf die Mikroorganismen des Bodens

Die Wirkung der Triazine in Bezug auf Bodenmikroorganismen ist von zahlreichen Autoren durch Versuche auf dem Feld wie auch im Laboratorium untersucht worden.

In den Versuchen auf dem Feld wurde sowohl der Einfluss der Triazine auf die Wichtigkeit der lebenden Pilze, Aktinomyzeten und Bakterien des Bodens, wie auch das Verhalten einiger funktioneller Bakteriengruppen beobachtet. Die erzielten Resultate widersprechen sich, wahrscheinlich infolge der Unterschiede in der Bodenbeschaffenheit. Die Gegenwart organischer Stoffe oder von Tonerde vermindert die Wirkung der Triazine. Es können Änderungen der Mikroflora unter dem direkten (eventuell eigenartiger Zerfall der Mikroflora) oder indirekten Einfluss der Unkrautvertilgungsmittel eintreten (kein Pflügen und das Verschwinden des Unkrauts), aber im allgemeinen wird nach mehr oder weniger langer Zeit alles wieder normal.

Die im Laboratorium angestellten Versuche auf reinem Nährboden und in künstlichem Kulturmedium zeigen, dass die Mikroorganismen eine je nach den Arten sehr veränderliche Empfindlichkeit oder Widerstandsfähigkeit den Triazinen gegenüber aufweisen. Über die Ursachen dieser mehr oder weniger grossen Empfindlichkeit wird noch diskutiert.

In Bezug auf die Viren und die Bakteriophagen werden einige Angaben gemacht.

Als Schlussfolgerung kann man sagen, dass die Mikroorganismen den unkrautvertilgenden Triazinen gegenüber sehr verschiedenartige Wirkungen zu haben scheinen, wie es auch den grösseren Pflanzen gegenüber der Fall ist.

References

ALIEV, A. M.: Deistvie simazina na zasorennost'i uorzhai kukuruzy (The action of simazine on weed infestation and the corn yield). Trudy Vsesoyuz. Nauch-Issledovatel. Inst. Udobrenii Agropuchy. **39**, 64 (1962).

AMANTAEV, E., A. ILYALETDINOV, and T. KYDYSHEV: Vliyanie simazina i atrazina na mikrofloru i soderzhanie nitratov v svetlo-kashtonovykh pochvakh Alma-Atinskoi oblasti (The effect of simazine and atrazine on the microflora and nitrate content in light chestnut soils of the Alma-Atinsk oblast). Agrobiologiya **3**, 462 (1963).

ARPAD, V., and M. GEZA: A Simazin es az Atrazin hatasa a kukoricavetesek talajanak mikroflorajara. Kukoricatermesztesi Kiserletek, p. 389 (1958–1960).

——, and V. JANOS: Simazin es dikonirt hatasanak vizsgalata a kukorica termeszetes uszogfertosodesere. Kukoricatermesztesi kiserletek (1958–1960).

---

* Übersetzung von den Autoren.

ATKINS, C. A., and Y. T. TCHAN: Study on soil algae: VI. Bioassay of atrazine and the prediction of its toxicity in soils using an algal growth method. Plant and Soil **27**, 432 (1967).

AVROV, O. E.: Vliyanie gerbitsidov na kluben'kovye bakterii i obrazovanie kluben-'kov u bobovykh rastenii (Effect of herbicides on nodule bacteria formation in legumes). Doklady Vsesoyuz Akad. Sel'skokhoz Naukin V. I. Lenina No. 3, p. 16 (1966).

BABAK, N. M., and L. M. PRESMAN: Vliyanie gerbitsidov na mikroorganismy pochv (Effects of herbicides on microorganisms in soils). Trudy Moldavskii Nauch-Isledovatel Inst. Oroshaemogo Zeml. Oroshehevidstva. **6**(1), 122 (1965).

BALICKA, N., and H. BILODUB-PANTERA: The influence of atrazine on some soil bacteria. Acta Microbiol. Polon. **13**, 149 (1964).

——, H. BILODUB, and L. SZUSZKIEWICZ: Der Einfluss des Atrazin auf die Mikroflora des Bodens. Zeszyty Nauk. Wysz. Skol. Rol. Wroclaw **17**, 281 (1964).

BARTHA, R., R. P. LANZILOTTA, and D. PRAMER: Stability and effects on some pesticides in soil. Applied Microbiol. **15**, 67 (1967).

BAUER, J. L.: The effect of repeated applications of the triazine herbicides on their decomposition in soils. Ph.D. Thesis, Auburn Univ., Ala. (1967).

BRYANT, J. B.: Bacterial decomposition of some aromatic and aliphatic herbicides. Ph.D. Thesis, Pa. State Univ. (1963).

BURNSIDE, O. C., E. L. SCHMIDT, and R. BEHRENS: Dissipation of simazine from the soil. Weeds **9**, 477 (1961).

——, C. R. FENSTER, and G. A. WICKS: Dissipation and leaching of monuron, simazine, and atrazine in Nebraska soils. Weeds **11**, 209 (1963).

CASELEY, J. C., and L. C. LUCKWILL: The effect of some residual herbicides on soil nitrifying bacteria. Ann. Rept. Agr. Hort. Research Sta., Long Ashton, Bristol, p. 78 (1964).

CASSINI, R.: Effets biologique et physico-chimique de dix annees de desherbage a la simazine dans une monoculture de mais a Grignon. Weed Research, (In Press) (1969).

——, and R. CASSINI: Influence de la simazine sur les champignons du sol. Symposium sur divers aspects de l'action des triazines. Bale—J. R. Geigy SA (1966).

CHANDRA, P., W. R. FURTICK, and W. B. BOLLEN: The effects of four herbicides on microorganisms in nine Oregon soils. Weeds **8**, 589 (1960).

CHARPENTIER, M., and J. POCHON: Bacteries telluriques cultivant sur amino-triazine (simazine). Ann. Inst. Pasteur, Paris **102**, 501 (1962).

CHOPRA, B. K., and E. A. CURL: Effect of prometryne on sporulation and fungistasis of *Fusarium oxysporum* f. sp. *Vasinfectum* in soil. Phytopathol. **58**, 1047 (1968).

CLARK, K. G., J. Y. YEE, F. O. LUNDSTOM, and T. G. LAMONT: A modified activity index procedure for determining the quality of the water insoluble nitrogen in mixed fertilizers containing urea-formaldehyde compounds. J. Assoc. Official Agr. Chemists **42**, 592 (1959).

COLE, H., D. R. MACKENZIE, and C. D. ERCEGOVICH: Maize dwarf mosaic. Interactions between virus-host-soil pesticides for certain inoculated hybrids in Pennsylvania field plantings. I. Main effects of virus and chemicals on yield. Plant Disease Reporter **52**, 545 (1968).

CURL, E. A., and H. H. FUNDERBUNK, JR.: Some effects of atrazine on *Sclerotium rolfsii* and inhibitory soil microorganisms. Phytopathol. **55**, 497 (1965).

DUBOVSKA, A., L. KOPLANOVA, and U. REHORKOVA: Vliyanie gerbitsidov na pochvenuyu mikrofloru kukuruzy (The effect of herbicides on the soil microflora of corn). Ochrana Rostlin 1 (4), 45 (1965).

DUKE, W. B.: The decomposition of atrazine and related s-triazine herbicides by soil microorganisms. Ph.D. Thesis, Oregon State Univ. (1964).

ENO, C. F.: The effect of simazine and atrazine on certain of the soil microflora and their metabolic process. Soil Sci. Soc. Florida Proc. **22**, 49 (1962).

FARMER, F. H., and R. E. BENOIT: The effect of simazine on nitrification. Bact. Proc. p. A 17 (1965).

—— ——, and W. E. CHAPPELL: Simazine, its effect on nitrification and its decomposition by soil microorganisms. Proc. N.E. Weed Control Conf. **19**, 350 (1965).

FINK, R. J., O. H. FLETCHALL, and O. H. CALVERT: Relation of triazine residues to fungal and bacterial colonies. Weed Sci. **16**, 104 (1968).

GHINEA, L.: L'influence des aminotriazines sur l'activite microbiologique du sol. Trans. Internat. Congress Soil Sci. 8th Congr. Bucharest **35**, III, 857 (1964).

GOGUADZE, V. D.: Vliyanie gerbitsidov na eazvitie Azotobaktera v nekotorykh pochvakh zapadnoi Gruzii (Effect of herbicides on *Azotobacter* development in some soils of Western Georgia). Agrokhimiya No. 3, 99 (1968).

GUILLEMAT, J.: Interactions entre la simazine et la mycoflore du sol. C. R. Acad. Sci. (Paris) **250**, 1343 (1960).

——, M. CHARPENTIER, P. TARDIEUX, and J. POCHON: Interactions entre une chloro-aminotriazine herbicide et la microflore fongique et bacterienne du sol. Ann. Epiphyt. **11**, 261 (1960).

HAUCK, R. D., and H. F. STEPHENSON: Nitrification of triazine nitrogen. J. Agr. Food Chem. **12**, 147 (1964).

HULEA A., G. ELIADE, and L. GHINEA: Recherches concernant l'influence de l'herbicide atrazine sur la microflore du sol (In Rumanian). Prob. Agr. (Bucharest) **13**(2), 57 (1961).

KAISER, P., and H. REBER: Unpublished data (1966).

KAMPF, R.: Untersuchungen über den Einfluss chemischer Pflanzenschutzmittel auf die Bodenfruchtbarkeit. Z. Pfl. Krankh. u. Schutz. **4**, 169 (1968).

KASZUBIAK, H.: The effect of herbicides on *Rhizobium*. I. Susceptibility of *Rhizobium* to herbicides. Acta Microbiol. Polon. **15**, 357 (1966).

KAUFMANN, D. D.: Effect of s-triazine and phenyluren herbicides on soil fungi in corn and soybean-cropped soil. Phytopathol. **54**, 897 (1964).

——, P. C. KEARNEY, and T. J. SHEETS: Microbial degradation of simazine. J. Agr. Food Chem. **13**,.238 (1965).

KELLER TH.: Über die Auswirkungen einiger Unkrautbekämpfungsmittel auf die mikrobiologische Tätigkeit des Bodens einer Kastanienselve. Mitt. Schweiz. Anst. f. Forsth. Verschw. **37**, 400 (1961).

KISS, A.: Microbiological examination of the action of herbicides used in vineyards in soils developed on loess. Agrokemia es Talajtan **16**, 11 (1967).

KLYUCHNIKOV, L. Y., A. N. PETROVA, and Y. A. POLESKO: Effect of simazine and atrazine on microflora of sandy soil. Mikrobiologiya **33**, 992 (1964).

KOLTCHEVA, B., and U. MARKOVA: The influence of simazine on the microflora in vineyard soils. Trans. Internat. Congress Soil Sci. 8th Congress, Bucharest III, 91 (1964).

MILLER, P. M., and J. E. AHRENS: Effect of an herbicide, a nematocide and a fungicide on *Rhizoctonia* infestation of Taxus. Phytopathol. **54**, 901 (1964).

MILLIKAN, D. F., and M. L. FIELDS: Influence of some representative herbicidal chemicals upon the growth of some soil fungi. Phytopathol. **54**, 901 (1964).

NEPOMILUEV, V.: The effect of simazine on microflora and microbiological processes in a peat podsolized soil cultivated by different methods. Chem. Abstr. **65**, 7923 (1966).

—— Die Wirkung von Herbiziden auf die Mikroflora eines Rasenpodsolbodens mit Ackerbohnen. Landw. Zbl. Abt. Pflanzl. Prod. **12**, 1641 (1967).

PANTOS, G.: Die Wirkung in der Praxis angewandten Herbizide auf einige Arten der Mikroflora und Mikrofauna des Bodens, auf einige Mykorrhiza-Pilze und die biologische Inaktivierung der Herbizide. Acta Agronomica Academiae Scientiarum Hungaricae **13,** 21 (1964).

—— Study of the effect on the soil microflora of herbicides used in practical farming. Agrokemia es Talajtan **13,** 63 (1964).

——, and P. GYURKO: Die Wirkung in der Praxis angewendeter Herbizide auf ide Bodenmikroflora sowie auf einige Arten der Mikrofauna, auf einige Mykorrhizapilze weiter einige Fragen der biologischen Inaktivierung der Herbizide. Erdeszettudomanyi Kozlemenyek (Forstwiss. Mitt.) No. 2, 3 (1962).

PETZOLDT, K.: Ist wiederholte Anwendung von Simazin unbedenklich? Gesunde Pflanzen **14,** 53 (1962).

POCHON, J., P. TARDIEUX, and M. CHARPENTIER: Recherches sur les interactions entre les aminotriazines herbicides et la microflore bacterienne tellurique. C. R. Acad. Sci. (Paris) **250,** 1555 (1960).

PORSCHNEVA N. S., A. I. LESOGOROVA, and E. T. MUZYCHIN: Einfluss von Herbiziden auf die Nährstoffdynamik und Mikroflora des Bodens im Obstgarten. Chimija w sel'sk. choz. No. 3, 45 (1966).

RANKOV, V., E. ELENKOV, P. SURLEKOV, and B. VELEV: Vlijanie nekotorvch gerbitsidov na razvitie azotfikeirujuscich bakterii. Agrokhimiya No. 4, 115 (1966).

REID, J. J.: Bacterial decomposition of herbicides. Proc. N.E. Weed Control Conf. **14,** 19 (1960).

——, H. CHMIEL, D. R. DILLEY, and P. FILNER: The increase of nitrate reductase activity and protein content of plants treated with simazine. Proc. Nat. Acad. Sci. **58,** 526 (1967).

RODRIGUEZ-KABANA R., E. A. CURL, and H. H. FUNDERBUNK, JR.: Effect of atrazine on growth response of *Fusarium oxysporum* f. *vasinfectum* in sterilized soil. Phytopathol. **57,** 463 (1967).

—— —— —— —— Effect of atrazine on growth response of *Sclerotium rolfsii* and *Trichoderma viride.* Can. J. Microbiol. **13,** 1343 (1967).

—— —— —— Effect of atrazine on growth activity of *Sclerotium rolfsii* and *Trichoderma viride* in soil. Can. J. Microbiol. **14,** 1283 (1968).

SCUDDER, W. T.: Persistence of simazine in Florida mineral and organic soils. Fla. Agr. Expt. Sta. Tech. Bull. No. 657 (1963).

SHAZKINA, T. P., and W. SHARKOW: Die Wirkung von Simazin und Heptachlor auf die Bodenfauna und den Gehalt an beweglichen Nährstoffen im Boden. Chimija w. sel'sk. Choz. No. 3, 50 (1966).

SIKKA, H. C., R. W. COUCH, D. E. DAVIS, and H. H. FUNDERBUNK, JR.: Effect of atrazine on the growth and reproduction of soil fungi. Proc. S. Weed Conf. **18,** 616 (1965).

SMIRNOVA, V. I.: The effects of herbicides on the development of microflora of the corn rhizosphere. Agrobiologiya L, 88 (1963).

——, and N. N. TRETTJAKOV: Vlijanie gerbitsidov na mikrofloru rizosfery kukuruzy i biologiceskuju aktivnost'poevy. Khimija w. sel'sk. Choz. No. 3, 52 (1965).

SOSNOVSKAYA, E. A., and P. D. PASCHENKO: The effect of herbicides on microflora of the soil under maize. Weed Abstr. **16,** 185 (1967).

SPIRIDONOV, Y. J., and A. I. JAKOVLEV: Effect of sym-triazines on soil cellulolytic microorganisms. Mikrobiologija **37,** 137 (1968).

STEINBRENNER, K., F. NAGLITSCH, and I. SCHLICHT: Der Einfluss der Herbizide Simazin und W 6658 auf die Bodenmikroorganismen und die Bodenfauna. Albrecht Thaer-Archiv. **4,** 611 (1960).

STEPANOVA, Z. A.: The effect of triazines on the microflora of pale-chestnut soils. Vest. sel'-Khoz. Nauki, Mosk. **12,** 41 (1967).

SWIETOCHOWSKI, B.: Untersuchungen im Laboratorium über das Tempo der Zersetzung des Simazins im Boden. Landw. Zbl. Abt. Pflanzl. Prod. **10,** 2654 (1965).

Szabo, I.: The effect of two herbicides on the soil microflora and root nodule formation in pea. Monomagy Agrartud Fosisk Kozl **7**, 33 (1964).

Tarlapan, M. I., and V. P. Zakhariya: Einfluss von Triazinen auf den NPK Gehalt im Boden. Agrokhimiya **11**, 115 (1966).

Teuteberg, A.: Der Einfluss der Herbizidanwendung auf die Bodenmikroflora. Erwerbs-Obstbau. **1**, 10 (1967).

—— Der Einfluss einer Herbizidbehandlung auf antibiotisch wirksame Mikroorganismen des Bodens. Z. Pfl. Krank. u. Schutz. **75**, 72 (1968).

Todorovic, M., and V. Crbic: Uticaj simazina i atrazina na mikrofloru zemljista u vinogradimae bez letnje obraede. Zemljiste i biljka **14**, 37 (1965).

Tsvetkova, S. D.: The action of simazine and atrazine on agrochemical proportion of soil. Vest sel-khoz Nauki Moskva **11**, 125 (1966).

Tulabaev, B., and N. Azimbegov: Deistvie proizvodnykh triazina i mocheviny na pochvennuyu mikrofloru. Khim. Sel'skom Choz. **5** (No. 3), 103 (1967).

Uhlig, S. K.: Untersuchungen über die Wechselwirkung zwischen Chlor-*bis*-äthylamino-*s*-triazin (Simazin) und mykorrhizabildenden Pilzen. Wiss. Zeitschr. tech. Univ. Dresden **15**, 639 (1966).

—— Über den Einfluss von Chlor-*bis*-äthylamino-*s*-triazin (Simazin) auf die Bildung ektotropher Mykorrhiza bei *Picea abies* L. Karsten und *Pinus silvestris* L. Archiv für Forstwesen **15**, 463 (1966).

Voderberg, V. K.: Abhangigkeit der Herbizid-Wirkung auf Boden-mikroorganismen vom Nährsubstrat. Nachrichtenblatt f. d. Deutsch Pflanzenschutzdienst **15**, 21 (1961).

De Vries, M. L.: Effect of biocides on biological and chemical factors of soil fertility. Ph.D. Thesis, Univ. of Wis., Madison (1962).

—— Influence of triazine-herbicides on soil organisms and the decomposition of these chemicals by microorganisms. New York: *Geigy Chemical Co.*, Ardsley (1963).

Zavarzin, V. I., and T. V. Beljaeva: Deistvie gerbitsidov na soderzanie pitatel-'nykh elementov v pochve. Khim. Sel'skom Choz. **4**(10), 34 (1966).

Microbial degradation of s-triazine herbicides

By

D. D. KAUFMAN* and P. C. KEARNEY*

Contents

I. Introduction

Biodegradation is a significant factor affecting the residual life and toxicity of many pesticides[1] in soils. Soil microorganisms may act upon a pesticide in several ways. One mechanism may involve degradation with ultimate detoxication and/or metabolism of the pesticide, whereas another mechanism may involve the activation or toxication of an initially nontoxic pesticide molecule. Still another mechanism may involve the transformation of a toxic molecule into a product which exerts some beneficial influence upon higher plants, soil fauna, or microorganisms. Such reactions have been observed during the microbial degradation of a number of pesticides. Because of the public health and environmental significance of pesticides and their residues, a thorough understanding of the chemical, physical, and microbial forces acting

---

* Crops Research Division, Agricultural Research Service, *U.S. Department of Agriculture*, Beltsville, Maryland.
   [1] Common and chemical names of pesticides mentioned in text are listed in Table III.

upon these chemicals is important. The purpose of this review is to discuss the parameters involved in the microbial degradation of s-triazine herbicides.

II. Evidence for the degradation of s-triazines in microbial systems

There is considerable evidence (BARTHA et al. 1967; BRYANT 1963; BURNSIDE 1959; BURNSIDE et al. 1961; BURSCHEL 1961; CHARPENTIER and POCHON 1962; COUCH et al. 1965; DUKE 1964; FARMER et al. 1965; GUILLEMAT 1960; GUILLEMAT et al. 1960; KAUFMAN 1969; KAUFMAN et al. 1963, 1964, 1965, and 1969; KONISHI and IMANISHI 1941; MACRAE and ALEXANDER 1965; MANORIK et al. 1968; McCORMICK and HILTBOLD 1966; MICKOVSKI and VERONA 1967; MURRAY and RIECK 1968; POCHON et al. 1960; RAGAB and McCOLLUM 1961; REID 1960; SCHOLL et al. 1937; SKIPPER 1966; SKIPPER et al. 1967; SPIRIDONOV and YAKOVLEV 1967; TERMAN et al. 1964; UHLIG 1966) indicating that soil microorganisms can utilize s-triazine compounds as a source of energy. Numerous s-triazine degrading microorganisms have been isolated and identified (Table I). The degradative capacity of these organisms has been demonstrated in several ways. Growth of microbial isolates in nutrient media containing s-triazines as the sole source of carbon and/ or nitrogen has been considered criterion for degradation. Evolution of $^{14}CO_2$ from (Fig. 1), or increased oxygen consumption (Fig. 2) in

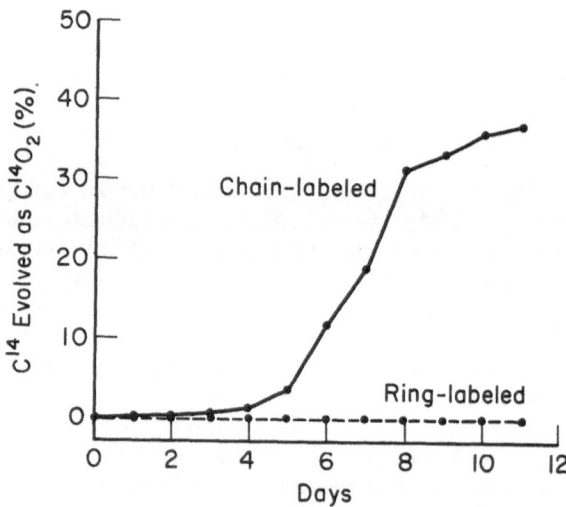

Fig. 1. $^{14}CO_2$ evolution from ring- and chain-labeled simazine treated culture solutions of *Aspergillus fumigatus* Fres (KAUFMAN et al. 1965); reprinted by permission from J. Agr. Food Chem. **13**, 238 (1965) as copyrighted by the American Chemical Society

Table I. *Soil microorganisms effective in degrading s-triazines*
(numbers indicate reference number from references section)

Organism	Simazine	Simetone	Simetryne	Atrazine	Atratone	Ametryne	Propazine	Prometone	Prometryne	Ipazine	Chlorazine	Trietazine	Dyrene
Actinomycetes													
Streptomyces sp.	13, 47, 49	—	—	—	—	—	—	—	—	—	—	—	—
Bacteria													
Achromobacter sp.	16	—	—	—	—	—	—	—	—	—	—	—	—
Acinetobacter sp.	16	—	—	—	—	—	—	—	—	—	—	—	—
Arthrobacter sp.	10, 47, 49	—	—	10	10	10	10	10	10	10	—	—	—
Bacillus sp.	—	—	—	23	—	—	—	—	—	—	—	—	—
Bacterium sp.	17	—	—	—	—	—	—	—	—	—	—	—	—
Bacterium globiforme	77	—	—	—	—	—	—	—	—	—	—	—	—
Corynebacterium sp.	77	—	—	—	—	—	—	—	—	—	—	—	—
Empedobacter sp.	17	—	—	—	—	—	—	—	—	—	—	—	—
Microbacterium sp.	17	—	—	10	10, 23	10	10	10	10	10	—	—	—
Pseudomonas sp.	—	—	—	10	10, 23	10	10	10	10	10	—	—	—
Fungi													
Aspergillus flavipes (Banier Sartory) Thom Church	47, 49	—	—	46	—	—	—	—	—	—	—	—	—
A. flavus	—	—	—	—	—	—	—	—	68	—	—	—	—
A. fumigatus Fres	44, 47, 49	44, 47, 49	44	44, 46	44	44	44	44	44	44	44	44	—
A. niger	65	—	—	65	—	—	—	—	65, 68	—	—	—	—
A. oryzae	—	—	—	—	—	—	—	—	68	—	—	—	—
A. repens	65	—	—	65	—	—	—	—	65	—	—	—	—
A. tamarii	—	—	—	—	—	—	—	—	68	—	—	—	—
A. ustus (Bain) Thom Church	13, 47, 49	—	—	46	—	—	—	—	—	—	—	—	—
Cephalosporium acremonium	65	—	—	65	—	—	—	—	65	—	—	—	—
Cladosporium herbarum	65	—	—	65	—	—	—	—	65	—	—	—	—
Curvularia lunta	—	—	—	—	—	—	—	—	68	—	—	—	—
Cylindrocarpon radicicola	29	—	—	—	—	—	—	—	—	—	—	—	—
Fusarium avenaceum	29	—	—	—	—	—	—	—	—	—	—	—	—
F. moniliforme Sheldon	47, 49	—	—	46	—	—	—	—	—	—	—	—	—
F. oxysporum	29, 47, 49	—	—	46	—	—	—	—	—	—	—	—	—
F. roseum (LK.) Snyder Hansen	20	—	—	20, 46	—	—	—	—	—	—	—	—	—
Geotrichum sp.	20	—	—	20	—	—	—	—	—	—	—	—	—
Penicillium sp.	20	—	—	20, 23	—	—	—	—	—	—	—	—	—
P. cyclopium	29, 65	—	—	65	—	—	—	—	65	—	—	—	—
P. decumbens Thom	—	—	—	46	—	—	—	—	—	—	—	—	—
P. frequentans	65	—	—	65	—	—	—	—	65	—	—	—	—
P. janthinellum Biourge	—	—	—	46	—	—	—	—	—	—	—	—	—
P. lanoso-coeruleum	29	—	—	—	—	—	—	—	—	—	—	—	—
P. luteum Zukal	—	—	—	46	—	—	—	—	—	—	—	—	—
P. purpurogenum (series) Stoll	13, 47, 49, 65	—	—	65	—	—	—	—	65	—	—	—	—
Rhizopus stolonifer (Ehr. ex Fr.) Lind	47, 49	—	—	46	—	—	—	—	—	—	—	—	—
Stachybotrys sp.	29, 47, 49	—	—	—	—	—	—	—	—	—	—	—	—
Torulopsis sp.	—	—	—	—	23	—	—	—	—	—	—	—	—
Trichoderma sp.	20	—	—	20	—	—	—	—	—	—	—	—	—
Trichoderma viride (Pers. ex Fr.)	47, 49	—	—	46	—	—	—	—	—	—	—	—	—

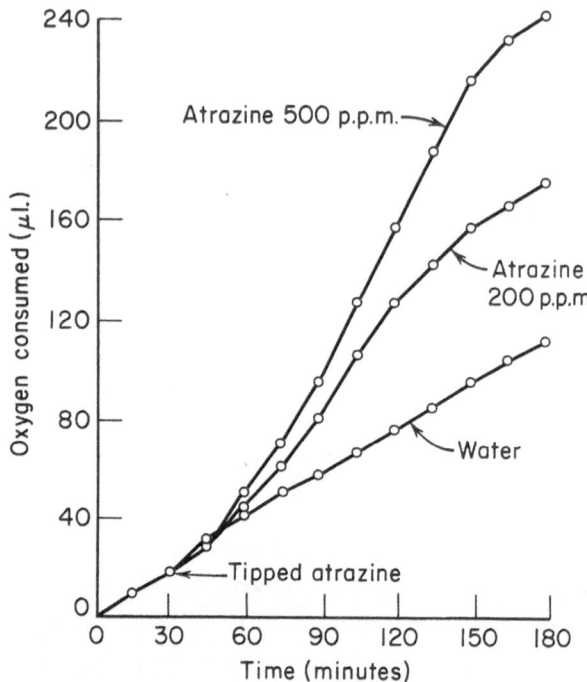

Fig. 2. Oxidation of atrazine by cells of *Bacillus* sp. grown in liquid inorganic
medium with atrazine as the carbon source (DUKE 1964)

s-triazine treated microbial systems has been interpreted by other
investigators as resulting from utilization of the s-triazine as a substrate.
Still other investigators (BURNSIDE *et al.* 1961, KAUFMAN *et al.* 1965 and
1969, MURRAY and RIECK 1968) have followed the progressive dissipation of triazines from microbial culture solutions by means of bioassays
(Fig. 3). Chemical analysis of s-triazine residues in these culture systems
indicated a direct correlation between the analytical techniques
(BRYANT 1963, CHARPENTIER and POCHON 1962, DUKE 1964, GRAMLICH
et al. 1964, GUILLEMAT 1960, GUILLEMAT *et al.* 1960, KAUFMAN *et al.*
1965, PANTOS *et al.* 1964, POCHON *et al.* 1960, RIED 1960, SKIPPER 1966,
SKIPPER *et al.* 1967, UHLIG 1966).

Some controversy exists in the literature as to whether the carbon,
nitrogen, or both, of the s-triazine molecule are utilized by soil microorganisms. Several investigators (GUILLEMAT *et al.* 1960, MICKOVSKI
and VERONA 1967, WAEFFLER 1961) have reported that although the
carbon of simazine is unavailable to soil microorganisms the nitrogen
could be utilized. GUILLEMAT *et al.* (1960) concluded that the nitrogen
in simazine could be utilized in the presence of an adequate carbon
source, whereas WAEFFLER (1961) concluded that simazine nitrogen

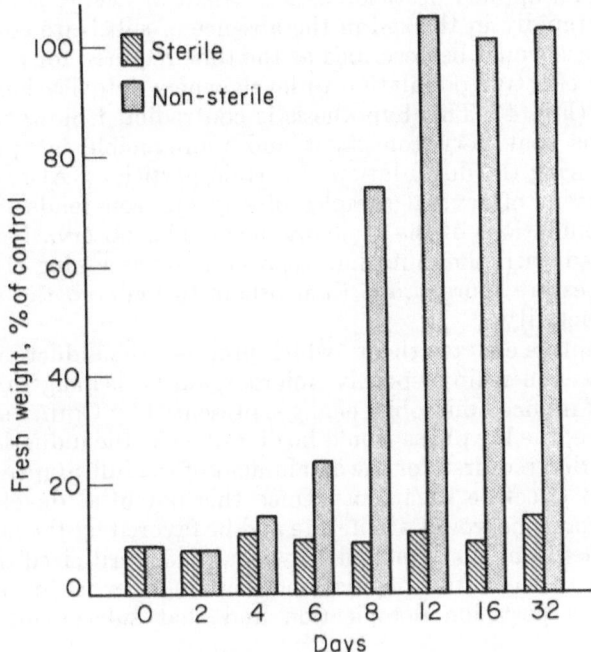

Fig. 3. Oat seedlings bioassay of sterile and nonsterile (inoculated with *A. fumigatus*) simazine-treated culture solutions added to a Hagerstown silty clay loam after various incubation periods (KAUFMAN *et al.* 1965); reprinted by permission from J. Agr. Food Chem. **13**, 238 (1965) as copyrighted by the American Chemical Society

could be utilized by soil organisms, but not as the sole source of nitrogen. Many soil microorganisms are indeed capable of utilizing the carbon of these compounds as demonstrated by the use of ^{14}C ring- and chain-labeled s-triazines (COUCH *et al.* 1965; KAUFMAN *et al.* 1963, 1964, 1965, and 1969; KEARNEY *et al.* 1964 and 1965). The ^{14}C-moiety of the side chains was used preferably as the energy source. Incorporation into protein, lipid, and to a lesser extent, nucleic acid fractions was observed primarily with ^{14}C-fragments originating from side chains (COUCH *et al.* 1965; KAUFMAN *et al.* 1963, 1964, 1965, and 1969). Other experiments have confirmed the utilization of s-triazine nitrogen by soil microorganisms (HAUCK and STEPHENSON 1964, KAUFMAN *et al.* 1965, KONISHI and IMANISHI 1941, MICKOVSKI and VERONA 1967, SCHOLL *et al.* 1937, TERMAN 1964).

The mechanism through which soil microbes develop the capacity to degrade pesticides is not completely understood. AUDUS (1960) discussed two major possibilities involving either (1) chance mutation or (2) adaptive enzymes. Mutants having the ability to degrade and/or metabolize an applied pesticide as a nutrient or energy source should proliferate rapidly in the soil in the absence of substrate competition. The lag phase would be described as the time required for the development of an effective population to levels where detoxication rates are detectable (Fig. 4). This hypothesis is contradicted, however, by the observations that very consistent and reproducible lag periods are observed during the degradation of certain pesticides. Also, independent enrichment of several samples of any one soil tends to produce effective populations of the same organism. This observation can only be reconciled with the mutation hypothesis by assuming that certain soil microbes are more prone than others to undergo the mutations leading to activity.

The adaptive enzyme theory which proposes the induction of adaptive enzymes in certain responsive microorganisms is in agreement with the views of induced microbial changes presented by Cohn and Monod (1953). Here, the lag phase would be described as the induction period, *i.e.*, that period required for the attainment of the full adaptive enzyme potential in effective organisms. Once this potential developed, the effective organisms would proliferate and be favored by the lack of substrate competition. Walker and Newman (1956) criticized this theory on the basis that effective populations would not persist in soils following complete pesticide detoxication and that subsequent pesticide

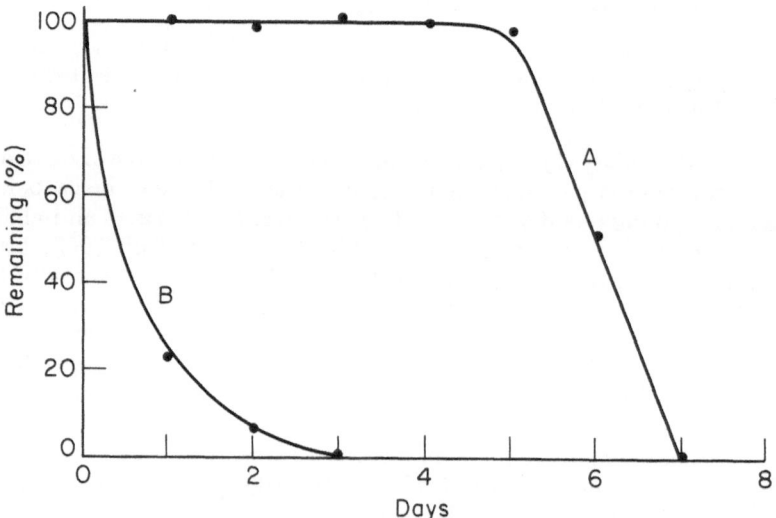

Fig. 4. Microbial degradation of a pesticide showing: A = a 5-day lag period followed by rapid dissipation of the pesticide, and B = the rapid dissipation of a second addition of the same pesticide, indicating the elimination of the initial lag phase (Kaufman 1969)

additions would also require a lag phase. Conversely, mutants would retain their effective characteristics in the absence of the pesticide substrate. The observations that pure cultures of pesticide degrading organisms quickly lose their effective ability to utilize certain pesticides when cultured on a less complex substrate such as glucose (HIRSCH and ALEXANDER 1960, KAUFMAN 1969) lends support to the induction or adaptive enzyme theory.

There is evidence that chloro-s-triazine herbicides are decomposed by a different mechanism, that of constitutive microbial systems or organisms which require no adaptation. Very little evidence has accumulated indicating that a lag phase occurs during the initial stage of chloro-s-triazine degradation. Although several investigators have used an enrichment method for the isolation of organisms effective in degrading chloro-s-triazines, few have seen the typical lag period followed by a period of rapid degradation. In contrast, most investigators have observed an immediate, but slow, continuous rate of degradation (Fig. 5), such as one might expect in a rate-limited, first-order reaction. Other factors must, therefore, limit the microbial degradation of chloro-s-triazine herbicides.

The build-up of effective organisms in soil by repeated applications of certain pesticides fails to occur with s-triazines. SHEETS and DANIELSON (1960) reported that rather than an increase in inactivation of these compounds, the soil retains the same inactivation intensity over a long period of time. Patterns of s-triazine herbicide persistence

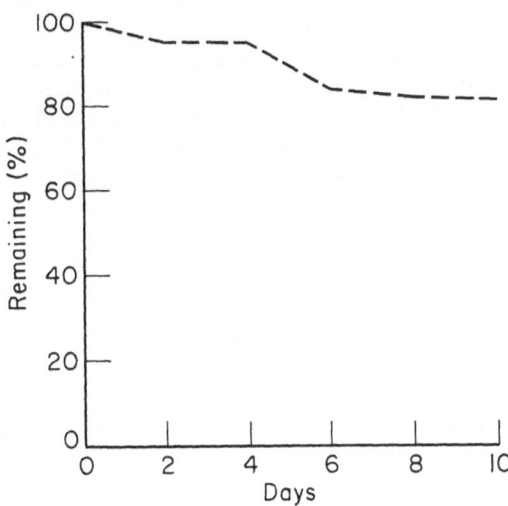

Fig. 5. Microbial degradation of simazine in a soil-solution enrichment culture system (KAUFMAN 1969)

and breakdown and its association with added organic matter suggests an incidental or nonpreferential involvement of these herbicides in microbial metabolism (McCORMICK and HILTBOLD 1966, SHEETS 1962, SHEETS and DANIELSON 1960, SHEETS and SHAW 1963). The presence of constitutive microbial systems operative in chloro-s-triazine degradation has been observed (KAUFMAN 1969). Certain soil microbes, repeatedly subcultured on nutrient media without triazines, readily metabolize chloro-s-triazines reintroduced to the growth medium as a sole or supplemental source of energy and carbon.

Adaptive mechanisms, however, may be involved in the microbial degradation of methylthio-s-triazines. DUKE (1964) followed the degradation of atrazine, atratone, and ametryne by measuring carbon dioxide evolution from treated and non-treated soils during a 25-day incubation period. Atrazine and atratone degradation occurred readily under all experimental conditions. Ametryne decomposition did not occur during the 25-day incubation period in previously untreated sterile soils treated with ametryne and sterile water (Fig. 6). When a viable soil extract, prepared from an ametryne enriched soil, was added to the previously untreated soils, ametryne degradation was rapid following an initial lag phase. Therefore adaptation and enrichment processes may be involved in the microbial degradation of methylthio-s-triazines.

Long lag periods are necessary for the development of microbial populations effective in degrading certain pesticides (AUDUS 1960). Extended lag periods, however, do not explain the persistence, or time concentration relations of certain persistent organic pesticides such as the s-triazines. BURSCHEL (1961) observed that the time required for inactivation and decomposition of simazine was a function of its soil concentration. The rate-limiting process is physical or chemical in nature rather than biological. The absence of a rapid proliferation of effective organisms and the occurrence of a slow rate of breakdown might be due to the effects of adsorption-desorption equilibria on availability of the s-triazines for uptake and metabolism by soil microorganisms. The significance of environmental factors in microbial degradation of s-triazine herbicides will be discussed in later sections. Whatever the mechanisms or parameters involved, ample evidence exists that soil microorganisms are capable of degrading and/or metabolizing s-triazine herbicides.

III. Mechanisms of chloro-s-triazine degradation

a) Dealkylation

Dealkylation appears to be the major mechanism involved in the microbial degradation of chloro-s-triazines. Indirect evidence for the oxidative dealkylation has been reported from experiments in which

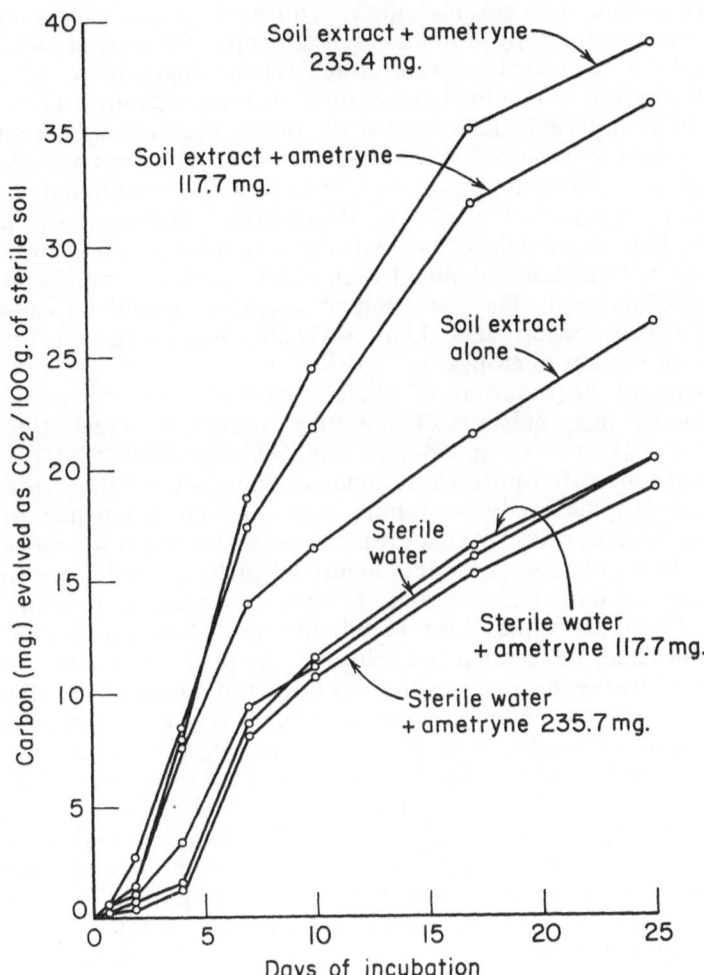

Fig. 6. Carbon dioxide evolution from sterile soil treated with ametryne and inoculated with soil extract or sterile water (DUKE 1964)

evolution of $^{14}CO_2$ occurred from microbial culture solutions containing ^{14}C-chain-labeled s-triazines (COUCH et al. 1965; KAUFMAN et al. 1963, 1964, 1965, and 1969; SKIPPER 1966; SKIPPER et al. 1967). Investigations utilizing pure cultures of the soil fungus *Aspergillus fumigatus* Fres have revealed the degradative pathway of simazine and atrazine. Two metabolites were found when this organism was cultured in the presence of ^{14}C-ring- and chain-, or ^{36}Cl-labeled simazine (KAUFMAN et al. 1963, 1964, and 1965; KEARNEY et al. 1965). The structure of one of these metabolites was identified as the 2-chloro-4-amino-6-ethylamino-s-triazine by thin-layer chromatography, infrared assay, nuclear magnetic resonance spectroscopy, and mass spectroscopy. The second metabolite retained the ^{36}Cl attached to the intact triazine ring, but was minus the two ethyl groups. The final structure is still undetermined (Kaufman et al. 1963 and 1965, Kearney et al. 1965). Preliminary chromatographic evidence indicated that the second metabolite was probably not the 2-chloro-4,6-diamino-s-triazine. Some evidence accumulated in the preceding experiments indicate dealkylation followed by partial deamination. Ammelide (2,4-dihydroxy-6-amino-s-triazine), was identified as a fungal metabolite by chromatographic comparison with authentic material. The detection of ammelide would be consistent with the observation that chloro-s-triazines can serve as a nitrogen source for certain microbes.

Microbial degradation of ^{14}C-ethylamino- and ^{14}C-ring-labeled atrazine by pure cultures of soil fungi occurs by much the same mechanism (Kaufman and Blake 1969). Evolution of ^{14}CO$_2$ occurred only from inoculated culture solutions containing chain-labeled atrazine. Chromatographic analysis of both ring- and chain-labeled atrazine culture solutions revealed that atrazine was metabolized to two or more degradation products. Products identified included 2-chloro-4-amino-6-isopropylamino-s-triazine and 2-chloro-4-ethylamino-6-amino-s-triazine. An unidentified third metabolite was also detected in both ring- and chain-labeled culture solutions; its presence in chain-labeled atrazine culture solutions precludes its being the completely dealkylated and dehalogenated end product ammelide, observed in the degradation of simazine. At least one additional metabolite, therefore, is involved if the degradation of atrazine is to proceed to the same product ammelide. Additional suspected intermediates would include 2-chloro-4-hydroxy-6-ethylamino-s-triazine, 2-hydroxy-4-amino-6-ethylamino-s-triazine, and 2,4-dihydroxy-6-ethylamino-s-triazine. On the basis of information obtained from microbiological investigations the following (Fig. 7) degradative pathway could be proposed for chloro-s-triazines.

Recently, Plimmer et al. (1968) demonstrated that the oxidative dealkylation of several s-triazine herbicides could be mediated by free radical reactions in model systems. These reactions may be analogous to reactions mediated by microbial enzymatic systems. Hydroxy or perhydroxy radicals generated from Fenton's reagent, Udenfriend's reagent, or the copper-ascorbate system caused loss of ^{14}CO$_2$ from chain-labeled (^{14}C-ethyl) atrazine or hydroxyatrazine, while the ring remained intact (Fig. 8). Three products in addition to atrazine were isolated and identified by mass spectral analysis. The two monalkyl derivatives of atrazine (2-chloro-4-ethylamino-6-amino-s-triazine and 2-chloro-4-amino-6-isopropylamino-s-triazine) plus the completely dealkylated 2-chloro-4,6-diamino-s-triazine were identified as major products. The distribution of products in the reaction mixture indicate that oxidative removal of the isopropyl group may be slower than the

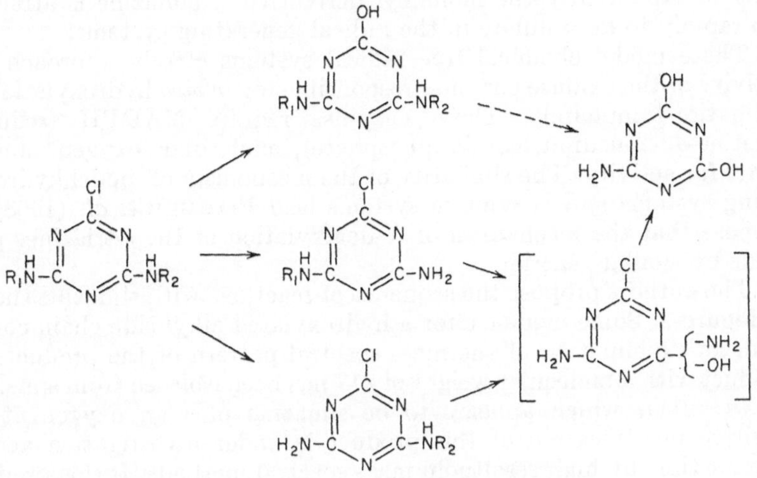

Fig. 7. Microbial degradation pathway proposed for chloro-*s*-triazines

ethyl group. This latter observation would correlate with observations made with several soil fungi (KAUFMAN and BLAKE 1969). *A. fumigatus* generally removed the ethyl side chain in preference to the isopropyl group, as did several other soil fungi. Several soil fungi, however, removed the isopropyl group in preference to the ethyl group.

Similar reactions (PLIMMER *et al.* 1968) performed with simazine disclosed only one major product, *i.e.*, the 2-chloro-4,6-diamino-*s*-

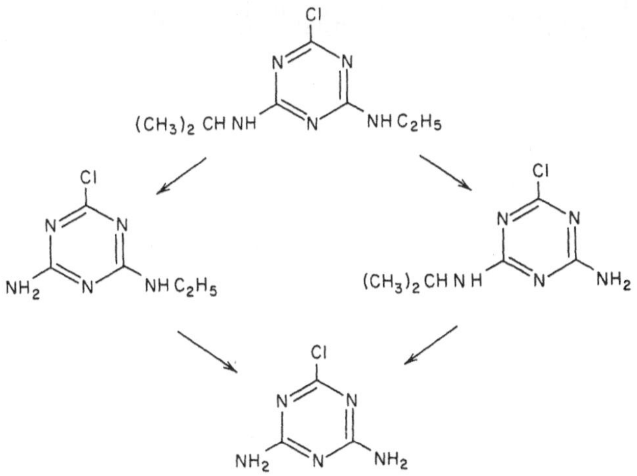

Fig. 8. Dealkylation of atrazine by hydroxyl radicals (PLIMMER *et al.* 1968)

triazine. Apparently the monalkyl derivative of simazine is attacked too rapidly to accumulate in the radical generating systems.

These model chemical free radical systems closely approach the activity of the oxidase enzymes responsible for *in vivo* hydroxylation of aromatic compounds. These enzymes require NADPH (reduced nicotine-adeninedinucleotide phosphate), molecular oxygen, and a metal for activity. The similarity of the mechanism of model hydroxylating systems and enzymatic systems lead PLIMMER *et al.* (1968) to propose that the mechanism of *N*-dealkylation of the *s*-triazines proceeds by similar reactions.

The authors propose the sequence of reactions with simazine shown in Figure 9. Some evidence for a hydroxylated alkyl side chain comes from an examination of the mass spectral pattern of the products. A product with a molecular weight of 217 has been isolated from simazine (m.wt. 211), which appears to be simazine plus an oxygen atom. Positive identification of this product is under investigation at the present time by high resoltuoin mass spectral methods. If this product can be confirmed, it would represent an intermediate between simazine and its dealkylated product.

Degradation of *s*-triazines by dealkylation in biological systems is not particularly surprising. Cyanuric acid (2,4,6-trihydroxy-*s*-triazine) was the principal urinary metabolite of mice injected with the alkylamino-*s*-triazine TEM (2,4,6-triethyleneimino-*s*-triazine) (NADKARNI *et al.* 1954). Formoguanamine (2,4-diamino-*s*-triazine) is converted in

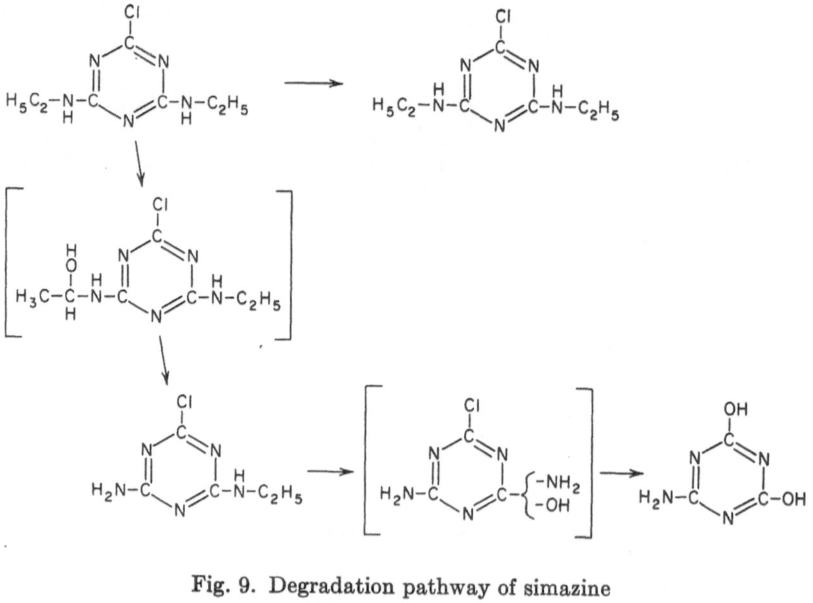

Fig. 9. Degradation pathway of simazine

orally dosed humans to ammelide (LIPSCHITZ and STOKEY 1948). Benzylidene biuret (2-phenyl-4,6-dihydroxy-s-triazine) is converted to cyanuric acid in dogs (WILLIAMS 1959). Dealkylation has also been observed with simazine, atrazine, and prometryne in higher plants (FUNDERBURK and DAVIS 1963, HURTER 1966, MUELLER and PAYOT 1965, SHIMABUKURO 1966 and 1967, SHIMABUKURO et al. 1966).

Dealkylation of s-triazine herbicides by microorganisms does not necessarily insure detoxication. Dealkylation of atrazine resulted in the formation of 2-chloro-4-amino-6-isopropylamino-s-triazine and 2-chloro-4-ethylamino-6-amino-s-triazine (Fig. 10) (KAUFMAN and BLAKE 1969). The former compound was almost as toxic to oats as atrazine at low equimolar concentrations ($4.64 \times 10^{-5}M$), whereas the latter compound was toxic to oats only at high concentrations ($23.2 \times 10^{-5}M$). The toxicity of these dealkylated metabolites has also been reported by others (GYSIN and KNUESLI 1960, KNUESLI et al. 1969, SHIMABUKURO 1966 and 1967, SHIMABUKURO et al. 1966). The presence of toxic dealkylated metabolites could account for the inability of some investigators to observe progressive dissipation of triazines when bioassaying inoculated culture solutions (Fig. 11) (BURNSIDE 1959, BURNSIDE et al. 1961, KAUFMAN and BLAKE 1969).

Monodealkylation of dialkylamino-s-triazines by soil microorganisms probably explains the increased phytotoxicity observed with time. SHEETS et al. (1962) observed an increased toxicity with time in soils

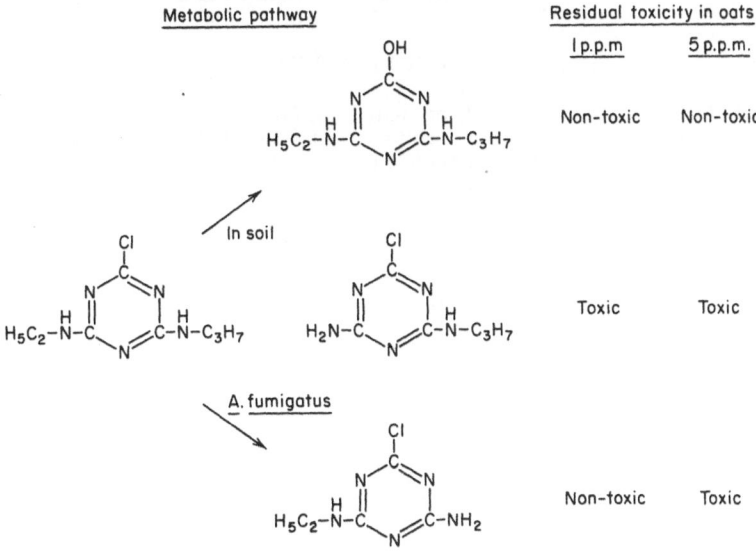

Fig. 10. Atrazine metabolism to toxic and nontoxic metabolites (KAUFMAN and BLAKE 1969)

treated with chlorazine. Progressive dealkylation of chlorazine could result in the formation of trietazine and, finally, simazine (Fig. 12). Similarly, the increased phytotoxicity of ipazine to pine seedlings 80 days after application could be explained by the conversion of ipazine to atrazine (Kozlowski 1965).

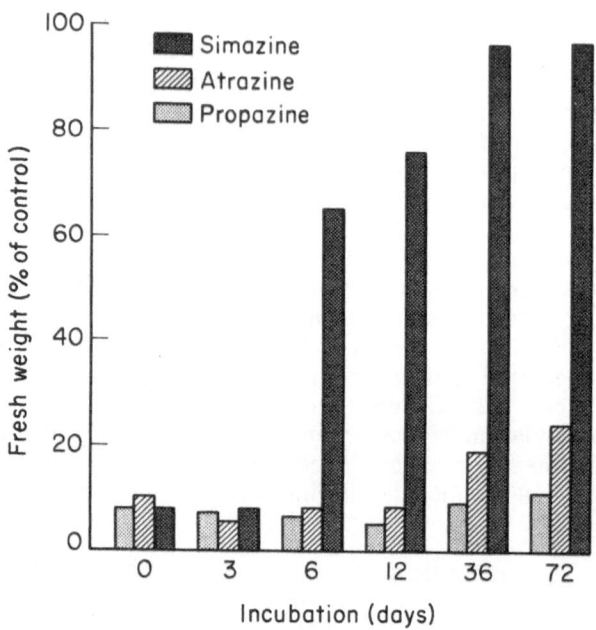

Fig. 11. Oat seedling bioassay of inoculated (*A. fumigatus*) simazine, atrazine, or propazine-treated culture solutions added to a Lakeland sandy loam soil after various incubation periods (Kaufman and Blake 1969)

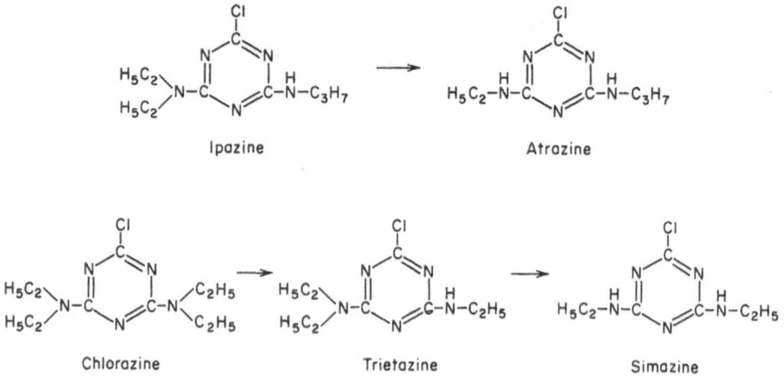

Fig. 12. Monodealkylation of several dialkylamino-s-triazines

b) Hydroxylation

Hydroxylation is a common degradative mechanism of microorganisms. Hydroxylation is frequently an initial reaction in the microbial degradation of halogenated pesticides. Although chlorine has no known essentiality in the nutrition of fungi (FOTHERGILL and ASHCROFT 1955), chlorinated metabolites are not uncommon (COCHRANE 1958).[2] Only one example is presently known where a soil microorganism, *Fusarium roseum* (LK, Snyder and Hansen), hydrolyzed the chloro-*s*-triazine atrazine to its corresponding hydroxy analogue. The results indicated a rapid degradation of atrazine to hydroxyatrazine (COUCH et al. 1965). Very little is known about the degradation of hydroxylated-*s*-triazines. A more rapid evolution of $^{14}CO_2$ from soil treated with ^{14}C-ring-labeled hydroxyatrazine than from soil treated with ^{14}C-ring-labeled atrazine has been observed (SKIPPER 1966, SKIPPER et al. 1967). Preliminary investigations in our laboratory indicate that hydroxy-*s*-triazine degradation by soil microorganisms may be more problematic than chloro-*s*-triazine degradation. Dealkylation of hydroxy compounds has been observed in higher plants (HURTER 1966).

c) Ring cleavage

Evolution of $^{14}CO_2$ from microbial systems treated with ^{14}C-ring-labeled *s*-triazines has been reported (COUCH et al. 1965; HAMILTON and MORELAND 1963; KAUFMAN et al. 1963, 1964, 1965, and 1969; MACRAE and ALEXANDER 1965; McCORMICK and HILTBOLD 1966; RAGAB and McCOLLUM 1961). These observations indicate that ring cleavage may occur, however limited it might be. In nearly all cases only low levels of $^{14}CO_2$ (zero to four percent) have been evolved from microbial systems treated with ring-labeled *s*-triazines. This would indicate that the *s*-triazine ring structure is somewhat resistant to microbial attack.

Environmental conditions, including soil type, temperature, moisture level, aeration, and supplemental energy sources influence the capacity of microbial systems to liberate $^{14}CO_2$. These factors may account for the apparent differences reported for the same *s*-triazines (HAMILTON and MORELAND 1963, MACRAE and ALEXANDER 1965, McCORMICK and HILTBOLD 1966, RAGAB and McCOLLUM 1961, WAGNER and CHAHAL 1966). The duration of the incubation period is also important when attempting to determine a soil's degradative capacity (KNUESLI et al. 1969). Extended experiments have indicated

---

[2] Editor's note: However, see HYLIN, SPENGER, and GUNTHER: Potential interferences in certain pesticide residue analyses from organochlorine compounds occurring naturally in plants. Residue Reviews **26**, 127 (1969).

that a long lag period follows the initial period of rapid $^{14}CO_2$ evolution, after which a second period of $^{14}CO_2$ evolution begins.

Complete mineralization of s-triazine compounds does occur, as evidenced in reports investigating their potential as slow-release nitrogen fertilizers (HAUCK and STEPHENSON 1964, KONISHI and IMANISHI 1941, SCHOLL et al. 1937). If dealkylation, deamination, or both occur in soil then ammelide, cyanuric acid, melamine (2,4,6-tri-amino-s-triazine), and ammeline (2-hydroxy-4,6-diamino-s-triazine) are potential degradative products of s-triazine herbicides.

The mechanism by which ring cleavage of s-triazines occurs is not known. Ring cleavage apparently occurs only after hydroxylation. Whether hydroxylation must accompany dechlorination, prior to ring cleavage, or whether hydroxylation subsequent to deamination would suffice prior to ring cleavage is not known. A pathway has been proposed for cleavage of the s-triazine ring, permitting the liberation of the number 2 carbon as carbon dioxide (GYSIN and KNUESLI 1960, MONTGOMERY and FREED 1964). Cleavage of the ring at this point would result in the formation of a substituted biguanide compound in addition to carbon dioxide. If the substituted biguanide were formed, it probably would undergo hydrolysis to a substituted biuret or a substituted guanidine and a substituted urea. Although certain biguanides, guanidines, and biuret compounds have been investigated for antibiotic possibilities, they also are vulnerable to microbial degradation.

Since little or no correlation has been observed between herbicide dissipation and carbon dioxide evolution, it seems likely that complete ring cleavage is not a major mechanism of s-triazine degradation in soil.

d) Conjugate formation

The degradative mechanisms previously described are those involving transformations of the parent compound. Conjugate formation occurs through the condensation of exogenous or endogenous substances with naturally occurring compounds. The naturally occurring compounds most commonly involved are carbohydrates, amino acids, or nucleoproteins. Conjugates are generally water-soluble, highly polar products, and are common to plant and animal systems. Few reports have indicated the role of conjugate formation in the inactivation of s-triazines. The formation of a conjugate with hydroxypropazine derived from propazine or prometryne in rats has been suggested (KNUESLI et al. 1969). The presence of water-soluble metabolites of atrazine in corn and sorghum could also be indicative of conjugate formation (SHIMABUKURO and SWANSON 1969). Although microbially produced s-triazine-conjugates have not been demonstrated the reactivity of the s-triazine fungicide dyrene with microbial metabolites containing aliphatic amino, aromatic hydroxyl, and sulfhydryl groups is well known (BURCHFIELD and STORRS 1957).

IV. Mechanisms of methoxy-s-triazine biodegradation

Knowledge concerning the biochemical hydrolysis of methoxy-s-triazines is very limited. Perhaps a reason for this is their limited use as nonselective weed control chemicals. Hydrolysis to hydroxylated compounds has been observed in higher plants (GYSIN and KNUESLI 1960, MUELLER and PAYOT 1965). In similar systems only 20 percent hydrolysis of methoxy compounds occurred as compared to 100 percent hydrolysis of chloro-s-triazines, indicating that the methoxy compounds were much more resistant to hydrolysis. KAUFMAN (1969) observed that the methoxy-s-triazines were most resistant to degradation by fungal systems used for studying chloro-s-triazine degradation.

DUKE (1964) examined the microbial degradation of atratone in soil and pure culture systems. Carbon dioxide evolution was used as an indication of degradation. In soil systems he observed significant interactions between atratone application rates, period of incubation and inoculum density introduced into the system. The quantity of carbon dioxide evolved from soils treated with low rates of atratone and inoculated with soil extract were significantly greater than all other treatments. Carbon dioxide evolution from soils treated with only soil extract was also significantly greater than treatments receiving high atratone application rates with soil extract. Similar results were observed in treatments receiving only low and high rates of atratone, i.e., a significantly greater carbon dioxide evolution was measured from soils receiving low rates of atratone. In pure culture investigations with a *Pseudomonas* sp., DUKE (1964) observed that peptone and yeast extract were more readily oxidized than atratone. Oxidation of peptone and yeast extract began immediately following their addition to the medium whereas atratone caused an immediate slight decrease in respiration. Therefore, there may exist some degree of atratone toxicity toward soil microorganisms. This inhibition may contribute to the long life of atratone in soils.

Whether or not methoxy-s-triazine degradation would proceed by demethoxylation, dealkylation, deamination, or ring cleavage, or all of these mechanisms, and in which order, is not known. Although degradation of methoxy-s-triazines by *Aspergillus fumigatus* Fres was observed (KAUFMAN 1969), no attempt was made to elucidate the mechanisms involved. Demethoxylation of pesticides is known to occur as a result of microbial activity (KEARNEY et al. 1967, KEARNEY and HELLING 1969). Oxidative demethylation and subsequent ring cleavage of vanillic acid are catalyzed by a cell-free system of *Pseudomonas fluorescens* (CARTWRIGHT and BUSWELL 1967, CARTWRIGHT and SMITH 1967). Demethylation required reduced glutathione (GSH) and NADPH. The methyl group was removed as formaldehyde and subsequently oxidized to formate and carbon dioxide.

V. Mechanisms of methylthio-s-triazine biodegradation

Although all available information strongly indicates that the chloro-s-triazines are chemically degraded in soils, there is good reason to believe that the methylmercapto-s-triazines are altered metabolically. DUKE (1964) obtained evidence indicating that microbial degradation of ametryne was an adaptive process. Bioassay of microbial culture solutions treated with prometryne indicated degradation of prometryne by Aspergillus niger, A. tamarii, A, flavus and A. oryzae (MURRAY and RIECK 1968). The methylthio (—S—CH₃) group is found in several insecticides as well as herbicides. Direct evidence for sulfoxidation by microsomal enzymes comes largely from animal and plant systems. Nevertheless, comparative biochemistry frequently holds true for reactions catalyzed by soil microorganisms.

Metabolic studies with pea plants (Pisum sativum) using ¹⁴C-ring-labeled prometryne revealed a number of oxidized products in addition to hydroxypropazine (MUELLER and PAYOT 1965). These oxidation products were subsequently identified as the corresponding sulfoxide and sulfone derivatives of prometryne and a pathway was proposed (Fig. 13). Therefore, instead of a direct displacement of Cl⁻, as in the

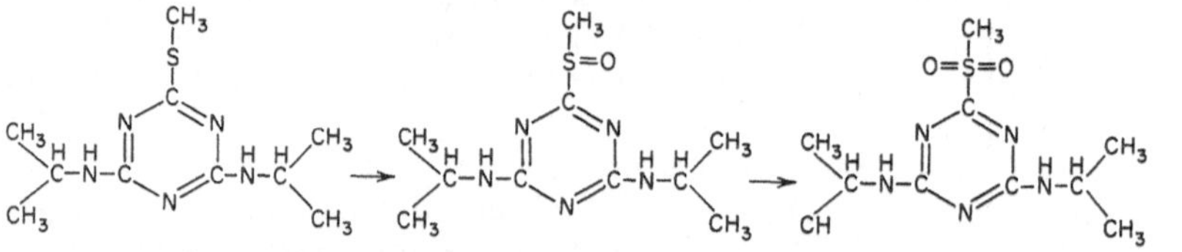

Fig. 13. Sulfoxidation of prometryne

case of propazine, prometryne apparently undergoes two intermediate oxidation steps prior to forming hydroxypropazine.

The hydrolytic properties of prometryne and its two oxidation products in acid, neutral, and basic conditions have been studied by DELLY (1963). His results are illustrated in Table II which shows the time required for hydrolysis of prometryne, the sulfoxide, and the sulfone in 0.1N hydrochloric acid, water, and 0.1N sodium hydroxide. The parent compound is very stable in water. By comparison, under acid, neutral, and alkaline condition, the sulfone is more rapidly hydrolyzed than the sulfoxide, six, 50, and ten times more rapidly, respectively.

In a comparison of the hydrolysis rates of a number of chloro-, methylthio-, and methoxy-s-triazines in normal acid, the ease of hydrolysis followed the order Cl > CH₃—S > CH₃—O. Therefore, the chloro compounds were the most readily displaced, followed by the methylthio-s-triazines (SHEETS and KEARNEY 1963); this will be dis-

Table II. *Hydrolytic behavior of prometryne and its derivatives*
(DELLY 1963)

Hydrolytic agent (25°C.)	Time required for hydrolysis of		
	—CH₃S	—CH₃SO	—CH₃SO₂
0.1N HCl	22 days	96 minutes	16 minutes
H₂O	About 500 years	150 days	3 days
0.01N NaOH	About 30 years	20 minutes	2 minutes

cussed more fully in relation to the soil persistence of the methylthio-s-triazines in subsequent sections.

WHITENBERG (1965) presented indirect evidence for the formation of the prometryne oxidation products in soils. Small amounts of hydroxy-propazine in cotton plants grown in ^{14}C-ring-labeled prometryne-treated soil, but none in plants grown in solution culture with prometryne. He postulated that hydroxypropazine, in soil-grown plants, may have been adsorbed from the soil, or the sulfoxide or sulfone may have been absorbed and further degraded to hydroxypropazine. In contrast to the findings in pea plants, neither oxidation product was detected in cotton plants. More indirect evidence for metabolism of prometryne in soils comes through consideration of the hydrolytic behavior of this compound reported by DELLY (1963) and illustrated in Table II, and the reported persistence of prometryne under field conditions. Hydrolysis at near neutral conditions is slow, whereas prometryne persistence in soil is fairly short by comparison to other triazines. As also pointed out previously the rate of chemical hydrolysis of propazine is more rapid than prometryne, and yet the reverse is true for soil persistence under certain conditions.

Direct evidence for prometryne oxidation products in soils has been found in our own laboratories. KEARNEY (1969) has studied the products resulting from ring-labeled prometryne degradation in Hagerstown silty clay loam maintained at constant moisture and temperature for 15 months. Trace amounts of the sulfoxide and sulfone were detected by gas and thin-layer chromatographic methods. No prometryne was recovered, but two additional unknowns were detected on thin-layer plates. Most of the radioactivity was recovered in acid and base extracts of soils previously extracted with four organic solvents. The humic acid fraction contained about 17 percent of the total activity recovered.

A number of pesticides contain methylthio-groups as part of their structure. The phosphate insecticide fenthion is metabolized to the sulfoxide as a major metabolite when applied to bean plants (FRANCIS and BARNES 1963, NIESSEN *et al.* 1962); the sulfone was also detected. The methyl carbamate insecticide Mesurol is also oxidized to the corresponding sulfoxide and sulfone in plants (NIESSEN and FREHSE 1963). Thus, parallels exist between the oxidation of prometryne in plants and soils and the oxidation of methylthio-insecticides in both systems.

Another route of metabolism of prometryne involves a reductive pathway. BOEHME and BAER (1967) have isolated the free mercapto derivative of prometryne as one of the urinary metabolites from rats and rabbits (Fig. 14). In addition to the mercaptan, the corresponding

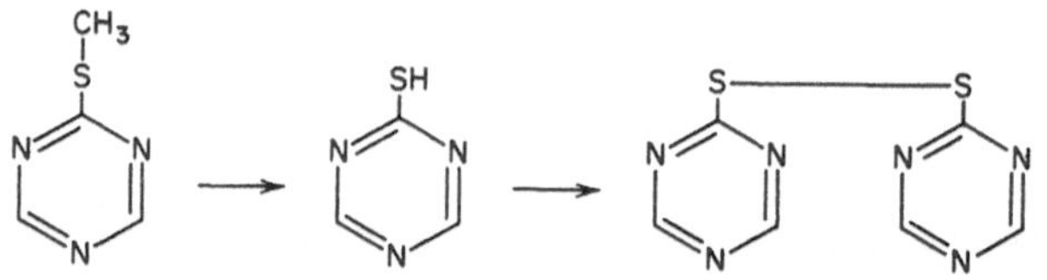

Fig. 14. Reductive metabolism of prometryne (BOEHME and BAER 1967)

disulfide was also detected. This reaction is apparently carried out by microorganisms in the intestine of the animal. The anaerobic metabolism of methionine, a thioether containing amino acid, by certain strains of *Pseudomonas* yields ammonia, α-ketobutyric acid, and methyl mercaptan. The enzyme methionase, which gives rise to methyl mercaptan, has been studied in cell-free systems and pyridoxal phosphate appears to be the coenzyme (MEISTER 1965). Another bacterial system converts methionine to α-amino-butyric acid and methyl mercaptan (MEISTER 1965). This system is present in *Escherichia coli* and requires the presence of adenosine triphosphate and pyridoxal. It should be noted with prometryne, however, that the sulfur atom remains bonded to the *s*-triazine ring in the case of prometryne and methyl mercaptan is not formed. No evidence for the mercaptan and disulfide derivatives of prometryne have been reported in soils.

VI. Effect of chemical structure on *s*-triazine biodegradation

Chemical structure affects the biodegradation of pesticides (ALEXANDER and ALEEM 1961, AUDUS 1960, KAUFMAN 1965). Chemical structure reportedly affects the nitrification rate of *s*-triazine nitrogen (HAUCK and STEPHENSON 1964, TERMAN 1964). Nitrification of cyanuric acid, ammeline, ammelide, and melamine varied inversely with the number of amino groups on the *s*-triazine ring, resulting in a decreased availability of nitrogen content of these compounds.

BRYANT (1963) observed a correlation between the alkylamino

substituent groups on the s-triazine ring and suitability as a substrate for various bacteria. Compounds containing a diethylamino group were preferred by a *Pseudomonas* sp., whereas an *Arthrobacter* sp. preferred the ethylamino substituted compounds. The isopropylamino group was utilized equally well by both genera. KAUFMAN (1969) observed a correlation between the size of the alkylamino substituent and the weight of oven dried mycelia produced by *Aspergillus fumigatus* when growing on several s-triazines as sole source of carbon. Mycelial weights increased as the number and size of substituents increased.

HAUCK and STEPHENSON (1964) suggested that the symmetrical resonating structure of the s-triazine molecule contributed to differences in degradation rates. Asymmetrical chlorotriazines were degraded more quickly than the corresponding symmetrically-substituted chlorotriazines. They suggested that polymerization resulting in an increased symmetry would lead to stability, whereas labile products of biochemical transformations tend toward asymmetry.

Methoxy- or methylthio-substituted triazines were degraded faster than the corresponding chloro-s-triazines (HAUCK and STEPHENSON 1964). Residual phytotoxicities of s-triazines, however, appears to follow the order of methoxy- > methylthio- > chloro-s-triazine (HARRIS et al. 1969). Although the 2-chloro derivatives have greater initial phytotoxicities, the 2-methoxy derivatives are less susceptible to degradation. The methylthio-derivatives appear intermediate in their residual phytotoxicities (SHEETS et al. 1962). If one assumes that microbial degradation is the only factor operative in s-triazine degradation in soil, the pattern of resistance to degradation would appear identical to the persistence pattern, *i.e.*, methoxy- > methylthio- > chloro-s-triazines. Such was the case when the utilization of methoxy-, methylthio- and chloro-analogues of simazine, atrazine, and propazine by *A. fumigatus* was examined (KAUFMAN 1969). Chloro-s-triazines were most susceptible to utilization, whereas methoxy-s-triazines were most resistant, and methylthio-s-triazines were intermediate in their susceptibility to utilization. In actual soil systems, however, one cannot consider a single operative factor to the complete exclusion of others. As indicated earlier, accumulating evidence increasingly points to a chemically mediated degradation of chloro-s-triazines and a biochemically mediated degradation of methylthio- and perhaps methoxy-s-triazines.

VII. Microbial degradation in soil systems

The ultimate question the microbiologist and biochemist must face is whether or not the degradative mechanisms observed *in vitro* are applicable or even relevant to *in vivo* systems. At first glance one might conclude that there is considerable evidence substantiating the biodegradation of s-triazine herbicides in soil systems. Degradation of s-triazines has been followed in soil by several methods, *i.e.*, extraction and chemical analysis, bioassays of residual phytotoxicity, oxygen uptake, and evolution of carbon dioxide. Studies comparing sterile vs. non-sterile soils have been conducted using these techniques. Generally, the results have been in agreement, indicating that soil factors influencing microbial activity also influence residual life of s-triazines in soil. When examining these environmental factors one must consider the effect these factors have on the soil microorganism and the chemical independently, as well as simultaneously. Chemical and physical factors which promote or inhibit microbial activity may also affect the availability or chemical degradation of the pesticide independent of the microbial effects. Thus, the increased decomposition of a pesticide in soil may not necessarily result from increased microbial activity *per se*, but merely reflect an increased solubility, reactivity, or availability of the chemical to the organism.

Soil microorganisms are generally more active in soils having high organic matter content than in soils having low organic matter contents. Increases in soil organic matter content decreased residual phytotoxicity of s-triazines (BURSCHEL 1961, HOLLY and ROBERTS 1963, SHEETS and SHAW 1963). Addition of organic matter to soils stimulated s-triazine degradation (McCORMICK and HILTBOLD 1966, SKIPPER 1966, WAGNER and CHAHAL 1966). Correlation between soil organic matter and percent simazine adsorbed is highly significant (HARRIS and SHEETS 1965) yet little is known concerning the effect simazine adsorption on soil organic matter has on the microbial degradation of simazine. Under pure culture conditions where adsorption is negligible, microbial degradation of s-triazines is increased in the presence of supplemental carbon sources (KAUFMAN *et al.* 1963 and 1965, SKIPPER 1966).

Microbial activity increases with increased temperature (ALEXANDER 1961). BURSCHEL (1961) observed that a temperature increase from 8.5° to 18°C. increased the rate of simazine decomposition by a factor of 3.5, whereas an additional temperature increase to 25°C. doubled the decomposition rate, the increase from 8.5° to 25°C. thus causing a sevenfold increase in the decomposition rate. AGUNDIS (1964) found simazine to persist in warm tropical soils for only 30 to 60 days, whereas persistence in colder Canadian and Russian soils was much longer (KOROLEV and STAROSEL'SKIY 1964, ROADHOUSE and BIRK 1961). HARRIS and WARREN (1964) reported that adsorption of atrazine increased with decreased temperature, the availability of atrazine, therefore, being greater at higher temperatures. Similar results were reported with both simazine and atrazine by TALBERT and FLETCHALL (1965). An additional indirect influence of temperature on adsorption may result from its affect on solubility (HARRIS and WARREN 1964). These two factors usually function simultaneously, *i.e.*, both lead to decreased adsorption as temperature rises. Thus the more rapid decomposition of s-triazine herbicides in warm soils may be the result of optimum conditions, *i.e.*, increased microbial activity coupled with increased availability and solubility of the chemical.

Soil pH, moisture content, and aeration also affect the activities of soil microorganisms. Soil bacteria and actinomycetes are most active in near neutral or alkaline, moist soils (WAKSMAN 1932), whereas soil fungi are generally more active in soils under low moisture conditions (MENON and WILLIAMS 1957) and at lower pH levels (WARCUP 1951). Since so many soil fungi appear to be involved in biodegradation of s-triazines one might expect decomposition to occur more rapidly in acid soils of low moisture content. Such is not the case, however. Persistence of several s-triazines is less in moist soils than in dry soils (DEWEY 1960, HARRIS and SHEETS 1965, HOLLY and ROBERTS 1963, SHEETS and HARRIS 1965, TALBERT and FLETCHALL 1964). Adsorption of atrazine and simazine varies inversely with pH (HARRIS and WARREN 1964, TALBERT and FLETCHALL 1965). The effect of soil pH and cation-exchange capacity on residual phytotoxicity and microbial degradation of s-triazines has not been adequately investigated, however.

Degradation of s-triazines under anaerobic conditions has been observed (AGUNDIS and BEHRENS 1966, KAUFMAN 1969, SKIPPER 1966). Anaerobic degradation observed by AGUNDIS and BEHRENS apparently involved hydroxylation, since SKIPPER (1966) observed that only a minute amount of $^{14}CO_2$ evolution occurred from chain-labeled atrazine treated soil under anaerobic conditions. The apparent lack of microbial degradation under anaerobic conditions probably accounts for the reduced residual phytotoxicity found to occur in surface soils when compared to subsoils containing equal amounts of simazine (KOROLEV and STAROSEL'SKIY 1964). A recent survey of 12 geographically separated soils by HARRIS *et al.* (1969) indicated that an average of 61 percent more atrazine was recovered from subsurface soils (15 inches) than from surface soils (three inches) treated with equal amounts of triazines under field conditions. Five northern soil samples contained more than twice as much atrazine residues as four southern samples.

Based on the obvious parallels between s-triazine persistence, environmental factors, and microbial activity it has frequently been concluded that microorganisms are primarily responsible for s-triazine degradation in soils. Evidence is accumulating, however, that chemical hydrolysis of chloro-s-triazines is more rapid in soils than microbial degradation. HARRIS (1965) found that the 2-hydroxy derivatives were the major degradation products occurring in soil. Sodium azide, a microbial inhibitor, did not appreciably affect the accumulation of the hydroxy derivatives of simazine, atrazine, and propazine in soil during an eight-week incubation period (HARRIS 1967). SKIPPER (1966) and coworkers (1967) followed $^{14}CO_2$ evolution from ring- and chain-labeled atrazine treated soils. Although 1.4 to 1.6 percent of the ^{14}C from chain-labeled atrazine, and 0.4 to 0.6 percent of the ^{14}C from ring-labeled atrazine, had been evolved as $^{14}CO_2$ at the end of two weeks, hydroxy-atrazine accounted for approximately 20 percent of the extracted ^{14}C-activity in both sterile and non-sterile soils. Similar results have been obtained in our laboratory with labeled simazine (Kaufman 1969).

These results support the conclusion that the first major reaction in the degradation of chloro-s-triazines in soils is a chemical hydroxylation, resulting in the formation of the corresponding hydroxy-s-triazine analogue. The significance of biodegradation of chloro-s-triazine in soil is thus subject to question. Under field conditions, the chemical vs. microbial degradation rate probably depends upon the soil environmental factors. Hydrolysis to hydroxy-s-triazines would probably be the dominant factor at high temperatures (over 30°C.) and low pH values (5.5). Attack by soil fungi may also be important at low pH levels. Neutral pH values may favor a limited bacterial attack. At basic pH values (8.5) the hydrolysis would again be dominant. Low temperatures and moisture levels are unfavorable for both microbial and chemical degradation of chloro-s-triazines in soils. Temperature and pH are only two of the factors involved, however. Physical characteristics, aeration, and inorganic and organic nutrient supply also affect both microbial and chemical degradation.

As indicated earlier (Duke 1964), the degradation of methylthio-s-triazines is probably a microbial process. The presence of a definite enrichment flora and a lag period would support this hypothesis. Their greater resistance to chemical hydrolysis but shorter soil persistence might also be indicative of microbial involvement. The significance of either chemical or microbial involvement in the soil degradation of methoxy-s-triazines is unknown.

In view of the observations discussed here it seems likely that a balance sheet considering the complete mineralization of s-triazine residues must be concerned with the degradation and fate of hydroxy-s-triazines in soils. Few investigations have considered the degradation of hydroxy-s-triazines. There is general agreement that dealkylation and evolution of $^{14}CO_2$ from chain-labeled s-triazines occurs more readily than ring cleavage and $^{14}CO_2$-evolution from ring-labeled s-triazines (Couch et al. 1965; Kaufman 1963, 1964, 1965, and 1969; Skipper 1966). Whether or not this radioactivity is actually derived from a hydroxylated analogue, the parent material, or both is not known. Skipper (1966) observed a two- to threefold increase in $^{14}CO_2$ evolved from ring-labeled hydroxyatrazine over ring-labeled atrazine, indicating a more rapid degradation of hydroxyatrazine than atrazine. Preliminary investigations of hydroxy-s-triazine degradation suggest a greater resistance of hydroxy-s-triazines to degradation by certain soil organisms. More definitive conclusions, however, can only be made with further investigations. Hydroxy-s-triazine degradation is currently under investigation at Beltsville.

Table III. *Common and chemical names of pesticides mentioned in text*

Common name	Chemical name
ametryne	2-methylthio-4-ethylamino-6-isopropylamino-s-triazine
ammelide	2,4-dihydroxy-6-amino-s-triazine
ammeline	2-hydroxy-4,6-diamino-s-triazine
atratone	2-methoxy-4-ethylamino-6-isopropylamino-s-triazine
atrazine	2-chloro-4-ethylamino-6-isopropylamino-s-triazine
chlorazine	2-chloro-4,6-bis(diethylamino)-s-triazine
cyanurate	2,4,6-trihydroxy-s-triazine
dyrene	2,4-dichloro-6-(0-chloroanilino)-s-triazine
fenthion	O,O-dimethyl-O-(3-methyl-4-methylmercaptophenyl) phosphorothioate
hydroxyatrazine	2-hydroxy-4-ethylamino-6-isopropylamino-s-triazine
hydroxypropazine	2-hydroxy-4,6-bis(isopropylamino)-s-triazine
ipazine	2-chloro-4-diethylamino-6-isopropylamino-s-triazine
melamine	2,4,6-triamino-s-triazine
Mesurol	4-(methylthio)-3,5-xylyl-N-methylcarbamate
prometone	2-methoxy-4,6-bis(isopropylamino)-s-triazine
prometryne	2-methylthio-4,6-bis(isopropylamino)-s-triazine
propazine	2-chloro-4,6-bis(isopropylamino)-s-triazine
simazine	2-chloro-4,6-bis(ethylamino)-s-triazine
simetone	2-methoxy-4,6-bis(ethylamino)-s-triazine
simetryne	2-methylthio-4,6-bis(ethylamino)-s-triazine
trietazine	2-chloro-4,6-diethylamino-6-ethylamino-s-triazine

Acknowledgments

The authors wish to acknowledge the helpful suggestions of Dr. Horace D. Skipper in the preparation of this manuscript.

Summary

Soil microorganisms are involved in the degradation of s-triazine herbicides. The degree of initial involvement of microorganisms in s-triazines may vary with the analogues in question. Chemical hydrolysis is probably the major detoxication method of chloro-s-triazines in soil systems, whereas methylthio-s-triazine detoxication is believed to be dependent upon microbial action. Three major degradative pathways are evident: hydrolysis at the number 2 carbon atom, N-dealkylation of side chains, and ring cleavage. Dealkylation, deamination, or both, is the major mechanism of chloro-s-triazine biodegradation by microorganisms, although dehalogenation and hydroxylation have been reported. Dealkylation, by microorganisms, if analogous to free radical systems, proceeds via hydroxylation of the carbon adjacent to the amino group. Further dealkylation and, or deamination and dechlorination has been observed with simazine, the product being ammelide. Additional investigations are needed to elucidate the mechanism of ring cleavage of s-triazines. Hydroxy-s-triazines appear to be important degradation products of the chloro-s-triazines. Since they are non-phytotoxic, their formation represents a detoxication mechanism. Although dealkylation of s-triazines occurs in soils, its importance has not been accurately ascertained. The lability of hydroxy derivatives to dealkylation should be examined.

Résumé*

Dégradation microbienne des desherbants dérivés de la triazine

Les microorganismes du sol interviennent dans la dégradation des desherbants dérivés de la s-triazine. L'importance de l'action initiale des microorganismes sur les s-triazines peut varier avec les analogues. L'hydrolyse chimique est probablement la principale voie de dégradation des chloro-s-triazines dans les sols, tandis que la disparition de la méthylthio-s-triazine paraît dûe à une action microbienne. Trois catabolismes principaux sont clairs: l'hydrolyse sur le carbone 2, la N-désalkylation des chaines latérales et le clivage du cycle. La désalkylation ou la désamination, ou les deux processus simultanément, sont les mécanismes fondamentaux de la biodégradation de la chloro-s-triazine par les microorganismes, bien que la déshalogénation et l'hydroxylation aient été signalées. Si la désalkylation par les microorganismes est analogue aux systèmes à radicaux libres, elle procède par hydroxylation du carbone adjacent au groupe aminé, Une désalkylation ultérieure, seule ou associée à une désamination et à une déchloration a été observée avec la simazine, avec formation finale de ammelide. De nouvelles recherches sont nécessaires pour élucider le mécanisme de clivage du cycle de la s-triazine. Les hydroxy-s-triazines paraissent être des produits importants de la dégradation des chloro-s-triazines. Puisqu'elles ne sont pas phytotoxiques, leur formation est un mécanisme de détoxication. Bien que la désalkylation des s-triazines se produise dans les sols, son importance n'a pas été déterminée avec exactitude. La facilité des dérivés hydroxylés à se désalkyler devrait être étudiée.

Zusammenfassung**

Mikrobische Degradierung der Triazin-Unkrautvertilgungsmittel

In die Degradierung der s-Triazin-Unkrautvertilgungsmittel sind Boden-Mikroorganismen verwickelt. Der Grad der anfänglichen Verwicklung von Mikroorganismen mit s-Triazinen kann sich mit den betreffenden Analogen ändern. Chemische Hydrolyse ist wahrscheinlich die Hauptentgiftungsmethode der Chlor-s-Triazine in Bodensystemen, während man annimmt, dass die Methylthio-s-Triazin-Entgiftung

---

* Traduit par R. Mestres.
** Übersetzt von M. Düsch.

von mikrobischer Tätigkeit abhängt. Drei hauptdegradierende Pfade sind offenkundig: Hydrolyse am Nummer 2 Kohlenstoffatom; N-Dealkylierung der Seitenketten und Ringspaltung. Dealkylierung, Deaminierung, oder beide, sind die Hauptmechanismen der Chlor-s-Triazin-Biodegradierung durch Mikroorganismen, obgleich Dehalogenierung und Hydroxylierung angegeben werden sind. Wenn es den freien fundamentalen Systemen entspricht schreitet die Dealkylierung durch Mikroorganismen über Hydroxylierung des angrenzenden Kohlenstoffes weiter zur Aminogruppe. Weitere De-alkylierung und (oder) Deaminierung und Dechlorierung sind bei Simazin, dessen Endprodukt Ammelid ist, beobachtet worden. Zusätzliche Untersuchungen sind notwendig, um den Mechanismus der Ringspaltung der Triazine aufzuhellen. Hydroxy-s-Triazine scheinen wichtige Degradierungsprodukte der Chlor-s-Triazine zu sein. Da sie nicht pflanzentoxisch sind, stellt ihre Bildung einen Entgiftungsmechanismus dar. Obgleich Dealkylierung der s-Triazine im Böden auftritt, ist ihre Wichtigkeit noch nicht genau ermittelt worden. Die Labilität der Hydroxy-Derivate zur Dealkylierung sollte geprüft werden.

References

1. Agundis, O.: The distribution and persistance of simazine and 2,4-D in tropical soils. Weed Soc. Amer. Abstr., p. 15 (1964).
2. —— Factors affecting atrazine inactivation in soils. Diss. Abstr. B27:4187B-(1967).
3. ——, and R. Behrens: Effect of aerobic and anaerobic atmospheres on the rate of atrazine disappearance from soil. Weed Soc. Amer. Abstr., p. 70 (1966).
4. Alexander, M.: Introduction to soil microbiology. New York: Wiley (1961).
5. ——, and M. I. H. Aleem: Effect of chemical structure on microbial decomposition of aromatic herbicides. J. Agr. Food Chem. 9, 44 (1961).
6. Armstrong, D. E., and R. F. Harris: Atrazine degradation as affected by soil and substrate composition. Agron. Abstr. 57, 81 (1965).
7. Audus, L. J.: Microbiological breakdown of herbicides in soils. In: E. K. Woodford and G. R. Sagar (eds.) Herbicides and the soil. Oxford: Blackwell (1960).
8. Bartha, R., R. P. Lanzilotta, and D. Pramer: Stability and effects of some pesticides in soil. Applied Microbiol. 15, 67 (1967).
9. Boehme, C., and F. Baer: Über den Abbau von Triazin-Herbiciden in tierischen organismus. Food Cosmetic Toxicol. 5, 23 (1967).
10. Bryant, J. B.: Bacterial decomposition of some aromatic and aliphatic herbicides. Ph.D. Thesis, Penn. State Univ. (1963).
11. Burchfield, H. P., and E. E. Storrs: Relative reactivities of 1-fluoro-2,4-dinitrobenzene and 2,4-dichloro-6-(o-chloroanilino)-s-triazine with metabolites containing various functional groups. Contrib. Boyce Thompson Inst. 19, 169 (1957).
12. Burnside, O. C.: The influence of environmental factors on the phytotoxicity and dissipation of simazine. Ph.D. Thesis, Univ. of Minn. (1959).
13. ——, E. L. Schmidt, and R. Behrens: Dissipation of simazine from the soil. Weeds 9, 477 (1961).

14. Burschel P.: Studies on the behavior of simazine in soil. Weed Research 1, 131 (1961).
15. Cartwright, N. J., and J. A. Buswell: The separation of vanillate o-demethylase from protocatechuate 3,4-oxygenase by ultracentrifugation. Biochem. J. 105, 767 (1967).
16. ——, and A. R. Smith: Bacterial attack on phenolic ethers, an enzyme system demethylating vanillic acid. Biochem. J. 102, 826 (1967).
17. Charpentier, M., and J. Pochon: Soil bacteria growing on aminotriazine (simazine) Ann. Inst. Pasteur 102, 501 (1962).
18. Cochrane, V. W.: Physiology of the fungi. New York: Wiley (1958).
19. Cohn, M., and J. Monod: Specific inhibition and induction of enzyme biosynthesis, p. 132. Symp. III. Soc. Gen. Microbiol., Comb. Univ. Press (1953).
20. Couch, R. W., J. V. Gramlich, D. E. Davis, and H. H. Funderburk, Jr.: The metabolism of atrazine and simazine by soil fungi. Proc. S. Weed Conf. 18, 623 (1965).
21. Delly, R.: Unpublished data, Anal. Lab., J. R. Geigy, S. A., Basel, Switzerland (1963).
22. Dewey, O. R.: Further experimental evidence of the fate of simazine in the soil. Proc. V. British Weed. Control Conf., Brighton I, 91 (1960).
23. Duke, W. B.: The decomposition of 2-chloro-4-ethylamino-6-isopropylamino-s-triazine (atrazine) and related s-triazine herbicides by soil microorganisms. M.S. Thesis, Ore. State Univ. (1964).
24. Farmer, F. H., R. E. Benoit, and W. E. Chappell: Simazine, its effect on nitrification and its decomposition by soil microorganisms. Proc. N.E. Weed Control Conf. 19, 350 (1965).
25. Fothergill, P. G., and R. Ashcroft: The nutritional requirements of Venturia inaequalis. J. Gen. Microbiol. 12, 387 (1955).
26. Francis, J. I., and J. M. Barnes: Studies on the mammalian toxicity of fenthion. Bull. World Health Org. 29, 205 (1963).
27. Funderburk, H. H., Jr., and D. E. Davis: The metabolism of C14 chain- and ring-labeled simazine by corn and the effect of atrazine on plant respiratory systems. Weeds 11, 101 (1963).
28. Gramlich, J. V., R. W. Couch, H. C. Sikka, D. E. Davis, and H. H. Funderburk, Jr.: Preliminary report on atrazine metabolism by microorganisms. Proc. S. Weed Conf. 17, 356 (1964).
29. Guillemat, J.: Interactions entre la simazine et la mycoflore du sol. Comptes Rendus Acad. Sci. Paris 250, 1343 (1960).
30. ——, M. Charpentier, P. Tardieux, and J. Pochon: Interaction entre une chloro-amino-triazine herbicide et la microflore fongigue et bactienne du sol. Annales des Epiphyties II, 261 (1960).
31. Gysin, H., and E. Knüsli: Chemistry and herbicidal properties of triazine derivatives. Adv. Pest Control Research 3, 289 (1960).
32. Hamilton, R. H., and D. E. Moreland: Fate of ipazine in cotton plants. Weeds 11, 213 (1963).
33. Harris, C. I.: Hydroxysimazine in soil. Weed Research 5, 275 (1966).
34. —— Fate of 2-chloro-s-triazines in soil. J. Agr. Food Chem. 15, 157 (1967).
35. ——, D. D. Kaufman, T. J. Sheets, R. G. Nash, and P. C. Kearney: Behavior and fate of s-triazines in soils. Adv. Pest Cont. Research 8, 1 (1969).
36. ——, and T. J. Sheets: Persistence of several herbicides in the field. Proc. N.E. Weed Control Conf. 19, 359 (1965 a).
37. —— —— Influence of soil properties on adsorption and phytotoxicity of CIPC, diuron and simazine. Weeds 13, 215 (1965 b).
38. ——, and G. F. Warren: Adsorption and desorption of herbicides by soil. Weeds 12, 120 (1964).

39. ——, E. A. Woolson, and B. E. Hummer: Dissipation of herbicides at three soil depths. Weed Sci **17**, 27 (1969).
40. Hauck, R. D., and H. F. Stephenson: Nitrification of triazine nitrogen. J. Agr. Food Chem. **12**, 147 (1964).
41. Hirsch, P., and M. Alexander: Microbial decomposition of halogenated propionic and acetic acids. Can. J. Microbiol. **6**, 241 (1960).
42. Holly, K., and H. A. Roberts: Persistence of phytotoxic residues of triazine herbicides in soil. Weed Research **3**, 1 (1963).
43. Hurter, J.: Abbauproducte von Simazine in Gramineen. Experientia **22**, 741 (1966).
44. Kaufman, D. D.: Unpublished data (1969).
45. ——, and J. Blake: Unpublished data (1969).
46. —— —— Atrazine degradation by soil fungi. Weed Sci. Soc. Amer. Abstr. No. 230 (1969).
47. ——, P. C. Kearney, and T. J. Sheets: Simazine: Degradation by soil microorganisms. Science **142**, 405 (1963).
48. —— —— —— Degradation of simazine by soil microorganisms. Weed Soc. Amer. Abstr., p. 12 (1964).
49. —— —— —— Microbial degradation of simazine. J. Agr. Food Chem. **13**, 238 (1965).
50. Kearney, P. C.: Unpublished data (1969).
51. ——, and C. S. Helling: Reactions of pesticides in soils. Residue Reviews **25**, 25 (1969).
52. ——, D. D. Kaufman, and M. Alexander: Biochemistry of herbicide decomposition in soil. In A. D. McLaren and G. H. Peterson (eds.): Soil biochemistry. New York: Dekker (1967).
53. —— ——, and T. J. Sheets: Metabolism of the alkyl moiety of 2-chloro-4,6-bis(ethylamino)-s-triazine. 147th Meeting Amer. Chem. Soc., Abstr. (1964).
54. —— —— —— Metabolites of simazine by *Aspergillus fumigatus*. J. Agr. Food Chem. **13**, 369 (1965).
55. Knuesli, E., D. Berrer, G. Dupuis, and H. Esser: s-Triazines. p. 51, In P. C. Kearney and D. D. Kaufman (eds.): Degradation of herbicides. New York: Dekker (1969).
56. Klozowski, T. T.: Variable toxicity of triazine herbicides. Nature **205**, 104 (1965).
57. Konishi, K., and A. Imanishi: Studies on the fertilizing value of melamines and guanidines. J. Sci. Soil Manure, Japan **15**, 564 (1941).
58. Korolev, L. I., and Ya. Yu. Starosel'skiy: After effect of simazine. Chem. Agr. (Engl. Trans.) **1**, 45 (1964); J.P.R.S.:28,023. Office of Technical Services, *U.S. Department of Commerce*, Washington, D.C. 20443 (Dec. 29, 1964).
59. Lipschitz, W. L., and E. Stokey: Diuretic action of formoguanamine in normal persons. J. Pharmacol. **92**, 131 (1948).
60. MacRae, I. C., and M. Alexander: Microbial degradation of selected herbicides in soil. J. Agr. Food Chem. **13**, 72 (1965).
61. Manorik, A. V., V. F. Vasil'chenko, M. M. Mandrovskaya, and S. M. Malichenko: Inactivation of herbicides of the triazine group by soil microorganisms. Agrokhimiya **4**, 123 (1968); through Chem. Abstr. **69**, 1959 f (1968).
62. McCormick, L. L., and A. E. Hiltbold: Microbiological decomposition of atrazine and diuron in soil. Weeds **14**, 77 (1966).
63. Meister, A.: Biochemistry of the amino acids, 2ed., vol. 2. New York: Academic Press (1965).
64. Menon, S. K., and L. E. Williams: Effect of crop, crop residues, temperature, and moisture on soil fungi. Phytopathol. **47**, 559 (1957).

65. MICKOVSKI, M., and O. VERONA: Decomposition of triazine herbicides by some soil fungi. Agr. Ital. (Pisa). **67**, 67 (1967); through Biochem. Abstr. **67**, 107564X (1967).

66. MONTGOMERY, M. L., and V. H. FREED: Metabolism of triazine herbicides by plants. J. Agr. Food Chem. **12**, 11 (1964).

67. MUELLER, P. W., and P. H. PAYOT: Fate of ^{14}C-labelled triazine herbicides in plants. Proc. I.A.E.A. Symp. "Isotopes in Weed Research," Vienna, p. 61 (1966).

68. MURRAY, D. S., and W. L. RIECK: Fungi isolates influencing phytotoxicity and degradation of a methyl mercapto triazine. Agron. Abstr. **60**, 95 (1968).

69. NADKARNI, M. V., E. I. GOLDENTHAL, and P. C. SMITH: The distribution of radioactivity following administration of triethylenimino-s-triazine-C^{14} in tumor-bearing and control mice. Cancer Research **14**, 559 (1954).

70. NIESSEN, H., and H. FREHSE: An infra-red-spectroscopic method for determining residues of N-methylcarbamates in plants. Pflanzenschutz-Nachrichten Bayer **16**, 205 (1963).

71. ——, H. TIETZ, and H. FREHSE: On the occurrence of biologically active metabolites of the active ingredient S1752 after application of lebaycid. Pflanzenschutz-Nachrichten Bayer **15**, 125 (1962).

72. PANTOS, G., P. GYURKO, T. TAKACS, and L. VARGA: The effect of herbicides used in practice on some species of the microfauna of the soil, on certain mycorrhiza fungi and the biological neutralization of herbicides. Acta. Agron. **13**, 21 (1964).

73. PLIMMER, J. R., and P. C. KEARNEY: Free radical oxidation of pesticides. Weed Sci. Soc. Amer. Abstr., p. 20 (1968).

74. —— ——, and J. R. ROWLANDS: Atrazine dealkylation by free radicals. 156th Meeting Amer. Chem. Soc., Abstr. A47 (1968).

75. POCHON, J., P. TARDIEUX, and M. CHARPENTIER: Recherches sur les interactions entre les aminotriazines et la microflore bacterienne tellurique. Comptes Rendes Acad. Sci. Paris **250**, 1555 (1960).

76. RAGAB, M. I. H.,; nd J. P. McCOLLUM: Degradation of C^{14}-labeled simazine by plants and soil microorganisms. Weeds **9**, 72 (1961).

77. REID, J. J.: Bacterial decomposition of herbicides. Proc. N.E. Weed Control Conf. **14**, 19 (1960).

78. ROADHOUSE, F. E. B., and L. A. BIRK: Penetration of and persistence in soil of the herbicide 2-chloro-4,6-bis(ethylamino)-s-triazine (simazine). Can. J. Plant Sci. **41**, 252 (1961).

79. SCHOLL, W., R. O. E. DAVIS, B. E. BROWN, and F. R. RIED: Melamine of possible plant food value. Ind. Eng. Chem. **29**, 202 (1937).

80. SHEETS, T. J.: Persistence of herbicides in soils. Proc. W. Weed Control Conf. **18**, 37 (1962).

81. ——, A. S. CRAFTS, and H. R. DREVER: Influence of soil properties on the phytotoxicities of the s-triazine herbicides. J. Agr. Food Chem. **10**, 458 (1962).

82. ——, and L. L. DANIELSON: Herbicides in soils. In: The nature and fate of chemicals applied to soils, plants, and animals (ARS-20). Washington, *U.S. Department of Agriculture*, p. 170 (1960).

83. ——, and C. I. HARRIS: Herbicide residues in soils and their phytotoxicities to crops grown in rotations. Residue Reviews **11**, 119 (1965).

84. ——, and P. C. KEARNEY: Unpublished data (1963).

85. ——, and W. C. SHAW: Herbicidal properties and persistence in soil of s-triazines. Weeds **11**, 15 (1963).

86. SHIMABUKURO, R. H.: Dealkylation of atrazine in root and shoot of pea plants and its significance. 152d Meeting Amer. Chem. Soc., Abstr. A34 (1966).

87. —— Significance of atrazine dealkylation in root and shoot of pea plants. J. Agr. Food Chem. **15**, 557 (1967).

88. ——, R. E. Kadunce, and D. S. Frear: Dealkylation of atrazine in mature pea plants. J. Agr. Food Chem. **14**, 392 (1966).
89. ——, and R. H. Swanson: Metabolism of root-applied versus foliarly-applied atrazine in corn. Weed Sci. Sco. Amer., Abstr. No. 197 (1969).
90. Skipper, H. D.: Microbial degradation of atrazine in soils. M.S. Thesis, Ore. State Univ. (1966).
91. Skipper, H. D., C. M. Gilmour, and W. R. Furtick: Microbial versus chemical degradation of atrazine in soils. Proc. Soil. Sci. Soc. Amer. **31**, 653 (1967).
92. Spiridonov, Y. Y., and A. I. Yakovlev: Infiltration and detoxication rate of soil herbicides under moist subtropical conditions. Khim. Sel. Khaz. **5**, 431 (1967); through Biochem. Abstr. **67**, 899812 (1967).
93. Talbert, R. E., and O. H. Fletchall: Inactivation of simazine and atrazine in the field. Weeds **12**, 33 (1964).
94. —— —— The absorption of some s-triazines in soils. Weeds **13**, 46 (1965).
95. Terman, G. L., J. D. DeMeret, C. M. Hunt, J. T. Cope, Jr., and L. E. Ensminger: Crop response to urea and urea pyrolysis products. J. Agr. Food Chem. **12**, 151 (1964).
96. Uhlig, S. K.: The interaction between chlorobisethylamino-s-triazine (simazine) and mycorrhiza-forming fungi. Wiss. Z. Tech. Univ. Dresden **15**, 639 (1966); through Biochem. Abstr. **66**, 85024w (1967).
97. Waeffler, R.: Soil microflora and triazine herbicides. *J. R. Geigy, S. A.,* Basle, Switzerland, (1961).
98. Wagner, G. H., and K. S. Chahal: Decomposition of carbon-14 labeled atrazine in soil samples from Sanborn field. Soil Sci. Soc. Amer. Proc. **30**, 752 (1966).
99. Waksman, S. A.: Principles of soil microbiology. Baltimore: Williams and Wilkins (1932).
100. Walker, R. L., and A. S. Newman: Microbial decomposition of 2,4-dichlorophenoxyacetic acid. Applied Microbiol. **4**, 201 (1956).
101. Warcup, J. H.: The ecology of soil fungi. Brit. Mycol. Soc. Trans. **34**, 376 (1951).
102. Whitenberg, D. C.: Fate of prometryne in cotton plants. Weeds **13**, 68 (1965).
103. Williams, R. T.: Detoxication mechanisms. New York: Wiley (1959).

Nonbiological detoxication of the s-triazine herbicides

By

L. S. JORDAN,* W. J. FARMER,** J. R. GOODIN,***
and B. E. DAY****

Contents

I. Introduction

Until recently, microbial decomposition was considered the main pathway for detoxication of the s-triazine herbicides in soils. Organisms have been isolated which can detoxify these herbicides (KAUFMAN et al. 1965). Factors such as temperature and soil organic matter which influence soil microbial activity have been shown to correlate with the rate of s-triazine loss from the soil (BURNSIDE et al. 1961, TALBERT and FLETCHALL 1964). More recent studies, however, indicate that microbial decomposition may play a less significant role in the detoxication process and various nonbiological pathways may be more important than previously thought.

Much of the research on nonbiological detoxication has been concentrated on photodecomposition, volatilization, soil adsorption, and two soil-associated chemical reactions: hydroxylation and dealkylation. In addition, codistillation has been suggested as a possible mechanism for loss of herbicide activity. Disagreement over whether codistillation

---

* Department of Horticultural Science, *University of California*, Riverside.
** Department of Soils and Plant Nutrition, *University of California*, Riverside.
*** Department of Agronomy, *University of California*, Riverside.
**** Associate Director, Citrus Research Center and Agricultural Experiment Station, *University of California*, Riverside.

occurs and to what extent it contributes to total loss of activity may have resulted from misinterpretation of data or a misunderstanding of the exact nature of codistillation.

Recent studies have demonstrated that degradation of *s*-triazine herbicides can occur in the absence of microorganisms. Correlations also have been established between nonbiological detoxication and soil properties. Since biological and some nonbiological reactions are both chemical in nature, the same correlations would be expected between soil properties and either microbially-catalyzed decomposition or detoxication reactions occurring in the absence of any biological agent.

This paper reviews the recent evidence for the non-biological detoxication of the *s*-triazine herbicides via photodecomposition, volatilization, hydroxylation, and dealkylation. Soil adsorption is discussed elsewhere in this volume, and no consideration will be given to codistillation.

II. Photodecomposition

Much speculative attention has been given to the influence of light in the deactivation of *s*-triazine herbicides on the soil, but little critical research has been conducted on actual losses from photodecomposition. Nevertheless, sufficient research has been carried out to demonstrate that photodecomposition does occur with some *s*-triazine herbicides.

GAST (1962) reported losses in activity for simazine and atrazine after exposure to ultraviolet (UV) and infrared light radiation, particularly if the herbicides were applied to a dry surface. DEWEY (1960) found a decrease in simazine activity after the herbicide was irradiated with a mercury-vapor high-pressure lamp. SHEETS and DANIELSON (1961) demonstrated inactivation of simazine by sunlight. MITCHELL (1961) irradiated simazine on filter paper with UV light, but his results were not conclusive.

JORDAN *et al.* (1965) studied the effect on simazine and atrazine of UV light sources with peak emission of 254, 311, and 360 mμ, respectively, representing far, middle, and near UV light. The two herbicides were deposited on aluminum planchets and irradiated for periods up to 500 hours. Temperature was maintained at 42°C. for irradiated samples and for samples kept in the dark to separate losses from volatility and photolysis. The absorption of simazine and atrazine under UV is greatest at 220 mμ (Fig. 1). Following irradiation, absorbance decreased. The greatest decrease occurred after irradiation with 254 mμ (far UV) and the least with 360 mμ (near UV). Loss of simazine and atrazine was rapid during the initial period of irradiation, but as time progressed the rate of loss decreased. The greatest overall loss occurred at 254 mμ and the least at 360 mμ of irradiation (Fig. 2).

Since the first law of photochemistry states that only light which is

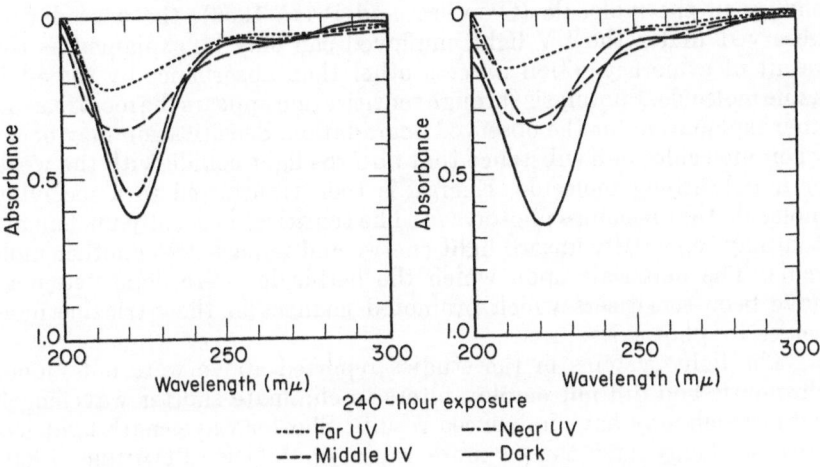

Fig. 1. Ultraviolet absorption spectra of simazine and atrazine deposited on aluminum planchets and kept in the dark or irradiated for 240 hours with far, middle or near ultraviolet light at 42 ± 2°C. (JORDAN *et al.* 1965); left = simazine, right = atrazine

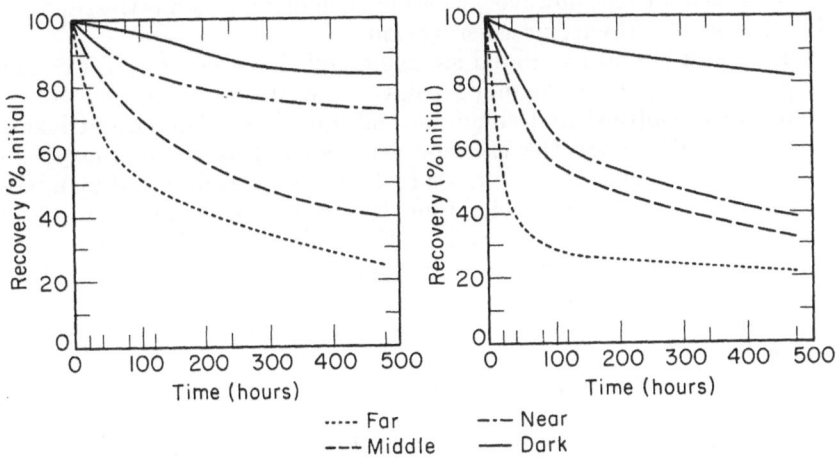

Fig. 2. Recovery of simazine and atrazine from aluminum planchets after storage in the dark or irradiated with near, middle, and far ultraviolet light for 20, 40, 80, 120, 240, and 480 hours at 42 ± 2°C.; the horizontal axis is divided into 100 hour increments and the intervals at which data were collected are designated by short vertical bars above the axis (JORDAN *et al.* 1965); left = simazine, right = atrazine

absorbed by a molecule can be effective in producing photochemical change in the molecule (CALVERT and PITTS 1966), the degradation observed under the UV light employed can only be explained as the result of some activation process other than absorbance by the s-triazine molecule. Photolysis through sensitization appears the most reasonable explanation for the observed degradation. Sensitization may occur when molecules of a substance that absorbs light collide with the weak or nonabsorbing molecule. Energy is then transferred to a receiving molecule that becomes photolyzed. The sensitizer is usually unchanged and may repeatedly absorb light energy and impart it to another molecule. The materials upon which the herbicides were deposited may have been sensitizers which promoted changes in the s-triazine molecules by photolysis.

The light systems in the studies reported above were not monochromatic and did not employ filters to eliminate shorter wavelength light, which may have influenced results. Shorter wavelength light has greater energy and causes more rapid photolysis. PLIMMER (1969) reported light of 220 mμ wavelength is necessary to bring about photochemical reaction in the s-triazines. Herbicide loss with light sources emitting UV at wavelengths longer than 290 mμ are of greater interest in investigating detoxication because their emission bands are in the sunlight spectrum that reaches the earth's surface. Detoxication of s-triazines under sunlight indicates that these herbicides may be degraded by light with wavelengths greater than 290 mμ (JORDAN *et al.* 1964). It is not clear, however, whether such degradation is the result of direct light absorbance or sensitization.

PLIMMER (1969) irradiated simazine and simetone with ultraviolet light (220 mμ) in methanolic solution. With simazine, the chloro group was displaced and simetone and 4,6-*bis*(ethylamino)-s-triazine were obtained. After the chlorine was replaced by the methoxyl, an alkylamino side chain was replaced by hydrogen, forming 4-ethylamino-2-methoxy-s-triazine. With simetone, loss of the side chain also resulted in formation of 4-ethylamino-2-methoxy-s-triazine. The photolysis rate of simetone was slower than that of simazine. Thus PLIMMER has demonstrated that an alteration of s-triazine molecules can be obtained *in vitro*. Such a demonstration has yet to be made in the field, however.

COMES and TIMMONS (1965) showed that atrazine and simazine on the soil may be detoxified by sunlight. They sprayed soil with simazine and atrazine and exposed it to sunlight during the spring and summer. Other soil was treated in the same manner but light was excluded. Bioassay analysis showed that atrazine loss from irradiation in the spring was 47 percent in 25 days and 73 percent in 60 days. Loss in the dark cultures was negligible. In the summer, 65 to 80 percent of the atrazine was rendered nontoxic to oats subjected to both the dark and light treatments. Simazine loss from irradiation during the first 25 days of spring was 25 percent. Soil temperatures in the summer ranged from 150° to 180°F. Thus, volatilization of the herbicides may have occurred, making it difficult to determine the extent of photodecomposition.

It is difficult to obtain reproducible results when studying the effect of light on soil-applied herbicides even when experimental conditions are controlled. It is not difficult to demonstrate that UV light reduces s-triazine, however. For example, a recent study by the senior author demonstrated that atrazine recovery from planchets was reduced by irradiation with UV light at 311 and 254 mμ (Table 1). Toxicity to oats

Table I. *Recovery of atrazine (percent) deposited on aluminum planchets and irradiated 235 hours with UV light (311 and 254 mμ) or kept in the dark at 38 ± 2°C.; growth of oats (mg. fresh weight) in soil treated with atrazine irradiated in planchets and soil treated with atrazine and irradiated under the same conditions as in the planchets; 500 mg. atrazine equivalent to 1 lb./acre as soil treatment*

Treatment	Recovery (%)	Oat weight (mg.)		
		Planchet irradiated	Soil irradiated	Control
Dark	79	44	24	115
311 mμ	38	77	24	114
254 mμ	17	113	29	118

was also decreased. However, when the herbicide was applied to the soil under the same conditions no loss of biological activity was observed. In other tests, loss of atrazine toxicity was observed on soil irradiated with 311 mμ of UV light for 240 hours (Table II). Atrazine activity was greater in soil maintained under the same conditions in the dark (38°C.). The highest volume of water reduced the effect of light on atrazine

Table II. *Growth of oats in soil treated with atrazine at one lb./acre in different volumes of water and irradiated for 240 hours with UV light (max. peak 311 mμ) or kept in the dark at 38°C.; data are presented as percent of growth of untreated control plants*

Treatment	Spray volume (gal./acre)				
	12.5	25	50	100	200
Irradiated	50	47	48	42	30
Nonirradiated	30	28	26	23	24

toxicity, probably by moving the herbicide into the soil deeper than the light penetrated.

Thus, the detoxication pathways resulting in losses of soil-applied *s*-triazine herbicides under sunlight remain in need of clarification. The variability of results from repeated uniform trials indicates the need for precisely controlled conditions for determining the effect of light on herbicides on the soil. No one has yet reported an experiment designed to properly determine the relative importance of light-induced detoxication of *s*-triazine herbicides on the soil under agricultural conditions.

III. Volatilization

Volatilization undoubtedly influences the loss of herbicidal *s*-triazine from the soil. The rate of loss is related to soil properties, moisture, temperature, and the physical and chemical nature of *s*-triazine.

DAVIS *et al.* (1959) reported that 50 percent of simazine volatilized from a metal surface at 160° to 165°F. Volatilization was less at lower

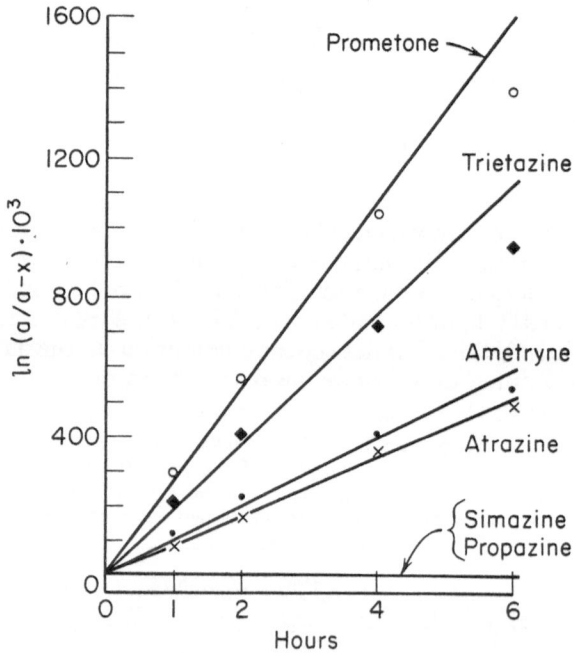

Fig. 3. First-order plots of volatility of six *s*-triazines from metal planchets at 25°C. (*a* = counts/minute at time zero assigned a value of 100 percent, *a* − *x* = percent counts/minute at each observation after time zero) (KEARNEY *et al.* 1964)

temperatures. Less simazine volatilized from sand than from metal and from clay than from sand.

Atrazine loss has been shown to be very rapid from metal planchets at 60°C. (DAVIS *et al.* 1963). At room temperature, loss from planchets was two percent in five hours, eight percent in seven days, and 61 percent in 36 days. A coating of a clear acrylic-ester resin substantially reduced the loss. JORDAN *et al.* (1965) noted rapid loss of simazine and atrazine from metal planchets at 43°C. in the dark. As time progressed, the rate of loss decreased (Fig. 2).

FOY (1964) studied the volatilization of eight *s*-triazine derivatives at 25° and 60°C. At 25°C. only trietazine and ipazine were lost. Trietazine loss was 85 percent in 12 hours and 99 percent in 48 hours. Ipazine loss was 60 percent in 12 hours and 95 percent in 48 hours. At 60°C. there was no loss of hydroxysimazine, hydroxypropazine, or prometone within 24 hours after deposit from water-hydrochloric acid solutions of pH 2 to 3. Loss of atrazine and propazine in 15 hours was about 95 percent. Simazine loss was 35 percent in 24 hours. Trietazine and ipazine were essentially volatilized in two hours.

FOY (1964) also made an interesting observation that prometone is quite volatile from chloroform solutions. He concluded that volatility may be affected by pH of aqueous solutions and by apolar solvents.

KEARNEY *et al.* (1964) studied the volatility of seven triazine herbicides. At 25°C. the volatility in descending order from nickel-plated planchets was prometone $\cong$ trietazine > atrazine $\cong$ ametryne $\cong$ prometryne > propazine $\cong$ simazine (Fig. 3). The vapor pressures of six of the seven compounds at different temperatures are shown in Table III. The comparison makes it obvious that volatility from metal planchets is correlated with vapor pressure of the herbicides. Volatility from planchets appeared to be first-order (Fig. 3). Volatility

Table III. *Vapor pressures of six commercial triazine herbicides calculated in mm. Hg.*[a]

Herbicide	Temperature			
	10°C.	20°C.	30°C.	50°C.
Prometone	5.9×10^{-7}	2.3×10^{-6}	7.9×10^{-6}	7.6×10^{-5}
Prometryne	2.4×10^{-7}	1.0×10^{-6}	4.0×10^{-6}	4.7×10^{-5}
Ametryne	1.9×10^{-7}	8.4×10^{-7}	3.3×10^{-6}	3.9×10^{-5}
Atrazine	5.7×10^{-8}	3.0×10^{-7}	1.4×10^{-6}	2.3×10^{-5}
Propazine	5.0×10^{-9}	2.9×10^{-8}	1.6×10^{-7}	3.4×10^{-6}
Simazine	9.2×10^{-10}	6.1×10^{-9}	3.6×10^{-8}	9.0×10^{-7}

[a] Data furnished by *Geigy Chemical Corp.*

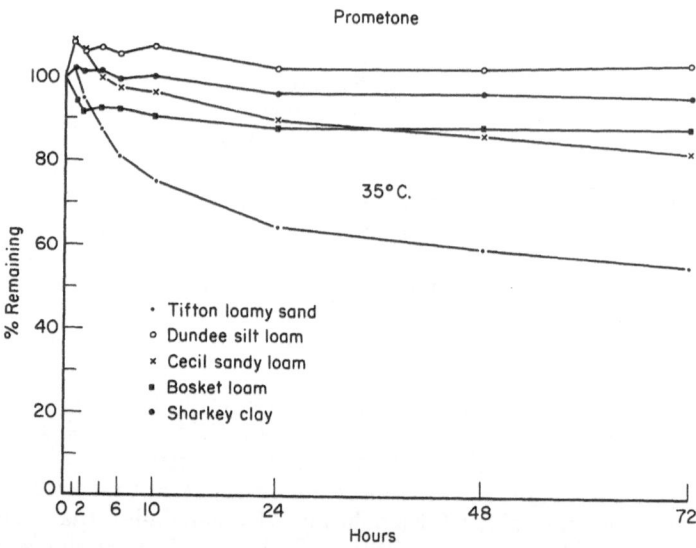

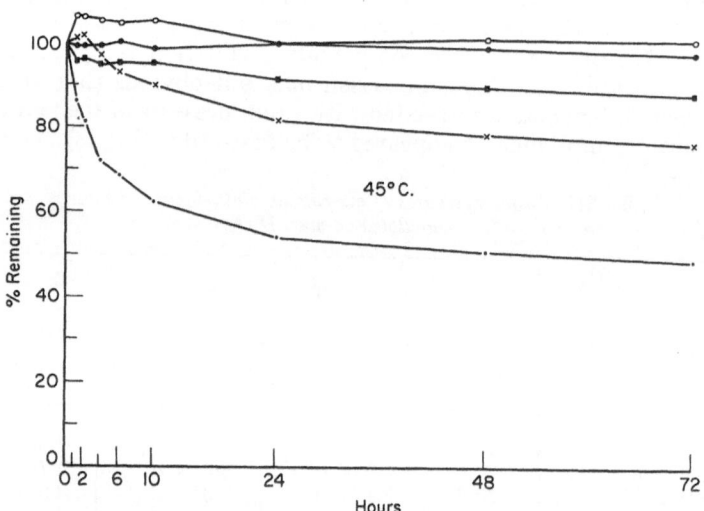

Fig. 4a. Volatility of prometone from five soils at 35° and 45°C. expressed as percent of the observed counts/minute immediately after application to soil (KEARNEY *et al.* 1964)

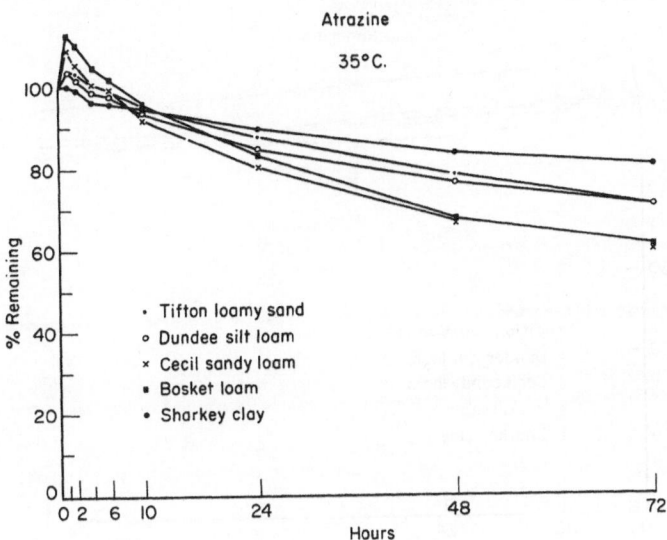

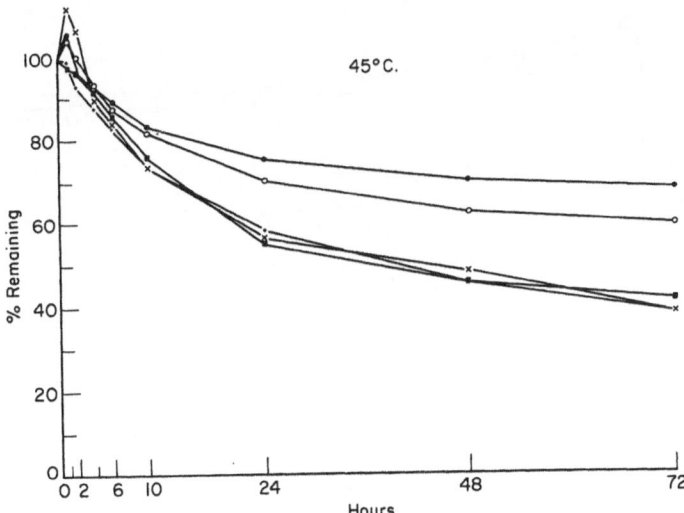

Fig. 4b. Volatility of atrazine from five soils at 35° and 45°C. expressed as percent of the observed counts/minute immediately after application to soil (KEARNEY *et al.* 1964)

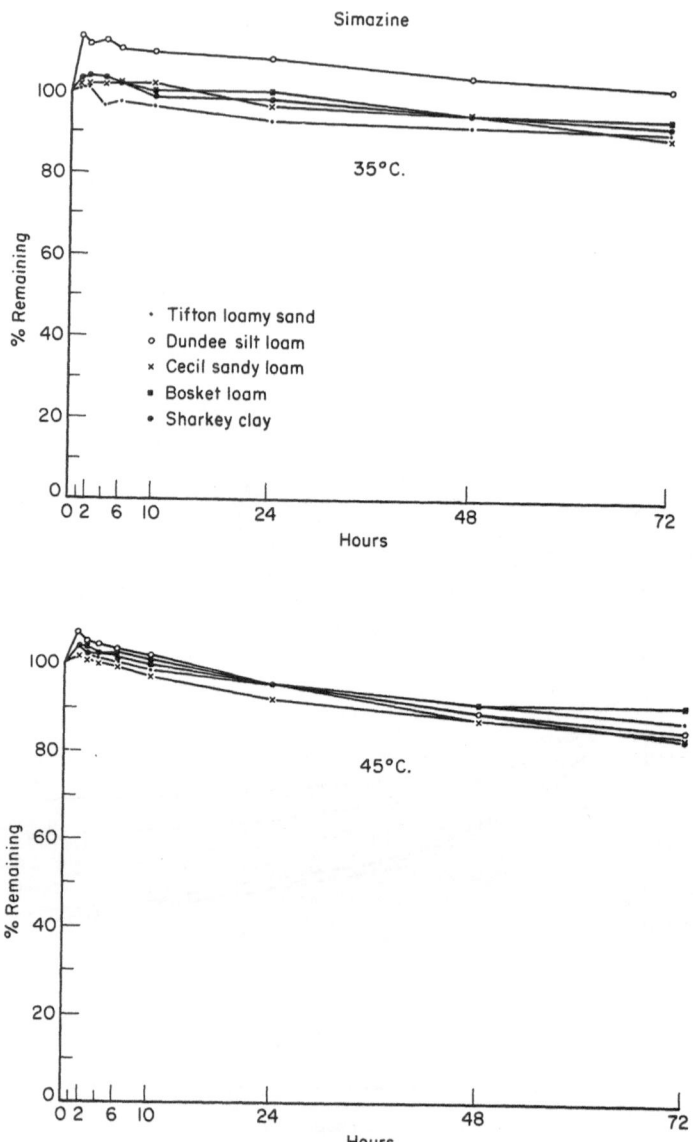

Fig. 4c. Volatility of simazine from five soils at 35° and 45°C. expressed as percent
of the observed counts/minute immediately after application to soil
(Kearney *et al.* 1964)

losses were greater at 35° than at 25°C. Vapor pressure increased as temperature increased (Table III), resulting in greater loss by volatilization. Atrazine loss essentially doubled with 10°C. temperature increase.

KEARNEY et al. (1964) observed that triazine herbicides volatilize more slowly from soils than from metal planchets. Soil type influenced vapor loss of prometone, had less influence on atrazine, and had little effect on loss of simazine (Fig. 4). From the limited amount of data presented, it appears that effect of soil properties on loss to volatility decreased as vapor pressure decreased. Volatility of prometone and atrazine increased with increased sand content and decreased with increased organic matter and clay content. An increase in temperature from 35° to 45°C. resulted in slightly increased prometone loss, had a greater effect on atrazine loss, and did not affect simazine loss from soil (Fig. 4).

KEARNEY et al. (1964) also compared the rate of volatility of seven s-triazine herbicides from moist and dry soils. Simazine appeared to be more volatile from dry soil. Prometryne and ametryne, and to a lesser extent prometone, atrazine, and trietazine, were less volatile from dry soil than from wet soil. Propazine volatility was not influenced by soil moisture.

IV. Hydroxylation reactions

The s-triazine herbicides can be detoxified through hydrolysis to the formation of nonphytotoxic hydroxyanalogs. Formation of such hydroxyanalogs has been reported for atrazine, simazine, and propazine in soils (ARMSTRONG et al. 1967, HARRIS 1967, ADAMS 1966, SKIPPER et al. 1967, ARMSTRONG and CHESTERS 1968). Considerable evidence has been accumulated showing the nonbiological nature of the reactions.

The work of SKIPPER et al. (1967) supported nonbiological detoxication as one of the major pathways of atrazine degradation. In greenhouse studies, they compared the loss in phytotoxicity of atrazine with $^{14}CO_2$ evolution from ring-labeled and chain-labeled atrazine from microbial isolates and soil. The half-life of atrazine as determined by $^{14}CO_2$ evolution was ten months. Bioassay data gave a half-life of one-half month. Hydroxyatrazine accounted for approximately 20 percent of the extracted ^{14}C activity after two to four weeks incubation of ^{14}C-atrazine in nonsterile or sterile soils.

Employing soil perfusion systems, ARMSTRONG et al. (1967) found that the conversion of atrazine to hydroxyatrazine occurred in the presence of soil. No microbial degradation of atrazine was detected after inoculation of a soil-free atrazine medium with perfusate. Atrazine hydrolysis occurred in sterilized soil at a pH of 3.9. The hydrolysis rate was tenfold greater in the presence of the soil than in the absence of soil at the same pH, thus indicating that atrazine hydrolysis is catalyzed by contact with soil.

HARRIS (1967) examined the effect of temperature on the loss of simazine, atrazine, and propazine in four soils. He identified the hydroxy derivatives as the degradation products in methanol extracts of the soils. Increasing soil temperatures to 95°C. greatly enhanced the conversion of atrazine to hydroxyatrazine, whereas increasing the temperature of an aqueous solution of atrazine without soil had only a slight effect. Since the 95°C. temperature is above that for normal microbial growth, this study provided evidence for nonbiological hydroxylation.

In addition, HARRIS (1967) showed that the presence in soil of 200 p.p.m. of a microbial inhibitor, sodium azide, had little effect on the

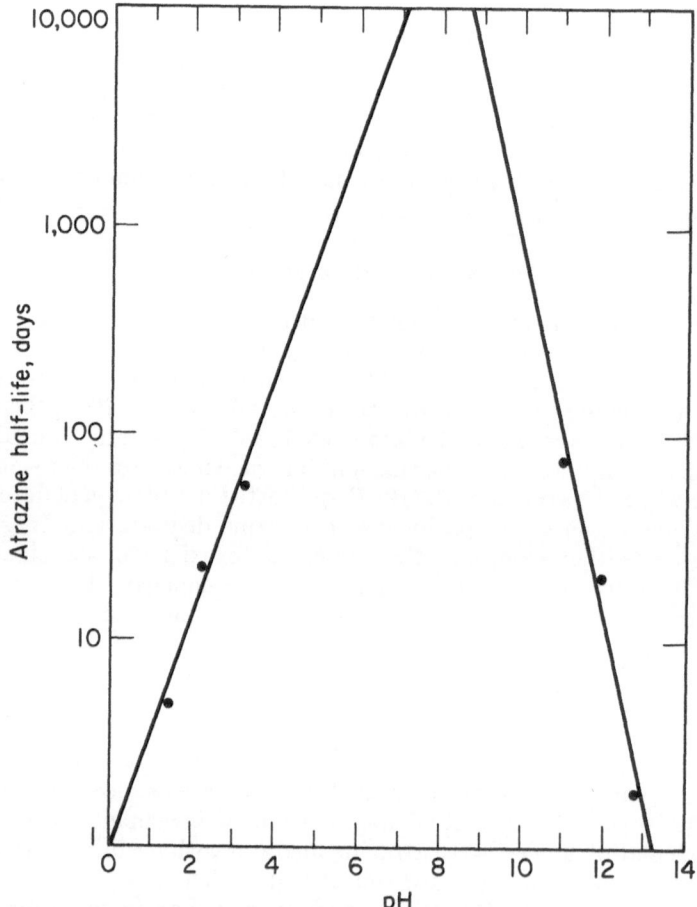

Fig. 5. Atrazine hydrolysis in aqueous solution as a function of pH (ARMSTRONG *et al.* 1967)

accumulation of hydroxy derivatives of simazine, atrazine, and propazine.

Chemical hydrolysis of atrazine (Fig. 5) occurs in strongly acid or basic solutions (ARMSTRONG *et al.* 1967). ARMSTRONG *et al.* (1967) postulated that the mechanisms proposed by HORROBIN (1963) for the hydrolysis of chlorotriazines would also hold for atrazine. Acid hydrolysis would result from protonation of a ring or chain nitrogen atom followed by cleavage of the C—Cl bond by water. Considerable evidence has been obtained to substantiate the mechanism postulated for adsorption-catalyzed hydrolysis of atrazine. This adsorption results from H-bonding between adsorbent and atrazine-ring N-atom and catalyzes hydrolysis by the mechanism proposed by HORROBIN (1963). The ring C-atom bonded to the Cl group, surrounded by electronega-

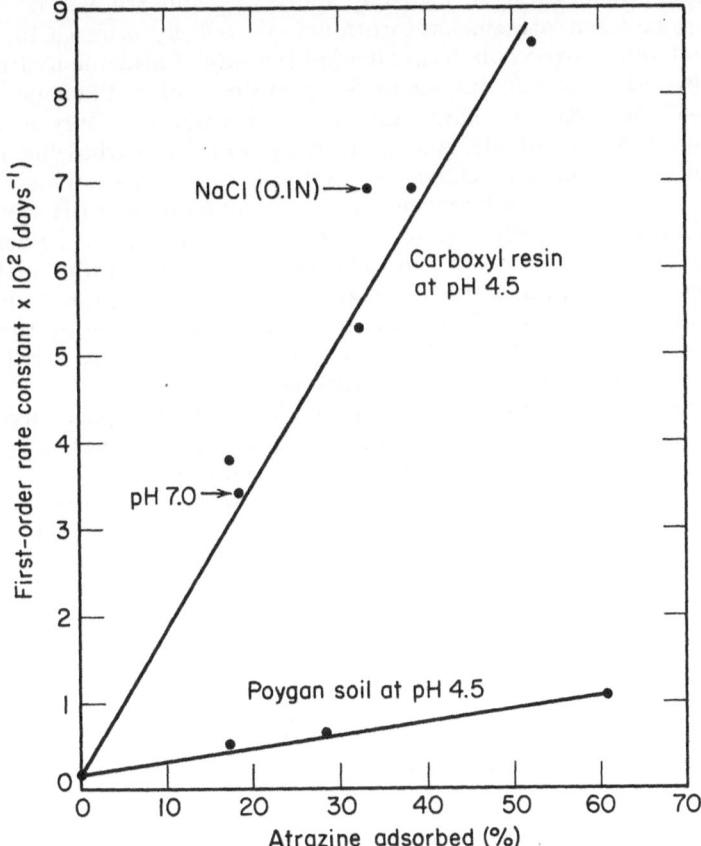

Fig. 6. Effect of atrazine adsorption on the first-order rate constant for atrazine hydrolysis (ARMSTRONG and CHESTERS 1968)

tive Cl and N atoms, is electron deficient and susceptible to displacement by strong nucleophilic agents such as OH ions, as evidenced by the susceptibility of atrazine to alkaline hydrolysis. In the neutral pH region, weak nucleophilic agents such as water are unable to replace the Cl group. However, H-bonding to the ring nitrogen atom causes further electron withdrawal from the electron-deficient C-atom, enabling the weak nucleophile water to replace the Cl group.

Several factors have been found to affect the rate of s-triazine hydroxylation in soils. ARMSTRONG et al. (1967) concluded that soil pH and organic matter content largely control the rate of atrazine hydroxylation. Atrazine degradation rates were greatest in soils of high organic matter and low pH. Increasing clay content of soils had only a slight effect on the degradation rate.

ARMSTRONG and CHESTERS (1968) examined the types of adsorption sites which catalyze atrazine hydrolysis and studied the relationship between atrazine adsorption and hydrolysis rates in soils. Only soil and a carboxylic resin affected the rate of atrazine hydrolysis (Fig. 6). Montmorillonite, cellulose acetate, and a phenolic resin exhibited high atrazine adsorption capacities but had only a slight effect on the rate of atrazine hydrolysis. For the carboxylic resin, atrazine adsorption was related closely to the degree of resin ionization as shown in Figure 7. Adsorption was at a maximum near pH 4, where the resin was completely in a protonated form, and decreased when the pH was increased or decreased. The decreasing adsorption with increasing pH was apparently a result of the decreased protonation of the resin and indicated that atrazine was adsorbed at the protonated carboxyl group. The decreased adsorption below pH 4 was attributed to protonation of the atrazine-ring nitrogen atom.

From the similarities of reaction kinetics between atrazine hydroxylation in solution and in soils, ARMSTRONG et al. (1967) have suggested that similar reaction mechanisms may be involved.[1]

From the results of the studies with soils of high clay content and high organic matter content, HARRIS (1967) concluded that montmorillonite protects simazine from hydrolysis through adsorption and that organic matter increases the degradation rate by acting as a catalytic agent. Simazine labeled with ^{36}Cl and ring-labeled ^{14}C produced a greater recovery of ^{36}Cl than ^{14}C. This was assumed to reflect the displacement of ^{36}Cl by a hydroxylation reaction.

WEBER et al. (1968) found that prometone was hydrolyzed to hydroxypropazine by a hydrogen sulfonic acid resin. He concluded that the prometone was adsorbed through proton association.

---

[1] In the discussion period following the presentation, Dr. M. M. Mortland, *Michigan State University*, pointed out that diffusion of the s-triazine to the adsorbing surface could also be responsible for the first-order reaction kinetics obtained for atrazine degradation.

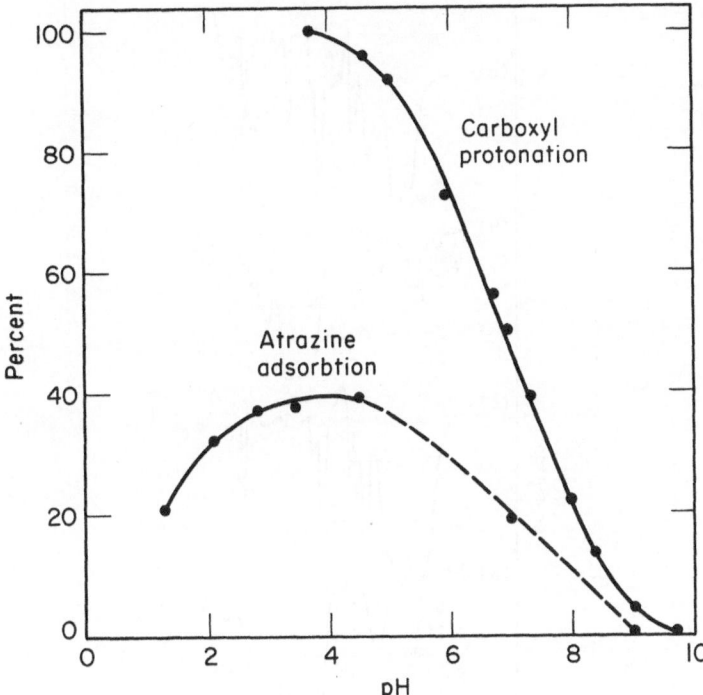

Fig. 7. Relationship between atrazine adsorption on carboxyl resin and the extent of carboxyl protonation (ARMSTRONG and CHESTERS 1968)

HANCE (1967) found that when high ratios of atrazine to adsorbing surfaces were used, nonbiological atrazine degradation was not significant.

The interaction of the *s*-triazines with silicate surfaces was examined by RUSSELL *et al.* (1968). They presented infrared data indicating protonation and hydroxylation when atrazine and propazine were adsorbed by montmorillonite (Fig. 8). Infrared spectra of atrazine-H-montmorillonite and propazine-NH_4-montmorillonite complexes were similar to the infrared spectra of the protonated hydroxy analogs. However, the spectra of adsorbed atrazine and propazine were not similar to those of the hydroxy analogs. They concluded that degradation of the chloro-*s*-triazines by interaction with the silicate surface results in the formation of the protonated hydroxy analogs of these compounds.

The presence of the keto form of the hydroxy analogs of atrazine and propazine was established on the basis of infrared spectra of the adsorbed compound. Two tautomeric forms of the hydroxy analogs are possible (Fig. 9). The appearance of a strong band at 1740 cm.$^{-1}$

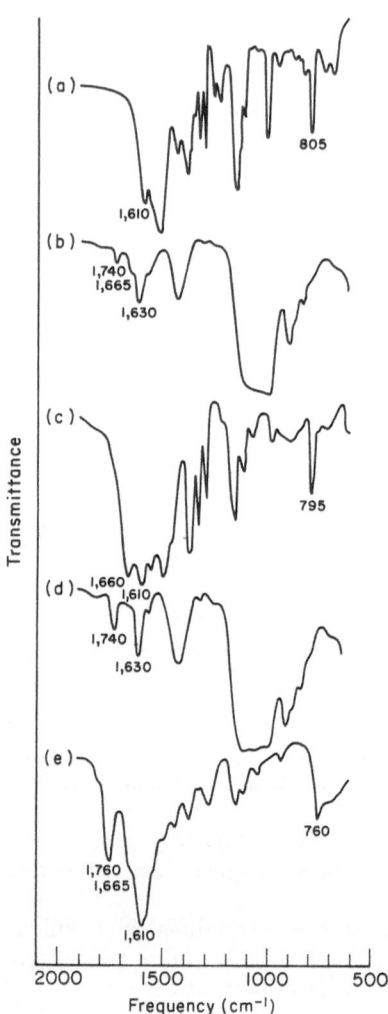

Fig. 8. Infrared spectra from 2,000 to 500 cm.$^{-1}$ of (a) propazine, (b) propazine-NH$_4$-montmorillonite clay complex, (c) hydroxypropazine, (d) hydroxypropazine-NH$_4$-montmorillonite clay complex, and (e) propazine + hydrochloric acid (Russell *et al.* 1968)

indicates that for the keto form (I) predominates in the protonated hydroxy species. This region of the infrared absorption spectra is characteristic of the carbonyl group.

In addition, Russell *et al.* (1968) discussed the chemical structure of the adsorbed, protonated hydroxytriazines, and on the basis of nuclear magnetic resonance data concluded that structure IV (Fig. 9) is most likely to predominate.

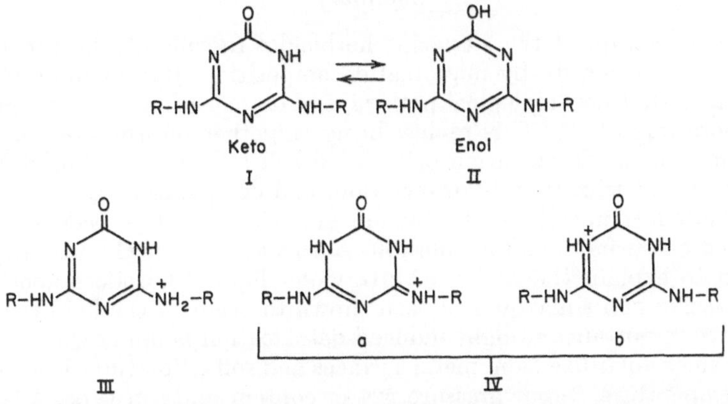

Fig. 9. Tautomeric structures (I and II) of the unprotonated and some possible tautomeric (III) and resonance (IV) structures of protonated hydroxy analogs of chloro-s-triazines (RUSSELL et al. 1968)

CRUZ et al. (1968), using infrared data, showed that the adsorption of propazine and prometone by montmorillonite was followed by protonation and hydroxylation on the montmorillonite surface.

BROWN and WHITE (1968) used similar data to show that montmorillonite was the effective mineral in the hydrolysis of 12 s-triazine herbicides by soil clays.

V. Dealkylation reactions

Because free radicals are known to occur in the soil as a result of biological production, it has been speculated that such free radicals might act as pesticide "scavengers." PLIMMER et al. (1968), who have carried out extensive studies of the N-dealkylation of s-triazines by soil fungi and higher plants, recently began employing electron-spin-resonance (ESR) measurements to study the interaction of pesticides with soil free radicals. Certain soils gave rise to complex hyperfine structures in their free radical ESR spectra, indicating that high concentrations of nonhumic acid fraction free radicals were present. They demonstrated that free radical reactions caused N-dealkalation of the s-triazines and have speculated that a similar mechanism is responsible for N-dealkalation of the s-triazines in soils as well as by soil fungi and higher plants.

PLIMMER et al. (1968) also made preliminary ESR measurements of atrazine and hydroxyatrazine interactions with a silt loam. In each case, changes in the ESR spectrum were observed. Hydroxyatrazine incorporated into the silt loam gave an ESR spectrum with a sharp signal characteristic of a free radical. Although the nature of these interactions is unknown, it appears from these measurements that free radical reactions do play a role in pesticide degradation processes.

Summary

Detoxication of the *s*-triazine herbicides in soils has been considered to be primarily by microbial decomposition. Recent information indicates that non-biological pathways of detoxication may be equally or more important. This review brings together information on four mechanisms of detoxication other than soil adsorption: photodecomposition, volatilization, hydroxylation, and dealkylation.

Photodecomposition of atrazine and simazine has been demonstrated on surfaces and in solution. A sensitization mechanism is proposed to explain the action of ultraviolet light. Photodecomposition of atrazine and simazine has been shown to occur in the field but the relative importance of light induced detoxication is unknown. Several *s*-triazines volatilize from metal surfaces and soils. Volatility is affected by temperature, vapor pressure, water content and soil type. Adsorption apparently influences the herbicide vapor pressure. Non-biological hydroxylation reactions of the *s*-triazines in soils are important. Adsorption catalyzes the reaction forming non-phytotoxic hydroxy-analogs. Several soil and environmental factors influence hydroxylation. The mechanism involves protonation of a ring nitrogen causing electron deficiency of the 2-carbon which is subjected to nucleophilic attack by water molecules. Nucleophilic replacement of chlorine by a hydroxyl group occurs at the 2-carbon. Free radical reactions in soils may cause detoxication by N-dealkylation.

Résumé*

La dégradation non biologique des desherbants type *s*-triazine

La disparition de l'activité des desherbants type *s*-triazine, dans les sols, a été considérée comme dûe principalement à la décomposition microbienne. Une information récente indique que des voies non biologiques de désintoxication peuvent avoir une importance égale ou supérieure. Cette mise au point rassemble les connaissances sur quatre mécanismes de désintoxication autre que l'adsorption par le sol: la photodécomposition, la volatilisation, l'hydroxylation et la désalkylation.

La photodécomposition de l'atrazine et de la simazine a été démontrée en surface et en solution. Un mécanisme de sensibilisation est proposé pour expliquer l'action des ultra-violets. Il a été démontré que la photodécomposition de l'atrazine et de la simazine se produit sur le terrain mais l'importance relative de la détoxification induite par le lumière est inconnue. Plusieurs *s*-triazines se volatilisent des surfaces métalliques et des sols. La volatilité est modifiée par la température, la tension de vapeur, l'humidité et le type du sol. L'adsorption influence apparemment la tension de vapeur du desherbant. Les réactions non biologiques d'hydroxylation des *s*-triazines dans le sol sont

---

* Traduit par R. Mestres.

importantes. L'adsorption catalyse la réaction de formation d'analogues hydroxylés non phytotoxiques. Plusieurs facteurs du sol et du milieu influencent l'hydroxylation. Le mécanisme implique une protonation d'un azote du noyau produisant une déficience en électrons du carbone 2 qui est sujet à une attaque nucléophile par des molécules d'eau. La substitution nucléophile d'un chlore par un groupe hydroxyle se produit sur le carbone 2. Les réactions radicalaires dans les sols peuvent provoquer la désintoxication par N-désalkylation.

Zusammenfassung*

Nicht-biologische Entgiftung der s-Triazin-Unkrautvertilgungsmittel

Man hat geglaubt, dass die Entgiftung der s-Triazin-Unkrautvertilgungsmittel in erster Linie durch mikrobische Zersetzung vor sich geht. Neuere Kenntnisse deuten darauf hin, dass nicht-biologische Entgiftungspfade gleich wichtig oder wichtiger sein können. Dieser Überblick bringt kenntnisse über vier Entgiftungsmechanismen zusammen, die von der Bodenadsorption verschieden sind: Photozersetzung, Verflüchtigung, Hydroxylierung und Dealkylierung.

Photozersetzung von Atrazin und Simazin sind auf Oberflächen und in Lösung gezeigt worden. Um die Wirkung des ultravioletten Lichtes zu erklären, ist ein Mechanismus zum Lichtempfindlichmachen vorgeschlagen worden. Es ist gezeigt worden, dass Photozersetzung von Atrazin und Simazin auf dem Feld vor sich geht, doch die relative Wichtigkeit der durch Licht verursachten Entgiftung ist unbekannt. Verschiedene s-Triazine verdunsten von Metalloberflächen und Böden. Die Verdunstung wird durch Temperatur, Dampfdruck, Wassergehalt und Bodenart beeinflusst. Die Adsorption beeinflusst offenbar den Dampfdruck des Unkrautvertilgungsmittels. Nicht-biologische Hydroxylierungs-Reaktionen der s-Triazine in Böden sind wichtig. Die Adsorption katalysiert die Reaktion indem sie nicht-pflanzentoxische Hydroxy-Analoge bildet. Der Mechanismus umfasst die Protonierung eines Ringstickstoffes und verursacht so einen Elektronenmangel des 2-Kohlenstoffes, der dadurch dem nucleophilen Angriff durch Wassermoleküle ausgesetzt wird. Nucleophile Ersetzung des Chlors durch eine Hydroxyl-Gruppe geht am 2-Kohlenstoff vor sich. Freie fundamentale Reaktionen in Böden können durch N-Dealkylierung Entgiftung verursachen.

References

ADAMS, JR., R. S.: Soxhlet extraction of simazine from soils. Soil Sci. Soc. Amer. Proc. **30**, 689 (1966).
ARMSTRONG, D. E., G. CHESTERS, and R. F. HARRIS: Atrazine hydrolysis in soil. Soil Sci. Soc. Amer. Proc. **31**, 61 (1967).

* Übersetzt von M. DÜSCH.

——, and G. CHESTERS: Adsorption catalyzed chemical hydrolysis of atrazine. Environmental Sci. Technol. **2**, 683 (1968).

BROWN, C. B., and J. L. WHITE: Reaction of 12 s-triazines with soil clays. Agron. Abstr., p. 89 (Nov. 1968).

BURNSIDE, O. C., E. L. SCHMIDT, and R. BEHRENS: Dissipation of simazine from the soil. Weeds **9**, 447 (1961).

CALVERT, J. C., and J. N. PITTS: Photochemistry. New York: Wiley (1966).

COMES, R. D., and F. L. TIMMONS: Effect of sunlight on the phytotoxicity of some phenylurea and triazine herbicides on a soil surface. Weeds **13**, 81 (1965).

CRUZ, MARIBEL, J. L. WHITE, and J. D. RUSSELL: Montmorillonite s-triazine interactions. Israel J. Chem. **6**, 315 (1968).

DAVIS, D. E., H. H. FUNDERBURK, JR., and N. G. SANSING: Absorption, transloca- tion, degradation, and volatilization of radioactive simazine. Proc. S. Weed Conf. **12**, 172 (1959).

——, D. R. ROBERTS, and H. H. FUNDERBURK, JR.: Radiochemical assay proce- dures for atrazine and atrazine degradation products. Proc. S. Weed Conf. **16**, 380 (1963).

DEWEY, R.: Further experimental evidence of the fate of simazine in the soil. Proc. Brit. Weed Control Conf. **5**, 91 (1960).

FOY, C. L.: Volatility and tracer studies with alkylamino-s-triazines. Weeds **12**, 103 (1964).

GAST, A.: Contributions to the knowledge of the behavior of triazines in soil. Presented 19th Ann. Symp. Crop Protection, Ghent, Belgium (1962).

HANCE, F. J.: Decomposition of herbicides in the soil by nonbiological chemical processes. J. Sci. Food Agr. **18**, 544 (1967).

HARRIS, C. I.: Fate of 2-chloro-s-triazine herbicides in soil. J. Agr. Food Chem. **15**, 157 (1967).

HORROBIN, S : The hydrolysis of some 1,3,5-triazines: Mechanism, structure, and reactivity. J. Chem. Soc. **1963**, 4130 (1963).

JORDAN, L. S., B. E. DAY, and W. A. CLERX: Photodecomposition of triazines. Weeds **12**, 5 (1964).

——, J. D. MANN, and B. E. DAY: Effects of ultraviolet light on herbicides. Weeds **13**, 43 (1965).

KAUFMANN, D. D., P. C. KEARNEY, and T. J. SHEETS: Microbial degradation of simazine. J. Agr. Food Chem. **13**, 238 (1965).

KEARNEY, P. C., T. J. SHEETS, and J. W. SMITH: Volatility of seven s-triazines. Weeds **12**, 83 (1964).

MITCHELL, L. C.: The effect of ultraviolet light (2537 Å) on 141 pesticide chemicals by paper chromatography. J. Assoc. Official Agr. Chemists **44**, 643 (1961).

PLIMMER, J. R.: Personal communication (Feb. 1969).

——, P. C. KEARNEY, and J. R. ROWLANDS: Free-radical oxidation of s-triazines: Mechanism of N-dealkylation. Presented Amer. Chem. Soc., Atlantic City, N.J. (1968).

RUSSELL, J. D., MARIBEL CRUZ, J. L. WHITE, G. W. BAILEY, W. R. PAYNE, JR., J. D. POPE, JR., and J. I. TEASLEY: Mode of chemical degradation of s-tri- azines by montmorillonite. Science **160**, 1340 (1968).

SHEETS, T. J., and L. L. DANIELSON: Herbicides in soils. In: The nature and fate of chemicals applied to soils, plants, and animals, pp. 170–181. *U.S. Depart- ment of Agriculture*, ARS 20-9, 221 pp. (1960).

SKIPPER, H. D., C. M. GILMOUR, and W. R. FURTICK: Microbial versus chemical degradation of atrazine in soils. Soil Sci. Soc. Amer. Proc. **31**, 653 (1967).

TALBERT, R. E., and O. H. FLETCHALL: Inactivation of simazine and atrazine in the field. Weeds **12**, 33 (1964).

WEBER, J. B., T. M. WARD, and S. B. WEED: Adsorption and desorption of diquat, paraquat, prometone, and 2,4-D by charcoal and exchange resins. Soil Sci. Soc. Amer. Proc. **32**, 197 (1968).

Persistence of triazine herbicides in soils

By

T. J. SHEETS*

Contents

I. Introduction

Although phytotoxic residues of the triazine herbicides are objectionable when, in some soils and under some environmental conditions, sensitive plants are injured the season after application, residual activity is essential for weed control and soil sterilization. Without residual activity, frequent applications of less persistent herbicides would be necessary, and costs of weed control would, therefore, be high. A good example of the advantages of residual phytotoxicity was described by HOROWITZ (1964), who suggested that a single application of simazine or atrazine at the beginning of winter rains in Israel would control winter weeds and maintain sorghum planted the next spring free of weeds until harvest.

The triazine herbicides exhibit varying degrees of persistence in soils. Within the group, persistence is related to chemical structure. Disappearance rates are dependent on several environmental and edaphic factors. That excessive persistence of atrazine and simazine,

---

* Pesticide Residue Research Laboratory, *North Carolina State University*, Raleigh, North Carolina.

as well as improper application, can cause problems has been recognized, and means have been devised to circumvent or eliminate injury to crops grown in rotation with corn. The best approach is to avoid planting sensitive crops on land sprayed the year before.

The persistence of the triazine herbicides was recognized early during their development, and research on this and related aspects has been extensive. Several reviews contain sections or paragraphs on the persistence of the triazines (HARRIS et al. 1968, SHEETS 1966, SHEETS and HARRIS 1965, VAN DER ZWEEP 1960, and others). The objectives of this paper are to review the research on the persistence of triazine herbicides in soils and to evaluate problems, real or potential, that are related to it.

II. Relation between structure and persistence

Triazine herbicides with a methoxy substituent in the *two* position of the ring are more persistent, generally, then those containing chloro or methylthio substituents (SHEETS et al. 1962, SHEETS and SHAW 1963, HOLLY and ROBERTS 1963, BUCHANAN and ROGERS 1963, SWITZER and RAUSER 1960). Prometone is one of the most persistent organic herbicides. In a Puerto Rican soil it was intermediate in persistence between fenac and picloram (DOWLER et al. 1968). In a greenhouse experiment where dilution and loss by leaching were eliminated and where losses to the atmosphere by volatilization were very low, prometone was more persistent than propazine and prometryne in four soils (Fig. 1). In one of the four soils residual activity of prometryne was greater than that of propazine, and in one soil propazine was more persistent than prometryne. When the residual activities of the 12 herbicides in four soils included in the experiment by SHEETS and SHAW (1963) were considered, the methylthio-derivatives appeared to be slightly more persistent than the chloro-derivatives (see Table I for results with two soils).

Results of experiments conducted under field conditions in Great Britain showed that the time required for 50 or 80 percent of a two-lb./acre application of prometone to disappear from two soils was more than twice the time required for disappearance of the same amount of propazine (HOLLY and ROBERTS 1963). In a clay soil at least 50 percent of a two-lb./acre application of prometone remained for 26 weeks or longer whereas 50 percent of the same rate of propazine disappeared from the same soil in about two weeks (HOLLY and ROBERTS 1963). Differences between the persistence of propazine and prometone were not as great in a sandy soil as in a clay soil.

Some differences in persistence have been observed among the 2-chloro-s-triazines. SWITZER and RAUSER (1960) ranked six triazines in order of activity and persistence in soils as follows: simazine ≥ atrazine ≅ propazine > ipazine > trietazine > chlorazine (Table I).

Propazine-bosket sandy loam   Propazine-tifton loamy sand

Prometone-bosket sandy loam   Prometone-tifton loamy sand

Prometryne-bosket sandy loam   Prometryne-tifton loamy sand

Propazine-cecil sandy loam   Propazine-sharkey clay

Prometone-cecil sandy loam   Prometone-sharkey clay

Prometryne-cecil sandy loam   Prometryne-sharkey clay

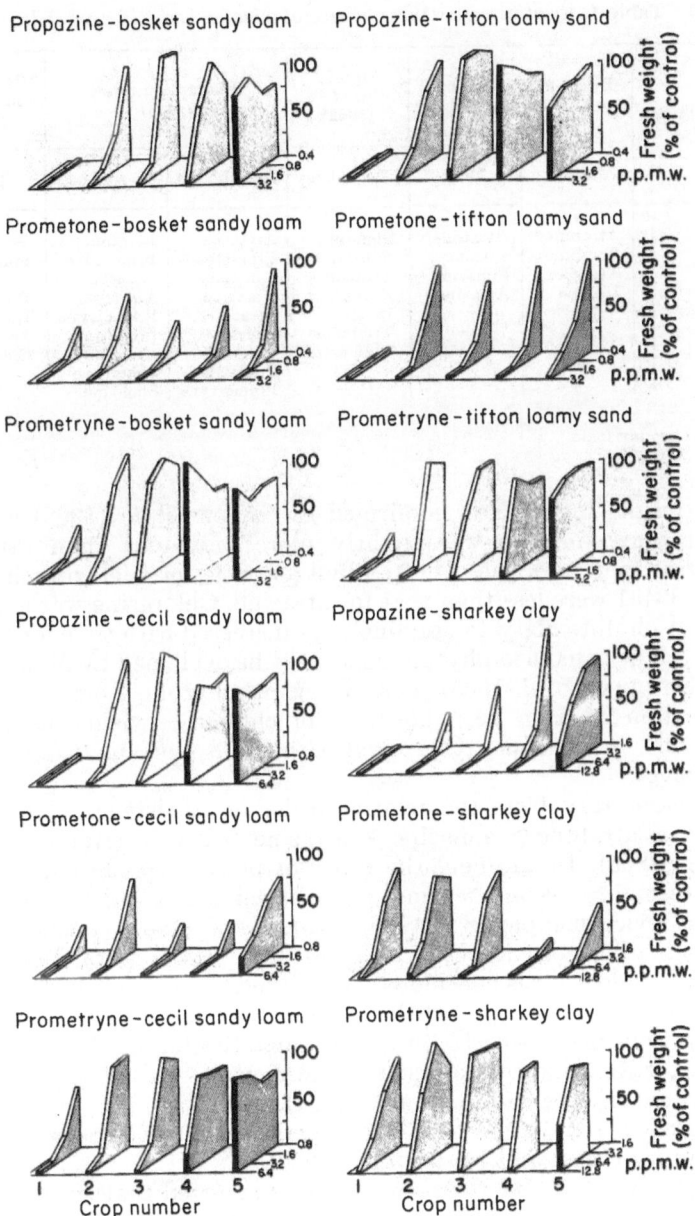

Fig. 1. Persistence of three s-triazines in four soils as shown by fresh weights of five successive crops of oats at four concentrations of the herbicides [reproduced by permission from Weeds **11**, 15–21 (1963)]; p.p.m.w. in the figure means parts per million by weight

Table I. *Relative persistences of several triazine herbicides in soils*

Persistence order		BUCHANAN & ROGERS (1963)		SHEETS et al. (1962)	SHEETS and SHAW (1963)		SWITZER and RAUSER (1960)
		Laboratory	Field	Greenhouse	Greenhouse[a]	Greenhouse[b]	Field
Persistence decreasing		Atratone	Atratone	Simetone	Atratone	Atratone	Prometone
		Simazine	Ipazine	Simazine	Prometone	Prometone	Simazine
		Atrazine	Simazine	Atrazine	Simetone	Simetone	Atrazine
		Ipazine	Atrazine	Propazine	Simazine	Ametryne	Propazine
		—	—	Ipazine	Atrazine	Prometryne	Ipazine
		—	—	Trietazine	Simetryne	Propazine	Trietazine
		—	—	Chlorazine	Ametryne	Atrazine	Chlorazine
		—	—	—	Propazine	Simetryne	—
		—	—	—	Prometryne	Simazine	—

[a] Bosket sandy loam.
[b] Cecil sandy loam.

These results were partly confirmed by SHEETS et al. (1962) who reported that simazine was slightly more persistent than atrazine, propazine, and norazine. The residual activities of trietazine, ipazine, and G-30451 were less than that for atrazine. Chlorazine was a special case; its phytotoxicity to seedling oats increased with time for several months. The enhanced phytotoxicity may be explained by N-dealkylation (KAUFMAN et al. 1965). Loss of one ethyl group from the *four* or *six* position or from both positions of chlorazine yields derivatives (trietazine or simazine, respectively) that are more phytotoxic than chlorazine.

In laboratory tests the relative order of persistence of four triazines was atratone > simazine ≅ atrazine > ipazine (BUCHANAN and ROGERS 1963). In the field the order of persistence shifted slightly, *i.e.*, atratone > ipazine > simazine > atrazine (Table I).

In a review completed in 1966, HARRIS et al. (1968) concluded that the residual phytotoxicities of atrazine and simazine were about equal when all evidence was considered collectively. A current evaluation of all the data leads to a slightly different conclusion; simazine often appears to be more persistent than atrazine. In Missouri several forage species planted immediately after treatment of the soil in the greenhouse were injured more by atrazine than by simazine, but in field plots seeded about one year after application, simazine caused the most injury (FINK and FLETCHALL 1963).

Recently FINK and FLETCHALL (1969) reported that simazine was more persistent than atrazine, and other work in Missouri (TALBERT and FLETCHALL 1964) showed that 0.16 lb./acre of atrazine and 0.31 lb./acre of simazine were present one year after application of two-lb./acre of each. In New Jersey the responses of wheat and barley to residues of simazine usually were more severe than the responses to residues of atrazine (ILNICKI 1969 a). In Lebanon, the residual activity of simazine was greater than that of atrazine (SAGHIR and CHOUDHARY 1967). In Israel, residual phytotoxicity of three triazines in soil was as follows: simazine > atrazine > propazine (HOROWITZ 1964). In Ontario, Canada, at low application rates atrazine disappeared more rapidly than simazine; but at five and six lb./acre the two were about equally persistent (SWITZER and RAUSER 1960). Occasionally, the two appear about equally persistent (MEGGITT 1969, ZEMANEK 1965), but data from many sources indicate that simazine usually persists a little longer than atrazine.

Other comparisons of differences in persistence under field conditions show that in England simazine was more persistent than propazine (HOLLY and ROBERTS 1963). At six locations in Nebraska residues of propazine were greater than those of atrazine in the upper three feet of soil (BURNSIDE et al. 1969), but in Israel residual activity of atrazine was greater than that of propazine (HOROWITZ 1964).

In field plots residual phytotoxicity of atrazine was greater than that of prometryne (BRYAND and ANDREWS 1967, HARRIS and SHEETS 1965); however, prometryne was less phytotoxic initially than atrazine (SHEETS and SHAW 1963). Whether prometryne is decomposed more rapidly than atrazine is open to question, but it appears that if allowance is made for differences in levels required to cause injury, prometryne may be more persistent than atrazine in some soils.

III. Variations in persistence among soils

Differences in persistence of triazine herbicides among soils are often observed (ADAMS 1968, AELBERS and HOMBURG 1959, HOLLY and ROBERTS 1963, SCUDDER 1963, SHEETS 1959, SHEETS and SHAW 1963, BURNSIDE et al. 1969). Residues in percent of simazine applied were greater in a sandy soil than in a clay or peat soil three, six, and 11 months after application (AELBERS and HOMBURG 1959). Residual toxicity of simazine was greater in a sand than in a peat (SCUDDER 1963). BURNSIDE et al. (1969) were able to conclude from studies conducted between 1962 and 1968 that soil texture had a greater effect on carryover of several herbicides, including propazine and atrazine, than climate in Nebraska. Residue carryover was greater in coarse than in fine-textured soils.

Experiments where soil-type comparisons are made under field conditions involve sites physically separated, often by great distances, unless soils are transported to adjacent sites. Such manipulations are seldom done; hence, soil-type effects on persistence per se are usually difficult to assess from field experiments due to the complicating influences of climate, especially rainfall and temperature. And while it is simple to show that persistence varies among soils (SHEETS 1959, SHEETS and SHAW 1963, SCHWEIZER and HOLSTUN 1966), determination of the causes for the differences is difficult in crude soil systems even under greenhouse or laboratory conditions.

The relation of triazine persistence (or conversely, decomposition) to soil properties and to soil temperature, soil moisture, and other soil environmental factors can best be assessed where conditions can be controlled. Laboratory experiments by Burnside et al. (1961), Near-pass (1965), and Armstrong et al. (1967) indicated that simazine loss from soils was greater at low than at neutral or high pH levels. Organic matter enhanced decomposition of 2-chloro-s-triazines (McCormick and Hiltbold 1966, Harris 1967, Roeth et al. 1969). Although adsorption to some clay soils may protect these herbicides against decomposition (Harris 1967, Armstrong et al. 1967), it functions to catalyze their hydrolysis in others (Armstrong et al. 1967, Armstrong and Chesters 1968).

The soil environment into which herbicides are introduced determines to a large extent the rate of disappearance, and many differences in persistence among soils can surely be attributed to variations of temperature and moisture levels. These influences on persistence will be discussed in the next section.

IV. Effects of weather and climate on persistence

Weather and climate influence the rate of disappearance of herbicides through their effects on volatilization, surface removal in runoff, and movement downward. Although these processes reduce herbicidal residues in soils, persistence is not altered since the herbicide molecule continues to exist in other segments of the environment.

Weather and climate alter persistence through their effects on degradation (photochemical, biological, and nonbiological decomposition). Microbial decomposition depends on favorable temperature and moisture conditions, and nonbiological decomposition also appears to be moisture and temperature dependent (Harris 1967). Warm, moist climates promote disappearance of triazine herbicides from soils (Dowler et al. 1968, Harris et al. 1969), and persistence is more prolonged in cold, dry climates (Adams 1968, Burnside et al. 1969, Roadhouse and Birk 1961).

In temperate and arctic regions triazine herbicides disappear from soil most rapidly during summer and least rapidly during winter (Burnside et al. 1961). Rates of disappearance during spring and fall months are usually intermediate between the extremely different rates of summer and winter. Even in the subtropical climate of the southern United States, atrazine, simazine, and ipazine were inactivated more rapidly during the summer than during winter or spring (Buchanan and Rogers 1963).

The halflife of simazine in soil varied indirectly with temperature (Burschel 1961). At 25°C., 50 percent of four p.p.m. disappeared in

20 days, at 18°C. in 39 days, and at 8.5°C. in 140 days. Simazine was not inactivated during six months in frozen soil (BURNSIDE *et al.* 1961). Others have shown that atrazine decomposition in soil increased with temperature between 10° and 35°C. (BURNSIDE 1965, McCORMICK and HILTBOLD 1966, ROETH *et al.* 1969). BUCHANAN and ROGERS (1963) reported that inactivation of simazine, ipazine, atrazine, and atratone proceeded more rapidly at 45°C. than at 25°, 30°, or 35°C. Results of field experiments confirm those obtained in the laboratory (BUCHANAN and ROGERS 1963, BURNSIDE 1965, HARRIS *et al.* 1969).

Within an area or region prevailing weather conditions alter persistence patterns. Atrazine and simazine are much more persistent during dry summers than during summers with abundant rainfall (BUCHHOLTZ 1965, HARRIS and SHEETS 1965, ILNICKI 1969 a, SWITZER and RAUSER 1960).

WILSON and COLE (1964) compared the effects of three watering schemes on the persistence of atrazine. Persistence was greatest in soil that was watered to field capacity once each week, intermediately persistent in soil that was watered to field capacity every 3.5 days, and least persistent in soil that was watered daily to field capacity. Soil moisture influences activity of soil microorganisms that degrade herbicides. Presumably, an optimum moisture level exists that extends over rather wide ranges in some soils. Excessive levels would retard decomposition by aerobes and enhance decomposition by anaerobes. Nonbiological decomposition might be slow if soils are extremely dry (HARRIS 1967).

Through well planned and well executed laboratory experiments, progress is being made toward understanding the factors that influence the persistence of the triazines, but this fact does not reduce the value of field and greenhouse experiments designed to study various aspects of persistence in soils. The results of the different approaches are complementary.

V. Effects of plant uptake on persistence

Uptake by plants and subsequent metabolic change or removal of harvested crops has been suggested as a means by which herbicides disappear. This is a logical assumption, but this path of herbicide removal from soils has not been studied as extensively as several others.

DEWEY (1960) observed that, at 16 months after application of 20 lb./acre simazine, residues were about ¼ as great (0.25 lb./acre *versus* one lb./acre) on a plot that was infested with *Cirsium* as on one close by without the weed. The author was not sure if the *Cirsium* infestation was the cause, or the result, of the low simazine residue.

BIRK and ROADHOUSE (1964) compared the persistence of atrazine in corn plots to that in fallow soil. At the end of the first season, residues were much greater in the soil that had been cropped than in the fallow soil. The corn plots were much drier during the growing season than the fallow plots, and the authors attributed the slow breakdown in corn plots to retarded microbial activity.

ASHLEY and RAHN (1965) investigated the effects of several cultural practices on the persistence of atrazine. They concluded that growing tolerant crops had little or no effect.

Atrazine residues were less in soil from pots cropped with corn, sorghum, or Johnsongrass than in soil from pots without crops (SIKKA and DAVIS 1966). The results from this greenhouse experiment were confirmed in a field test. Soil samples taken one, two, three, and six months after application from plots of corn and Johnsongrass sprayed with two, four, or 15 lb./acre of atrazine contained smaller residues of atrazine in the zero to six-inch depth than samples from fallow plots sprayed at the same rates.

Aside from the direct effects of plant uptake on residue levels, indirect effects of shading, of different water retention and percolation rates, and of altered microbial activity in cropped soils probably also influence the rate of dissipation of atrazine. Some of these factors tend to hasten dissipation, others tend to retard it.

In Nebraska, atrazine was more persistent buried at three, nine, or 15 inches under sod than at the same depths in bare soil (HARRIS et al. 1969). Plants differed from those grown by SIKKA and DAVIS (1966), and in Nebraska the atrazine was buried in aluminum pipes where root penetration was surely restricted to some extent. The more rapid disappearance from the bare soil was attributed to a higher soil temperature in bare soil than in the soil with sod (HARRIS et al. 1969). Thus, corn and Johnsongrass may have been even more effective in removing atrazine than the data of SIKKA and DAVIS (1966) indicate. Many other speculative statements are possible. Obviously, demonstration of the importance of plant uptake per se as a factor or process in herbicide loss from soils is elusive.

VI. Formulation and persistence

Granular atrazine persisted longer in soil than a wettable powder formulation applied as an aqueous suspension (BUCHHOLTZ 1965, MEGGITT 1964, 1969, WISK and COLE 1965); however, in a few experiments a difference between the two forms was not observed (MEGGITT 1962). Atrazine applied as a spray at three lb./acre in the spring injured oats slightly the following fall, but oats and soybeans planted the next spring appeared normal (WISK and COLE 1965); in contrast, atrazine applied as granules at the same time and rate injured oats and soybeans severely the next spring. When atrazine was formulated as the water-soluble ammonium sulfate granule, it was more persistent than when it was formulated on insoluble Attaclay (KURATLE and COLE 1964). Differences in persistence between granular and wettable powder formulations of atrazine are no longer of practical significance because the granular formulation is not marketed.

A granular formulation of prometryne was investigated recently (JORDAN et al. 1968), and at desert locations the granular form performed better than the spray form. The reverse was true at coastal and intermediate locations.

Oil-water emulsions have been used as carriers for atrazine applied postemergence (LeBARON 1966, SLIFE 1967, SYLWESTER 1966, WRIGHT 1966), and questions have been raised about the effect of the oil on persistence of the herbicide (SYLWESTER 1966). SLIFE (1967) suggested that the potential for problems with residues in the soil might be less with an oil-water mixture as the carrier than when water is used alone.

VII. Persistence at different soil depths

Subsurface placement of herbicides often enhances and extends weed control. The extended period usually can be attributed to longer persistence of herbicidally active residues from subsurface than from surface applications. Effectiveness of prometryne was increased by soil incorporation and subsurface placement (SCHWEIZER and HOLSTUN 1966, JORDAN et al. 1968, WIESE and HUDSPETH 1968). SCHWEIZER and HOLSTUN (1966) reported that residues from four lb./acre of prometryne, applied subsurface but not incorporated, were almost twice as toxic to oats 25 weeks after application as residues from the same rate applied to the soil surface.

Although rates of decomposition of most herbicides are sufficiently rapid that residues are seldom detected below the upper eight-inch layer, low concentrations of atrazine and simazine have been detected below the plow layer (BURNSIDE et al. 1963, MEGGITT 1969, BURNSIDE et al. 1969). In the investigations of BURNSIDE et al. (1963) bioassay of soil samples collected 16 months after application showed that the amounts of the two herbicides below 12 inches were greater than the amounts in the surface six inches of some plots.

One year after applying nine lb./acre to a Puerto Rican forest soil, DOWLER et al. (1968) found low concentrations of prometone in samples of soil taken from depths of zero to six, six to 12, 12 to 24, 24 to 36, and 36 to 48 inches. On the other hand, the triazine herbicides often cannot be detected below the plow layer. For example, on plots sprayed with one or three lb./acre during each of six years in a low rainfall area, simazine was not detected in the eight to 12-inch depth but two lb./acre remained in the upper eight inches of the three-lb./acre plots (DAWSON et al. 1968).

Detection of the triazine herbicides below the plow layer raises a question about the rate of disappearance under conditions existing at those soil depths. At least two groups have worked on this problem (HARRIS et al. 1969, ROETH et al. 1969). HARRIS et al. (1969), in cooperation with several associates at different locations in the continental United States and Puerto Rico, buried atrazine within horizontally-positioned aluminum tubes at three depths (two, nine, and 15 inches) and in 12 soil profiles (ten locations). The tubes were removed between three and 7.5 months after treatment and the soil analyzed for atrazine. Data from several locations are shown in Table II. Although the data were variable and recoveries from different depths of some locations were not different, the mean recovery (all soils) for the 15-inch depth was 61 percent greater than that for the three-inch depth. The authors pointed out that the organic matter content, temperature, and aeration were more favorable for degradation at the three-inch depth than at the other levels.

Table II. *Atrazine recoveries from soil after burial for various times at three, nine, and 15 inches* [from data published in Weed Sci. **17**, 27–31, (1969); reference HARRIS *et al.* (1969) in this paper]

Location	Exposure time (days)	Atrazine recovered from three depths		
		3 inches (%)	9 inches (%)	15 inches (%)
Mayaguez, Puerto Rico	146	2	7	6
Ft. Lauderdale, Florida	115	<1	<1	<1
Beltsville, Maryland	110	14	9	16
Lincoln, Nebraska	138	7	3	16
Lincoln, Nebraska	137	3	14	26
Hayes, Kansas ·	105	36	51	58
Rosemount, Minnesota	149	20	27	36
Stoneville, Mississippi	135	18	27	26

ROETH *et al.* (1969) removed samples from depths of zero to nine, 14 to 24, and 36 to 48 inches in two soils (Sharpsburg silty clay loam and Keith silt loam) and measured atrazine adsorption, microbial populations, and organic matter and clay contents in addition to atrazine degradation. Atrazine adsorption, microbial populations, and soil organic matter decreased with increasing depth. Paralleling these trends, atrazine degradation, determined under laboratory conditions, was two or three times more rapid in soil from the zero to nine-inch depth than in that from lower depths.

Decomposition of atrazine buried at depths of three, six, 12, 24, and 48 inches in four soil types varying with respect to texture and drainage was determined in another investigation at Beltsville, Maryland (WOOLSON 1969). Procedures were identical to those described above (HARRIS *et al.* 1969). Percent recoveries 135 days after application (WOOLSON 1969) varied with depth in all soils except one (Table

III), but an additional trend that was not apparent from the results of HARRIS et al. (1969) was observed. In the three soils where recoveries were different among depths, the percent recovery was greatest at intermediate depths. Recoveries from samples buried at 48 inches were less than peak recoveries observed at 12 or 24 inches.

Table III. *Atrazine recoveries from four Maryland soils after burial for 135 days at three, six, 12, 24, and 48 inches* (data from WOOLSON 1969)

Soil depth (inches)	Christiana clay loam (%)	Mattapex silt loam (%)	Othello silt loam (%)	Sassafras sandy loam (%)
3	7	12	19	23
6	12	15	30	30
12	48	25	44	20
24	40	50	35	19
48	38	35	13	35

In spite of the several factors that favor slower degradation as soil depth increases, significant amounts of atrazine were lost in three to 7.5 months at all locations included in the experiment by HARRIS et al. (1969). The results of WOOLSON (1969) showing greater rates of loss of atrazine at 48 inches than at 12 and 24 inches are puzzling. Obviously, some factor other than favorable organic matter level, temperature, and aeration must be involved in the more rapid decomposition (low recoveries) at the 48-inch depth. WOOLSON (1969) commented on this point as follows: "poor drainage increased atrazine retention at 12 inches but decreased its retention at lower depths."

VIII. Disappearance curves

The hypothesis has been advanced that the disappearance of the triazine herbicides, specifically simazine and atrazine, from soils conforms to a first-order reaction (the rate of loss is proportional to the concentration), and data have been presented to support this concept (BURSCHEL 1961, ARMSTRONG et al. 1967). Curves in Figure 2 were reproduced from BURSCHEL (1961).

Several investigators have determined residues of simazine and atrazine at intervals which permit construction of curves showing the relation between residue and time. In a nonsterile silt loam incubated in the laboratory, BURNSIDE et al. (1961) obtained the disappearance curve for simazine that is shown in Figure 3. In contrast to the results of BURSCHEL (1961), the curve of BURNSIDE et al. (1961) appears to consist of three phases. A lag period, during which no appreciable loss occurred, was followed by a rapid decomposition phase. The concen-

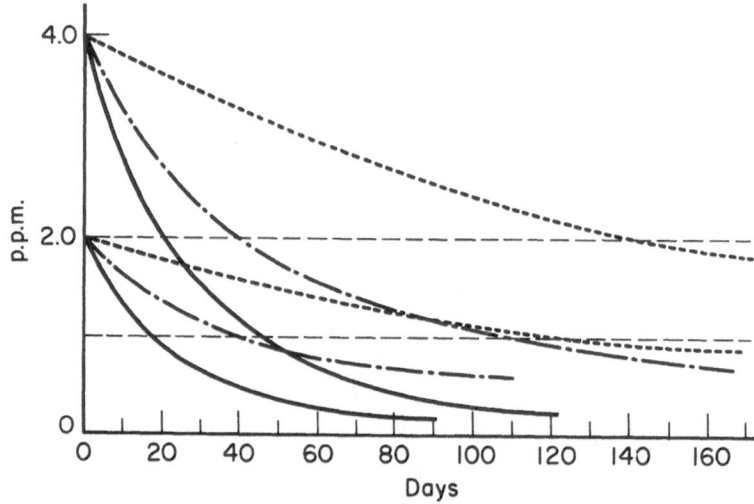

Fig. 2. The breakdown of two and four p.p.m. of simazine at two temperatures on
soil (löss) containing 10 percent organic matter [reproduced by permission
from Weed Research **1**, 131–141 (1961); reference BURSCHEL (1961) in
this paper]; ----- 8.5°C., - — - —-18°C., and ———25°C.

tration of simazine decreased rapidly during the second phase. During
the third phase, decomposition continued at a slower, ever decreasing
rate. Others have also shown the existence of a lag phase in the disappearance of simazine (HOLLY and ROBERTS 1963) and atrazine (BURN-
SIDE 1965). Lag periods in herbicide decomposition are usually attributed to a requirement for adaptation of microorganisms that are
effective for degradation; however, lags might occur in the decomposition of herbicides under field conditions due to unfavorable weather
and its effects on several processes.

AELBERS and HOMBURG (1959) and ROADHOUSE and BIRK (1961)
have presented data that, when the percent remaining was plotted as
a function of time, did not show an obvious lag phase (Fig. 4). Also,
curves of SIKKA and DAVIS (1966) for atrazine (not reproduced here)
and those of TALBERT and FLETCHALL (1964) for simazine and atrazine
(not reproduced here) show rapid disappearance, apparently without
a lag phase. However, in these experiments the amount of herbicide
remaining at the first sampling date was less than 50 percent. A slight
inflection in some curves could be interpreted as evidence for a lag
phase (BURSCHEL 1961, TALBERT and FLETCHALL 1964). Volatilization
as well as nonbiological decomposition appears to account for significant losses of atrazine from soil (FOY 1964, KEARNEY et al. 1964,
JORDAN et al. 1965); rapid loss by these processes could prevent the
appearance of lag phases in disappearance curves even when an

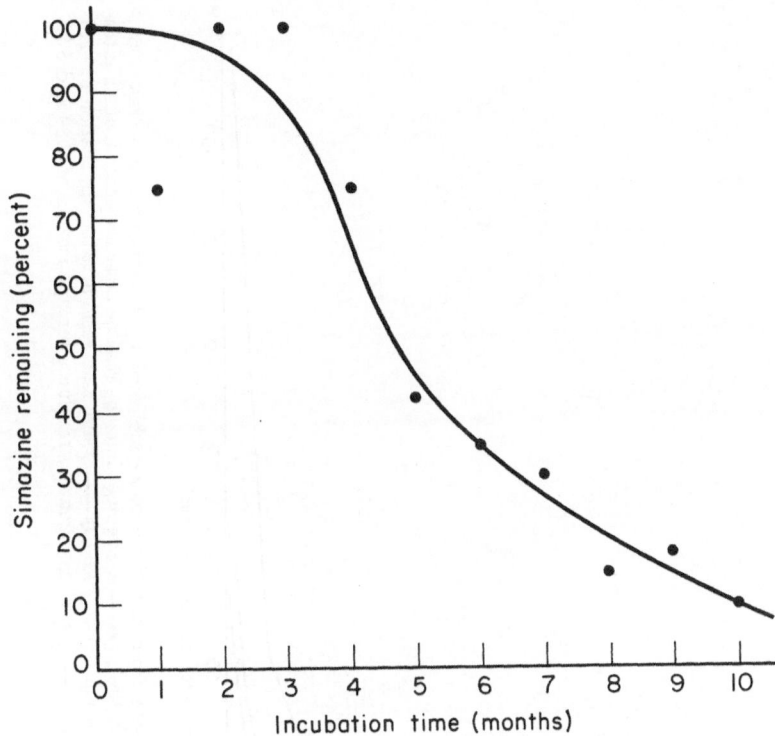

Fig. 3. Inactivation of simazine (four p.p.m.) in non-sterile Waukegan silt loam at 85°F. [from data published in Weeds **9**, 477–484 (1961); reference BURNSIDE *et al.* (1961) in this paper]

adaptation of effective microorganisms occurs. However, when all the evidence is considered, especially recent data on microbial and non-microbial decomposition (HARRIS 1967, HARRIS *et al.* 1968, ARMSTRONG *et al.* 1967, ARMSTRONG and CHESTERS 1968, SKIPPER *et al.* 1967, WEBER *et al.* 1968), a lag phase due to the adaptation of effective microorganisms appears to be the exception rather than the rule. Most investigators have observed immediate and continuous degradation that may be interrupted by adverse weather under field conditions. After the rapid phase, decomposition in field soils slows; and in some cases low levels persist for several months (Fig. 4, see also AELBERS and HOMBURG 1959, HOLLY and ROBERTS 1963).

IX. Problems associated with persistence in soils

Injury to susceptible crops grown in rotation with resistant crops, particularly corn and sorghum, which receive atrazine, simazine, or

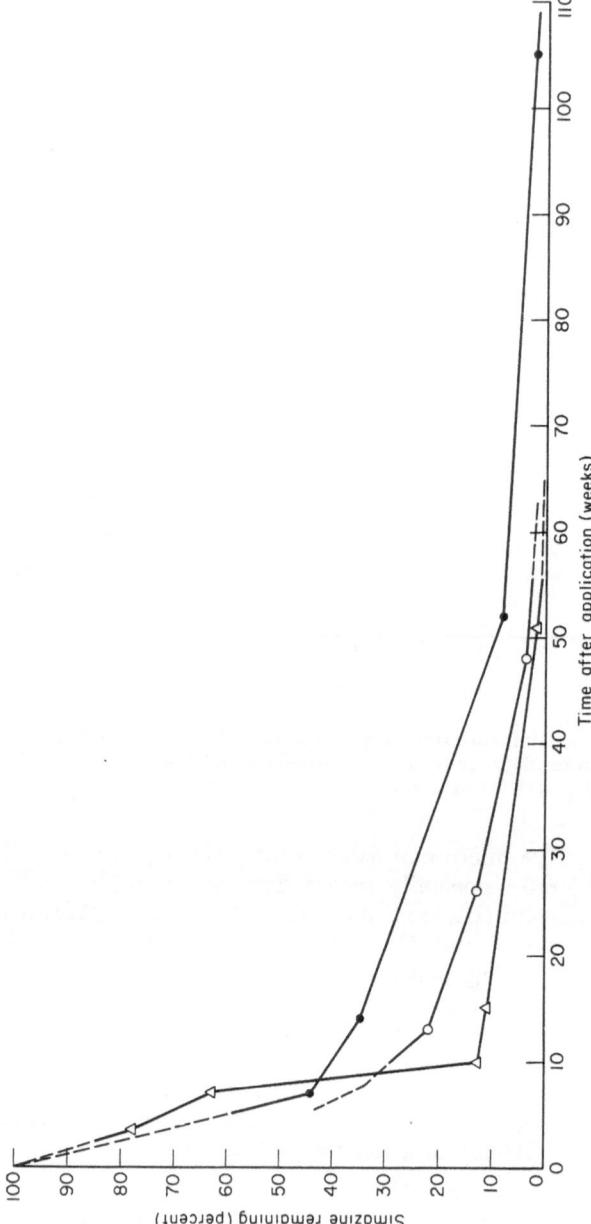

Fig. 4. Curves showing the disappearance of simazine from soil (from data published by Roadhouse and Birk 1961, Aelbers and Homburg 1959, and Burschel 1961); ●——● average simazine residue after application of six rates ranging from 0.5 to 20 lb./acre (Roadhouse and Birk 1961), O——O simazine residue after application of 4.46 lb./acre (Aelbers and Homburg 1959), and △——△ average simazine residue after application of 0.89 and 8.92 lb./acre (Burschel 1961)

propazine applications, is the major problem attributable to persistence of the triazines in soils (HARRIS *et al.* 1968, SHEETS and HARRIS 1965, SPLITTSTOESSER and DERSCHEID 1962). Atrazine injury to susceptible crops grown in rotation with corn has been encountered far more frequently in Canada and in the Midwest and Great Plains areas than in the southeastern United States. Compare the results of SCUDDER (1963) and BUCHANAN and ROGERS (1963) to those of FRANK (1966), BURNSIDE *et al.* (1963), and SPLITTSTOESSER and DERSCHEID (1962). Sugar beets are extremely sensitive to atrazine and under some conditions in Canada and the northern United States are injured two years after the season of application (ADAMS 1968, FRANK 1966). Problems have also been encountered in eastern Europe with residues of simazine, atrazine, and propazine (NEURURER 1962).

A survey by FRANK (1966) in Ontario, Canada showed that in one case involving 483 acres sprayed with atrazine in 1961 and planted to sugar beets in 1962, 16 percent (78 acres) was damaged severely, slight to moderate injury was observed on 38 percent, and 46 percent was apparently unaffected. Surveys in 1963 and 1964 gave similar results.

In Illinois, carryover problems from atrazine applied preemergence declined from a peak number in 1962 as farmers practiced more accurate and precise application methods (SLIFE 1964 and 1966). However, there was a recurrence of problems in the corn belt during the 1968 season from residues of atrazine applied in 1967 (SLIFE 1968, SYLWESTER 1969). The marked increase in frequency and amount of phytotoxic residues resulted from several factors. First, farmers were advised to:

"Use (Atrazine) where corn follows corn, since there may be some 'carryover' effect in the next season. If possible, follow corn with corn where Atrazine has been used, and if possible, do not use Atrazine where the area is to go into soybeans, small grain or small-seeded legumes the next year. The year before the area goes back into small grain, legumes or soybeans, use cultural control only or another herbicide which does not have 'carryover' action." (*Iowa State University* 1968.)

However, in most seasons crops are not damaged severely from atrazine applied the year before, and many farmers apparently plant and take the risk (SYLWESTER 1969). At least two other factors contributed to the injury encountered in 1968. Unusually wet weather in the spring and early summer of 1967 greatly altered normal weed control practices (harrowing, rotary hoeing, cultivation, and early postemergence applications of atrazine), and some atrazine was applied postemergence during late June or early July as a rescue effort or last resort (SLIFE 1968, SYLWESTER 1969). The wet period was followed by a dry summer, dry fall, dry winter, and dry spring in some states. The late postemergence applications and the dry weather that followed both tended to augment carryover of atrazine residues. Preplanting or preemergence applications that would have been applied four to six weeks earlier probably would not have caused injury of the same extent and severity.

Although carryover poses some hazard in North Carolina and other southeastern states, the only residue carryover from one spring to the next resulting in injury even to the most susceptible crops has usually been attributed to improper application (Worsham 1967). Cereals seeded in the fall after simazine or atrazine application, even in mild humid areas, may be injured.

Since the carryover hazard was recognized, studies have been conducted in many places on several aspects of the problem. Residue levels of simazine or atrazine and plant injury from their residues at any specific time are related to the rate of application (Ashley and Rahn 1965, Austin et al. 1968, Buchholtz 1965, Fink and Fletchall 1969, Frank 1966, Harris and Sheets 1965, Ilnicki 1969 a, Jones and Andrews 1964, Meggitt 1962 and 1969, Scudder 1963, Stroube and Bondarenko 1960). Twelve months after treatment Stroube and Bondarenko (1960) were unable to detect simazine in soil sprayed at a rate of two lb./acre but found residues of 0.46 and 2.8 lb./acre in the surface six inches of plots sprayed at a rate of four and eight lb./acre, respectively, and Buchholtz (1965) and Meggitt (1969) observed greater injury to oats and soybeans from four than from two lb./acre of simazine or atrazine about one year after application.

The hazard to rotational crops is greatest in fields which have been sprayed two or three years in succession with simazine or atrazine (Fink and Fletchall 1969, Frank 1966, Ilnicki 1969 a, Jones and Andrews 1964, Meggitt 1969). Meggitt (1969) observed no effect on oats, soybeans, and field beans one year after application of two lb./acre of atrazine; however, one year after the second or third annual application of two lb./acre, the sensitive crops were severely injured. At a more southern location Austin et al. (1968) found no evidence of accumulation or increased injury to sensitive rotational crops from three successive annual applications. Climatic and edaphic conditions vary sufficiently from season to season so that phytotoxic residues may remain during one season and not another. Also, increased injury as annual sprayings are repeated is more likely to occur in regions that are cold and dry than in those that are warm and moist (Burnside et al. 1965).

Although carryover has increased with the number of applications, accumulation to levels that sterilize the soil has not occurred even at rates of application four or five times as great as the rate needed for weed control (Austin et al. 1968, Bryand and Andrews 1967, Meggitt 1969, Scudder 1963). In New Jersey there was no difference in yields of sweet corn from annual applications of from one to eight lb./acre of simazine or atrazine over a seven-year period (ILNICKI 1969 a).

Planting cereals the fall after spring or summer use of simazine or atrazine has been recognized for several years as a hazardous practice. ILNICKI (1969 a) determined the response of wheat and barley to residues of these two herbicides over a six-year period. The spray schedule was devised so that the effect of continuous annual application (one each year) could be compared to that of two or more annual applications followed by a year without herbicide. The results showed that injury was less the year without herbicide. When an omission of an annual application preceded an application by one year or more, injury to wheat and barley was about as severe as when the herbicides were applied each year. Only by omitting an application during the year of seeding was the severity of response markedly reduced.

ILNICKI (1969 b) stated:

"An examination of species tolerances indicates that a majority of the crops which follow corn in rotation as practiced in New Jersey could be seriously affected by carryover of triazine herbicides as a result of successive applications to corn. The danger from carryover of triazines can be avoided or lessened if the application rates used are not in excess of those necessary for commercial weed control and if the application schedule employed allows for a periodic omission. If carryover is suspected, a crop more tolerant to the residue should be chosen for planting that season."

MEGGITT (1969) indicated that one crop of corn (one growing season) without atrazine was sufficient to permit growing susceptible crops on land sprayed once with four pounds per acre of atrazine or once/year for two or three years with two lb./acre. Advice against planting sensitive crops the year after atrazine use is printed on the manufacturer's label.

At excessive rates, residues persisted for two or three years in some regions. Residues from eight lb./acre of atrazine were not eliminated by growing one unsprayed crop of corn (MEGGITT 1969). At ten lb./acre atrazine and simazine injured oats in Nebraska the third growing season after application (BURNSIDE et al. 1965).

The results of SCHIRMAN and BUCHHOLTZ (1966) are especially interesting. Oat yields in 1961 on plots sprayed in 1960 with one, two, or four lb./acre of atrazine for quackgrass control exceeded those of the untreated control. Thus severe weed infestations reduced yields more than residues of atrazine.

Reports from Tennessee indicated that, at rates of two lb./acre, prometryne persisted in sufficient amounts to injure crops grown the next year (AUSTIN et al. 1968, BRYAND and ANDREWS 1967), but in Michigan prometryne at four lb./acre did not injure sensitive plants one year after a single application (MEGGITT 1969). Some injury developed one year after application of six lb./acre.

Data provided by WIESE (1967) showed the persistence and carry-over in soils of propazine used for weed control in grain sorghum in the southern Great Plains of the United States. Residues of propazine from two successive annual applications of two and four lb./acre were sufficient to injure wheat planted 1.5 years after the last application. Injury was much more severe on wheat planted six months after the last of three annual sprayings. Forage and grain yields of wheat planted six months after the third annual application of one lb./acre were reduced by 70 to 75 percent. Soybean yields were reduced by 61 percent on a plot sprayed two and three years before with four lb./acre of propazine. Yield of soybeans planted one year after the third annual application of two lb./acre was reduced by 73 percent. MARTIN (1967) reported that residues of atrazine and propazine used by farmers for weed control in grain sorghum on the high plains of west Texas persisted and injured cotton grown the next season. Problems with propazine appear similar to those encountered with atrazine.

Varietal and species differences in response are an important consideration in a discussion of injury from residues of herbicides in soil. ILNICKI (1969 b) observed that:

"Each investigator who has reported crop tolerance to residual triazine herbicides has established a relative tolerance for the crop with respect to other crops grown in the test. This relative tolerance is somewhat similar for simazine and atrazine; however, when the relative tolerance is established by researchers under different climatic and edaphic conditions and when different varieties are compared, the results are often in disagreement. It is important to realize that readings of relative crop tolerances are useful only as guides and that they do not have universal application."

Other potential or suspected problems should be watched and carefully investigated as the acreage sprayed with atrazine or other triazine herbicides increases. Low levels of atrazine can be carried in surface water moving off sprayed fields during and after rain storms (WHITE et al. 1967). Therefore, data are needed on the persistence of atrazine in water and aquatic soils. One such study with simazine shows that only 50 percent of the total applied remained one week after the fourth weekly addition of 0.25 or 0.5 p.p.m. to pond water (SUTTON et al. 1966).

Uptake by resistant crops grown in rotation with treated crops, resulting in illegal residues in the unsprayed crop, has been posed as a possible problem. Since a high percentage of each of the triazines that are used for selective weed control disappears before a rotational crop is planted and since the triazines are degraded in plants, the likelihood of illegal residues appearing at harvest in an unsprayed, resistant plant seems nil.

Summary

Literature on the persistence of triazine herbicides in soils is reviewed from the standpoint of herbicide-structure relations, variations among soils, effects of soil properties and characteristics, influences of climate, weather, and the soil environment, effects of uptake by tolerant plants, differences attributable to formulation, characteristics of disappearance curves, and problems associated with persistence in soils.

Methoxy triazines are usually much more persistent than chloro or methylthio triazines. Simazine persists slightly longer than atrazine in most soils, but in some soils the two are about equally persistent. Persistence is greatest in cool, dry climates. Atrazine and simazine disappear rapidly during the summer and slowly during the winter. Decomposition of chloro-substituted triazines occurs more rapidly in acid than in neutral or alkaline soils. Growing tolerant plants apparently hastens disappearance of atrazine from soil, but some results contrary to this view have been published. Granular forms of atrazine persist longer than wettable powder forms.

Generally, persistence of atrazine increases with soil depth to 24 inches. Below 24 inches the rate of loss (probably dilution) in some soils may increase somewhat.

Disappearance of atrazine and simazine apparently conforms to a first-order reaction. Although a lag phase followed by a period of rapid loss is sometimes observed, under favorable conditions degradation usually begins immediately after application and continues at an ever decreasing rate.

Injury to susceptible crops grown in rotation with resistant crops which receive atrazine, simazine, or propazine is the major problem associated with persistence of triazine herbicides. In areas where a hazard to sensitive plants exists the season after application, growing a tolerant crop for one season without use of a persistent herbicide usually allows residues to drop to nonphytotoxic levels. Although carryover of residues causes greater injury after two or three annual applications, accumulation to levels that sterilize the soil has not occurred.

Résumé*

Persistance des desherbants dérivés de la triazine dans les sols

Les travaux sur la persistance, dans les sols, des desherbants dérivés de la triazine ont été dépouillés dans l'objectif de définir les relations structure—pouvoir desherbant, les différences selon les sols,

---

* Traduit par R. Mestres.

les effets des propriétés et des caractéristiques des sols, les influences du climat, du temps et de l'écologie, les effects de l'absorption par les plantes non sensibles, les différences imputables à la formulation, les caractéristiques des courbes de dégradation et les problèmes associés avec la persistance dans les sols.

Les méthoxy triazines sont habituellement plus persistantes que les chloro ou méthylthio triazines. La simazine persiste légèrement plus que l'atrazine dans la plupart des sols, mais dans certains cas les deux produits subsistent pratiquement de façon identique.

La persistance est maximale dans les climats froids et secs. L'atrazine et la simazine disparaissent rapidement en été et lentement pendant l'hiver. La décomposition survient plus rapidement dans les sols acides que dans les sols alcalins. La pousse des plantes non sensibles accélère apparemment la disparition de l'atrazine du sol, mais quelques résultats contraires ont été publiés. Utilisée sous forme granulés, l'atrazine persiste plus longtemps que sous forme poudres mouillables.

En général, la persistance de l'atrazine augmente avec la profondeur du sol jusqu'à 60 cm. Au dessous de 60 cm. les pertes peuvent augmenter légèrement dans quelques sols.

La disparition de l'atrazine et de la simazine correspond apparemment à une réaction du premier ordre. Bien qu'une phase stationnaire suivie d'une perte rapide soit parfois observée, dans des conditions favorables, la dégradation commence généralement aussitôt après l'application et se poursuit à une vitesse toujours décroissante, sauf interruption par le mauvais temps.

Les dégats aux plantes susceptibles cultivées en assolement avec des cultures résistantes traitées à l'atrazine, la simazine ou la propazine sont les seuls problèmes importants soulevés par la persistance des desherbants du type triazine. Dans les zones où existe un danger pour les plantes sensibles pendant la saison qui suit le traitement, la culture d'une plante résistante pendant cette saison, sans emploi d'herbicide persistant, permet aux résidus d'atteindre des concentrations non phytotoxiques.

Bien que le cumul des résidus produise des dégats plus importants après deux ou trois traitements annuels, des teneurs capables de stériliser le sol n'ont pas été atteintes.

Zusammmenfassung*

Die Hartnäckigkeit der Triazin-Unkrautvertilgungsmittel in Böden

Literatur über die Hartnäckigkeit der Triazin-Unkrautvertilgungsmittel in Böden wurde vom Standpunkt der Unkrautvertilgungsmittelstruktur-Verhältnisse, der Abweichungen unter den Bodenarten,

---

* Übersetzt von M. Düsch.

der Folgen der Bodeneigentümlichkeiten und charakteristischen Merkmale, der Einflüsse des Klimas, des Wetters und der Bodenumgebung, der Erfassungsergebnisse durch widerstandsfähige Pflanzen, der Formulierung zuschreibbaren Verschiedenheiten, der charakteristischen Merkmale ihrer Abbaukurven und der Probleme, die mit der Hartnäckigkeit in Böden verbunden sind, revidiert.

Methoxy-Triazine sind gewöhnlich hartnäckiger als Chlor- oder Methylthio-Triazine. In den meisten Böden bleibt Simazin etwas länger bestehen als Atrazin, aber in einigen Böden sind die beiden ungefähr gleich hartnäckig. Die Hartnäckigkeit ist in kühlen trockenen Klimas am grössten. Atrazin und Simazin verschwinden im Sommer schnell und im Winter langsam. Zerfall tritt in sauren Böden schneller auf als in alkalischen. Wachsende widerstandsfähige Pflanzen beschleunigen anscheinend das Verschwinden von Atrazin aus dem Boden; es sind jedoch einige Resultate, die im Gegensatz zu dieser Ansicht stehen, veröffentlicht worden. Körnige Formen von Atrazin sind beharrlicher als Pulverformen, die fähig sind Feuchtigkeit aufzunehmen.

Im allgemeinen nimmt die Hartnäckigkeit von Atrazin mit Bodentiefe bis zu 60 cm. zu. Unterhalb 60 cm. kann die Verlustrate in einigen Böden etwas zunehmen.

Das Verschwinden von Atrazin und Simazin entspricht anscheinend einer Reaktion erster Ordnung. Obgleich manchmal eine langsame Phase, der eine Periode schnellen Verlustes folgt, beobachtet wird, beginnt die Degradierung gewöhnlich unter günstigen Bedingungen sofort nach der Anwendung and schreitet in einer immer mehr abnehmenden Rate fort, ausgenommen wenn sie durch ungünstiges Wetter unterbrochen wird.

Das einzige bedeutende Problem, das mit der Hartnäckigkeit der Triazin-Unkrautvertilgungsmittel verbunden ist, ist der Schaden, der empfindlichen Ernten zugefügt wird, die in Rotierung mit widerstandsfähigen Ernten, welche Atrazin erhalten, angebaut werden. In Landstrichen, wo eine Gefahr für empfindliche Pflanzen in der Saison nach Anwendung existiert, erlaubt das Anpflanzen einer widerstandsfähigen Ernte für eine Saison ohne Gebrauch eines hartnäckigen Unkrautvertilgungsmittels, ein Absinken der Rückstände auf nicht pflanzengiftige Ebenen. Obgleich Übertragungen von Rückständen nach zwei oder drei jährlichen Anwendungen grösseren Schaden zufügen, ist eine Anhäufung von solchen Mengen, die den Boden sterilisieren würden noch nicht vorgekommen.

References

ADAMS, R. S., JR.: Soil factors contributing to atrazine carryover. Minnesota Sci. **25** (1), 9 (1968).

AELBERS, E., en K. HOMBURG: De inactivering en penetratie van simazine in de grond. Mededel. Landbouwhogeschool en Opzoekingsst. Staat Gent **24**, 893 (1959).

Armstrong, D. E., and G. Chesters: Adsorption catalyzed chemical hydrolysis of atrazine. Environ. Sci. Technol. **2**, 683 (1968).

—— ——, and R. F. Harris: Atrazine hydrolysis in soil. Soil Sci. Soc. Amer. Proc. **31**, 61 (1967).

Ashley, R. A., and E. M. Rahn: Persistence of atrazine and diphenamid soil residues and tolerance of these residues by certain crops. Proc. N.E. Weed Control Conf. **19**, 362 (1965).

Austin, D., H. Andrews, and L. N. Skold: Disappearance of atrazine, DCPA, diphenamid, diuron, linuron, norea, prometryne, and trifluralin from soils treated for three years. Proc. S. Weed Conf. **21**, 314 (1968).

Birk, L. A., and F. E. B. Roadhouse: Penetration of and persistence in soil of the herbicide atrazine. Can. J. Plant Sci. **44**, 21 (1964).

Bryand, T. A., and H. Andrews: Disappearance of diuron, norea, linuron, trifluralin, diphenamid, DCPA, and prometryne from the soil. Proc. S. Weed Conf. **20**, 395 (1967).

Buchanan, G. A., and E. G. Rogers: Role of temperature in the inactivation of some s-triazine herbicides. Proc. S. Weed Conf. **16**, 393 (1963).

Buchholtz, K. P.: Factors influencing oat injury from triazine residues in soil. Weeds **13**, 362 (1965).

Burnside, O. C.: Longevity of amiben, atrazine, and 2,3,6-TBA in incubated soils. Weeds **13**, 274 (1965).

——, C. R. Fenster, and G. A. Wicks: Dissipation and leaching of monuron, simazine, and atrazine in Nebraska soils. Weeds **11**, 209 (1963).

—— —— ——, and J. V. Drew: Effect of soil and climate on herbicide dissipation. Weed Sci. **17**, 241 (1969).

——, E. L. Schmidt, and R. Behrens: Dissipation of simazine from soil. Weeds **9**, 477 (1961).

——, G. A. Wicks, and C. R. Fenster: Herbicide longevity in Nebraska soils. Weeds **13**, 277 (1965).

Burschel, P.: Untersuchungen über das Verhalten von Simazin im Boden. Weed Research **1**, 131 (1961).

Dawson, J. H., V. F. Bruns, and W. J. Clore: Residual monuron, diuron, and simazine in a vineyard soil. Weed Sci. **16**, 63 (1968).

Dewey, O. R.: Further experimental evidence on the fate of simazine in the soil. Proc. 5th Brit. Weed Control Conf., p. 91 (1960).

Dowler, C. C., W. Forestier, and F. H. Tschirley: Effects and persistence of herbicides applied to soil in Puerto Rican forests. Weed Sci. **16**, 45 (1968).

Fink, R. J., and O. H. Fletchall: Forage crop establishment in soil containing atrazine or simazine residues. Weeds **11**, 81 (1963).

—— ——: Soybean injury from triazine residues in soil. Weed Sci. **17**, 35 (1969).

Foy, C. L.: Volatility and tracer studies with alkylamino-s-triazines. Weeds **12**, 103 (1964).

Frank, R.: Atrazine carryover in production of sugar beets in southwestern Ontario. Weeds **14**, 82 (1966).

Harris, C. I.: Fate of 2-chloro-s-triazine herbicides in soil. J. Agr. Food Chem. **15**, 157 (1967).

——, D. D. Kaufman, T. J. Sheets, R. G. Nash, and P. C. Kearney: Behavior and fate of s-triazines in soils. Adv. Pest Control Research **8**, 1 (1968).

——, and T. J. Sheets: Persistence of several herbicides in the field. Proc. N.E. Weed Control Conf. **19**, 359 (1965).

——, E. A. Woolson, and B. E. Hummer: Dissipation of herbicides at three soil depths. Weed Sci. **17**, 27 (1969).

Holly, K., and H. A. Roberts: Persistence of phytotoxic residues of triazine herbicides in soil. Weed Research **3**, 1 (1963).

HOROWITZ, M.: The use of triazine herbicides to replace cultivations in dryland sorghum in Israel. Proc. 7th Brit. Weed Control Conf., p. 913 (1964).

ILNICKI, R. D.: Unpublished data, *Rutgers—The State University of New Jersey* (1969 a).

——: Personal communication, *Rutgers—The State University of New Jersey* (1969 b).

Iowa State University of Science and Technology (with Cooperative Extension Service, *U.S. Department of Agriculture*): Pre-emergence and pre-plant weed control in corn and soybeans. Weed Control Series, Pm—371 (Rev.), 6 pp. (1968).

JONES, L. B., and H. ANDREWS: Effect of rate of application on the persistence in the soil of several preemergence herbicides. Proc. S. Weed Conf. **17**, 374 (1964).

JORDAN, L. S., J. M. LYONS, W. H. ISOM, and B. E. DAY: Factors affecting performance of preemergence herbicides. Weed Sci. **16**, 457 (1968).

——, J. D. MANN, and B. E. DAY: Effects of ultraviolet light on herbicides. Weeds **13**, 43 (1965).

KAUFMAN, D. D., P. C. KEARNEY, and T. J. SHEETS: Microbial degradation of simazine. J. Agr. Food Chem. **13**, 238 (1965).

KEARNEY, P. C., T. J. SHEETS, and J. W. SMITH: Volatility of seven s-triazines. Weeds **12**, 83 (1964).

KURATLE, H., and R. H. COLE: Effects of granular atrazine formulation on weed control and persistence. Proc. N.E. Weed Control Conf. **18**, 363 (1964).

LeBARON, H. M.: Oil water emulsions as carriers for atrazine. Proc. N.C. Weed Control Conf. **21**, 46 (1966).

MARTIN, A. G.: Herbicide residue analysis in the Texas High Plains. Proc. S. Weed Conf. **20**, 122 (1967).

McCORMICK, L. L., and A. E. HILTBOLD: Microbiological decomposition of atrazine and diuron in soil. Weeds **14**, 77 (1966).

MEGGITT, W. F.: Herbicide residues in the soil and the effect on subsequent crops. Abstr. Weed Soc. Amer., p. 15 (1964).

—— Persistence of herbicide residues in soils. *Michigan State University*, East Lansing, unpublished report (1969).

—— Residual studies on triazine herbicides. Proc. N.C. Weed Control Conf. **19**, 82 (1962).

NEARPASS, D. C.: Effects of soil acidity on the adsorption, penetration, and persistence of simazine. Weeds **13**, 341 (1965).

NEURURER, H.: Zur Kenntnis der Auswirkung von Herbiziden im Boden. 1. Mitteilung: Untersuchung über die Nachwirkung von Bodenherbiziden auf die Fruchtfolge. Pflanzenschutzber. **28**, 145 (1962).

ROADHOUSE, F. E. B., and L. A. BIRK: Penetration of and persistence in soil of the herbicide 2-chloro-4,6-bis(ethylamino)-s-triazine (simazine). Can. J. Plant Sci. **41**, 252 (1961).

ROETH, F. W., T. L. LAVY, and O. C. BURNSIDE: Atrazine degradation in two soil profiles. Weed Sci. **17**, 202 (1969).

SAGHIR, A. R., and A. H. CHOUDHARY: Triazine herbicides on maize and their residual effect on following crops. Weed Research **7**, 272 (1967).

SCHIRMAN, R., and K. P. BUCHHOLTZ: Influence of atrazine on control and rhizome carbohydrate reserves of quackgrass. Weeds **14**, 233 (1966).

SCHWEIZER, E. E., and J. T. HOLSTUN, JR.: Persistence of five cotton herbicides in four southern soils. Weeds **14**, 22 (1966).

SCUDDER, W. T.: Persistence of simazine in Florida mineral and organic soils. *University of Florida* Agr. Exp. Sta. Tech. Bull. 657, 23 pp. (1963).

SHEETS, T. J.: The comparative toxicities of monuron and simazine in soil. Weeds **7**, 189 (1959).

—— Problems in the persistence of herbicides in plants and soils. Proc. 8th Brit. Weed Control Conf., p. 842 (1966).

——, A. S. Crafts, and H. R. Drever: Influence of soil properties on the phytotoxicities of the s-triazine herbicides. J. Agr. Food Chem. **10**, 458 (1962).

——, and C. I. Harris: Herbicide residues in soils and their phytotoxicities to crops grown in rotations. Residue Reviews **11**, 119 (1965).

——, and W. C. Shaw: Herbicidal properties and persistence in soils of s-triazines. Weeds **11**, 15 (1963).

Sikka, H. C., and D. E. Davis: Dissipation of atrazine from soil by corn, sorghum and Johnsongrass. Weeds **14**, 289 (1966).

Skipper, H. D., C. M. Gilmour, and W. R. Furtick: Microbial versus chemical degradation of atrazine in soils. Soil Sci. Soc. Amer. Proc. **31**, 653 (1967).

Slife, F. W.: Soil residue problems with herbicides. Illinois Custom Spray Operators' Training School **16**, 6 (1964).

—— Personal communication, *University of Illinois*, Urbana. (1966, 1968).

—— Atrazine-oil and other postemergence treatments for corn. Illinois Custom Spray Operators' Training School **19**, 6 (1967).

Splittstoesser, W. E., and L. A. Derscheid: Effects of environment upon herbicides applied preemergence. Weeds **10**, 304 (1962).

Stroube, E. W., and D. D. Bondarenko: Persistence and distribution of simazine applied in the field. Proc. N.C. Weed Control Conf. **17**, 40 (1960).

Sutton, D. L., T. O. Evrard, and S. W. Bingham: The effects of repeated treatments of simazine on certain aquatic plants and residue in water. Proc. N.E. Weed Control Conf. **20**, 464 (1966).

Switzer, C. M., and W. E. Rauser: Effectiveness and persistence of certain herbicides in soil. Proc. N.E. Weed Control Conf. **14**, 329 (1960).

Sylwester, E. P.: Use of atrazine in water plus oil for weed control in corn. Proc. N.C. Weed Control Conf. **21**, 45 (1966).

—— Personal communication, *Iowa State University*, Ames (1969).

Talbert, R. E., and O. H. Fletchall: Inactivation of simazine and atrazine in the field. Weeds **12**, 33 (1964).

Van der Zweep, W.: The persistence of some important herbicides in soil. In E. K. Woodford and G. R. Sagar (eds.): Herbicides and the soil, p. 79. Oxford: Blackwell (1960).

Weber, J. B., T. M. Ward, and S. B. Weed: Adsorption and desorption of diquat, paraquat, prometone, and 2,4-D by charcoal and exchange resins. Soil Sci. Soc. Amer. Proc. **32**, 197 (1968).

White, A. W., A. P. Barnett, B. G. Wright, and J. H. Holladay: Atrazine losses from fallow land caused by runoff and erosion. Environ. Sci. Technol. **1**, 740 (1967).

Wiese, A. F.: Unpublished results, *Texas A & M University* and *U.S. Department of Agriculture*, Southwestern Great Plains Research Center, Bushland, Texas (1967).

——, and E. B. Hudspeth, Jr.: Subsurface application and shallow incorporation of herbicides on cotton. Weed Sci. **16**, 494 (1968).

Wilson, H. P., and R. H. Cole: Effect of formulation, rate and soil moisture on atrazine persistence. Proc. N.E. Weed Control Conf. **18**, 358 (1964).

Wisk, E. L., and R. H. Cole: Persistence in soils of several herbicides used for corn and soybean weed control. Proc. N.E. Weed Control Conf. **19**, 356 (1965).

Woolson, E. A.: Personal communication and unpublished data, Crops Research Division, Agricultural Research Service, *U.S. Department of Agriculture*, Beltsville, Maryland (1969).

Worsham, A. D.: Safe use of herbicides. Proc. W. Weed Control Conf. **21**, 20 (1967).

Wright, W. G.: Atrazine-oil for post emergence weed control in corn and sorghum. Proc. N.C. Weed Control Conf. **21**, 44 (1966).

Zemanek, J.: The study of the influence of triazine herbicides on the subsequent crops. Ochrana Rostlin **2**, 9 (1965).

Ways and means to influence the activity and the persistence of triazine herbicides in soils

By

Homer M. LeBaron*

Contents

I. Introduction

There are many interacting factors which influence the activity and persistence of the triazine herbicides in soils. A brief review of these would include soil pH, texture, organic matter, cation exchange, prevalent cations or other minerals, chemical reactions, and mechanical

---

* Geigy Agricultural Chemicals, *Geigy Chemical Corporation*, Ardsley, New York.

analysis. Environmental factors such as temperature, moisture, leaching, sunlight, microbial activity, plant growth, and rhizosphere are also highly important. Weed species and their depth of germination may affect triazine activity. The physical and chemical characteristics of the herbicide itself determine how it responds to all the above factors.

It is obvious that, under usual agricultural practices, most soil and environmental factors are not within the ability of man to manipulate or control. The above title, however, suggests that the activity and persistence of the triazine herbicides are, at least to some extent, within our capability to regulate. As with all soil-applied herbicides, it is difficult for the user to achieve 100 percent effectiveness or predictability with the triazines.

The fact that these herbicides perform as consistently as they do testifies that they are effective on many weed species and under a relatively wide range of soil and climatic conditions. This remarkable consistency is, however, in large measure due to the extensive research and use experience accumulated over the past 12 years. In the case of atrazine[1] in particular, its history is a success story based on a happy union between knowledge of how and why the product performs under specific conditions and its amazing biological selectivity and flexibility.

Nevertheless, the two most serious problems associated with the use of these herbicides for selective weed control in crop production have always been and will likely continue to be improvement in activity and reduction in persistence. The purpose of this paper is to review the research on the various means of influencing the activity and persistence of triazine herbicides in soils.

II. Selection of triazine herbicides

At the present time there are about a dozen triazine herbicides which are available for commercial use or are in advanced stages of development in the United States. Several additional triazines are used for weed control on a limited commercial basis elsewhere. While there are certain properties (both chemical and biological) which are quite similar between these herbicides, they all differ sufficiently to warrant their use in situations where no other triazine will do as well. Even with the chloro-triazines, simazine and atrazine, which are probably as much alike as any two triazines, we find that simazine can be used safely in many crops where atrazine would not be as safe. When the comparison is between different types, such as chloro-, methylthio-, and methoxy-triazines, the differences in performance become much greater.

---

[1] Herbicides and other pesticides mentioned in text and their common or trade names are given in the Foreword and in Table I.

In most crops and weed control situations, therefore, the user does not have a choice between two or more triazine herbicides. Even where more than one triazine is labeled and recommended for the same crop, the selection is usually not difficult to make. The label restrictions, clarifications, and local experience will often determine that only one of the choices is possible or most desirable.

In corn, for example, where the greatest selection between triazine herbicides exists, atrazine will usually be the obvious choice because of its wide adaptability and flexibility. Simazine, however, may be preferred for improved control of crabgrass (*Digitaria* spp.), *Panicum* spp., shattercane (*Sorghum bicolar*), and other hard-to-control annual grasses. DAVID (1962) reported that simazine at four lb./acre applied to sandy soil in pots gave almost complete kill of *Digitaria* spp., while atrazine was relatively ineffective. Where soil carry-over from chloro-triazines represents a serious limitation, Primaze® may be a good alternative. Ametryne has given good control of emerged weeds when applied as a post-directed spray after the corn is at least 12 inches high.

In sorghum, propazine and atrazine can be used preplant incorporated and preemergence, depending upon the area and soil type, while atrazine can also be applied postemergence. Terbutryn (GS-14260) is in the late stages of development for preemergence use in sorghum. While atrazine is usually preferred to propazine because of improved grass control and greater flexibility, propazine is frequently the best choice in the Southern part of the sorghum belt because of higher crop tolerance. Soil residue from terbutryn does not represent as much risk to rotational crops as from the chloro-triazines.

Sugarcane can be treated preemergence or postemergence directed with ametryne, atrazine, or simazine. GS-14254, the first methoxy-triazine to be developed for selective uses, will also likely be labeled for similar treatments in the future. Since soil persistence and crop tolerance from triazines are not normally problems in this crop, the relative importance of these herbicides is based on specific advantages in weed control or crop tolerance.

While neither atrazine nor simazine generally shows any effect or inhibition on growth of corn, a few studies have shown greater inhibition in crop growth when simazine was mixed with the soil, especially at high rates, at pH above neutral, and in mineral soils compared to atrazine (BURNSIDE and BEHRENS 1961, PLAISTED and PANTON 1961, ANDERSON 1964, SAGHIR and CHOUDHARY 1967). Soil incorporation or dependable precipitation is also more important for simazine to be effective (SPLITTSTOESSER and DORSCHEID 1962).

III. Method and timing of application

Depending upon weed species, soil characteristics, and environmental conditions, atrazine can be successfully applied either preplant, preemergence, or postemergence in corn. The preplant application can be made in the fall or spring, sprayed on the surface or incorporated to variable depths with several types of equipment, and sprayed in water or combined with a fluid fertilizer application. It has become very difficult to suggest a uniform order of preference due to the many factors which influence atrazine activity. Any preferential listing for methods of atrazine application would have to be given on a regional basis since any one of the methods might be preferred in some area. Even local differences in preference between farmers are quite common and may vary from year to year with climatic conditions.

a) Factors influencing the type of application

Herbicide performance is not the only criterion for determining the method of application. Convenience and necessity are also very important. Rainfall or poor weather may, for example, interfere at the time the farmer would prefer to apply the herbicide. The rate of atrazine required and the risk of soil carry-over may differ with the method of application. The prevalent weed species, depth and time of germination, and other factors will help determine which method of application a farmer uses.

Nevertheless, in the final analysis the major consideration in most cases is which method will most likely give the best and most consistent weed control and crop safety. Fortunately, crop safety in the instance of atrazine in corn, is usually not a consideration due to its wide safety margin.

b) Preemergence applications

From the standpoint of the best or most complete weed control preemergence applications of atrazine or other chloro-triazine herbicides would normally be the preferred method. In most areas of the country where corn is grown some precipitation can be expected within two weeks after planting and application. From ¼ to one inch of rainfall within this period, depending on soil type, organic matter, weed species, and other factors, is usually sufficient to "activate" atrazine. Research and field experience have conclusively demonstrated that for control of most annual weeds in most soils, preemergence applications of atrazine and most other triazine herbicides provide better results than other methods of application when sufficient rainfall occurs within the first week or two (LOVELY 1964, BURNSIDE and WICKS 1964, McGLAMERY 1967).

Unless the farmer or applicator using a preemergence application performs the planting and application in one operation, it may be impossible for him to complete spraying before the weeds have emerged. He may not want to perform both operations at one time, however, as it slows the planting operation and requires special equipment. For this reason, as well as for certain performance and soil carry-over advantages, other methods of triazine herbicide applications have been used.

c) Preplant applications

Preplant incorporation of triazines reduces somewhat the dependence upon rainfall or irrigation soon after application. However, it often results in a soil dilution which reduces its effectiveness, especially if incorporated too deeply (KNAKE et al. 1967). While preplant incorporation of some triazines has given good results and has become an alternate method of application which is preferred by many farmers, it is most advantageous where rainfall is low or undependable. In these areas, such as the Western corn belt states, the choice is whether to apply atrazine preemergence and hope for rain, which will often provide complete season-long weed control, or to apply it preplant with incorporation and expect less than perfect, but more consistent, weed control.

Other major advantages of preplant incorporation include the convenience of application, extending the period when the herbicide can be applied, and improved control of certain problem weeds, such as deep germinating species. Even applications during the previous fall have given promising results with little loss in activity (TALBERT and FLETCHALL 1964, BUCHANAN 1962, BEHRENS et al. 1968). However, for best or most consistent results, atrazine should be applied no more than a few weeks prior to planting. SANTELMANN et al. (1968) found that prometryne gave best weed control when used no more than 21 days prior to planting, and that propazine could be incorporated up to two weeks before planting.

Another feature of preplant incorporation of triazines which could be important is the soil carryover compared to preemergence applications. Since this method of application eliminates essentially any loss of triazine from volatility or photodecomposition, this could result in greater carryover in rotational crops (BURNSIDE 1964). Some research indicates this to be so. KNAKE (1966) reported that a bio-assay of soil taken seven months after three lb./acre of atrazine was applied preemergence to the soil surface showed no significant reduction in oat growth, while the dry weight of oats was reduced about 40 percent when atrazine was incorporated about ½ inch deep with a Gandy row wheel or 1.5 inches deep with a power incorporator. SCHWEIZER and HOLSTUN (1966) found that prometryne residues from a four lb./acre sub-surface application were nearly twice as toxic to oats after 25 weeks as were residues from the same rate applied as a surface pre-emergence treatment. From extensive studies on the Geigy Research Farms and elsewhere we would conclude that there is occasionally greater carry-over of triazines where the herbicide is mechanically incorporated soon after application than is found when compared to surface treatments. Any significant difference, however, would be the exception rather than the rule. Soil incorporation normally will result in a longer period for triazine degradation depending on how soon before planting the application is made, and more immediate contact with soil moisture and processes involved in detoxification. These factors may compensate for the greater volatilization and photodecomposition which take place with surface applications.

d) Postemergence applications

Of the chloro-triazine herbicides in commercial use, only atrazine has sufficient foliar activity to be recommended for postemergence weed control. This very important advantage appears to be related to its greater water solubility. Even when applied after the weeds emerge, however, soil interactions and root uptake are often necessary for consistent results (BURNSIDE et al. 1964, THOMPSON and SLIFE 1969). Weeds should generally be treated with atrazine before they exceed 1.5 inches in height. Grasses, especially, become tolerant of atrazine as they become well established.

Although there are several important advantages in being able to wait until the weeds emerge before applying a herbicide, until recent years the postemergence use of atrazine has been more from necessity than for other reasons. In sorghum, for example, the margin of selectivity is not sufficient for atrazine to be safely applied before the crop is established, except in the heavier soils of the Northeastern High Plains area or in the Southwest where heavy rainfall is not likely.

A significant improvement in the effectiveness of postemergence applications of atrazine has been in the addition of highly refined nonphytotoxic oils to the spray tank. Much research has been and is being done to determine the relationships between oil specifications and effectiveness. While differences are small and not always consistent, one gallon/acre of a nonphytotoxic paraffin oil having from 70- to 110-second viscosity (SUS at 100°F.) and containing at least one percent of a suitable emulsifier appears to be near optimum (LeBARON 1966). Surfactants and other additives have also been promising, but have not given as consistent enhancement of atrazine activity compared to the oils (THOMPSON & SLIFE 1969).

These spray additives have improved the flexibility of atrazine sufficiently so that, under certain conditions, postemergence sprays may be the preferred method of atrazine application (STRITZKE and PARKER 1967, BUCHHOLTZ 1967). Such a preference will likely be limited to special weed problems, high organic matter soils, or where low rates of application are used because of soil carry-over or economics. The delay in time between preplant or preemergence and postemergence applications, however, may at least partly compensate for a possible reduction in residue carryover from using a lower rate of atrazine postemergence. Studies have shown that even a few weeks' delay in time of application can result in a significantly greater carryover. This is particularly noticeable when the delay is in the spring at which time the level of moisture, decomposition of organic matter, and other soil changes are at a peak (PETERS and KELLY 1964, FRANK 1966, SAGHIR and CHOUDHARY 1967, BURNSIDE 1968). BUCHHOLTZ (1965 a and b) found that variables other than date of application will influence the comparative carryover of triazines when applied in different ways.

Formulation or spray additives can have a significant effect on herbicide movement or activity in the soil (BAYER 1967). McCowan (1968) reported that oils enhanced the preemergence activity of herbicides, including triazines. The reasons for occasional responses of spray additives on preemergence herbicides are not fully understood, and results are too unpredictable at present to justify the addition of surfactant or oil in soil applications of triazines (WICKS et al. 1968).

In spite of the potential advantages of atrazine-oil-water emulsions as a postemergence spray, the preemergence or preplant applications of atrazine alone will be superior under most conditions for consistent weed control in corn. Furthermore, under extremely wet conditions or when the crop is under stress, injury to corn and sorghum has been observed when spray additives have been used. Atrazine plus oil should be considered, generally, as a last resort or salvage weed control program to be used when earlier applications were impossible or unsuccessful. The user should also consider supplemental cultivation following the postemergence application in the event these treatments fail to give complete control due to weather conditions or size of weeds.

Compared to the chloro-triazines, most methylthio- and methoxy-triazines are much more active as postemergence herbicides. In contrast to atrazine, however, their foliar activity is generally non-selective (SAGHIR and CHOUDHARY 1967). When used after the crop emerges, therefore, these triazines (e.g., ametryne, prometryne, terbutryn, and GS-14254) must usually be applied as directed sprays. Nevertheless, except when used at very low rates (e.g., prometryne as a chemical hoe treatment in cotton), these herbicides are used for dual purpose control of both emerged and later germinating weeds.

e) Other modifications

Various modifications or combinations of the preplant, preemergence, and postemergence applications of triazine herbicides have been used or are being developed for special weed problems or to improve efficiency. Probably the most frequently used combination of methods has been the split application of atrazine for control of quackgrass. This consists of a pre-plow treatment followed by a preemergence treatment. Under some conditions, atrazine can be applied pre-plow either in the fall or spring, but fall treatments have tended to be slightly superior (BUCHHOLTZ 1963, FLANAGAN 1963, VENGRIS 1964). In most of the Northeast, however, where corn land can be plowed in the spring, the spring pre-plow treatment has been consistently superior (RALEIGH 1961, FERTIG 1962 and 1963).

In some areas (KNAKE and SLIFE 1964), the single pre-plow treatment of atrazine at four lb./acre in the fall or spring has been equal or superior to the split application for quackgrass control. However, results have usually favored a split treatment of two lb./acre pre-plow and two lb./acre preemergence, partly because the two lb./acre of atrazine preemergence also controls most annual weeds in addition to the quackgrass.

The combinations of pre-plow followed by postemergence applications, and two postemergence applications about a week apart have been promising. For example, pre-plow followed by postemergence applications of atrazine in oil-water emulsions have been very effective on Canada thistle and other perennial weeds which emerge prior to plowing.

Other methods of triazine herbicide application which have been used effectively and are on the increase include low volume aerial and ground applications. The most obvious advantage of both methods is the reduced amount of water or other carrier needed. There are also considerable savings in time, wear and tear on machinery, and other advantages. One of the major reasons for the increasing aerial applications of triazine herbicides is because of the large amount of acreage that can be covered in a given time, even when fields are too wet for ground equipment to be used. Such a situation usually coincides with actively germinating or growing weeds when a timely herbicide application is most critical and when atrazine is most effective. BOVEY and BURNSIDE (1965), BOVEY and FENSTER (1964), and WIESE and OWEN (1967) have reported equal or superior weed control and crop yield from aerial applications of triazine herbicides compared to ground applications. Further research and development are needed before the low volume applications of wettable powder herbicides by ground equipment become popular, but the present problems will eventually be resolved.

IV. Rate of application

Within the scope of this review it is not possible to give more than brief mention of the importance of application rate on the activity and persistence of triazine herbicides. It should go without saying that the rate of herbicide used is determined primarily by the amount required to give relatively consistent control of the prevalent weeds. This would imply that for a given weed problem in a specific soil and climate, the rate is fixed. In the final analyses, however, the use rate will often be a compromise between this and relative crop tolerance to the herbicide, risk of soil carryover involved, and economics. Use of various techniques such as methods of application, supplemental cultivation, and band vs. broadcast applications may also modify the rate required. It is in this context, therefore, that rate may be considered as a variable which can be used to influence activity and persistence of triazine herbicides.

Obviously, to the corn farmer in the Southeastern United States who has never been preoccupied with carryover of atrazine and knows that crop tolerance is not a factor, the rate of atrazine to be applied should be a simple problem (AUSTIN et al. 1968). He would be influenced largely by economics and a desire for optimum and consistent weed control. The corn farmer in the Dakotas, on the other hand, is much concerned about soil residue carryover to his oats next year, and may compromise somewhat on weed control or use supplemental cultivation in order to keep the rate of triazine as low as possible. Under no condition should a herbicide be applied at an excessive rate.

The rate of application has often been closely correlated with the carryover of triazine herbicides in soils (FRANK 1966, BUCHHOLTZ 1965 a and b, BRYANT 1967). Since the dissipation of triazine herbicides is a first-order reaction, the amount remaining after a given time should be directly related to the amount applied. There are so many additional factors which govern the soil carryover of triazines, however, that rate sometimes appears relatively unimportant. Even within the same area in the same year, carryover may be only loosely related to rate applied. For this reason, limited evaluations on the carryover effect of repeated applications are subject to error. There are frequently such great differences between years that the amount of triazine remaining for more than one year cannot be detected. In three English soils, for example, the time required for 80 percent disappearance of simazine at two lb./acre varied from seven to 27 weeks during the same year, while even greater variation occurred between years on the same soil (HOLLY and ROBERTS 1963).

The major reason for reducing the rate of application of most triazine herbicides below what is considered optimum for consistent weed control is fear of carryover. BUCHHOLTZ (1965 a) found that significant injury to oats occurred in about half of the trials where atrazine had been applied at two lb./acre to the previous corn crop, and in all applications at four lb./acre. While up to four lb./acre of atrazine may be required for control of quackgrass and certain other weeds (BUCHHOLTZ 1963, MEGGITT 1960), BUCHHOLTZ and DOERSCH (1968) reported that annual weed control and corn yield from a combination control program consisting of atrazine at a reduced rate early in the season followed by one or two cultivations have been equal or superior to atrazine without cultivation at higher rates.

As noted before, the use of atrazine plus oil postemergence in some areas has been for the purpose of reducing the rate. FRANK (1966) found that a reduction from two lb./acre to 1.5 lb./acre of atrazine was important in reducing carryover injury to sugar beets in Southwestern Ontario. At 1.5 lb./acre no sugar beet injury was reported when applied preemergence broadcast, while serious crop injury was sometimes noted at two lb./acre preemergence broadcast and at 1.5 lb./acre postemergence broadcast. The difference in carry-over between the preemergence and postemergence applications was believed caused by weather conditions during the few weeks after preemergence applications which favored leaching. Band applications reduced the occurrence of crop injury from all rates or methods. Only at relatively high rates (three lb./acre) was injury to sugar beets serious when planted two years following atrazine application.

Uniform and proper application of triazine herbicides go hand-in-hand with rate of application. The need for accurate, well-maintained, and properly calibrated application equipment is agreed to by all when it comes to selective use of herbicides. It has been emphasized by BUCHHOLTZ (1967), FRANK (1966), and others that more cases of atrazine injury to rotational crops result from faulty application than from any other cause. For maximum activity and minimum carryover, overlapping of spray swaths by double spraying of headlands or when turning at ends of rows, and changes in speed while climbing hills or crossing ditches should be avoided. Spraying should take place when the wind is blowing at less than five miles/hour. Spray drift can be reduced by using wide angle nozzles to lower the boom height, by increasing spray volume or droplet size, by lowering pressure, and by using a spray shield. Other spraying errors which should be avoided include cleaning or flushing the sprayer in the field, and improper width when band spraying.

V. Formulations

There are many factors which must be considered in herbicide formulation research and development. The effects of formulation on herbicide activity and persistence are only two aspects which must be evaluated. These characteristics, nevertheless, are of extreme importance in any changes or improvements which are made in formulations of triazine herbicides.

Because of the solubility limitations of triazine herbicides in common commercial solvents, most triazines have been formulated as wettable powders. Prometone and other methoxy-triazines are sufficiently soluble to be formulated as two lb./gallon emulsifiable solutions (2E). Ametryne and some of the methylthio-triazines can also be made into emulsifiables, usually as 2E formulations. Economics, however, have favored the use of wettable powder formulations of ametryne and prometryne. It has not yet been possible to produce economical emulsifiable formulations of the chloro-triazines.

While most of the triazine herbicide formulations used in European and other countries are 50 percent wettable powders (50W), all commercial wettable powder triazines used in recent years in the United States are 80W's. Comparative tests between these two concentrations or other changes made over the past years have shown no significant or consistent difference in biological activity, handling qualities, or effect on soil carryover (FRANK 1966).

Between 1960 and 1963, granular formulations of atrazine were manufactured and sold for weed control in corn. Not only did these formulations provide somewhat less consistent control of early weeds, but their most serious handicap was the increased possibility of soil carryover compared to wettable powder formulations (BUCHHOLTZ 1965, FEENEY and COLE 1964, KNAKE 1963, LOVELY 1964, PETERS and KELLEY 1964, FRANK 1966). For these reasons, it was decided in 1963 to discontinue further sales of atrazine granules for use in selective weed control. Simazine, however, is used in a four percent granular formulation (4G) for both selective and non-selective weed control. Although economics favor the 80W formulation, the 4G is often preferred where ease of application or longer residual weed control is important. Atrazine and prometone are available in granular or pelleted form for non-selective use only.

Further research is being conducted in an effort to improve the handling, performance, and residual characteristics of triazine herbicide formulations. Results at this time appear promising that significant improvements in triazine formulations are possible.

VI. Combinations of herbicides

In order to improve the spectrum or consistency of weed control and/or to reducet he risk of carryover or crop injury from either of two herbicides, herbicide combinations have been studied extensively in recent years. Most such studies in corn have included atrazine at reduced rates with other herbicides which would reduce the risk of soil carryover. Improvement in activity over atrazine alone has often been of secondary interest except for specific problems, such as sod planting or minimum tillage corn with atrazine plus paraquat, and control of nutsedge (*Cyperus esculentus*) with atrazine plus butylate. However, it should be mentioned that one of the arguments used in favor of herbicide combinations or rotations is to prevent the shift to monocultures or tolerant weed species which have been noted following repeated use of the same herbicide.

To review the research results, advantages, and short-comings of all combinations which have included a triazine herbicide could result in a large manuscript in itself. It will not be possible, therefore, to do

more than summarize briefly the current status of a few promising
combinations, with emphasis on corn. No mention will be made here
of split applications where each herbicide is applied separately.

The combination of atrazine plus butylate has been used as a
preplant incorporated application at rates of one to two lb./acre and
two to four lb./acre, respectively. This treatment has been especially
promising for the control of nutsedge, which has been increasing in
many areas. It often provides wider spectrum control of broadleaf
weeds and grasses compared to either herbicide alone. However, while
butylate is not likely to cause carryover injury to rotational crops, its
use may result in some stunting to the corn.

Atrazine plus propachlor is another of the more compatible combinations for corn. It is applied preemergence at rates of one to two
lb./acre and two to four lb./acre, respectively. Propachlor is a short
residual herbicide which gives very good early control of annual
grasses, but often fails, when applied alone, to control broadleaf weeds
or late germinating grasses. One special feature about propachlor is
that it performs best in medium to high organic matter soils and under
relatively light rainfall, which is in direct contrast to atrazine.

Atrazine plus prometryne (Primaze® 80W) is the only commercial
mixture of two triazine herbicides available in the U.S.A. for selective
use at the present time. It must be applied only preemergence to corn
at rates of ¾ to 1.5 lb./acre active ingredient of each triazine. The
primary advantage of this product over atrazine is the reduced carryover to sensitive rotational crops (SHEETS and SHAW 1963, WISK and
COLE 1965, HARRIS and SHEETS 1965, SAGHIR and CHOUDHARY 1967).
COLBY and HARRIS (1966) reported, however, that the combination
did not appear to reduce the soil residues after several months in a
greenhouse study compared to atrazine alone. While prometryne
residues often remain as long in the soil compared to atrazine, its
carryover effect is usually much less due to stronger soil adsorption
and the somewhat greater tolerance of many rotational crops (BUCH-
HOLTZ 1965 b). Primaze 80W may give better control of certain annual
grasses compared to atrazine alone in some cases (COLBY and HARRIS
1966), particularly under heavy rainfall. However, some crop injury
has been experienced with the combination on high pH soils which are
eroded or low in organic matter.

Atrazine plus linuron, like Primaze 80W, has its major advantage
in reducing the carryover possibility from atrazine alone. The combination is, in most respects, very similar to atrazine plus prometryne.

The combination of atrazine plus paraquat serves a very specialized
need. It provides rapid knock-down of established sod or growing
vegetation where no or very little tillage is planned. Paraquat is a very
fast acting nonselective foliar herbicide, but it has no residual or soil
activity. Atrazine is applied to control later emerging weeds. The use
of such a treatment appears limited to hilly and high rainfall areas

where normal tillage and erosion would present difficulties. The combination is usually applied at or soon after corn planting, at rates of two to three lb./acre of atrazine and ¼ to one lb./acre of paraquat. Because of the possible toxic effects of paraquat to the applicator, caution must be taken when applying it.

Combinations of atrazine plus simazine have been used to a limited extent as a tank mix in corn, although the advantages of this mix over atrazine alone are minor. BUCHHOLTZ (1965 a) and others have found little or no difference in the comparative carryover effect from atrazine and simazine when applied at the same rates. FINK and FLETCHALL (1963) observed that simazine residues resulted in substantially lower yields of nine rotational forage crops than did atrazine residues. While simazine residues usually tend to persist slightly longer than those of atrazine (SAGHIR and CHOUDHARY 1967), oats and several other bio-assay or rotational crops apparently have a greater sensitivity to atrazine residues. Simazine does, however, provide more consistent control of several problem annual grasses compared to atrazine, as mentioned earlier, particularly in areas of high rainfall or light soils.

The combination of atrazine plus 2,4-D has been used extensively as a postemergence treatment. Studies of this combination have been aimed at an effective but low-cost treatment which provides control of grasses as well as broadleaves without fear of soil carryover. It has often been applied with oil as a spray additive. While this combination appears to have synergistic activity for the control of small weeds, crop injury has quite commonly been observed. Furthermore, late germinating weeds must often be controlled by cultivation or other means.

Combinations including triazine herbicides are also gaining interest in crops other than corn. In sorghum, atrazine plus norea, propazine plus norea, propazine plus terbutryn, and propazine plus linuron have shown some promise for preemergence weed control. The only significant advantage in these combinations appears to be the reduction of carryover from chloro-triazines. None of them shows consistently superior weed control compared to propazine or atrazine alone. In some instances, the risk of crop injury is also increased.

Combinations of prometryne with the organic arsenicals (e.g., MSMA) have been promising in cotton. This treatment is used for increasing the spectrum of activity with a low cost post-directed application to emerged weeds. Depending on the rate of prometryne used, some residual control may also be achieved. There also has been interest in combinations of prometryne for improved control of broadleaf weeds with herbicides such as trifluralin and nitralin which is most effective on grasses. Such a combination, however, would usually be a split application with prometryne alone being applied as a preemergence "over-shot" treatment or as a post-directed spray.

Similar combinations or split applications including prometryne have been studied in soybeans. There is a real need for improved control of broadleaf weeds in this major crop. The margin of soybean selectivity does not appear to be great enough to allow use of prometryne preemergence without extensive precautions and limitations. Its use post-directed in soybeans is more promising, and is being investigated.

For the control of established as well as germinating weeds in perennial crops, such as orchards, nurseries, and vineyards, combinations of simazine with paraquat or amitrole have been very useful. Since each of these herbicides is included for its specific contributions, each is usually applied at the normal rate for the crop and soil type. Reduction of soil carryover is usually not an important consideration in perennial crops. For increased weed control spectrum or the control of specific problem weeds in sugarcane, combinations of ametryne plus atrazine, ametryne plus 2,4-D, atrazine plus 2,4-D, simazine plus fenac, or simazine plus trifluralin have been promising.

VII. Interactions with other pesticides

Among the side effects or unexpected biological responses from triazine herbicides, we should consider possible interactions with other pesticides. Two chemicals can possibly depress or enhance the biological activity compared to when the chemicals are used alone. Such interactions could occur when the materials are applied together or separately.

CORBIN and SHEETS (1968) found that Dexon (p-dimethylamino-benzenediazo sodium sulfonate) was strongly antagonistic to atrazine activity. Dry weights of cucumber and wheat were greater when atrazine and Dexon were mixed together in the soil than when atrazine was applied alone. Recent studies (WEBSTER and SHEETS 1969) indicate that the antagonistic responses of seedlings to combinations of atrazine and Dexon in soil are caused by a decrease in atrazine uptake in the presence of Dexon. NASH and HARRIS (1969) found that Dexon caused similar interference with root absorption or translocation of 14 s-triazine herbicides. Cotton injury from prometryne in a very light soil was also observed to be greatly reduced when an application of Fumazone (dibromochloropropane) had been previously made.

Little has been done to study the interaction between residue carryover from a previous herbicide and the herbicide used for weed control in the present crop. ADAMS and ESPINOZA (1969) studied the effects of sublethal amounts of atrazine alone and in combination with CDAA, linuron, amiben, and trifluralin on the growth of soybeans. Sublethal concentrations of atrazine did not significantly depress soybean growth. The effect of atrazine in combination with CDAA or linuron was not significant, but dry matter production of soybeans was consistently greater with these combinations than with atrazine, CDAA, or linuron alone. When damping-off organisms were not controlled, severe injury to soybeans occurred in treatments containing amiben or trifluralin alone, apparently by stimulating emergence of soybean seedlings one to two days earlier. When damping-off organisms were controlled with a fungicide, the growth of soybean seedlings was depressed by treatments containing combinations of trifluralin and atrazine in proportion to the amount of atrazine present. However, combinations of atrazine and amiben resulted in the best growth of soybeans in all instances.

VIII. Effect of soil nutrients

Research has shown that various nutrients and cations in the soil influence the interactions and, therefore, the activity and persistence of triazines in soils. In addition, it has been demonstrated that the triazine herbicides may influence the plant uptake of certain nutrients from the soil. This may affect triazine activity indirectly.

DeVries (1963) reported that simazine increased significantly the uptake of nitrogen by corn in all soils studied. The uptake of magnesium and phosphorus was increased in limed soil, and potassium uptake was greatest in acidified soil. Aluminum sulfate decreased injury from simazine. When lime was mixed into the soil, a synergistic effect on the activity of simazine resulted in the death of all pine seedlings. This was believed caused by an excess of calcium ions and not a pH effect. Nearpass (1965) found that simazine penetrated to lower layers and was more persistent in all soils studied when the soil pH was increased by adding lime. Weber et al. (1968) and others have reported that any treatment which decreased soil pH tended also to reduce triazine phytotoxicity. It has generally been assumed that this effect was caused by greater protonation of the triazine at lower pH, resulting in increased soil adsorption. Armstrong et al. (1967) observed that soil pH and organic matter content largely controlled the rate of atrazine hydrolysis. Uhlig (1964) observed that under conditions of adequate soil moisture, the susceptibility of Lepidium sativum to simazine was increased by supplements of nitrogen, phosphorus, potassium, and magnesium. Upchurch et al. (1963) observed greater phytotoxicity of simazine at high levels of phosphorus than at low levels.

Observations from county weed control demonstrations in Minnesota suggested that atrazine gave better weed control on soils high in available phosphorus than on soils of low phosphorus (Adams 1965). Adams (1965) and Adams and Espinoza (1969) subsequently conducted extensive research on the interrelationships between the susceptibility of rotational crops to injury from triazine residues and the level of phosphorus and other minerals in the soil. They found that high levels of phosphorus tended to suppress plant growth in the presence of simazine or atrazine, but that this relationship varied with soil type. An interaction between triazines and phosphorus also affected the uptake of some minerals by plant shoots. Based on the observation that the phosphorus-atrazine treatments generally acted to reduce the accumulation of manganese by soybeans, the similarity between manganese deficiency and atrazine injury, and the importance of manganese in photosynthesis, ADAMS and coworkers (1969) are studying possible interactions among phosphorus, manganese, and atrazine.

In an unpublished report on the influence of nitrogen, phosphorus, and potassium levels on atrazine carryover injury to oats, BUCHHOLTZ (1966 b) observed that oat injury was lowest on plots fertilized with high rates of nitrogen and little or no phosphorus. Conversely, injury was greatest on plots which had received low to moderate rates of nitrogen and moderate to high rates of phosphorus. Variable rates of potassium showed no consistent effect, either alone or in interactions with nitrogen or phosphorus. He stated, however, that the differences observed were not generally great enough to be of major importance in explaining variations in injury from atrazine residue that are observed in the field. ASHLEY and RAHN (1967) also observed that an application of nitrogen at 50 lb./acre reduced the injury to an oat cover crop from atrazine residues. DOLL and MEGGITT (1969) on the other hand, reported that atrazine caused greater injury to oats as the nitrogen level increased, but no significant differences were found for potassium and phosphorus.

McCUTCHEN and ANDREWS (1967) reported increased herbicidal activity and higher cotton yields from prometryne when a nitrogen solution containing 15 percent urea was used as the carrier for pre-emergence applications compared to NH_4NO_3 fertilization, followed by water as the herbicide carrier. GRAVES et al. (1968) found that both urea and $NaNO_3$ increased the risk of corn injury from soil applications of atrazine plus linuron combinations. Crop injury was also noted from urea with no herbicide.

MINSHALL (1967 and 1969) found that soil applications of urea or KNO_3 increased the rate of exudation from the stem stump of tomatoes by 100 to 300 percent. Even more remarkable, the triazine concentration in the augmented exudate was as much as 40 percent higher than it was where no urea was applied. KNO_3 increased the concentration of atrazine as much as 26 percent. MINSHALL concluded that urea influenced the rate of metabolic root pressure which resulted in active movement of the triazine herbicides from the soil solution to the plant xylem. Recent work with intact plants has demonstrated that urea increased the atrazine concentration in shoots of tomato plants by about 20 percent in the light when transpiration would be active and about 17 percent in the dark when root pressure would dominate.

Some evidence suggests that nitrogen or other nutrients have an indirect effect on triazine activity and persistence in soils. ARMSTRONG (1966) reported that additions of nitrogen as NH_4NO_3, $NaNO_3$, or KNO_3 to soil perfusion systems showed an increased rate of atrazine degradation. While it appeared that microorganisms might be using atrazine as a carbon source, efforts to obtain degradation in nitrogen-amended enrichment trials were unsuccessful. Only a small amount of atrazine which disappeared could be accounted for in increased hydroxyatrazine formed. Also, C^{14}-atrazine studies failed to show any degradation products in the nitrogen-containing perfusate after 60 days.

BRENCHLEY (1969) found that atrazine toxicity in tomatoes was influenced by magnesium levels. At high magnesium rates plants showed decreased atrazine toxicity and more rapid recovery than those receiving no magnesium. DAVIS (1967), however, failed to find any correlation between atrazine toxicity and magnesium concentration with *Lemna minor* (duckweed).

IX. Use of water

The artificial use of water to influence the activity and carryover of triazine herbicides is, unfortunately, relatively limited. This can, however, be a very effective technique where irrigation water and proper methods of application are available. In order to activate triazines, overhead or sprinkler irrigation is generally most effective with a minimum amount of water. Furrow and subirrigation have also been useful. The objective is simply to get the herbicide to the weed at or soon after germination. While rainfall is usually unpredictable, many growers will keep the need for moisture in mind when deciding when to plant or apply the herbicide.

The amount of rainfall or irrigation needed for optimum weed control from preemergence applications of triazines will depend on the solubility and other properties of the chemical, soil texture, organic matter content, and weed species and their depth of germination. The range would vary from about ¼ inch under conditions of light mineral soils, where the triazine is used for control of shallow germinating weeds, to at least one inch of precipitation needed on heavier, medium organic matter soils for control of deep germinating grasses (SPLITT-STOESSER and DERSCHEID 1962, BURNSIDE and BEHRENS 1961). Some studies indicate that triazine herbicides are effective when applied to a wet soil, as well as prior to precipitation (STICKLER *et al.* 1969, THOMPSON and SLIFE 1969). It has also been observed that excessive or heavy rainfall may shorten the period of weed control.

Water can also be very useful in reducing or eliminating soil carry-over from triazines. SWITZER and RAUSER (1960) found that two lb./acre of atrazine was detoxified in eight weeks under conditions of high moisture and high temperature, while under dry conditions, the same rate caused oat injury 12 months later. Leaching of the herbicide out of the plow layer is often assumed to be the process involved when water is applied to reduce residue problems. BURNSIDE et al. (1969) found that residue carryover from atrazine and propazine tended to be concentrated in the top six inches of the six soils studied, but significant leaching did occur under unusually heavy rainfall on a very fine sandy loam soil. Leaching is very likely an important factor in light textured soils low in organic matter. However, the effect of water on microorganism and soil processes which stimulate triazine degradation is often more important. ROETH et al. (1969) found that increasing the moisture content from 40 percent to 80 percent of field capacity caused up to six times as much metabolism of atrazine. Atrazine was also degraded two to three times faster in the topsoil compared to the subsoil.

Even after crop injury has developed due to triazine residues, it can often be reduced or delayed by rainfall or irrigation (FRANK 1966). At the same time, however, water activates or increases the phytotoxicity of triazines. GROVER (1966) reported that as the moisture level in the soil decreased, simazine was less effective in reducing the top growth of oats. ED_{50} (concentration resulting in 50 percent stand reduction or effective dose) values were almost three times as high when the moisture level was 30 percent compared to 60 percent. At all simazine concentrations, on the other hand, oat growth was increased as the amount of water increased, confirming that while water is needed to activate the herbicide, plants that survive will grow best with adequate moisture.

Triazine herbicides are quite resistant to leaching in soils which contain significant amounts of organic matter. Residue analysis of soil which had been treated annually with triazines for several years have seldom shown significant levels below the eight- to 12-inch depth. The exceptions have been in light-textured soils with little organic matter and high rates of triazines applied annually. Even in these instances, however, there has not been a triazine accumulation at lower depths.

X. Use of tillage

When the modern herbicides were in their early stages of development, their potential was so remarkable that many observers expected cultivation or other mechanical methods to be replaced completely by use of chemicals. Indeed, this has been a practical possibility in certain instances. Atrazine has made this hope become more of a reality on more crop acres than has any other herbicide. Observations indicate that at least on some soil types, substituting herbicides for cultivation is not only practical but may benefit corn or other crops by avoiding root damage (HINESLY et al. 1967, BUCHHOLTZ and DOERSCH 1968).

In recent years, however, there has been an increasing realization that the use of herbicides and tillage should often supplement each other. Even with atrazine, reports continue to confirm that a timely cultivation or rotary hoeing, particularly under conditions of low rainfall or minimum moisture, improves rather than lessens its effectiveness (BUCHHOLTZ 1966, BURNSIDE 1964, BURNSIDE et al. 1964, BURNSIDE and WICKS 1964, SPLITTSTOESSER and DERSCHEID 1962, KNAKE et al. 1965). STANIFORTH and LOVELY (1964) emphasized that the failure of preemergence herbicides to control weeds under dry conditions should not be a major shortcoming if cultivation is begun when weeds begin to grow. BURNSIDE et al. (1964) observed that combinations of tillage, narrow row spacings, and atrazine applied preemergence gave more dependable weed control in sorghum than any one alone. They noted that even the weeds surviving the atrazine treatments appeared to be reduced in vigor so that tillage was more effective in destroying them. Reductions in labor, tillage, crop injury, herbicide cost, and herbicide residue, as well as increases in sorghum yield and weed control were cited as advantages of combining weed control methods. BUCHHOLTZ and DOERSCH (1968) reported that while broadcast applications of triazines with no cultivation yielded as much as cultivated plots, the herbicides plus one cultivation increased the average corn grain yield by six percent. Since this increase was probably due to improved weed control, cultivation would not likely have been justified except where weeds were not controlled satisfactorily. HINESLY et al. (1967) concluded, based on a three-year study at seven locations in Illinois, that little or no benefit is obtained for cultivation of corn in most soils if weeds can be controlled by herbicides. Yields in some soils were even higher without supplemental cultivation. Only on light-colored soils that tend to crust was there an occasional advantage for cultivation.

While supplemental tillage will usually be beneficial when low rainfall or other adverse conditions have reduced the effectiveness or residual control from triazine herbicides, an early incorporation immediately after application may be undesirable. Effectiveness of weed control has often been reduced from mechanical incorporation of triazine herbicides, especially with prometryne or other methylthiotriazines (WEISE and HUDSPETH 1968). Compared to surface applications, incorporation may also increase crop injury, particularly on light soils. There are, however, significant exceptions which have been reported. JORDAN et al. (1968) found that at three locations in Southern California, prometryne as well as other herbicides gave better weed control when shallowly but thoroughly incorporated compared to surface applications. It is not possible to determine from most incorporation studies what contribution was made by the incorporation treatment separately from the herbicide.

Tillage is not only often useful to supplement weed control with herbicides but various tillage methods can also influence significantly the residual life or carryover of the triazine herbicides. BUCHHOLTZ

(1965 a) reported that plowing the seedbed for oats resulted in less carry-over injury from atrazine than discing. Cultivation of the corn between the rows after treatment, however, did not influence the injury to oats the next spring. WICKS and BURNSIDE (1965) reported similar findings with atrazine and propazine in sorghum. Where atrazine is applied as a band treatment BUCHHOLTZ (1966 a) suggested that discing thoroughly, followed by plowing, is the most satisfactory practice to use in order to reduce carryover. LYNBENOV (1966) observed that atrazine at three kg./hectare for weed control in corn caused injury to wheat the following year when pre-sowing tillage for the wheat was not deeper than six to eight cm. He recommended that in dry years when pre-sowing tillage to at least 18 to 20 cm. is impossible, the area should be resown to corn instead of to wheat.

Deep plowing beyond the normal plow layer has been suggested as a means to reduce the risk of carryover from triazines. While this could be useful on a limited scale when the soil depth and structure is such that deep plowing would have other beneficial effects, such extensive tillage would be of doubtful justification when the sole purpose is to reduce the risk of herbicide carryover. HARRIS et al. (1969) found atrazine to persist longer as depth of placement was increased. From 12 locations in the United States and Puerto Rico, atrazine recovery averaged 61 percent higher at the 15-inch than at the three-inch depth. The advantage of either shallow or deep plowing as compared to discing, in reducing the carryover effects of triazines, therefore, appears to be due to dilution or removal of the residues out of the root or germination zone of sensitive plants.

Oat injury was less intense from three repeat applications of three lb./acre atrazine residues in Illinois (KNAKE 1966) as the amount of tillage (discing for corn seedbed preparation and cultivations of corn crop) was increased. Yields were not significantly affected. BUCHANAN (1962), on the other hand, reported that weekly cultivations had no effect on the rate of inactivation of several triazine herbicides.

Fall plowing vs. spring plowing has also been evaluated for the comparative effect on triazine carryover. BUCHHOLTZ (1965 a) found that while one study resulted in better oat yields from spring-plowed plots, in another experiment the fall-plowed plots tended to have slightly less reduction in oat yield. FRANK (1966) observed decreased sugar beet injury from atrazine residues with fall plowing compared to spring plowing when the atrazine was applied preemergence. This relationship tended to be reversed, however, when atrazine was applied postemergence.

XI. Selection of rotational crops

Crops which are considered sensitive to triazine residues vary greatly in their degrees of tolerance. Depending on the triazine used and the possibility of some soil residue remaining, the proper selection of a rotational crop has been an important means of reducing or preventing injury from occurring. Even though economics, type of farm, and farmer preference will largely determine which crop rotation is followed, knowledge about relative crop tolerance to triazine residues can be helpful.

It is not yet possible to state with confidence what concentration of a triazine herbicide in a given soil will be tolerated by specific crops without yield reduction under field conditions. Several good studies have been made, however, to determine the GR$_{50}$ values (level of herbicide required to give a growth reduction of 50 percent) for atrazine and other triazines in certain soils. TALBERT and FLETCHALL (1964) found that the series of crops studied ranged in decreasing tolerance to atrazine and simazine from F-6 inbred corn, GT-112 inbred corn, sorghum, soybeans, wheat, cucumbers, oats, and mustard. FRANK (1966) reported the decreasing order of tolerance in crops studied to be navy beans, soybeans, oats, and sugar beets.

BURNSIDE (1959) reported that the relative tolerance of eight crops to simazine in decreasing order was: corn, sorghum, flax, soybeans, rye, wheat, barley, and oats. He found that the concentrations of simazine in soil which could be determined by bioassay using various crops were as follows: oats—from $\frac{1}{4}$ to one p.p.m., soybeans—from $\frac{1}{2}$ to eight p.p.m., and corn—from eight to 256 p.p.m. FINK and FLETCHALL (1963) observed that of nine forage crops tested in the greenhouse and field in soil previously treated with atrazine or simazine, tall fescue and alfalfa were the most tolerant crops, timothy the most susceptible, with red clover, Korean lespedeza, ladino clover, sweetclover, orchardgrass, and bromegrass being intermediate. BUCHHOLTZ (1967) suggested that planting crops such as sudangrass and proso millet could avoid the effects of atrazine residues.

KNAKE et al. (1966) have reported that repeated broadcast applications of atrazine for three years on five different soil types in Illinois resulted in no yield reduction to soybeans planted the fourth year, even though early symptoms of injury occasionally appeared. Oats which were planted on a similar series of plots, however, yielded only 78 percent as much as oat plots with no history of atrazine. Not only was the difference due to greater atrazine tolerance of soybeans, but the standard tillage practices of plowing and discing prior to planting soybeans, compared to discing only for oats, were considered important factors.

FINK and FLETCHALL (1969) in Missouri observed no injury to soybeans planted about one year following atrazine at five lb./acre or less. Following two years of treatment, however, substantial reductions in yield of soybean forage occurred at five lb./acre or more of atrazine when soil samples were bioassayed in the greenhouse on May 4. Soybeans planted in the field about one month later (June 10) showed no

yield reduction of soybeans following two consecutive years of treatment, even though sufficient triazine residue remained to reduce the
number of broadleaf weeds. The authors believed that the difference
was caused by the delay in planting of the field trials. They suggested
that a delay in soybean planting might be justified for further dissipation of atrazine or simazine residues where these herbicides have
been used.

ADAMS and NELSON (1967) have conducted simulated atrazine
residue studies at three locations in Minnesota over two years to
determine what "residue" rates are tolerated by soybeans and oats.
The amount of atrazine resulting in a 50 percent reduction in grain
yield of oats varied from about $\frac{1}{2}$ to more than 1.5 lb./acre depending
on location and weather conditions. On the basis of two years' data, it
was concluded that approximately 40 percent reduction in stand
would not lower the yield of oat grain significantly. No reductions in
the yield of soybeans were observed at rates up to one lb./acre. This
ability of oats and other crops to "grow-out-of" or survive early injury
from triazine residues with no or minor reductions in final yield has
been observed frequently. ADAMS (1969) has claimed that equations
can be worked out to predict accurately whether crop yield might be
affected by atrazine residues based on early crop injury and weather
records. The predictability varies with the crop. Soybeans, for example,
have shown no or little correlation between early stand evaluations
and seed yields. BURNSIDE (1967 and 1968) reported that the ability of
oats to tiller profusely compensated for early stand reductions from
both simazine and atrazine residues.

Several researchers have suggested that crop selectivity to herbicides or herbicide residues might be improved by exploiting varietal
differences or by altering the genetic structure of the crop. While most
research on triazine selectivity has shown small variations within
species, some genetic work has shown great differences because of the
effects of single major genes.

GROGAN et al. (1963) showed that, although most varieties of corn
are resistant to atrazine and simazine, complete susceptibility occurs
in at least one inbred line (GT-112) because of the action of a single
recessive gene. CHAPLIN and ALBAN (1960) found differences in simazine resistance among 90 varieties of sweet corn. Based on differential
responses between several corn inbreds to simazine and atrazine,
ANDERSON (1964) concluded that there was a relationship between
corn resistance to the European corn borer, stalk rot, and triazine
herbicides, all of which seemed to be dependent upon the activity
of 2,4-dihydroxy-7-methoxy-1,4-benzoxazin-3-one(Dimboa), its 2-glu-
coside, or similar cyclic hydroxamates naturally present in most corn
varieties. PALMER and GROGAN (1965) claimed, however, that susceptibility of GT-112 to triazine herbicides was not caused by the
absence of the 2-glucoside or by its inability to hydrolyze atrazine.

The tolerance of a large number of sorghum varieties also varied considerably when treated with propazine, terbutryne, and other herbicides (MILLER and BOVEY 1969). The degree of tolerance tended to be directly related to vigor and productivity, rather than simple inheritance. WEISE and QUINBY (1969) observed considerable differences in the resistance of 60 sorghum varieties to propazine, with some evidence that resistance to this triazine was dominant in inheritance.

OSGOOD et al. (1969) found that interspecific hybrids within the genus *Saccharum* (sugarcane) differed substantially in their tolerance to diuron and ametryne.

Even within a relatively sensitive crop such as oats, significant though marginal differences in varietal response to atrazine have been observed. BUCHHOLTZ (1965 a and 1966 a) suggested that the degree of oat injury from atrazine carryover could be reduced somewhat by planting less sensitive varieties such as Garland and Minhafer.

From an extensive screening of all 1,541 accessions of flax in the World Collection, ANDERSON and BEHRENS (1967) reported that significant differences in atrazine susceptibility were found, with C.I. 719 showing the most resistance. Breeding work was continued with a selection of C.I. 719 (C.I. 2484) which appeared to be the most tolerant of all genotypes evaluated (COMSTOCK and ANDERSON 1968). When it was crossed with "Koto," a susceptible variety, atrazine tolerance in the F_3 and backcross populations behaved as a quantitative character rather than as the simple inheritance reported in corn by GROGAN et al. (1963). It was concluded that selection for atrazine tolerance in flax populations would be difficult.

A similar study (ANDERSON 1969) of about 2,500 strains of soybeans confirmed recently that varietal differences in atrazine susceptibility could also be important in this crop. Careful investigation indicated, however, that seed size may be more important than genetic variation. KARIM and BRADSHAW (1968) further observed varietal differences in survival and final weight of wheat, rape, and mustard when treated with simazine. The progenies from mustard plants which survived 10 oz./acre of simazine showed a very marked increase in resistance over the controls, suggesting that genetic variations in crops could be valuable in extending herbicide selectivity.

Another area that is under investigation is that of developing crop plants having a higher photosynthetic efficiency, as this is closely related to productivity. This development is being attempted by breeding for this characteristic, by using various bioassay techniques, or combinations of both. MOSS (1968) observed that a close relationship existed between plants which had a high rate of photosynthesis and their relative tolerance to atrazine. Corn, sorghum, Johnsongrass, and Bermudagrass all have the high photosynthetic efficiency characteristic. Recent studies indicate that this characteristic is directly dependent upon a special highly organized type of chloroplast around the plants' vascular system, in addition to the usual leaf chloroplasts. The possibility of such a relation between plant tolerance to triazines and efficient photosynthesis is supported by observations that vigorous, rapidly photosynthesizing plants are able to escape or to overcome triazine injury better as compared to weak or chlorotic plants. BRENCHLEY (1968) reported, for example, that high chlorophyll tomato plants were able to recover better from injurious effects of atrazine, indicating more detoxification.

XII. Cropping as a means of triazine dissipation

Some reports have indicated that the rate of triazine dissipation in soil can be enhanced by cropping or allowing weeds to grow (SIKKA and DAVIS 1966). In theory this appears sound as we know that plants growing in treated soil will absorb some of the herbicide through their roots and will break it down. The chloro-triazines can be metabolized readily by corn, sugarcane, Johnsongrass, and Bermudagrass (GYSIN and KNÜSLI 1960). Fields treated with atrazine or simazine for quackgrass control and planted to corn have been observed to be less injurious to sensitive crops the next year than were treated fields that were fallowed (KNAKE and SLIFE 1964). WICKS and BURNSIDE (1965) observed that more oats grew in the old sorghum rows on plots that had been treated with atrazine and propazine the previous year. BURNSIDE (1959) claimed that the amount of simazine removed from soil by soybean plants was proportional to the length of time the plants grew in the soil. SIKKA and DAVIS (1966) reported that corn, sorghum, and Johnsongrass were effective in reducing the amount of atrazine remaining in both field and greenhouse soils. Uptake by corn, Johnsongrass, and sorghum within three months in pot culture accounted for losses of approximately 25, 25, and 20 percent, respectively, of the atrazine applied.

Several researchers have failed, however, to find any significant advantage in a crop of corn to enhance atrazine dissipation. NALEWAJA (1966) found that at two locations in North Dakota the injury to wheat from atrazine applied in 1964 was greater where corn had been grown compared to fallowed plots. ROGERS (1968) did not observe a significant reduction in the soil concentration of ametryne and prometryne as a result of oats growing in the soil for 56 days.

While these observations represent direct contradictions, such widely differing results should not be unexpected. It is obvious that, under field conditions, many variables exist at the same time. Compared to fallowed soil, a cropped soil is often cooler, dryer, and shadier, which would contribute to increased soil carryover. Differences in temperature, moisture, photodecomposition, and volatility could more than compensate for the small amount of triazine taken up and metabolized by plants.

XIII. Use of microorganisms

There have been numerous reports claiming the importance of a wide range of soil microorganisms in the detoxification and degradation of triazine herbicides (KAUFMAN et al. 1965). Some claims have even been made of an enrichment or adaptive effect in which triazines were degraded more quickly in soil previously treated with the herbicide compared to the dissipation rate from the initial application (AHRENS 1967 b). Since such a phenomenon has been demonstrated in the case of 2,4-D and other herbicides, it would have very important ramifications for both weed control and carryover if this adaptive mechanism occurred with a soil-applied herbicide such as atrazine. ENO and CASELEY (1965) observed, however, no indication of any soil enrichment over a four-month period with organisms assumed to be capable of degrading simazine.

Several soil microbiologists have been optimistic that an adaptive organism could be found or developed which could be used under field conditions to speed up the rate of triazine degradation and prevent undesirable carryover. MAXFIELD (1969) has tried to coat wheatgrass seed with selected microorganisms to reduce the time required before reseeding rangeland following cheatgrass control with atrazine. He has isolated several microorganisms from corn fields treated with high rates of atrazine. Apparently one species of *Stagonospora* can degrade atrazine quite rapidly, but when this organism is used to coat the seed of wheatgrasses to provide protection against the small amounts of atrazine residues, the seeds fail to germinate. The inhibition is not parasitic and work is continuing to resolve this problem.

Most recent research raises serious questions as to the practical significance of microbial degradation of triazines. While not discounting soil microorganisms as possible participants in atrazine detoxification, evidence is accumulating in support of the theory that either the ability of microorganisms to act directly on the triazine molecule is limited or else its ultimate contribution is evidenced over a long period of time. Recent studies by McCORMICK and HILTBOLD (1966) revealed that not only was there no enrichment effect following subsequent applications of triazine herbicides, but only a small part of the molecule was evolved as $^{14}CO_2$ over a period of three months. About 0.02 percent of the ^{14}C-ring-labeled prometryne was evolved, while about five percent of the ^{14}C from the methylthio position of prometryne, 0.06 percent of ^{14}C-ring-labeled atrazine and up to two percent of ^{14}C side chain-labeled simazine were evolved as $^{14}CO_2$. Decomposition of triazine herbicides was directly related to the breakdown of soil organic matter. These researchers are of the opinion that triazines are passively degraded, incidental to the metabolism of soil-derived substances, and are nonstimulatory to growth or enzymatic capability of the microflora. Microorganisms cannot be expected to proliferate in response to triazines in soils unless the herbicide provides some advantages such as an exclusive energy source or growth factor.

Microorganisms have also been used in an attempt to degrade large quantities of waste pesticides. In a recent report on a pesticide disposal project, STOJANOVIC (1968) stated that atrazine not only resisted microbial decomposition but also inhibited breakdown of soil organic matter.

It is possible that reports of an enrichment effect from previous applications of triazines may be due to greater metabolism of organic matter or other factors which stimulate microbial activity. That microbial activity and degradation of triazine herbicides in most soils are interrelated appears certain. The point of controversy is the relative importance of microbial detoxification of triazines as a direct or specific reaction under field conditions.

XIV. Use of charcoal and other additives

Although the long-term weed control from triazine herbicides was a characteristic looked for by their developers and has been one of their greatest assets, the inability to "shut them off" when they are no longer needed has been one of their most serious liabilities. It is natural, therefore, that research on ways to control artificially or to remove their residual effectiveness has also been conducted.

CASTELFRANCO and DEUTSCH (1962) postulated that nucleophilic chemicals such as calcium polysulfide added to the soil should be effective in enhancing triazine detoxification. While this concept was not found to be of practical value, other means have been more successful. Probably the most promising additive to find commercial acceptance has been activated charcoal. By mixing finely ground activated carbon with contaminated soil, the farmer could have a wider latitude in his rotational crops. GAST (1962) reported that simazine could be completely deactivated by additions of activated charcoal, while in soil treated with manure or wood ash the simazine was still biologically active. He estimated that 50 to 200 parts of activated charcoal were necessary for detoxification of one part of simazine.

Research in Oregon (ANONYMOUS 1965) showed that when 1.6 lb./acre of atrazine were applied on June 30, 25 lb./acre of charcoal incorporated on July 30 prevented injury to oats planted on August 4. Fifty lb./acre of charcoal deactivated most of the 3.2 lb./acre of atrazine but was less effective at the 4.8 lb./acre rate. Treating the seed with the maximum amount of charcoal that would adhere to the seedcoat was not an effective method of deactivating the atrazine residues in the soil. A timing study (ANONYMOUS 1967) showed that 200 lb./acre of charcoal was adequate to overcome the effects of 1.6

lb./acre rate of atrazine applied on March 15 or earlier when oats were planted June 20 of the same year. The charcoal also prevented oat injury from 4.8 lb./acre of atrazine applied on December 14 or earlier the previous year.

The amount of carbon required to detoxify soils containing residues of atrazine or other triazines obviously will depend on several factors. AHRENS (1965) suggested that a safe rule to follow is to use approximately 200 lb./acre of finely ground activated charcoal for one lb./acre of triazine remaining in the soil. At higher levels of residue or when planting very sensitive rotational crops such as beets, ratios of 400:1 may be required. BOVEY and MILLER (1969) found that greater ratios were required to detoxify propazine and terbutryn under tropical conditions. Even 600 lb./acre of activated carbon failed to completely prevent injury to oats, cucumbers, or beans by 2.5 lb./acre of propazine. ANDERSON (1968) observed that the soil organic matter content and the efficiency of incorporation influenced the amount of charcoal required with a range of 200:1 to 400:1 of charcoal to simazine. The charcoal was also less efficient at lower rates of simazine. He pointed out that further work was needed to learn how to apply and incorporate the charcoal, how long the inactivation of triazine persists, whether or not harmful quantities of herbicide might become available later, and what effect the charcoal has on later herbicide treatments.

In developing a bioassay for atrazine residues in soils, BUCHHOLTZ (1967) observed that as much as 2,000 lb./acre of carbon allowed normal growth of oats, while 250 lb./acre of carbon gave apparent protection from up to two lb./acre of atrazine. Without carbon an application of $\frac{1}{8}$ lb./acre of atrazine decreased oat weights substantially, while $\frac{1}{2}$ lb./acre killed all the plants.

The texture or fineness of the carbon is very important. AHRENS (1966) found that while charcoal granules showed no effect on simazine-treated nursery stock, powdered activated carbon provided protection against simazine injury. COFFEY and WARREN (1969) reported that a variation of more than 50 percent in the efficiency of herbicide adsorption was observed in three powdered activated carbons even though they were all steam activated, from lignite sources, and 70 percent passed through a 325-mesh screen. The sizes of the individual particles, however, were claimed to account for the differences.

Various modifications of charcoal applications have been tried in order to increase the effectiveness or reduce the amount of adsorbent needed. While seed coatings have not been very promising, selected placement of the charcoal around or over the seeds or plant roots has been effective and economical. KRATKY and WARREN (1969) reported recently that sensitive crops were protected against two lb./acre of simazine by a small amount of activated carbon-vermiculite mixture directly over the seed. LINSCOTT and HAGIN (1967) found that narrow bands of activated carbon over the rows were successful in protecting alfalfa from injury when the seedbed had been treated with 1.5 lb./acre of methoprotryne (G-36393), while the same rate of atrazine caused crop injury.

The use of a powdered carbon slurry has been used commercially to some extent as a root dip prior to transplanting of strawberries or other plants into triazine treated soil or when a triazine herbicide would be applied following transplanting. CRANDALL (1966) observed that strawberry plants whose roots were dipped in a powdered carbon slurry before planting showed no injury from simazine at one lb./acre, while considerable loss of plants occurred without the root dip. The root dip also protected the strawberry plants against atrazine at one lb./acre to a large extent. Weed control was not affected. SCHUBERT (1967) found that root dipping provided protection to strawberry plants from twice the weed control rate of simazine. ROBINSON (1965) reported that the ability of strawberries to survive treatment with simazine depended on the nature of the root dip. By order of effectiveness and crop safety, the adsorbents tested were as follows: activated carbon > unactivated charcoal = soot > vermiculate > kieselguhr > dried farmyard manure = dried grass.

Tomato transplants were able to survive better when treated with 0.8 g. of activated carbon in 100 cu. cm. of water at planting into soil treated one week earlier with one to three lb./acre of atrazine, compared to plants not treated with carbon (HOPEN 1967). However, some injury from atrazine occurred at all rates.

Activated carbon root dips have also been promising for nursery plants and conifers (AHRENS 1967 a). AHRENS (1966) found that under severe drought stress, white spruce transplanted into sod was not only protected against simazine and other herbicides, but more spruce survived in untreated plots when root dips were used.

While the commercial use of activated charcoal for soil detoxification of herbicides is still limited, its long-range potential appears promising. Economics and assurance of the desired effect are the major problems. Where activated charcoal is used as a soil treatment, a very fine powder should be distributed uniformly and mixed thoroughly into the upper few inches of soil for maximum effectiveness. The cost of such a treatment may restrict its use to relatively high-value land, special crops, limited areas of misapplications, or as band applications, root dips, and similar methods where less charcoal is required.

It is probable that, as we make use of rapid methods for determining triazine residues in soils and gain more knowledge on the relationship between the triazine residue remaining and the extent of crop injury, we will be able to apply just enough charcoal to bring the triazine residue below the "threshold" level for a specific crop in a specific soil. When we can be sure of a definite need and beneficial response from rates of charcoal below 50 lb./acre, such a treatment could be of commercial value on a large scale.

Spent charcoal is available in relatively large quantities as a waste product from industries such as sugar refineries and breweries where activated charcoal is used for purification. Although such material could be extremely cheap, preliminary studies indicate that such deactivated carbon is relatively ineffective at practical rates for detoxification of triazines in soils.

A report from BORTELS and FRICKE (1966) claiming that deposits of industrial flue dust have increased the adsorptive capacity of the soil and account largely for the rapid detoxification of simazine in the Cologne Basin is of interest. By raising the soil temperature the effect of the dust tended to be nullified. They concluded that repeated applications of simazine on these soils would be inadvisable because possible residual accumulations of the chemical may become activated by changes in climate and/or cultural conditions.

Recent studies at Auburn University (HILTBOLD 1969) have been conducted on the use of processed garbage compost for pesticide degradation or detoxification from soils. While results have demonstrated very striking effects on tying up some pesticides, the compost has not been very promising in detoxification of triazine herbicides.

There have been claims that incorporating plant residues such as corn stalks or corn cobs back into the soil has aided in reducing triazine residue carry-over. LAVY et al. (1965) found, however, that up to 10 percent corn cobs in soil were required to prevent soybean injury from four lb./acre of atrazine. In addition to the effect of plant residues on adsorbing the triazine herbicides and degrading them by stimulation of microorganism activity, there is also a possibility that the carbohydrates released by fresh organic matter decomposition could decrease the inhibiting effect of triazines on photosynthesis. The addition of two percent sucrose to a nutrient solution has been found to decrease atrazine inhibition from 70 percent to 26 percent (DAVIS 1967). ASHLEY and RAHN (1967) reported that the addition of 100 lb./acre of sugar offered some protection to an oat cover crop from atrazine residues.

The addition of animal manures or organic matter of any kind should be helpful in reducing undesirable triazine residues from soils, especially soils naturally low in organic matter. At the same time, however, increases in soil organic matter may require slightly higher rates of triazine herbicides for weed control. ARMSTRONG (1966) found that the addition of leaf compost to a sandy loam soil increased the rate of simazine decomposition. While no decomposition was observed over 3.5 months when no compost was added, the half-life of simazine for additions of one percent, five percent, and 10 percent leaf compost were about 140, 60, and 40 days, respectively. McCORMICK and HILTBOLD (1966) found that the decomposition of atrazine paralleled the response of soil organic matter decomposition and was closely associated with repeated additions of energy-containing material (e.g., glucose).

Roeth *et al.* (1969) suggested that the kind and amount of organic matter may govern atrazine degradation to a greater extent than does clay content. Doherty and Warren (1969) also observed that type of organic matter is important. Sphagnum peat was not nearly so adsorptive for prometryne, simazine, and other herbicides as fibrous peak or muck soil. Prometryne showed no adsorption to clay when one percent bentonite was mixed with quartz sand, while simazine adsorption was increased. Fibrous peat was 13 and three times more adsorptive while muck soil was seven and two times more adsorptive than bentonite for prometryne and simazine, respectively. While prometryne adsorption was closely related to percent organic matter, it was also influenced by chemical properties of the organic matter, possibly based on the number of reactive groups available for hydrogen bonding. Grover (1966) found that peat moss added to soil reduced the

Table I. *Chemical designations of herbicides and other pesticides mentioned in text*

Common name	Trademark	Chemical name
ametryne	Evik	2-ethylamino-4-isopropylamino-6-methylmercapto-*s*-triazine
amiben	Amiben, Vegiben	3-amino-2,5-dichlorobenzoic acid
amitrole	Weedazol Cytrol	3-amino-*s*-triazole
atrazine	AAtrex	2-chloro-4-ethylamino-6-isopropylamino-*s*-triazine
butylate	Sutan	*S*-ethyl diisobutylthiocarbamate
CDAA	Vegadex	*N,N*-diallyl-2-chloroacetamide
diuron	Karmex	3-(3,4-dichlorophenyl)-1,1-dimethylurea
fenac	Fenac	(2,3,6-trichlorophenyl)acetic acid
GS-14254	Sumitol	2-*sec.* butylamino-4-ethylamino-4-methoxy-*s*-triazine
methoprotyne (G-36393)	—	2-methylthio-4-isopropylamino-6-γ-methoxypropylamino-*s*-triazine
linuron	Lorox	3-(3,4-dichlorophenyl)-1-methoxy-1-methylurea
MSMA	(varied)	monosodium methanearsonate
nitralin	Planavin	4-(methylsulfonyl)-2,6-dinitro-*N,N*-dipropylaniline
norea	Herban	3-(hexahydro-4,7-methanoindan-5-yl)-1,1-dimethylurea
paraquat	Paraquat	1,1'-dimethyl-4,4'-bipyridinium salts
prometone	Pramitol	2,4-bis(isopropylamino)-6-methoxy-*s*-triazine
prometryne	Caparol	2,4-bis(isopropylamino)-6-methylthio-*s*-triazine
propachlor	Ramrod	2-chloro-*N*-isopropylacetanilide
propazine	Milogard	2-chloro-4,6-bis(isopropylamino)-*s*-triazine
simazine	Princep	2-chloro-4,6-bis(ethylamino)-*s*-triazine
terbutryn (GS-14260)	Igran	2-*tert.* butylamino-4-ethylamino-6-methylthio-*s*-triazine
trifluralin	Treflan	α,α,α-trifluoro-2,6-dinitro-*N,N*-dipropyl-*p*-toluidine
2,4-D	(varied)	(2,4-dichlorophenoxy)acetic acid
—	Primaze	1:1 ratio between (prometryne) 2,4-bis(isopropylamino)-6-methylthio-*s*-triazine + (atrazine) 2-chloro-4-ethylamino-6-isopropylamino-*s*-triazine

simazine injury to oats. The ED_{50} value increased by approximately 0.16 p.p.m. with an increase of 2.2 percent in soil organic matter.

Ion-exchange resins have been suggested as another antidote to absorb unwanted pesticide residues from soil. The economics and practical value of such methods are still in doubt.

XV. Conclusions

The many factors which influence the activity and persistence of triazine herbicides in soils include temperature, moisture, leaching, sunlight, microbial activity, rhizosphere, soil organic matter, pH, texture, mechanical analysis, cation-exchange capacity, prevalent ions, chemical reactions, species, depth of germination, and stage of growth. It is obvious that under practical field conditions the majority of these factors is beyond the capacity of man to control, but we have learned how several characteristics of the soil, plant, and environment can be modified in order to enhance the consistency of weed control, crop safety, and detoxification of the triazine herbicides.

The choice of the triazine herbicide itself is an important means of influencing the dependability of weed control and soil carry-over in some instances. In corn, for example, atrazine, simazine, and Primaze® can all be used, and ametryne may be labeled for special weed problems. Atrazine, because of its wide adaptability and flexibility, will usually be the preferred triazine in this crop. However, simazine may be used for improved performance where crabgrass (*Digitaria* spp.), *Panicum* spp., shattercane (*Sorghum bicolor*) and other hard to control annual grasses predominate. Primaze® may be useful in areas where the soil persistence of atrazine and simazine causes serious concern. Ametryne shows promise as a postemergence directed spray for control of emerged weeds after the corn is at least 12 inches high. In sorghum, atrazine is usually preferred to propazine, where it can be used safely, because of improved grass control. Propazine, however, is frequently the best choice in the southern part of the sorghum belt because of higher crop tolerance. Terbutryn has also been promising in sorghum and represents a much reduced soil carryover problem. Sugarcane can be treated with atrazine, ametryne, or simazine, and a new methoxy-triazine (GS-14254) is being developed for use in this crop. At least in limited areas, a number of other crops can be treated with more than one of the several commercial triazine herbicides.

The proper timing and method of application can be the most important means of influencing the biological performance of the triazine herbicides. For example, atrazine can be used in corn by any one of three methods of application: (1) preplant (with or without mechanical incorporation), (2) preemergence, and (3) postemergence (with or without spray additives). For special problems, such as quackgrass (*Agropyron repens*), the application may be split with part of the herbicide being applied preplow and preplant in the fall or spring and the rest being applied preemergence or postemergence. Prometryne in cotton, ametryne in sugarcane, and other uses of triazine herbicides can also be applied at different times or by various methods. The objective of a specific timing or method of application is usually to improve weed control performance; however, it may also influence the soil persistence of a triazine herbicide. The earlier the application is made, the more time the herbicide has to be degraded. On the other hand, mechanical incorporation into the soil may prolong the carry-over slightly by reducing volatilization and photodecomposition.

That the rate of application of triazine herbicides will influence their activity and persistence is well understood. However, since no one should recommend or use more herbicide than required for dependable weed control, the rate of application will not usually provide much flexibility. Nevertheless, by including other variables such as method of application, supplemental cultivation, and band vs. broadcast application, the choice of rate can have a significant influence on performance and soil carry-over.

Between 1960 to 1963 when atrazine was commercially available in both wettable powder and granular formulations, it became apparent that the wettable powder was often superior in performance and resulted in less carry-over. The granular formulations of atrazine were discontinued, therefore, except for non-selective uses. Simazine is available in both 80W and 4G formulations. While the 80W is often preferred because of performance and economics, the granules may be best where ease of application or extended weed control is important. Ametryne can be applied as an 80W or 2E formulation. Economics favor the 80W, but the emulsifiable formulation has given improved performance for certain post-emergence uses and may be preferred for handling qualities. Research to improve the handling and performance characteristics of all triazine herbicides is continuing, and formulation will likely have an important influence on their activity and persistence where more than one formulation is available.

The use of herbicide combinations is one of the most promising means of influencing weed control, crop tolerance, and soil carryover. While a few combinations consisting of more than one triazine herbicide (e.g., Primaze® in corn and ametryne plus atrazine in sugarcane) provide some significant improvement over either of the triazines alone in certain situations, the combination of a triazine with a non-triazine herbicide has often been more promising. Most such combinations have demonstrated reduced soil carryover compared to atrazine alone. The more promising combinations have often improved the control of certain weeds, particularly, under conditions of heavy-soil, high organic matter, and low rainfall. These improvements, however, are not without limitations such as requirement for tank mixes, more risk of crop injury, handling problems, and frequent failures.

The interaction of triazine herbicides with other pesticides has not, to the present time, demonstrated an important influence on their activity or soil carryover. Reliable evidence exists from both basic studies and field observations, however, to suggest that significant interactions can occur. The increased use of pesticides separately, in combination, or in rotation, further indicates that such interactions may increase and need to be considered.

While the modification of soil nutrients is not likely a practical means of influencing the biological properties of triazine herbicides, they can have important effects on triazine activity and persistence. Some nutrients may simply influence the pH or other soil factors, which in turn affect triazine adsorption, movement, and uptake by plants. Other nutrients may alter the plant uptake and translocation of triazines in a more direct way. Nitrogen, for example, appears to enhance triazine uptake by affecting the metabolic root pressure of some plants.

Water has an extremely important influence on the activity and .persistence of triazine herbicides. The triazine is usually dependent upon water in order to be dissolved and moved into the soil solution where it is absorbed by the weed roots or shoot. Water will also greatly enhance the dissipation of triazine herbicides in soils. Therefore, the herbicide will usually be applied as early as possible in order to make maximum use of rainfall. From a practical point of view, however, water is not within our ability to control, except in the limited areas where irrigation is available.

Tillage is an important means of influencing the activity and soil carry-over from triazines. It is one of the variables most universally available and easiest to apply. Mechanical incorporation or supplemental cultivation, especially under conditions of low rainfall, tolerant weeds, or marginal rates of application, will improve the weed control from most triazine herbicides. Also, plowing and thorough mixing of the soil following the crop harvest will reduce the risk of triazine injury to rotational crops.

Within limits, the selection of rotational crops is a very important means of controlling the extent of triazine carry-over problems. While it is not always practical to choose between several crops, it is often possible to select the more tolerant of two or three crops. Soybeans, for example are much more tolerant of triazine residue compared to oats and most vegetable crops. Based on research and experience, it is usually possible to estimate whether or not a triazine application will result in a carryover problem and which would be the best crop to select. Bioassay or chemical assay prior to planting the rotational crop may also be useful.

While cropping has been shown to reduce the amount of triazine soil residues in a few instances, it is not considered important or practical under most conditions. There is good evidence that compensating factors such as increased temperature, moisture, and sunlight under fallow conditions will more than make up for the amount of triazine degraded by the crop.

The use of microorganisms as a treatment has not, as yet, been found to be a significant value in degrading triazines in soils. Microbial degradation of triazines in soil appears to be slow and without adaptive characteristics. Nevertheless, some research is continuing in the hope that this technique may be useful under specific conditions.

The use of charcoal and other additives has been very promising as a means of reducing the carryover effect of triazine herbicides. By adsorbing the triazines preferentially, activated carbon and other adsorbents prevent the triazines from being released into the soil solution. The soil is thereby detoxified. When used as a general soil treatment, the amount of finely ground charcoal needed has varied from 50 to 400 lb. per lb./acre of triazine residue. While charcoal has been used to some extent and appears to have potential, it is presently limited to special problems on high value crops or where root dipping and band applications can be utilized. The most important factors which limit the practical use of this technique include economics, predictability of the desired effect, difficulties in obtaining a uniform application, and thorough mixing with the soil.

Summary

Most of the factors which influence the activity and persistence of triazine herbicides in soil are not within man's capacity to control under practical field conditions. We have learned, however, that several of the soil characteristics and environmental factors which determine triazine response can be modified. By utilizing the inherent flexibility of the triazine herbicides, combined with various techniques now available and yet to be developed, we can markedly enhance their consistency of weed control, crop safety, and detoxification in soils.

Various means which are being used or have been tried to improve weed control are discussed. In the case of some of the longer residual triazine herbicides, prevention or reduction of carry-over injury to sensitive rotational crops must be considered as a companion problem to weed control. The optimum solution varies with the type of agriculture, weeds, soil, and climatic conditions, but will often include more than one technique.

In corn, sorghum, sugarcane, and a few other crops, more than one triazine herbicide can be used depending on the weed problem, soil type, rotational crops, and other factors.

Method and timing of application offer considerable flexibility in the use of several triazine herbicides. For example, atrazine can be applied in corn as (1) preplant (with or without mechanical incorporation), (2) pre-emergence, and (3) postemergence (with or without spray additives). Prometryne in cotton and ametryne in sugarcane can also be applied at different times or by various methods. The choice is usually based on weed control, but method and timing can also affect soil carry-over.

While rate of application is not usually very flexible, when variables such as method of application and supplemental cultivation are included, the choice of rate can have an important influence on soil persistence.

Formulation has had a major effect on both weed control and persistence from triazine herbicides. While economics have favored the wettable powder formulations (80W), some triazines are available in granular form for perennial crops or nonselective uses. Emulsifiable formulations (2E) of certain triazines offer some advantages.

Combinations of triazines with other herbicides offer one of the most promising means of reducing soil carry-over, although there are certain risks and disadvantages which must be realized. The greatest interest has been in combinations with atrazine for weed control in corn. Some combinations have also given improved performance depending on weed species, soil, or environmental conditions.

Other means of influencing the activity or persistence of triazine herbicides have been studied. Some of these methods which have been very useful include tillage (supplemental cultivation, plowing, etc.), water (especially where irrigation is available), and careful selection of rotational crops. Variables or treatments which have been studied but found to be of limited or no practical value at the present time include cropping vs. fallow, triazine-pesticide interactions, use of adaptive microorganisms, soil nutrients, charcoal, and other additives.

Certain chemicals or soil amendments may yet be found to have important influences on the activity and persistence of triazine herbicides.

Résumé*

Voies et moyens d'influencer l'activité et la persistance des desherbants type triazine dans les sols

En pratique, dans les conditions que l'on retrouve en plein champ, la plupart des facteurs qui influent sur l'activité et la persistance d'action dans le sol des herbicides de la série des triazines sont indépendants de la volonté de l'homme. On a trouvé, cependant, que certaines caractéristiques du sol et des facteurs environnants qui déterminent les résultats pouvant être obtenus au moyen des triazines sont susceptibles de modification. En tirant parti de la souplesse d'utilisation inhérente aux herbicides triaziniques ainsi que de diverses techniques qui ont été ou qui peuvent être mises au point, il est possible de favoriser nettement la constance d'action de ces produits contre les

---

* Traduit par l'auteur.

mauvaises herbes et pour la défense des récoltes et la désintoxication des sols.

On discute divers moyens qui ont été mis en oeuvre ou en essai pour améliorer l'efficacité de la lutte contre la végétation adventice. Dans le cas de certaines des triazines à action résiduelle plus ou moins prolongée, la prévention ou la réduction des dommages par répercussion sur les cultures assolées sensibles doit être considérée comme un problème qui va de pair avec la lutte contre les plantes adventices. La solution optimale varie selon le genre de culture, les mauvaises herbes en cause, le sol, les conditions climatiques; souvent, elle demande l'application de plusieurs techniques.

Pour le maïs, le sorgho, la canne à sucre ainsi que certaines autres cultures, on peut utiliser plus d'un herbicide triazinique, selon la nature des plantes adventices, le type de sol, les cultures assolées, ainsi que d'autres facteurs.

Plusieurs des herbicides triaziniques offrent une grande souplesse d'utilisation quant à la méthode et au moment choisi pour l'application de cette dernière. Par exemple, l'atrazine peut être appliquée au maïs, soit: (1) en période de pré-plantage (avec ou sans incorporation mécanique); (2) à la phase souterraine de la croissance; (3) ou à la phase aérienne de la croissance (avec ou sans vaporisation complémentaire). La prométryne dans les cultures de coton et l'amétryne dans les champs de canne à sucre peuvent également être appliquées à différents moments ou par des méthodes diverses. Le choix est généralement basé sur la nature des mauvaises herbes. Ici aussi, la méthode et le moment d'application peuvent influencer l'entraînement dans le sol, et ces facteurs doivent également entrer en ligne de compte.

En général, la quantité appliquée est plus ou moins constante. Cependant, quand on introduit des variables représentées par exemple par la méthode d'application et les cultures supplémentaires, le choix de la dose peut avoir une influence importante sur la persistance dans le sol.

La forme sous laquelle les produits sont présentés a un effet considérable à la fois sur l'action contre la végétation adventice et sur la persistance des herbicides. La formules à poudre mouillable (80W) sont en principe plus avantageuses, mais on offre certaines triazines sous forme de granulés, soit pour les plantes vivaces, soit pour le désherbage non sélectif. Les formes émulsionnables de certaines triazines offrent des avantages.

Les associations de triazines et d'autres herbicides offrent un des moyens les plus prometteurs pour réduire l'entraînement en profondeur, bien que l'on doive tenir compte de certains risques et inconvénients. Le plus grand intérêt a été manifesté à l'égard des associations à l'atrazine pour lutter contre les mauvaises herbes dans les cultures de maïs. D'autres associations ont d'ailleurs donné des résultats améliorés, selon l'espèce de plante adventice en cause, le sol, ou les conditions du milieu.

On a étudié d'autres moyens pour chercher à influencer l'activité ou la persistance des triazines. Parmi ces méthodes, qui se sont révélées fort utiles, on compte les travaux culturaux (culture complémentaire, labourage), l'apport d'eau (surtout dans les régions disposant de moyens d'irrigation) et le choix judicieux des cultures alternées. Les variables ou les traitements qui ont été étudiés mais qui jusqu'à ce jour se sont avérés en pratique d'une utilité faible ou nulle comprennent la substitution d'une mise en culture pour la jachère, les interactions triazines-pesticides, le recours aux micro-organismes adaptatifs, aux produits nutritifs pour sols, au carbone, ainsi qu'à d'autres produits d'addition.

Il nous reste peut-être encore à découvrir une influence importante de certains produits chimiques ou de certains amendements sur l'activité et sur la persistance des herbicides triaziniques.

Zusammenfassung*

Wege und Mittel um die Wirksamkeit und Hartnäckigkeit von Triazin-Unkrautvertilgungsmittel in Böden zu beeinflussen

In der Praxis sind die meisten Umstände, die die Wirksamkeit und Lebensdauer der Triazin-Unkrautvertilgungsmittel im Boden bestimmen, menschlicher Beeinflussung nicht zugänglich. Wir haben jedoch gelernt, dass einige der Bodeneigenschaften und Umweltsfaktoren, die für das Ansprechen auf die Triazine verantwortlich sind, beeinflusst werden können. Dadurch dass wir uns die den Triazin-Unkrautvertilgungsmitteln innewohnende Anpassungsfähigkeit zunutze machen und uns gleichzeitig verschiedener bereits verfügbarer und noch zu entwickelnder Verfahren bedienen, können wir die Beständigkeit ihrer Wirkung in Bezug auf Unkrautbekämpfung, Ernteschutz und Entgiftung im Boden erheblich verbessern.

Verschiedene Mittel zur Verbesserung der Unkrautbekämpfung, die im Gebrauch sind oder versuchsweise eingesetzt worden sind, werden besprochen. Bei einigen der herbiziden Triazine mit längerer Verbleibzeit muss die Verhütung oder Verminderung von Schäden an empfindlichen Wechselernten, die durch diese Nachwirkung verursacht werden können, als ein Begleitproblem der Unkrautbekämpfung angesehen werden. Die optimale Lösung richtet sich nach der Art der Kultur, des Unkrauts und des Bodens sowie nach den klimatischen Verhältnissen, wird aber häufig mehr als ein Verfahren umfassen.

Bei Mais, Sorghum, Zuckerrohr und einigen anderen Kulturen kann mehr als ein Triazin-Unkrautmittel Verwendung finden, je nach den Unkrautverhältnissen, der Bodenart, Wechselernten und anderen Umständen.

Zeitpunkt und Methoden der Anwendung sind bei einigen Triazin-Unkrautvertilgungsmitteln äusserst elastisch. Atrazin z.B. kann im Falle von Mais (1) vor der Aussaat (mit oder ohne mechanische Ein-

* Übersetzt vom Autor.

verleibung), (2) vor oder (3) nach Keimdurchbruch (mit oder ohne spritzbare Zusatzmittel) eingesetzt werden. Bei Baumwolle kann Prometryne and bei Zuckerrohr Ametryne ebenfalls zu verschiedenen Zeiten und nach verschiedenen Methoden angewandt werden. Die Wahl wird im allgemeinen aufgrund der Erfordernisse der Unkrautbekämpfung getroffen, doch kann die Methode und der Zeitpunkt auch einen Einfluss auf die Boden-Nachwirkung haben.

Bei Berücksichtigung veränderlicher Grössen wie das Anwendungsverfahren und Zusatzbebauung ist die Anwendungsfrequenz zwar im allgemeinen nicht sehr variierbar, kann aber einen wesentlichen Einfluss auf die Verweilzeit im Boden ausüben.

Die Formulierung der Präparate kann sich auf die Unkrautbekämpfung sowie auch auf die Verweilzeit der Triazin-Unkrautmittel bedeutend auswirken. Während aus wirtschaftlichen Rücksichten die benetzbaren Pulverformulierungen (80W) vorgezogen werden, sind einige Triazine zum Gebrauch für Dauerkulturen oder zum nichtelektiven Gebrauch in granulierter Form erhältlich. Emulgierbare Formulierungen (2E) bestimmter Triazine bieten gewisse Vorteile.

Kombinationen von Triazinen mit anderen Unkrautvertilgungsmitteln ergeben eine der vielversprechendsten Möglichkeiten zur Verminderung der Boden-Nachwirkung, obwohl damit Risiken und Nachteile verbunden sind, die man sich vor Augen halten muss. Das grösste Interesse besteht für Kombinationen mit Atrazin für Unkrautbekämpfung bei Mais. Je nach Unkrautart, Bodenbeschaffenheit, und Ortsverhältnissen sind auch mit anderen Kombinationen Leistungsverbesserungen erzielt worden.

Andere Mittel zur Beeinflussung der Wirkung oder Verweildauer von Triazin-Unkrautmitteln sind ebenfalls untersucht worden. Zu solchen Methoden, die sich als sehr nützlich erwiesen haben, gehören Bestellung (zusätzliche Bebauung, Pflügen usw.), Bewässerung (besonders wenn eine Rieselanlage zur Verfügung steht), und sorgfältige Auswahl von Wechselkulturen. Zu Varianten oder Behandlungen, die geprüft wurden, sich aber bis heute als nur beschränkt praktisch nutzbar oder nutzlos erwiesen haben, gehören Ernte im Gegensatz zu Brachernte, Wechselwirkungen zwischen dem Triazin und einem Schädlingsbekämpfungsmittel, Verwendung anpassungsfähiger Mikroorganismen, von Bodennährstoffen, Holzkohle und anderen Zusatzstoffen.

Es mag sich schliesslich doch noch herausstellen, dass manche Chemikalien oder Bodenverbesserungsmittel die Wirkung und Verweildauer von Triazin-Unkrautvertilgungsmitteln stark beeinflussen.

References

ADAMS, R. S., JR.: Phosphorus fertilization and the phytotoxicity of simazine. Weeds **13**, 113 (1965).
—— Personal communications. Univ. of Neb. (1969).

——, and W. G. Espinoza: The effect of phosphorus and atrazine on mineral composition of soybeans. J. Agr. Food Chem. **17**, 818 (1969).

——, and S. E. Nelson: Effect of simulated atrazine residues on oat yield. Research Rept. N.C. Weed Control Conf. **24**, 4 (1967).

Ahrens, J. F.: Detoxification of simazine and atrazine treated soil with activated carbon. Proc. N.E. Weed Control Conf. **19**, 364 (1965).

—— Summary of research conducted on weed and grass control in Christmas tree plantings. Univ. Conn. Unpublished rept. (1966).

—— Improving herbicide selectivity in transplanted crops with root dips of activated carbon. Proc. N.E. Weed Control Conf. **21**, 64 (1967 a).

—— The persistence of simazine in soil. Weed Soc. Amer. Abstr., p. 75 (1967 b).

Anderson, A. H.: The inactivation of simazine and linuron in soil by charcoal. Weed Research **8**, 58 (1968).

Anderson, R. N.: Differential response of corn inbreds to simazine and atrazine. Weeds **12**, 60 (1964).

—— A search for atrazine resistance in soybeans. Weed Sci. Soc. Amer. Abstr., 157 (1969).

——, and R. Behrens: A search for atrazine resistance in flax (*Linum usitatissimum* L.). Weeds **15**, 85 (1967).

Anonymous: Deactivation of a triazine herbicide in the soil. S. Ore. Expt. Sta. Medford, Proj. 186 (1965).

—— Use of activated charcoal to reduce the effect of soil residue and/or increase crop tolerance. 1966–1967. Farm Crops Dept., Ore. State Univ. (1967).

Armstrong, D. E.: Atrazine degradation in soil. Ph.D. thesis, Univ. of Wisc. (1966).

——, G. Chesters, and R. F. Harris: Atrazine hydrolysis in soil. Soil Sci. Soc. Amer. Proc. **31**, 61 (1967).

Ashley, R. A., and E. M. Rahn: Persistence of atrazine at two locations as affected by soil incorporation and certain additives. Proc. N.E. Weed Control Conf. **21**, 557 (1967).

Austin, D., H. Andrews, and L. N. Skold: Disappearance of atrazine, DCPA, diphenamid, diuron, linuron, norea, prometryne, and trifluralin from soils treated for three years. Proc. S. Weed Conf. **21**, 314 (1968).

Bayer, D. E.: Effect of surfactants on leaching of substituted urea herbicides in soil. Weeds **15**, 249 (1967).

Behrens, R., A. L. Darwent, S. D. Evans, and W. W. Nelson: Fall vs. spring applications of atrazine for weed control in corn. Research Rept. N.C. Weed Control Conf. **25**, 148 (1968).

Bortels, H., and E. Fricke: The phytotoxic effect of simazine in relation to its adsorption in soil. Nachr Bl. dt. PflSchutzdienst., Stuttgart. **18**(5), 65 (1966); through Weed Abstr. **15**, 1786 (1966).

——, and O. C. Burnside: Aerial and ground applications of preemergence herbicides in corn, sorghum, and soybeans. Weeds **13**, 334 (1965).

Bovey, R. W., and C. R. Fenster: Aerial application of herbicides on fallow land. Weeds **12**, 117 (1964).

——, and F. R. Miller: Effect of activated carbon on the phytotoxicity of herbicides in a tropical soil. Weed Sci. **17**, 189 (1969).

Brenchley, R. G.: The response of tomatoes to atrazine as affected by magnesium and photoperiod. Ph.D. thesis, Ore. State Univ. (1969).

Bryant, T. A.: Disappearance of atrazine, DCPA, diphenamid, diuron, linuron, norea, prometryne, and trifluralin from the soil. MS thesis, Univ. of Tenn. (1967).

Buchanan, G. A.: Role of temperature in the inactivation of some s-triazine herbicides. MS thesis, Univ. of Fla. (1962).

Buchholtz, K. P.: Use of atrazine and other triazine herbicides in control of quackgrass in corn fields. Weeds **11**, 202 (1963).

—— Factors influencing oat injury from triazine residues in soil. Weeds **13**, 362 (1965 a).

—— Studies on weed control, corn yields and soil residues using triazine herbicides in 1964. Unpublished rep't (1965 b).

—— Herbicides on corn can harm next crop. Crops and Soils **19**(No. 3), 20 (1966 a).

—— Atrazine injury to oats on plots having various fertility levels. Unpublished rep't., Univ. of Wisc. (1966 b).

—— Kill weeds, minimize residues. A simple test for atrazine residues. Crops and Soils **20**(No. 3), 7 (1967).

——, and R. E. DOERSCH: Cultivation and herbicides for weed control in corn. Weed Sci. **16**, 232 (1968).

BURNSIDE, O. C.: The influence of environmental factors on the phytotoxicity and dissipation of simazine. Ph.D. thesis, Univ. of Minn. (1959).

—— The effect of herbicide incorporation on weed control in corn and sorghum. Proc. N.C. Weed Control Conf. **21**, 26 (1964).

—— Effect of herbicides used for wild cane control in corn on subsequent oat yield. Research Rep't. N.C. Weed Control Conf. **24**, 3 (1967).

—— Oat yields one year after preemergence and postemergence atrazine applications on sorghum. Research Rep't. N.C. Weed Control Conf. **25**, 1 (1968).

——, and R. BEHRENS: Phytotoxicity of simazine. Weeds **9**, 145 (1961).

——, C. R. FENSTER, G. A. WICKS, and J. V. DREW: Effect of soil and climate on herbicide dissipation. Weed Sci. **17**, 241 (1969).

——, and G. A. WICKS: Cultivation and herbicide treatments of dryland sorghum. Weeds **12**, 307 (1964).

——, G. A. WICKS, and C. R. FENSTER: Influence of tillage, row spacing, and atrazine on sorghum and weed yields from non-irrigated sorghum across Nebraska. Weeds **12**, 211 (1964).

CASTELFRANCO, P., and D. B. DEUTSCH: Action of polysulfide ion on simazine in soil. Weeds **10**, 244 (1962).

CHAPLIN, D., and E. K. ALBAN: Varietal response of sweet corn to simazine. Proc. N.C. Weed Control Conf. **17**, 52 (1960).

COFFEY, D. L., and G. F. WARREN: Inactivation of herbicides by activated carbon and other absorbents. Weed Sci. **17**, 16 (1969).

COLBY, S. R., and C. I. HARRIS: Atrazine-prometryne combinations: herbicidal activity, fate in corn, and soil persistence. Proc. N.E. Weed Control Conf. **20**, 115 (1966).

COMSTOCK, V. E., and R. N. ANDERSON: An inheritance study of tolerance to atrazine in a cross of flax. Crop Sci. **8**, 508 (1968).

CORBIN, F. T., and T. J. SHEETS: Antagonistic effects of certain pesticide combinations on plants. Weed Sci. Soc. Amer. Abstr., p. 22 (1968).

CRANDALL, P. C.: 1966 herbicide trials. Unpublished rep't, Wash. State Univ., Vancouver (1966).

DAVID, R.: Influence of simazine and atrazine on *Digitaria:* investigation of the residual effect of these compounds in soil. Phytiat. Phytopharm. **11**, 191 (1962).

DAVIS, D. E.: Annual report for calendar year 1966. Unpublished rep't, Auburn Univ. (1967).

DEVRIES, M. L.: The effect of simazine on Monterey pine and corn as influenced by lime, bases, and aluminum sulfate. Weeds **11**, 220 (1963).

DOLL, J. D., and W. F. MEGGITT: Uptake and effectiveness of herbicides at different nutrient levels and combinations. Weed Sci. Soc. Amer. Abstr., 212 (1969).

DOHERTY, P. J., and G. F. WARREN: Adsorption of four herbicides by different types of organic matter and a bentonite clay. Weed Research **9**, 20 (1969).

ENO, C. F., and J. C. CASELEY: Soil enrichment with organisms capable of metabolizing herbicides. Rep't. Fla. Expt. Sta., p. 186 (1965).

FEENEY, R. W., and R. H. COLE: Comparison of spray and granular formulations of several herbicides. Proc. N.E. Weed Control Conf. **18**, 369 (1964).

FERTIG, S. N.: A summary of quackgrass control studies for 1961. Proc. N.E. Weed Control Conf. **16**, 286 (1962).
—— A summary of four years of quackgrass studies. Proc. N.E. Weed Control Conf. **17**, 304 (1963).
FINK, R. J., and O. H. FLETCHALL: Forage crop establishment in soil containing atrazine or simazine residues. Weeds **11**, 81 (1963).
—— —— Soybean injury from triazine residues. Weed Sci. **17**, 35 (1969).
FLANAGAN, T. R.: Quackgrass survival following fall and spring herbicide treatments before corn. Proc. N.E. Weed Control Conf. **17**, 297 (1963).
FRANK, R.: Atrazine carryover in production of sugar beets in Southwestern Ontario. Weeds **14**, 82 (1966).
GAST, A.: Contributions to the knowledge of the behavior of triazines in soil. Paper read at XIV Annual Symposium for Crop Protection, Gent, Belgium (1962).
GRAVES, C. R., D. AUSTIN, and H. ANDREWS: Effects of linuron and atrazine with sources and rates of nitrogen on selected corn varieties in the seedling stage. Proc. S. Weed Conf. **21**, 84 (1968).
GROGAN, C. O., E. F. EASTIN, and R. D. PALMER: Inheritance of susceptibility of a line of maize to simazine and atrazine. Crop Sci. **3**, 451 (1963).
GROVER, R.: Influence of organic matter, texture, and available water on the toxicity of simazine in soil. Weeds **14**, 148 (1966).
GYSIN, H., and E. KNÜSLI: Chemistry and herbicidal properties of triazine derivatives. Adv. Pest Control Research **3**, 289 (1960).
HARRIS, C. I., and T. J. SHEETS: Persistence of several herbicides in the field. Proc. N.E. Weed Control Conf. **19**, 359 (1965).
——, E. A. WOOLSON, and B. E. HUMMER: Dissipation of herbicides at three soil depths. Weed Sci. **17**, 27 (1969).
HILTBOLD, A. E.: Personal communication. Auburn Univ. (1969).
HINESLY, T. D., E. L. KNAKE, and R. D. SEIF: Herbicide versus cultivation for corn with two methods of seedbed preparation. Agron. J. **59**, 509 (1967).
HOLLY, K., and H. A. ROBERTS: Persistence of phytotoxic residues of triazine herbicides in soils. Weed Research **3**, 1 (1963).
HOPEN, H. J.: 1967 vegetable crop, and gladiolus herbicide field-plot test results. Univ. of Ill. (1967).
JORDAN, L. S., J. M. LYONS, W. H. ISOM, and B. E. DAY: Factors affecting performance of preemergence herbicides. Weed Sci. **16**, 457 (1968).
KARIM, A., and A. D. BRADSHAW: Genetic variation in simazine resistance in wheat, rape and mustard. Weed Research **8**, 283 (1968).
KAUFMAN, D. D., P. C. KEARNEY, and T. J. SHEETS: Microbial degradation of simazine. J. Agr. Food Chem. **13**, 238 (1965).
KNAKE, E. L.: Liquid vs. granular preemergence herbicides. Agr. Chemicals **18**(No. 5), 52 (1963).
—— Effect of tillage on atrazine residue and oat yield. Unpublished rep't, Univ. of Ill. (1966).
——, A. P. APPLEBY, and W. R. FURTICK: Soil incorporation and site of uptake of preemergence herbicides. Weeds **15**, 228 (1967).
——, D. L. MULVANEY, and P. E. JOHNSON: The effect of atrazine applications on subsequent soybean and oat yields. Research Rept. N.C. Weed Control Conf. **23**, 2 (1966).
——, and F. W. SLIFE: 1963 Research summary-Quackgrass research. Unpublished annual rep't, Univ. of Ill. (1964).
—— ——, and R. D. SEIF: The effect of rotary hoeing on performance of preemergence herbicides. Weeds **13**, 72 (1965).
KRATKY, B. A., and G. F. WARREN: Activated carbon vermiculite mixture for increasing herbicide selectivity. Weed Sci. Soc. Amer. Abstr., 48 (1969).
LAVY, T. L., J. J. LINSCOTT, and F. W. ROETH: Use of activated carbon and corncobs in detoxifying atrazine treated soils. Proc. N.C. Weed Control Conf. **22**, 114 (1965).

LeBaron, H. M.: Oil-water emulsions as carriers for atrazine. Proc. N.C. Weed Control Conf. **23**, 46 (1966).

Linscott, D. L., and R. D. Hagin: Protecting alfalfa seedlings from a triazine with activated charcoal. Weeds **15**, 304 (1967).

Lovely, W. G.: Soil incorporation of liquid and granular herbicides for weed control in corn. Proc. N.C. Weed Control Conf. **21**, 18 (1964).

Lynbenov, Ya.: Effect of pre-sowing tillage of soil on residual effect of some triazine herbicides. Rast. Nauki **2**, 141 (1965); through Weed Abstr. **15**, 1360 (1966).

Maxfield, J. E.: Personal communication. Univ. of Nev. (1969).

McCormick, L. L., and A. E. Hiltbold: Microbial decomposition of atrazine and diuron in soil. Weeds **14**, 77 (1966).

McCowen, F. H.: Turf herbicide Rx: Add oil. Agr. Chemicals **23**(No. 4), 18 (1968).

McCutchen, T., and H. Andrews: Nitrogen solution vs. water as a carrier for prometryne, trifluralin, C-2059 and SD-11831 applied preemergence on cotton. Proc. S. Weed Conf. **20**, 57 (1967).

McGlamery, M. D.: Preplant incorporation of corn herbicides. Proc. N.C. Weed Control Conf. **24**, 13 (1967).

Meggitt, W. F.: Herbicide combinations for quackgrass control. Proc. N.C. Weed Control Conf. **17**, 83 (1960).

Miller, F. R., and R. W. Bovey: Tolerance of Sorghum bicolor (L.) Moench to several herbicides. Agron. J. **61**, 282 (1969).

Minshall, W. H.: Metabolic root pressure and the uptake of triazine herbicides. Weed Soc. Amer. Abstr., p. 61 (1967).

—— Effect of nitrogenous materials on the uptake of triazine herbicides. Weed Sci. **17**, 197 (1969).

Moss, D. N.: Relation in grass of high photosynthetic capacity and tolerance to atrazine. Crop Sci. **8**, 774 (1968).

Nalewaja, J.: Summary of 1966 weed control trials—field crops. Unpublished rep't N.D. State Univ. (1967).

Nash, R. G., and W. G. Harris: Dexon fungicide antagonism toward herbicial activity of s-triazines. Weed Sci. Soc. Amer. Abstr., 240 (1969).

Nearpass, D. C.: Effect of soil acidity on the absorption, penetration and persistence of simazine. Weeds **13**, 341 (1965).

Osgood, R. V., R. R. Romanowski, and H. W. Hilton: Differential tolerance of Hawaiian sugarcane varieties to diuron and ametryne. Weed Sci. Soc. Amer. Abstr., 31 (1969).

Palmer, R. D., and C. O. Grogan: Tolerance of corn lines to atrazine in relation to content of benzoxazinone derivative, 2-glucoside. Weeds **13**, 219 (1965).

Peters, R. A., and P. E. Kelley: Evaluation of herbicides for annual grass control in corn. Proc. N.E. Weed Control Conf. **18**, 278 (1964).

Plaisted, P. H., and C. Panton: The susceptibility of different corn varieties to atrazine, propazine, simazine and trietazine. Report No. 7 to Geigy Chem. Co. from Boyce Thompson Inst., Nov. 24 (1961).

Raleigh, S. M.: Quackgrass control. Proc. N.E. Weed Control Conf. **15**, 315 (1961).

Robinson, D. W.: The use of adsorbents and simazine on newly planted strawberries. Weed Research **5**, 43 (1965).

Roeth, F. W., T. L. Lavy, and O. C. Burnside: Atrazine degradation in two soil profiles. Weed Sci. **17**, 202 (1969).

Rogers, E. G.: Leaching of seven s-triazines. Weed Sci. **16**, 117 (1968).

Saghir, A. R., and A. H. Choudhary: Triazine herbicides on maize and their residual effects on following crops. Weed Research **7**, 272 (1967).

Santelmann, P. W., H. A. L. Greer, and I. L. Six: Factors influencing the activity of soil incorporated herbicides. Okla. State Univ. Bull. B-658 (1968).

Schubert, O. E.: Can activated charcoal protect crops from herbicide injury? Crops and Soils **19**(No. 9) (1967).

SCHWEIZER, E. E., and J. T. HOLSTUN: Persistence of five cotton herbicides in four southern soils. Weeds **14**, 22 (1966).

SHEETS, T. J., and W. C. SHAW: Herbicide properties and persistence in soils of s-triazines. Weeds **11**, 15 (1963).

SIKKA, H. C., and D. E. DAVIS: Dissipation of atrazine from soil by corn, sorghum and Johnsongrass. Weeds **14**, 289 (1966).

SPLITTSTOESSER, W. E., and L. A. DERSCHEID: Effects of environment upon herbicides applied preemergence. Weeds **10**, 304 (1962).

STANIFORTH, D. W., and W. G. LOVELY: Preemergence herbicides in corn production. Weeds **12**, 131 (1964).

STICKLER, R. L., E. L. KNAKE, and T. D. HINESLY: Soil moisture and effectiveness of preemergence herbicides. Weed Sci. **17**, 257 (1969).

STOJANOVIC, B.: 35th Ann. Meeting Nat. Agr. Chem. Assoc., Unpublished report (1968).

STRITZKE, J. F., and B. PARKER: Summary of atrazine and oil work in South Dakota. Proc. N.C. Weed Control Conf. **24**, 59 (1967).

SWITZER, C. M., and W. E. KAUSER: Effectiveness and persistence of certain herbicides in soil. Proc. N.E. Weed Control Conf. **14**, 329 (1960).

TALBERT, R. E., and O. H. FLETCHALL: Inactivation of simazine and atrazine in the field. Weeds **12**, 33 (1964).

THOMPSON, L., JR., and F. W. SLIFE: Foliar and root absorption of atrazine applied postemergence to giant foxtail. Weed Sci. **17**, 251 (1969).

UHLIG, S. K.: Investigations and experiences on the effect of various environmental factors on the activity of simazine. Tag. Ber. Symp. über den Einfluss von Umweltbedingungen auf die Wirkung von chemischen Pflanzenschutzmitteln, Biol. Zent. Anst. Dt. Akad. Landw. Wiss. Berlin, Magdeburg, 1962. **62**, 121 (1964); through Weed Abstr. **15**, 636 (1966).

UPCHURCH, R. P., G. P. LEDBETTER, and F. L. SELMAN: The interaction of phosphorus with the phytotoxicity of soil applied herbicides. Weeds **11**, 36 (1963).

VENGRIS, J.: Summary of quackgrass control studies with atrazine in Massachusetts. Proc. N.E. Weed Control Conf. **18**, 382 (1964).

WEBER, J. B., P. W. PERRY, and K. IBARAKI: Effect of pH on the phytotoxicity of prometryne applied to synthetic soil media. Weed Sci. **16**, 134 (1968).

WEBSTER, H. L., and T. J. SHEETS: A physiological explanation of the atrazine-Dexon interaction. Weed Sci. Soc. Amer. Abstr., 239 (1969).

WICKS, G. A., and O. C. BURNSIDE: Residues in soil one year after herbicides were applied to sorghum. Weeds **13**, 173 (1965).

—— ——, and C. R. FENSTER: Effect of atrazine and oil on weed control in sorghum across Nebraska in 1968. Proc. N.C. Weed Control Conf. **25**, 14 (1968).

WIESE, A. F., and E. B. HUDSPETH, JR.: Subsurface application and shallow incorporation of herbicides on cotton. Weed Sci. **16**, 494 (1968).

——, and D. OWEN: Oil carriers and aircraft for applying propazine and atrazine. S. Weed Proc. Conf. **20**, 109 (1967).

——, and J. R. QUIMBY: Inheritance of 2,4-D and propazine resistance in grain sorghum. Weed Sci. Soc. Amer. Abstr., p. 29 (1969).

WISK, E. L., and R. H. COLE: Persistence in soils of several herbicides used for corn and soybean weed control. Proc. N.E. Weed Control Conf. **19**, 356 (1965).

Quantitative determination of triazine herbicides in soils by bioassay

By

Richard Behrens*

Contents

I. Introduction

Man in his efforts to achieve a better understanding of the world around him has continually faced the necessity of finding suitable techniques of study. In many instances, he has utilized the responses of living organisms to gain the desired knowledge through biological assay methods. According to FINNEY (1947), "The term biological assay in its widest sense should be understood to mean the measurement of the potency of any stimulus, physical, chemical or biological, physiological or psychological, by means of the reactions which it produces in living matter." A substantial portion of present-day research involves, under this definition, the use of the biological assay. The purpose of this review is to consider the use of bioassays specifically for the quantitative determination of triazine herbicides in soil.

---

* Department of Agronomy and Plant Genetics, *University of Minnesota,* St. Paul.

II. History and development of the biological assay

In the late 1800's, Ehrlich made the first attempts to use the biological assay in a quantitative manner in his efforts to standardize diphtheria antitoxin (Finney 1952). During the 1920's, a major advance in biological assay methodology came with the introduction of standard preparations of a compound rather than animal units, as the reference for estimating the concentration of unknowns or the relative activity of other substances. Went (1928) first utilized a plant response—*Avena* coleoptile curvature—in his famous quantitative bioassay for auxin. This extremely sensitive test is still frequently used in the study of plant hormones. Further progress in biological assay technique came in the 1940's with the introduction of experimental designs and statistical techniques that provided means for evaluating the reliability and precision of bioassay results (Finney 1952).

Crafts (1935) first developed a "plant-indicator" test to study herbicides. He examined the relative toxicity, persistence, and degree of leaching of sodium arsenite and sodium chlorate in several California soils using oats as the test plant.

The use of bioassay for s-triazine determination in soils preceded the development of reliable techniques for the extraction of these compounds from soils for chemical assay. Gysin and Knüsli (1960) attributed to Gast prior to 1958 the first use of the biological assay for the determination of a s-triazine herbicide. Gast used mustard and oats to measure simazine levels in soil. Van der Zweep (1958) published an account of the use of a biological assay for simazine using rye plants to measure from 0.2 to 5.0 p.p.m. of simazine in sand. In the past ten years numerous studies of the s-triazines in soils have utilized bioassays.

The fact that quantitative results can be obtained by bioassay methods is well known. Yet the assumptions made and conditions that must be met to obtain reliable data are often not clearly understood. Freed (1964) states: "The validity of biological assay procedure rests on two basic assumptions. They are: (a) that a given plant response to a chemical increases in an order related to the dose and (b) that within the limits of sample variation, these responses are reproducible when the plant material and the environmental conditions are the same."

To insure the validity of the results of a biological assay, certain conditions must be met: (1) the response of the plant to the unknown and standard treatments included in the assay must be obtained under identical environmental conditions; (2) the unknown samples must not contain some component or factor that will modify the response of the test plant, and thus, invalidate the comparison with the standard sample; (3) the plant material must be assigned to treatments in a random manner; and (4) it must be possible to determine the statistical limits of reliability for the bioassay at some desired level of probability, normally at the five percent or one percent level (EMMONS 1950).

Biological assays have a number of limitations that should be recognized if a suitable assay is to be properly developed. (1) Biological assays are not well suited for the determination of a compound present in extracts of biological materials. Almost invariably there are interfering components in the extract that influence the response of the test organism. (2) Relatively narrow ranges of environmental conditions and concentrations of the test compound are necessary if the biological assay is to be successful. Wide variances in environmental conditions from those that are normal for the test organism places stresses on it that are likely to restrict or prevent the desired response. Dosages of the test compound that induce an intermediate degree of plant response (i.e., 50 percent growth reduction) provide the best estimate of its potency (FINNEY 1952). (3) Some biological assays require a considerable length of time to grow the test organism to a stage where responses will be significant. For example, in herbicide bioassays with plants, a test period of 30 days is common. During this period, the plant may be exposed to a continually changing concentration of the test compound which is being deactivated or lost from the soil throughout the test. (4) The response of organisms in a biological assay are an indirect measurement of the test compound. When the bioassay involves plants grown in the soil, significant amounts of the test compound may be adsorbed by the soil so that it is not available to the test plant. This may be beneficial or detrimental depending on the objective of the experiment. If the objective is to determine the amount of test compound available to affect plant growth, it is beneficial; while if the objective is to determine the total quantity of the test compound, it is detrimental.

III. Plant biological assay methods for herbicides in the soil

Procedures for biological assay of herbicides in soil are too numerous to mention individually. There are many variations depending on the herbicide, the particular aspect under study, the test plant, and the type of response utilized. Despite these variations, the essential components are much the same, and are similar to those first described by CRAFTS (1935) when he used Kanota oats as the indicator plant. He used the bioassay to determine the relative potency of soil sterilants and also studied the persistence of sodium arsenite and sodium chlorate in four soils.

In his procedure, a weighed amount of air-dry soil was moistened with a dilute solution of the test chemical. He obtained uniform distribution of the toxicant and water by placing ⅓ of the test soil in a metal can and moistening it with ⅓ of the test solution. Then, he alternately added ⅓ portions of soil and chemical solution to the can.

After standing overnight the cultures were seeded with 13 oat seeds/can, which were thinned to ten uniform seedlings after emergence. The soil was watered to field capacity when necessary. Plant height and the fresh weight of top growth were determined 30 days after planting. These data were plotted against the concentration of herbicide applied to the soil to show the relationship between herbicide concentration and the growth response of the oats.

The data obtained in biological assays is usually quantitated by means of standard response curves which are constructed by plotting the growth response of test plants against the concentrations of the herbicide that were added to the soil in which the plants were grown. Then, unknown soil concentrations of the herbicide are estimated by comparing the response of test plants with the growth responses plotted on the standard curve. The inclusion of standards in each bioassay is essential to obtain reliable quantitative estimates of unknown herbicide concentrations. Unpredictable variations in test plant sensitivity and in environmental conditions between bioassay runs makes this necessary.

The manner of expressing data obtained in bioassays varies considerably. Plant growth responses have been expressed as direct measurements (i.e., dry or fresh weight) or, more commonly, as percentages of the untreated control. These values are plotted against the herbicide concentration graphed on an arithmetic or logarithmetic scale. The transformation of herbicide dose to a log-scale is desirable because the relationship between log dose and mean plant response often approaches linearity, which greatly simplifies statistical analysis of the assay data.

AGUNDIS (1966), BURNSIDE (1959), JANSEN et al. (1958), and TALBERT and FLETCHALL (1964) have used analysis of variance to provide a measure of the sensitivity of their bioassay techniques. However, more frequently no attempt is made to determine the variability in a bioassay. The common practice is to estimate from a standard curve the amount of herbicide in soil samples collected from the various treatments of an experiment. Some workers report these estimates with no statistical analysis, while others conduct an analysis of variance on the estimates. FINNEY (1952) strongly advocates the utilization of more advanced statistical techniques than have been used in herbicide bioassays. It seems probable that the application of more appropriate statistical methods to data from biological assays would greatly improve the value of the results obtained.

The biological assay has been used frequently to determine the influence of soil factors such as organic matter, cation exchange capacity, or pH on the phytotoxicity or detoxification of a herbicide. In studies of this nature, the results have often been expressed as the concentration of herbicide required to induce a certain level of response by the test plant. For example, a 50 percent reduction in fresh weight of tops, ED$_{50}$ (SHEETS and SHAW 1963), dry weight of tops, GR$_{50}$ (UPCHURCH and MASON 1962), or reduction in stand, NR$_{50}$ (UPCHURCH *et al.* 1966) have been utilized to compare the potency of herbicides in soils with different characteristics. The relative importance of the various soil factors is then evaluated by means of correlation and regression analyses.

The precision of a bioassay in the determination of a herbicide concentration depends to a great extent on the recognition and control of the many factors that can introduce variability or a bias into the data obtained. Some of these factors are discussed below.

a) The test organism

The test organism, usually a plant, must grow well under the environmental conditions imposed in the bioassay. It should develop a good gradation in response over a wide range of herbicide concentrations. The test plant should not be diseased or highly susceptible to diseases. Plant-pathogen-herbicide interactions can introduce variability or bias into a biological assay. ESPINOSA *et al.* (1968) demonstrated that 3-amino-2,5-dichlorobenzoic acid (amiben) and $\alpha,\alpha,\alpha,$-trifluoro-2,6-dinitro-N,N-dipropyl-p-toluidine (trifluralin) increased the susceptibility of soybean seedlings to damping off. On the other hand, atrazine appeared to have some fungicidal action that reduced the soybean susceptibility to disease organisms. SHULDT and WOLF (1956) also reported that a number of s-triazines had fungicidal properties. If ESPINOSA *et al.* (1958) had been making potency comparisons between amiben, trifluralin, and atrazine using soybeans as the test plant, the pathogen effect would have caused amiben and trifluralin to appear more potent in relation to atrazine than is actually the case.

The seed of the test plant to be used in the bioassay should be carefully selected (MITCHELL and LIVINGSTON 1968). The plant material used should be as genetically homozygous as possible. Variability in response to a herbicide within a species has been reported frequently. JACOBSOHN and ANDERSEN (1968) summarized these reports and described significant differences in the response of wild oat lines to several herbicides. Only seeds of uniform size can be expected to produce the seedlings of similar size and vigor that are so necessary in keeping the variability within the bioassay at a low level. Further, if the mode of action of the test herbicide is the suppression of photosynthesis, the seed size is the major determinant of the food supply available for growth of the bioassay plant. ANDERSEN (1969) demonstrated the seed size effect in studies of the tolerance of soybean genotypes to atrazine. He found a significant correlation between soybean seed size and atrazine tolerance with the large-seeded types being more tolerant than the small-seeded types. Seed that is low in vigor should be avoided, also. Seedlings from seed with low vigor may differ greatly in time of emergence and rate of growth. This variability may seriously decrease the precision of a bioassay.

b) Environmental conditions

A uniform environment is a necessity in maintaining a high level of precision in a bioassay. Properly designed growth chambers have an advantage over a greenhouse for conducting bioassays. Sunlight in the greenhouse varies considerably from day to day and throughout the season, while the fluorescent lights of growth chambers provide a relatively constant light source. Greenhouse lighting can be considerably improved by providing supplemental fluorescent lighting. Supplemental lighting is especially important in the winter to maintain a good growth rate and reduce plant etiolation. In considering growth chamber lighting, uniformity of intensity should not be taken for granted. Unless the chamber walls are covered with reflective material a substantial drop in light intensity occurs from 12 to 18 inches from the wall. If plants are placed in this area, considerable variability will be introduced into the assay.

Precise temperature control is not often attainable in greenhouses. Fluctuations arise from variations in sunlight as well as from the cycling of heaters and ventilating systems. However, the maintenance of temperature uniformity within the greenhouse area where the test is located is possible. Care must be taken to eliminate temperature gradients caused by the colder outside walls or the heating units. Also, temperature differences are common at the edges of greenhouse benches so bioassay plants should not be placed near the edge of the bench. In growth chambers, temperature gradients are often found. If the light intensity is high in a growth chamber, vertical temperature gradients are common. Temperature differences and variations in air velocity are often great near air ducts in growth chambers. Since air velocity has a significant effect on the transpiration rate of plants, it may effect the uptake and translocation of the test herbicide and, thereby, introduce variability into the bioassay.

The moisture content of the soil (GROVER 1966, UPCHURCH and MASON 1962) can greatly change the phytotoxicity of herbicides to plants. Care must be taken to maintain a uniform soil moisture throughout the bioassay run or bias as well as variability may be introduced. The larger plants, normally those growing in soil containing the lower concentrations of herbicide, will often transpire more water resulting in a drier soil in those pots. GROVER (1966) has shown that the ED_{50} of simazine for oats grown at 30 and 60 percent of field capacity is 1.3 and 0.4 p.p.m.w., a very substantial difference.

Other soil factors exert a considerable influence on the phytotoxicity of a herbicide. Therefore, every precaution should be taken to use exactly the same soil throughout the bioassay. In bioassays where soil dilution is used, bias as well as variability would be introduced if the soil used in making the dilutions differed from the undiluted treated soil. Increasing dilutions would contain greater proportions of the untreated soil and, thereby, would exert a greater influence on the bioassay results. GROVER (1966) has shown that organic matter differences 2.89 to 7.68 percent, changed the ED_{50} of simazine for oats from 0.43 p.p.m.w. to 0.76 p.p.m.w. While organic matter is generally the most important soil factor in influencing herbicide phytotoxicity, pH, various cations, clay content, and cation exchange capacity may exert effects, also (SHEETS et al. 1962, UPCHURCH and MASON 1962, UPCHURCH et al. 1966, BOUCHET 1967, GROVER 1966).

The stage of growth of the bioassay plant at harvest can affect the results of the test. HOLLY and ROBERTS (1963) reported that ryegrass harvested in the two- to 2½-leaf stage, did not give valid estimates of simazine when these results were compared with those obtained with plants harvested in the three- to 3½- and four-leaf stages of growth. It seems probable that the early growth of the ryegrass seedlings, at the expense of food reserves in the seed, was much the same whether or not simazine was present in the soil. Differences arose as growth stages were reached in which photosynthesis became the source of food rather than the seed. Then, simazine inhibition of photosynthesis gave rise to growth inhibition of the ryegrass that was proportional to the simazine concentration.

IV. Biological assays for the triazine herbicides in soils

The oat (*Avena sativa* L.) has been the plant species most frequently used in biological assays of soil for the s-triazine herbicides. SHEETS and SHAW (1963) used the oat as test plant in an extensive series of bioassays which established the relative potency of 14 s-triazine herbicides in four soil types. The oat is relatively sensitive to the s-triazines and is often used to determine low concentrations of these herbicides. The quantitation by oat bioassay of soil concentrations of s-triazines from a low of 0.04 lb./acre of atrazine (DARWENT and BEHRENS 1968) to a high of 2.5 lb./acre of simazine (STRAUBE and BONDARENKO 1960) has been reported. MCCORMICK (1965) obtained the most precise estimate of atrazine in a Decatur clay loam soil within a concentration range of 0.12 to 0.18 p.p.m. BURNSIDE (1959) could detect significant differences at the five percent level of probability between simazine concentrations of 0.00, 0.25, 0.50 and 1.0 p.p.m.w., MCCORMICK (1965) and SIKKA and DAVIS (1966) diluted atrazine-treated soil with herbicide free soil to obtain herbicide concentrations in the range where the oat bioassay was most sensitive. In persistence studies, where the s-triazine concentration required to give a 50 percent growth reduction was determined, soil dilution also has been used (HOLLY and ROBERTS 1963).

In studies on the s-triazines in soil, the soybean (*Glycine max* L.) is the second most frequently used bioassay plant. Most of the bioassays using this test plant have been concerned with the persistence of simazine or atrazine in soil. BURNSIDE (1959) noted that soybeans were less sensitive to simazine than oats but were especially suitable for measuring simazine concentrations in the range used for selective weed control in normal field applications. Soybeans give the best degree of response to s-triazine concentrations in a range from 0.20 to 8.0 p.p.m.w. AGUNDIS (1966) could detect significant differences at the five percent level between atrazine concentrations of 0.00, 0.50, 1.0, 2.0, 4.0 and 8.0 p.p.m.w. in a soil containing 8.5 percent organic matter. In a soil similar in clay content but containing six percent organic matter, he found that there was a loss in bioassay sensitivity at the higher concentrations and a gain in sensitivity at the lower concentrations. That is, the dry weight of soybeans was not significantly different when grown in the 4.0 and 8.0 p.p.m.w. concentrations of atrazine but their weight differed significantly when grown in 0.25 and 0.50 p.p.m.w. TALBERT and FLETCHALL (1964) used soybeans to detect simazine and atrazine concentrations ranging from 0.25 to 1.50 p.p.m. BURNSIDE *et al.* (1963) reported a linear relationship between soybean response and the concentration of simazine in the soil from zero to one p.p.m.w. The fact that soybean growth responses to simazine and atrazine are optimum in the concentration ranges often encountered in the field eliminates the need for soil dilution as has been necessary when oats is used as the test plant.

A number of other plant species have been used occasionally as test plants in bioassays for the s-triazine herbicides. UPCHURCH *et al.* (1966) used cotton as a test plant to determine the amount of simazine and other herbicides required to reduce the dry shoot weight of cotton 50 percent (GR_{50}) at six field locations during two growing seasons. GR_{50} values ranged from 0.3 to 22.9 lb./acre of simazine in this study. These values did not differ greatly from those obtained for soybeans in the same study, which indicates that the growth response range of cotton and soybeans to simazine does not differ greatly. Cucumber (CORBIN and UPCHURCH 1967) and mustard (GYSIN and KNÜSLI 1960) are other broadleaved species that have been used as bioassay plants in studies of the s-triazines in soils. Data on their effective response ranges were not reported.

Among the grasses, rye (VAN DER ZWEEP (1958), Italian ryegrass (HOLLY and ROBERTS 1963), and *Alopecurus agrestis* (BOUCHET 1967) have been used occasionally in bioassays for the s-triazines. HOLLY and ROBERTS (1963) stated that 0.05 to 0.15 lb./acre of triazine herbicides reduced the growth of the Italian ryegrass 50 percent and indicate that this sensitivity compared favorably with that reported by VAN DER ZWEEP for the rye bioassay.

JANSEN *et al.* (1958) and ADDISON and BARDSLEY (1968) have

explored the possibilities of using algae in bioassays for evaluating the phytotoxicity of herbicides, including some s-triazines. JANSEN et al. (1958) observed considerable suppression of algal colonial growth on nutrient-moistened perlite by seven s-triazines at concentrations of 0.06 to 4.0 p.p.m. ADDISON and BARDSLEY (1968) evaluated a bioassay procedure using *Chlorella vulgaris* Berzerinck as an assay organism. In this procedure, concentrated nutrient media and an algal suspension were added to a water extract of the test soil. Standard concentrations of the herbicide in water were prepared in the same manner. After incubation in light under controlled conditions for 144 hours, the light transmittance of the algal suspension was determined. Comparison of transmittance of unknowns with that of knowns gave an estimate of the herbicide concentration in the soil. A prometryne residue of 0.044 ± 0.005 p.p.m. was detected by the *Chlorella* bioassay. The major difficulty with this bioassay is the necessity of extracting the herbicide from the soil. Quantitative extraction with water is not likely for most s-triazines. Furthermore, the proportion extracted is likely to vary with each soil type used.

PARKER (1965) described a bioassay method for photosynthetic inhibitors that shortens the time required to complete the bioassay from three to four weeks to three or four days. He made use of paraquat, a herbicide that is activated to a toxic form in the plant by photosynthesis. Paraquat is less toxic to plants treated with photosynthetic inhibitors such as the s-triazines. In this procedure, duckweed (*Lemna minor* L.) fronds were floated on a soil slurry containing atrazine. After 24 hours in the light, the culture was sprayed with one lb./acre of paraquat and then returned to the light for 16 to 24 hours. The *Lemna* fronds were then rated for injury. The *Lemna* fronds growing on the cultures containing the lowest concentrations of the triazines developed the greatest degree of injury. This bioassay will detect atrazine concentrations of one to eight oz./acre.

V. Aspects of the triazine herbicides studied by bioassay

Bioassays have been used to examine many aspects of the behavior of the triazine herbicides in the soils. There is no doubt that the biological assay has been the major means of conducting investigations in this research area. The significant findings of this research have been discussed in detail by other symposium participants and need not be repeated here. However, a brief summary of the areas covered by means of biological assay seems appropriate.

Perhaps the greatest effort with bioassays has been that of investigating the persistence of various triazine herbicides following single and repeated applications under field conditions which has been reported by BURNSIDE (1965), BURNSIDE et al. (1963), BRYANT and ANDREWS (1967), DARWENT and BEHRENS (1968), DAWSON et al.

(1968), SHEETS and SHAW (1963), and TALBERT and FLETCHALL (1964). Related studies examined, by means of biological assay, a number of factors influential in the dissipation or deactivation of triazines in the soil. The degree of leaching of several triazines in a number of soils has been determined by ASHTON (1961), BURNSIDE et al. (1963), and DARWENT and BEHRENS (1968). Bioassay techniques established the importance of temperature, moisture content and oxygen concentration of the soil in the deactivation of the triazines (AGUNDIS 1966, BURNSIDE 1965). Significant information has been obtained on the removal of triazines from the soil by several plant species (BURNSIDE 1959, SIKKA and DAVIS 1966) and insights into the importance of microbiological and chemical means of triazine deactivation has been gained (AGUNDIS 1966, McCORMICK 1965).

Another area of study where biological assay techniques have assumed a major role is in the determination of the importance of various soil factors on the phytotoxic activity of the triazines. The primary importance of organic matter and the soil water was established (GROVER 1966, HARRIS and SHEETS 1965, UPCHURCH and MASON 1962, UPCHURCH et al. 1966). Also, the influence of clay content, pH, cation-exchange capacity and various individual cations on triazine toxicity received attention through bioassay techniques (HARRIS and SHEETS 1965, SHEETS et al. 1962, UPCHURCH et al. (1966).

There can be no question of the great importance of the information obtained in these studies, and it seems certain that the biological assay will continue to be a valuable technique in further studies of this nature.

VI. Comparison of chemical and bioassay methods

Possible alternatives of biological assay methods for the quantitative determination of the triazine herbicides in soils are chemical methods of analysis. Yet, a survey of the literature discloses that chemical assay for triazines in soils has been rarely used. VAN DER ZWEEP (1958) compared chemical and bioassay methods for the determination of simazine in sand. Both methods were satisfactory in determining simazine concentrations as low as 0.2 p.p.m. and the estimates of simazine concentrations were very similar by both methods. Several years later ERCEGOVITCH (1962) stated that chemical methods by which triazine residues can be determined in soils were not available because in heavier soils the recovery of simazine and atrazine was very poor. He indicated little success in correlating chemical assays, using recovery by chloroform extraction with bioassay results. SIKKA and DAVIS (1966) used absolute methanol for extraction of atrazine from a Norfolk sandy loam soil and obtained recoveries of 90 percent of atrazine concentrations ranging from 0.5 to five p.p.m.w. They reported a close correlation of results obtained with chemical and biological assay methods. HARRIS *et al.* (1969) also used methanol extraction methods and chemical assay to determine atrazine persistence in 12 soils. They recovered approximately 60 percent of the atrazine from field treated soil samples with recovery percentages ranging from 26 to 92 percent. Recovery from the same soils treated under laboratory conditions was much better, generally ranging from 70 to 90 percent. The wide variations in the recovery of atrazine from soils by solvent extraction which arise from differences in the soils and in treatment techniques greatly restricts the value of chemical assay methods in triazine persistence studies.

Biological assay methods are of particular value in studies of the influence of edaphic and climatic factors on the phytotoxicity of the triazines. In experiments of this nature, plant responses can provide the only meaningful estimate of the relationships between triazine herbicide activity and edaphic or climatic factors. Elaborate, expensive equipment is not necessary in conducting bioassays; furthermore, they are relatively simple to conduct. These are additional significant advantages over present chemical assay techniques.

Summary

Biological assays make use of the responses of living organisms to evaluate the potency of a stimulus. Bioassays of a quantitative nature were first developed about 70 years ago. Plants were first used as the test organism in 1928 when auxin concentrations were determined by means of the *Avena* coleoptile curvature response. Bioassays for herbicides were reported eight years later in 1935.

A proportional relationship between dose and response and the reproducibility of this relationship are essential in assuring the validity of results obtained by bioassay methods. The conditions under which the bioassay is conducted must provide an unbiased comparison of the unknown and standard samples. Environmental uniformity and test plant homogeneity are required to attain reliable results.

Triazine herbicides in the soil are under intensive study by means of the biological assay. Oat and the soybean have served most frequently as the test plant. The persistence of the triazines in soils has been of greatest concern and has received the most attention in the bioassay studies. Important information has been obtained on the influence of climatic and soil factors on the potency and dissipation of many of the triazine herbicides by this method of study. The several biological assay methods that were used in this research are described in this paper. Some limitations of the biological assay and precautions regarding its use are considered. The relative usefulness of chemical and bioassay methods of triazine determination in soils is discussed.

Résumé*

Détermination quantitative des desherbants type triazine dans les sols par essai biologique

Les essais biologiques utilisent les réponses d'organismes vivants pour évaluer la force d'un stimulus. Des essais biologiques quantitatifs ont été développés voici environ soixante-dix ans. Les plantes furent utilisées les premières comme organismes d'essai en 1928 lorsque les concentrations en auxines ont été déterminées au moyen de la courbure d'*Avena* coleoptile. Des essais biologiques pour les desherbants ont été décrits huit ans plus tard, en 1935.

Une relation de proportionnalité entre la dose et la réponse et la reproductibilité de cette relation sont essentielles pour assurer la validité des résultats obtenus par les méthodes biologiques. Les conditions dans lesquelles l'essai biologique est conduit doivent fournir une comparaison non erronée de l'inconnu et des étalons. L'uniformité du milieu et l'homogénéité des plantes d'essai sont nécessaires pour obtenir des résultats valables.

Les desherbants du type triazine sont actuellement soumis à une étude approfondie par des essais biologiques. L'avoine et le soja sont utilisés le plus souvent comme plantes d'essai. La persistance des triazines dans les sols a été très étudiée et a fait l'objet des soins les plus attentifs dans les études biologiques. Cette méthode d'étude a permis d'obtenir des informations importantes relatives à l'influence des facteurs climatiques et du sol sur l'activité et la disparition de nombreux desherbants du type triazine.

Les nombreuses méthodes d'essais biologiques qui ont été utilisées dans cette recherche sont décrites dans ce texte où sont également considérées les limites des essais biologiques et les précautions à prendre pour leur mise en oeuvre. L'utilité relative des méthodes chimiques et biologiques de dosage de la triazine dans les sols est discutée.

Zusammenfassung**

Quantitative Bestimmung von Triazin-Unkrautvertilgungsmitteln in Böden durch Bio-Analyse

Bio-Analysen machen von den Reaktionen lebender Organismen Gebrauch, um die Durchschlagskraft eines Reizmittels auszuwerten. Bio-Analysen quantitativer Natur wurden zum ersten Mal vor ungefähr 70 Jahren entwickelt. 1928 wurden zum ersten Mal Pflanzen als

---

* Traduit par R. MESTRES.
** Übersetzt von M. DÜSCH.

Testorganismen verwendet, als man Auxin-Konzentrationen mit Hilfe der *Avena* coleoptile curvature-Reaktion bestimmte. Bio-Analysen für Unkrautvertilgungsmittel wurden acht Jahre später, 1935, erwähnt.

Ein proportionales Verhältnis zwischen Dosis und Reaktion und die Reproduzierbarkeit dieses Verhältnisses sind für das Sicherstellen der Gültigkeit der Resultate, die durch bio-analytische Methoden erhalten wurden, wesentlich. Die Bedingungen, unter denen die Bio-Analysen durchgeführt werden, müssen einen unparteiischen Vergleich der Standard- und der unbekannten Proben liefern. Umgebungsbedingte Einheitlichkeit und Testpflanzen-Gleichartigkeit sind notwendig, um zuverlässige Resultate zu erhalten.

Triazin-Unkrautvertilgungsmittel werden mit Hilfe biologischer Analysen eingehend studiert. Hafer and Sojabohnen haben am häufigsten als Testpflanzen gedient. Die Hartnäckigkeit der Triazine in Böden ist von grösster Wichtigkeit gewesen und hat die meiste Aufmerksamkeit in den bio-analytischen Studien erhalten. Es wurden wichtige Informationen über den Einfluss von klimatischen und Boden-Faktoren auf die Stärke und Verflüchtigung vieler Triazin-Unkrautvertilgungsmittel durch diese Untersuchungsart gewonnen. Die verschiedenen biologisch-analytischen Methoden, die in dieser Forschung angewandt wurden, werden in dieser Arbeit beschrieben. Einige Einschränkungen der biologischen Analysen und Vorsichtsmassnahmen, ihre Verwendung betreffend, sind in Betracht gezogen worden. Die relative Nützlichkeit der chemischen und bio-analytischen Methoden für die Triazin-Bestimmung in Böden wird diskutiert.

References

ADDISON, D. A., and C. E. BARDSLEY: *Chlorella vulgaris* assay of the activity of soil herbicides. Weeds **16**, 427 (1968).

AGUNDIS, O.: Some factors that influence the deactivation of atrazine in the soil. Ph.D. thesis, Univ. of Minn., St. Paul (1966).

ANDERSEN, R. N.: A search for atrazine resistance in soybeans. Weed Sci. Soc. Amer., Abstr. no. 157 (1969).

ASHTON, F. M.: Movement of herbicides in soil with simulated furrow irrigation. Weeds **9**, 612 (1961).

BOUCHET, F.: Étude de l'influence de la nature du sol sur l'action herbicide de la simazine. Weed Research **7**, 102 (1967).

BRYANT, T. A., and H. ANDREWS: Disappearance of diuron, norea, linuron, trifluralin, diphenamid, DCPA and prometryne from the soil. Proc. S. Weed Conf., p. 395 (1967).

BURNSIDE, O. C.: Influence of environmental factors on the phytotoxicity and dissipation of simazine. Ph.D. thesis, Univ. of Minn., St. Paul (1959).

—— Longevity of amiben, atrazine, and 2,3,6-TBA in incubated soils. Weeds **13**, 274 (1965).

——, C. R. FENSTER, and G. A. WICKS: Dissipation and leaching of monuron, simazine, and atrazine in Nebraska soils Weeds **11**, 209 (1963).

——, E. L. SCHMIDT, and R. BEHRENS: Dissipation of simazine from the soil. Weeds 9, 477 (1961).

CORBIN, F. T., and R. P. UPCHURCH: Influence of pH on detoxification of herbicides in soil. Weeds 15, 370 (1967).

CRAFTS, A. S.: Toxicity of sodium arsenite and sodium chlorate in four California soils. Hilgardia 9, 459 (1935).

DARWENT, A. L., and R. BEHRENS: Dissipation and leaching of atrazine in a Minnesota soil after repeated applications. Proc. N.C. Weed Control Conf. 20, 66 (1968).

DAWSON, J. H., V. F. BRUNS, and W. J. CLORE: Residual monuron, diuron and simazine in a vineyard soil. Weeds 16, 63 (1968).

EMMONS, C. W.: Hormone assay. New York: Academic Press (1950).

ERCEGOVICH, C. D.: Unpublished communication, Geigy Chemical Corp., Ardsley, N.Y. (Aug. 22, 1962).

ESPINOZA, W. G., R. S. ADAMS, JR., and R. BEHRENS: Interaction effects of atrazine and CDAA, linuron, amiben or trifluralin on soybean growth. Agronomy J. 60, 183 (1968).

FINK, R. J., and O. H. FLETCHALL: Soybean injury from triazine residues in soil. Weed Sci. 17, 35 (1969).

FINNEY, J. J.: Probit analysis. London: Cambridge Univ. Press (1947).

—— Statistical method in biological assay. New York: Hafner (1952).

FREED, V. H.: Determination of herbicides and plant growth regulators. In L. J. Audus (ed.): Physiology and biochemistry of herbicides, p. 39. New York; Academic Press (1964).

GROVER, R.: Influence of organic matter, texture and available water on the toxicity of simazine in soil. Weeds 14, 148 (1966).

GYSIN, H., and E. KNÜSLI: Chemistry and herbicidal properties of triazine derivatives. Adv. Pest Control Research 3, 289 (1960).

HARRIS, C. I., and T. J. SHEETS: The influence of soil properties on adsorption and phytotoxicity of CIPC, diuron and simazine. Weeds 13, 215 (1965).

——, E. A. WOOLSON, and B. E. HUMMER: Dissipation of herbicides at three soil depths. Weed Sci. 17, 27 (1969).

HOLLY, K., and H. A. ROBERTS: Persistence of phytotoxic residues of triazine herbicides in soil. Weed Research 3, 1 (1963).

JACOBSOHN, R., and R. N. ANDERSEN: Differential response of wild oat lines to diallate, triallate and barban. Weeds 16, 491 (1968).

JANSEN, L. L., W. A. GENTNER, and J. L. HOLTON: A new method for evaluating potential algicides and determination of the algicidal properties of several substituted urea and s-triazine compounds. Weeds 6, 390 (1958).

McCORMICK, L. L.: Microbiological decomposition of atrazine and diuron in soil. Ph.D. thesis, Auburn Univ. (1965).

MITCHELL, J. W., and G. A. LIVINGSTON: Methods of studying plant hormones and growth regulator substances. Agricultural Handbook No. 336. U.S. Government Printing Office, Washington, D.C. (1968).

PARKER, C.: A rapid bioassay method for the detection of herbicides which inhibit photosynthesis. Weed Research 5, 181 (1965).

SHEETS, T. J., A. S. CRAFTS, and H. R. DREVER: Influence of soil properties on the phytotoxicity of the s-triazine herbicides. J. Agr. Food Chem. 10, 458 (1962).

——, and W. C. SHAW: Herbicidal properties and persistence in soils of s-triazines. Weeds 11, 15 (1963).

SHULDT, P. H., and C. N. WOLF: Fungitoxicity of substituted s-triazines. Contrib. Boyce Thompson Inst. 18, 377 (1956).

SIKKA, H. C., and D. E. DAVIS: Dissipation of atrazine from soil by corn, sorghum and Johnsongrass. Weeds 14, 289 (1966).

STRAUBE, E. W., and D. D. BONDARENKO: Persistence and distribution of simazine applied in the field. Proc. N.C. Weed Control Conf. 17, 40 (1960).

TALBERT, R. E., and O. H. FLETCHALL: Inactivation of simazine and atrazine in the field. .Weeds **12**, 33 (1964).

UPCHURCH, R. P., and D. D. MASON: The influence of soil organic matter on the phytotoxicity of herbicides. Weeds **10**, 9 (1962).

——, F. L. SELMAN, D. D. MASON, and E. J. KAMPRATH: The correlation of herbicidal activity with soil and climatic factors. Weeds **14**, 42 (1966).

VAN DER ZWEEP, W.: De bepaling van simazin in grondmonsters. Mededelingen van de Landbouwhogeschool en de Opzoekingsstations van de Staat te Gent **23**, 1000 (1958).

WENT, F.: Wuchsstoff und Wachstum. Rec. Tran. Bot. Ne'erl. **25**, 1 (1928).

Quantitative determination of triazine herbicides in soils by chemical analysis

By

A. M. Mattson*, R. A. Kahrs*, and R. T. Murphy*

Contents

I. Introduction

Reliable methods for the chemical analysis of soils for the presence of triazine herbicide are of great importance in determining their persistence in the soil. A large amount of material has been published about the determination of residues in a variety of kinds of samples. Although all of the techniques reported have not been applied to the analysis of soils, they could be. Therefore, this discussion will attempt to review the literature in regard to the broader aspects of the determination of triazine residues but the emphasis will be on methods which are applicable to the analysis of weathered soil samples having residues in amounts that would arise from use of triazine under normal agricultural practice. There are no practical analytical methods avail-

---

* Analytical Chemistry Department, Geigy Agricultural Chemicals, Division of *Geigy Chemical Corporation*, Ardsley, N.Y.

able for the analyses of soil for hydroxy-triazines or other degradation products which have sufficient sensitivity or which can be applied to many types of soils. Therefore this report will be limited to the determination of unchanged triazines. The *s*-triazines generally referred to are those of present commercial importance: atrazine, simazine, propazine, ametryne, prometryne, and prometone.

Following this review, details of the analytical method used in our laboratory for determining atrazine residues in soil will be presented. Data comparing efficiency of the various extraction procedures will also be given.

II. Literature review

The three important steps in the chemical analysis of soils are extraction, cleanup of the extract, and determination of the amount of triazine present. A discussion of information available under each of these topics will be given with an attempt to evaluate this information.

a) Extraction procedures

A variety of solvents has been used for extraction of triazine herbicides from soil. Chloroform (CHILWELL and HUGHES 1962, GYSIN and KNÜSLI 1960, MAJOR 1962, MONTGOMERY and FREED 1961, SHEETS and KEARNEY 1964), carbon tetrachloride (KNÜSLI *et al.* 1964), dichloromethane (BENFIELD and CHILWELL 1964, CHILWELL and HUGHES 1962, KNÜSLI *et al.* 1964), dioxane (KNÜSLI *et al.* 1964), methanol (BENFIELD and CHILWELL 1964, NEGI *et al.* 1964, SHEETS and KEARNEY 1964, DELLEY *et al.* 1967), and 8*M* urea solution (SHEETS and KEARNEY 1964) have been used. Procedures for carrying out the extractions have used agitation of the soil and solvent for various periods of time at various temperatures and Soxhlet extractions. McGLAMMERY *et al.* (1967) compared the effectiveness of 12 solvent systems for the extraction of atrazine from soil. They found that Soxhlet extraction gave better recoveries than shaking at room temperature. Chloroform, methanol, acetone, and methylene dichloride gave comparable results.

No one "best" method for extracting soils can be selected from the literature. Early workers often used the less polar solvents. It now is recognized generally that water-miscible solvents give better extractions and that moisture content of the soil is an important factor. Therefore, water should be used as one component of the extracting solvent or the moisture content of the soil should be adjusted. Such extractions with solvent-water mixtures make for a more difficult analytical procedure because manipulatory difficulties are increased by the more polar solvents and usually more interfering materials are extracted which increases the need for better cleanup. A study of the comparative efficiencies of various extraction procedures applied to weathered atrazine residues in soil which was done in our laboratory is presented in section III.

b) Cleanup procedures

Cleanup procedures are generally needed for most soil types in order to attain a satisfactory limit of detection such as 0.05 p.p.m. Interferences can arise not only from natural soil constituents but also from other chemicals added to the soil. GYSIN and KNÜSLI (1960) used acid and basic aqueous solvent washing and partitioning for cleanup of extracts of various triazine herbicides. These methods can be used if residues are high and interferences are minimal. Column chromatography using alumina has been described by KNÜSLI et al. (1964). Alumina has been used successfully in our laboratory for a number of years. By regulating the activity of the alumina this cleanup has been adapted for use with a number of triazines [Geigy Analytical Bulletins No. 7 (1965) and No. 10 (1965)]. DELLEY et al. (1967) have described cleanup procedures involving alumina columns followed by a sodium bisulfate cleanup column. These procedures were designed to give general methods for the cleanup of various kinds of samples and triazines.

c) Determination of amount of residue

There are three major methods of determination available for the quantitative determination of triazine residues. These are (a) ultraviolet spectrophotometry, (b) colorimetric spectrophotometry, and (c) gas chromatography.

1. Ultraviolet spectrophotometry.—The determination of the triazine depends on the hydrolysis of the compound to the hydroxyderivative, which is then measured spectrophotometrically. It is applicable to chloro-, thiomethyl-, and methoxy-substituted triazines. The hydroxy-derivatives are obtained by acid hydrolysis. They all have an absorption maximum near 240 mμ and minima near 225 and 255 mμ. Quantitative measurements for low amounts of triazines are made by determining the absorption at the maximum and minima wavelengths and calculating the net absorbance at 240 mμ by a baseline technique.

This method was one of the earliest methods developed. GYSIN and KNÜSLI (1960 and 1964) described this method and DELLEY et al. (1967) reported on its application to a number of chloro-, thiomethyl-, and methoxy-substituted triazines. In our laboratory it has been used successfully for large numbers of various kinds of samples including soils with a limit of detectability of 0.04 p.p.m. Normally its use was preceded by cleanup using an alumina column [Geigy Analytical Bulletins No. 7 (1965) and No. 10 (1965)].

2. Colorimetric spectrophotometry.—The Zincke reaction has been applied to the determination of chlorotriazines. BURCHFIELD and SCHULDT (1958) first showed that this reaction was applicable to pesticides containing active halogen and applied it to the determination of chloro-triazine. This method depends on the reaction of pyridine in alkaline solution with chloro-triazines to produce a somewhat unstable yellow color. RAGAB (1959) and RAGAB and McCOLLUM (1968) investigated various parameters of the reaction as applied to chloro-triazines and gave several methods for producing more stable chromophores. KNÜSLI *et al.* (1964) reviewed the method and gave procedures for its application to the analysis of soil samples. RADKE *et al.* (1966) evaluated the pyridine-alkali method and reinvestigated conditions relating to color stability and reproducibility and applied it to the analysis of soil perfusates for atrazine. This method has not found wide applicability because of the difficulty in controlling the color formation and because it can be used only for chloro-triazines.

3. Gas chromatography.—Gas chromatography is being used increasingly for the determination of triazine residues. Early work was done using the flame ionization detector for residue analysis (CHILWELL and HUGHES 1962, BENFIELD and CHILWELL 1964, HENKEL and EBBING 1964) and for analysis of technical materials (STAMMBACH *et al.* 1961). This detector is in general not used currently for residue determinations because of its lack of specificity. It can be used in special cases where residues are large in relation to the interferences encountered. In general, as for all residue determinations, detectors of enhanced selectivity are preferable for use in analyzing for triazines.

The microcoulometric titration detectors have enhanced specificity for chlorine or sulfur (COULSON *et al.* 1960) or nitrogen (MARTIN 1966). These detectors are commercially available (*Dohrmann Instruments Co.*) and are widely used for pesticide residue analysis. The use of this detector for determining residues of chloro- and thiomethyl-triazines in plant materials was described by MATTSON *et al.* (1965). Since that time large numbers of soil samples have been analyzed for chloro- and thiomethyl-triazines by this detection system in our laboratory. Generally an alumina column was used for cleanup and a limit of detectability of 0.05 p.p.m. was achieved. As little as 20 ng. of triazines give peaks suitable for quantitative determinations. The nitrogen-specific microcoulometric titration detector is a more recent development. It is applicable to all triazines including the methoxy-substituted compounds. It appears to have sensitivity similar to that found for the chloro- and sulfur-sensitive detection systems. In certain instances nitrogen selectivity appears to avoid more interferences than encountered with the other two detectors.

The flame photometric detector was first developed by BRODY and CHANEY (1966) and is available in a commercial unit (*Tracor Instruments, Inc.*). It has enhanced specificity for sulfur or phosphorus. The

principle on which the detector is based is the photometric detection
of flame emission of phosphorous and sulfur compounds in a hydrogen-
rich air flame. It is more selective for phosphorous- than for sulfur-
containing compounds in the ratio of approximately five to one.
However it has been used in our laboratory successfully for the
determination of thiomethyl-triazines. Peaks suitable for quantitative
determinations can be obtained with five ng. of prometryne, for
example. Reagents must be carefully selected or purified to avoid
interference because of the presence of sulfur.

The electrolytic conductivity detector was described by COULSON
(1965 and 1966) and is now commercially available (*Tracor Instruments, Inc.*). It has selectivity toward halogen-, sulfur-, or nitrogen-
containing compounds. The effluent from the gas chromatography
column is combusted and the elements are quantitatively determined
by electrolytic conductance. A comparison of this detector and the
microcoulometric detector in determining chlorinated pesticides was
reported by CRAMMER and CARROLL (1967). The conductivity detector
appears to have some advantages in simplicity, economy, sensitivity,
and peak shape. It has been used successfully for the determination of
triazine herbicide residues using the nitrogen sensitive mode in our
laboratories in Basle, Switzerland (EBERLE 1969).

Electron capture detectors have not been used to any extent because of lack of selectivity toward triazine compounds. The chloro-
triazines have been reported to require from 100 to 300 ng. to produce
½ full-scale deflection asc ompared to five ng. of p,p'-DDT (FDA
Manual 1968) when using electron capture detectors with a tritium
radioactive source and a direct current mode. It is possible that newer
electron capture detectors could enhance the selectivity of some of the
triazine herbicides.

The alkali-flame thermionic detector has been reported to have
selectivity toward nitrogen compounds 1,000 times that evinced for
carbon compounds (TINDLE et al. 1968). These authors reported excellent recovery data at levels of one, ten, and 100 p.p.m. for atrazine
and simazine added to corn and soil. They also reported a minimum
detectable limit of 0.2 to 0.5 ng. for the standard triazines and minimum detectable residue concentrations of 0.05, 0.02, and 0.001 p.p.m.
for residues in corn, soil, and water without any cleanup. The detectable limits apparently were extrapolated from data obtained at higher
levels of triazine concentration because no data was presented at these
lower levels.

Microwave emission detection gas chromatography for the determination of atrazine in soil has been reported by BACHE and LISKE
(1968) following chloroform extraction and alumina column cleanup
at residue levels near 0.05 p.p.m. This detection system is not available
commercially.

4. Thin-layer chromatography.—DELLY et al. 1967 have presented an excellent review of thin-layer chromatography methods for the detection of triazine herbicides. These authors also present their method for the identification of eight triazine compounds by thin-layer chromatography. They state that by their technique 0.05 μg. or less of triazines can be determined and that a sensitivity of at least 0.1 p.p.m. can be obtained if a one-g. sample aliquot is spotted on the plate. Semi-quantitative values (±20 percent) can be obtained by comparing spots with known amounts of standards.

    5. Polarography.—The triazine herbicides with chloro- or methyl-thio-groups at the 2-position have been shown (THOMPSON 1965) to be polarographically active. HAYES et al. (1967) reported work done using conventional d.c. polarography and single-sweep polarography with a number of triazine herbicides. They found the lower limit of detection to be about 1.5 to 2.0 μg. using a conventional d.c. polarograph and about 0.15 μg. using single-sweep polarography. In general, it was found that it was not possible to distinguish between the various triazines when more than one was present. The methoxy-triazines were not polarographically reducible in aqueous solutions at either acid or alkaline pH values. The authors suggested that polarography could provide a rapid and accurate method for determination of triazine herbicide providing suitable extraction and cleanup methods could be developed.

III. Detailed analytical method

    This analytical method has been developed over a period of time in our laboratory and has been used to analyze a large number of soil samples on a routine basis. Extraction is done using an acetonitrile-water mixed solvent, cleanup is by use of an alumina column, and final determination is by microcoulometric titration detector gas chromatography. It is written specifically for atrazine and micro-coulometric gas chromatography detection but its use for other triazine herbicides and with ultraviolet spectrophotometric deter-mination of the triazines will be described. At the present time gas chromatographic determination is recommended but where this is not available the ultraviolet spectrophotometric method can be used successfully.

a) Sampling of the soil

    Soil samples are received in a frozen state and held frozen until analysis. The soil is thawed and large stones are removed. It is then mixed thoroughly and sub-sampled for chemical analysis and for moisture determinations. Moisture is determined by air drying at ambient temperature to a constant weight.

b) Extraction procedure

The extraction is done at reflux temperature using ten percent water-acetonitrile as the solvent. Place a sample of 100 g. of soil in a 500-ml. round-bottom boiling flask fitted with a 24/40 joint. Add 300 ml. of ten percent water-acetonitrile (v/v). Fit the flask with a Liebig condenser with 300-mm. jacket and heat to reflux using a heating mantle and continue the refluxing for one hour. Cool the mixture to room temperature and allow the soil to settle.

The supernatant liquid is filtered through two kinds of filter paper. Place one sheet of Reeves-Angel Grade 802 (32 cm.) paper in a long-stem glass funnel (148 mm. diam.) and follow this with one sheet of Whatman 2V paper (32 cm.). Pour the supernatant solution into the funnel, collect the filtrate, and mix thoroughly. Transfer a 60-ml. aliquot (equivalent to 20 g. of soil) to a 500-ml. separatory funnel. Dilute this aliquot with 300 ml. of water and add about 20 ml. of a saturated solution of sodium sulfate. Extract the aqueous solution with 25 ml. of methylene chloride. Allow the phases to separate and filter the methylene chloride through a one-inch pad of anhydrous sodium sulfate (about 70 g.) into a 250-ml. Erlenmeyer flask (equipped with a 24/40 joint). Repeat the extraction with 25 ml. of methylene chloride and filter the methylene chloride through the same sodium sulfate tube. Wash the sodium sulfate with 25 ml. of methylene chloride into the Erlenmayer flask. Evaporate the solvent to dryness using a flash evaporator (bath at 40°C.).

c) Column cleanup

The cleanup is done using alumina. The alumina used is Woelm Aluminum Oxide, Basic (*Alupharm Chemical Co.*, New Orleans, La.) which must be adjusted to Activity V as described on the package. The column used is 22 mm. OD and is equipped with a perforated support (similar to *Scientific Glass Apparatus Co.*, Catalog No. JC-2100, Size No. II, 22 mm. O.D.); to this column a 100-ml solvent reservoir has been joined. Place a glass wool plug at the bottom and add 12.5 g. of the alumina and tap the column gently to settle the packing. Place a glass wool plug on top of the alumina. Transfer the sample to the column with two five-ml portions of carbon tetrachloride. When the last five-ml. portion has just penetrated into the alumina, rinse the Erlenmeyer flask with 65 ml. of carbon tetrachloride and add this to to the cleanup column. Discard the eluate.

When the last of the carbon tetrachloride has reached the top of the column add 50 ml. of five percent ethyl ether in methylene chloride. Collect this eluate which contains the atrazine in a 250-ml. Erlenmeyer flask (24/40 joint). Evaporate the solvent just to dryness using a flash evaporator. Quantitatively transfer the residue into vials (two dram) with successive one- to two-ml. portions of methylene chloride. Evaporate the solvent in the vials with a gentle stream of air while heating in a water bath at 40°C.

d) Determination of atrazine by gas chromatography

The gas chromatograph used is a Micro Tek Model MT 220 equipped with a Dohrmann Microcoulometric Detector using a chloride sensitive titration cell (T-300). The column is three-percent Reoplex on Chromosorb W, HP, or Gas Chrom "Q" 60/80 mesh packed in borosilicate glass tubing (1/4 in. × two ft.) Nitrogen is used as the carrier and purge gas with flows of 130 ml./minute and 12 ml./minute, respectively. The oxygen flow is 68 ml./minute.

The temperatures used are injection port, 235°C.; column, 190°C.; transfer line, 255°C.; furnace, 870°C. The bias voltage is 250 mv, the attenuation is 200 ohms, and the chart speed is 1/2 inch/minute.

1. Standardization of gas chromatography response.—Dissolve 0.100 g. of standard atrazine in 100 ml. of chloroform in a volumetric flask. This stock solution contains one μg./μl. Dilute a portion of the

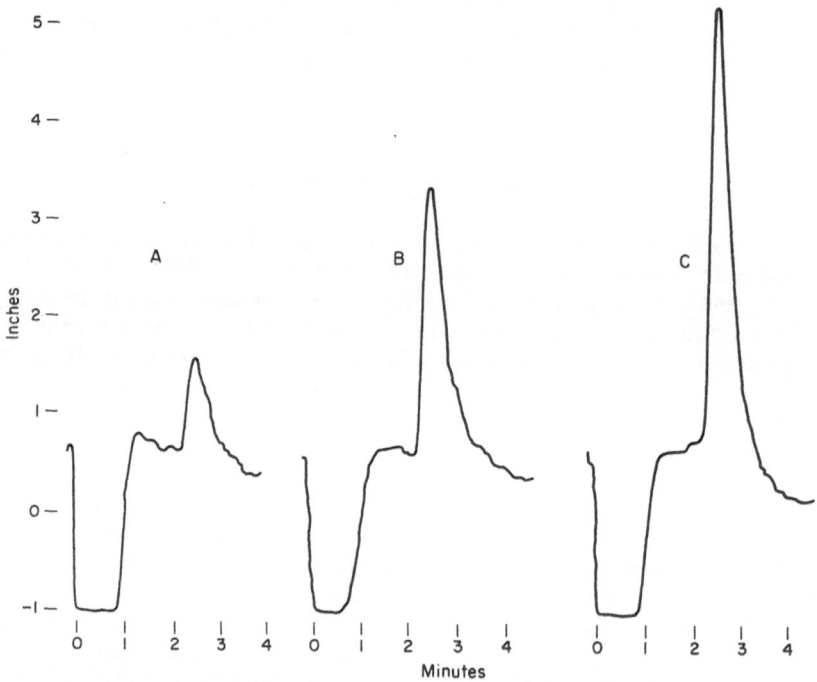

Fig. 1. Gas chromatograms of standard atrazine; A = 20 ng., B = 40 ng., and C = 60 ng. (see Table I for data)

stock solution 100x with benzene to obtain a standard solution containing ten ng./μl. Inject volumes ranging from two to six μl. (20 to 60 ng. of atrazine) into the gas chromatograph. The amounts of standards injected may have to be adjusted to suit the sensitivity of the particular gas chromatograph used. Peak areas are calculated by triangulation by multiplying the width at ½ the height (inches) times the height of the peak. Typical gas chromatograms are shown in Figure 1 for 20, 40, and 60 ng. and numerical values obtained are shown in Table I. The solvents (benzene and chloroform) did not

Table I. *Standardization of GLC response with standard atrazine showing daily standardizations*

Set no.[a]	Atrazine injected (ng.)	Peak response			Slope	Intercept
		½Base (in.)	Height (in.)	Area (sq. in.)		
1	20[b]	0.21	1.10	0.23	49.9	9.3
	40[b]	0.20	3.20	0.64		
	50	0.20	3.72	0.74		
	60[b]	0.21	5.00	1.05		
2	20	0.18	1.23	0.22	42.9	10.7
	30	0.19	2.10	0.40		
	40	0.20	3.54	0.71		
	50	0.21	4.00	0.84		
	60	0.22	5.32	1.17		
3	20	0.18	1.36	0.25	48.0	8.9
	40	0.19	3.40	0.65		
	50	0.20	4.11	0.82		
	60	0.21	5.20	1.09		
4	20	0.20	1.70	0.34	45.4	5.0
	30	0.18	3.00	0.54		
	40	0.22	3.40	0.75		
	50	0.21	4.80	1.01		

[a] Set no. refers to daily standardizations.
[b] Chromatograms shown in Figure 1.

produce any blank values. The areas so obtained may be plotted against the nanograms of atrazine to obtain a standard graph (Figure 2). A preferable procedure is to calculate the slope and intercept without plotting the points by means of a desk top computer (Olivetti-Underwood Programma 101). Computation of the intercept and slope is faster, more objective, and is more suitable for record keeping. The line shown (Fig. 2) was drawn using the value calculated for the slope and intercept.

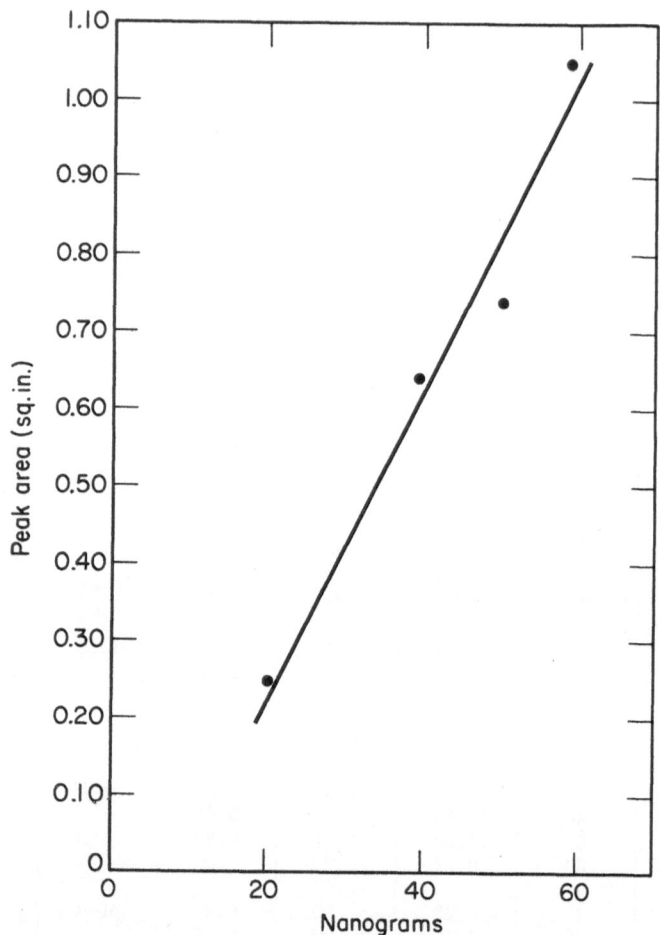

Fig. 2. Standard graph for atrazine determination (values from Table I, Set 1)

2. Determination of atrazine in the sample.—Add various amounts of benzene to the sample residues in the vials depending on the amount of atrazine expected. The following volumes of benzene would be suitable for the conditions given when injecting five μl:

             0.04 p.p.m. atrazine, 0.200 ml. benzene
             0.25 p.p.m. atrazine, 0.500 ml. benzene
             0.50 p.p.m. atrazine, 1.00 ml. benzene
             1.00 p.p.m. atrazine, 2.00 ml. benzene

For residue values lying between those given above, adjustments can be made in the injection volumes to obtain peak areas falling within the standardization values.

The peak areas are calculated and the nanograms may be determined from the applicable standard graph or from the computed slope and intercept. The residue concentrations are calculated in terms of p.p.m. A correction is made for the volume increase in the extracting solvent due to water in the soil. Another correction is made to obtain the p.p.m. on the basis of air dried soil:

$$\text{p.p.m.} = \frac{\text{atrazine found (ng.)}}{\text{mg. soil injected}} \times \frac{V + (W \times M/100)}{V} \times \frac{100}{(100 - M)}$$

where V = volume of extraction solvent (300 ml.), W = weight of soil sample (100 g.), M = percent moisture in soil, and 100 = conversion factor.

A computer program will calculate the ng. of atrazine using the slope and intercept derived from the standards, make the necessary corrections, and print out the p.p.m. of atrazine.

Known amounts of atrazine are added to control samples of the soil to establish the validity of the analytical procedure. Atrazine is added to approximate the range of residues expected in the soil samples to be analyzed. These additions are made to the soil in the round-bottom boiling flask before addition of the extraction solvent. Recovery of this atrazine is not a measure of the efficiency of extracting atrazine from weathered soils but does validate the general analytical procedure. Typical data are shown in Table II for 0.1, 1.0, and 2.0 p.p.m. of added atrazine. The chromatograms obtained from unfortified soil and 0.1 p.p.m. of added atrazine are shown in Figure 3 A and B. The data given in Table II are correlated with the standardizations given in Table I.

Typical data found for treated samples are shown in Table III and typical chromatograms are shown in Figure 3 B and C. Again the data are correlated with the standardizations given in Table I.

3. Application to other s-triazine herbicides.—This method has also been applied extensively to simazine, propazine, ametryne and prometryne. Prometone has been determined to a limited extent. The same extraction procedure is used for all of these compounds. Some changes in the activity of the alumina are needed for the best cleanup and recovery. Activity IV alumina is used for propazine, ametryne, and prometryne and Activity V alumina is used for atrazine, simazine, and prometone. The eluants remain the same as described above for atrazine.

Gas chromatographic determination of thiomethyl-triazines has been done using a sulfur-sensitive microcoulometric titration cell or by flame photometric detection. Prometone would require a nitrogen sensitive detector such as the microcoulometric or electrolytic conductivity cell for gas chromatographic determination.

A. M. MATTSON *et al.*

Table II. *Recovery of atrazine added to soil before extraction*

Set[a] no.	Atrazine added (p.p.m.)	Final volume (μl.)	Injected		Peak			Atrazine found		
			Volume (μl.)	Crop (mg.)	½ Base (in.)	Height (in.)	Area (sq. in.)	ng.	p.p.m.	Recovery (%)
1	0.10[b]	200	5	500	0.19	4.10	0.779	49	0.10	98
2	0.05	200	6	600	0.21	2.35	0.494	34	0.057	114
3	1.0	2000	3	30	0.22	2.18	0.480	32	1.1	110
4	2.0	2000	2	20	0.22	2.70	0.594	32	1.6	80

[a] Corresponds to Set no. in Table I for standardizations used.
[b] Chromatogram shown in Figure 3.

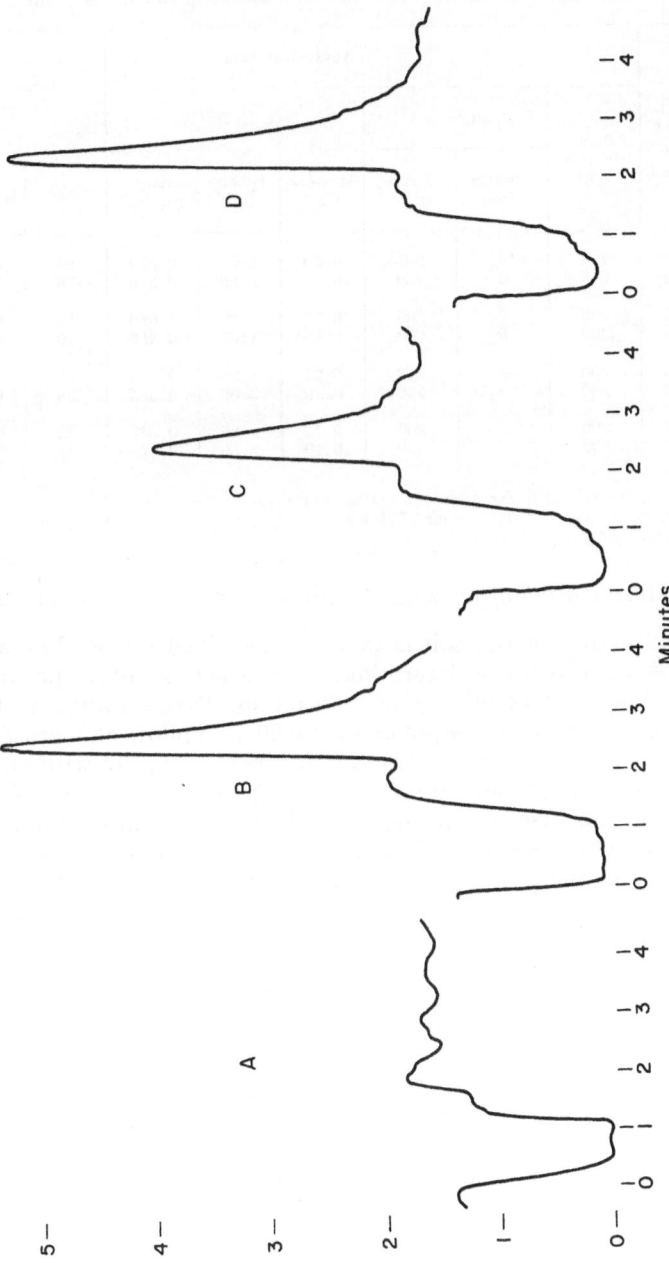

Fig. 3. Gas chromatograms for atrazine determination in soil; A = control, 500 mg. of soil injected, B = control plus 0.10 p.p.m. of atrazine added, 500 mg. of soil injected (Table II), C = treated soil, 0.10 p.p.m. of atrazine found, 400 mg. of soil injected (Table III), and D = treated soil, 0.25 p.p.m. of atrazine found, 200 mg. of soil injected (Table III)

Table III. *Typical raw data obtained in analysis of soil for atrazine*[a]

Set no.[b]	Soil depth (in.)	Final volume (μl.)	Injected		G. C. Peak			Atrazine found	
			Volume (μl.)	Crop (mg.)	½ Base (in.)	Height (in.)	Area (sq. in.)	(ng.)	(p.p.m.)
1	0–3[c]	200	2	200	0.20	4.05	0.810	50	0.25
	3–6[c]	200	4	400	0.22	2.58	0.568	38	0.10
2	0–3	200	2	200	0.24	4.50	1.080	57	0.29
	3–6	200	5	500	0.18	1.98	0.356	26	0.05
3	0–3	200	2	200	0.21	3.88	0.815	48	0.24
	3–6	200	2	200	0.19	2.08	0.395	28	0.14
4	0–3	200	2	200	0.22	3.85	0.847	43	0.22
	3–6	200	3	300	0.19	2.50	0.475	27	0.09

[a] Atrazine applied at four lb./acre; analyzed after 123 days.
[b] Set No. refers to daily standardization (Table I).
[c] Chromatograms shown in Figure 3.

e) Determination of triazines by ultraviolet spectrophotometry

The extraction of the soil is done as described above. For ultraviolet spectrophotometric determination a larger sample is needed to achieve a limit of detectability of 0.05 p.p.m. Take an aliquot of the acetonitrile-water extract equivalent to 50 g. Follow the procedure described previously for the extraction of the soil: extract with methylene chloride making the proportionate adjustments in the volumes of water and methylene chloride used. For the cleanup column, use 25 g. of alumina instead of 12.5 g. and prepare the column as described above. Transfer the residue remaining after evaporation of the methylene chloride to the cleanup column with ten ml. of carbon tetrachloride and rinse the flask with two additional five-ml. portions. Add 80 ml. of carbon tetrachloride to the column as the first eluant. Follow this with five percent ethyl ether in carbon tetrachloride but use 100 ml. in this instance and collect the second eluate.

The column just described is adequate for use with most crops. However, for alfalfa and similar crops when using a 50-g. sample, a larger column is required (25 × 400 mm.) and 60 g. of the alumina is used. The amounts of eluates are increased to 200 ml. for the carbon tetrachloride and 150 ml. for the five percent ethyl ether in carbon tetrachloride.

Hydrolysis of the chloro-triazine to hydroxy-triazine is done at room temperature with 50 percent sulfuric acid. Heat the ethyl ether-carbon tetrachloride eluate on a steam bath to boil off the ethyl ether and if necessary reduce the volume to about 100 ml. Transfer the

solution quantitatively to a 125-ml. separatory funnel. Add exactly
one ml. of 50 percent sulfuric acid (V/V) to the funnel and shake the
mixture vigorously. Repeat the shaking every 15 minutes for two
hours. Then add exactly nine ml. of water and shake the funnel to mix
the contents. Allow the phases to separate and discard the carbon
tetrachloride. Wash the acid solution with 25 ml. of 20 percent ethyl
ether in chloroform. Transfer the acid solution to a second clean
separatory funnel and wash it again with 25 ml. of ethyl ether.

The thiomethyl-triazines (ametryne and prometryne) and the
methoxy-triazine (prometone) require an elevated temperature for
hydrolysis. Evaporate the ethyl ether-carbon tetrachloride eluate to
approximately 15 ml. on the steambath. Transfer this to a test tube
(30 × 140 mm.) with a small amount of carbon tetrachloride and
evaporate the solvent to dryness using a gentle stream of air and a
water bath at 40°C. Add ten ml. of $1N$ H_2SO_4 and heat the test tube in
a 100°C. bath for three hours. Cool the test tube and transfer the acid
solution to a 125-ml. separatory funnel and wash as described above.

Transfer the acid solution to a one-cm. silica absorption cell for
ultraviolet spectrophotometry. Measure the absorbancy at 225, 240,
and 255 mμ against a reagent blank. Determine the net absorbance (E)
at 240 mμ using a baseline technique according to the following
equation:

$$E = A^{240} - \left[\frac{A^{225} + A^{255}}{2} \right]$$

One μg. of hydroxy-triazine per ten ml. of final acid solution will
give an absorbance of about 0.009. The exact value must be determined
in each laboratory. Standardize the method with known amounts of
standard atrazine carried through the hydrolysis procedure. Two μg.
of atrazine can be detected. With a 50-g. soil sample this is equivalent
to 0.04 p.p.m. Some kinds of ethyl ether have been found to cause
high reagent blanks. It is best to wash the ethyl ether with $1N$ sulfuric
acid and water before use. The excess water should be removed with
anhydrous sodium sulfate. The ethyl ether can be checked by measuring the absorbance produced by the ethyl ether wash as described
above.

IV. Comparison of extraction methods

A study in our laboratory compared various extraction solvents and
conditions for recovering weathered atrazine residues from soil. The
soil used was a silty clay loam (Webster) from Iowa which had been
treated with atrazine 284 days before sampling.

A large sample was air dried for two weeks at room temperature to
constant weight. It was then sieved through three screens, No. 8 mesh

(2.38 mm. opening), No. 10 mesh (1.68 mm. opening), and No. 16 mesh (1.0 mm. opening). The soil was mixed thoroughly and samples were taken for the different extractions. The analytical method described above was used with gas chromatagraphic determination of the atrazine.

Table IV. *Comparison of extraction procedures for weathered atrazine residues in soil*

| Solvent | Extraction | | Sample no.[a] | Atrazine found (p.p.m.) |
	Conditions	Time[b] (hours)		
Chloroform	Soxhlet	I 24	1	0.05
		II 24		0.00
		T 48		0.05
		I 24	2	0.70
		II 24		0.37
		T 48		1.07
10% Water-methanol	Soxhlet	I 24	1	0.05
		II 24		0.00
		T 48		0.05
		I 24	2	1.95
		II 24		0.05
		T 48		2.00
Methanol	Reflux	1	1	0.03
		1	2	1.09
10% Water-methanol	Reflux	1	1	0.05
		1	2	0.80
10% Water-chloroform	Reflux	1	1	0.06
		1	2	1.55
10% Water-acetonitrile	Reflux	1	1	0.07
		1	2	1.95
Methanol	Room temp.	18[c]	1	0.06
		18[c]	2	0.60

[a] Sample no. 1 = low residue sample and Sample no. 2 = high residue sample.
[b] I = first 24-hour extraction, II = second 24-hour extraction of the same sample, and T = total for both I and II.
[c] 18 hours standing at room temperature followed by ½-hour mechanical shaking.

The various extraction procedures used are described below:

A. *Soxhlet extraction*—two consecutive 24-hour periods
   (a) Chloroform
   (b) Methanol containing ten percent water
B. *Refluxing with the soil*—one-hour period
   (c) Methanol
   (d) Methanol containing ten percent water
   (e) Chloroform plus ten percent water
   (f) Acetonitrile containing ten percent water
C. *Room temperature*—Standing overnite (18 hours) at room-temperature followed by ½-hour shaking.
   (g) Methanol

Two weathered soils were chosen for analysis. By preliminary analysis one had been found to contain about 0.08 p.p.m. of atrazine (Sample No. 1) and the other had approximately 24 times this residue or 1.9 p.p.m. (Sample No. 2). The results obtained are shown in Table IV. The highest recovery for both the high- and the low-level residue samples was found by the 24-hour water-methanol-Soxhlet extraction and the water-acetonitrile one-hour reflux extraction. The water-chloroform extraction gave a recovery of about 80 percent of the best results. All of the other methods gave much poorer results (30 to 55 percent of the best results). The water-acetonitrile reflux procedure for one hour is preferable to the water-methanol-Soxhlet procedure because it is less time consuming and requires less complicated apparatus.

Summary

The quantitative chemical analysis of soils for s-triazine herbicides is of importance in evaluating the persistence of these compounds in soils treated by normal agricultural practices. The three main steps in analyzing soils are extraction, cleanup of the extract, and final determination of the amount of herbicide present. Various procedures which have been reported in the literature for each of these steps are discussed.

A detailed method used in our laboratory for routine analysis of soils for atrazine is presented. Extraction of the soil is done by refluxing with an acetonitrile-water mixture. Cleanup is accomplished by use of an alumina column and determination of the atrazine is done by gas chromatography. A limit of detection of 0.05 p.p.m. is readily attained.

The efficiency of the extraction of atrazine from weathered soils by acetonitrile-water is compared with other extraction systems. The use of water as one of the components of the extracting system appears to be essential.

388 A. M. Mattson *et al.*

Résumé*

Dosage des desherbants du type triazine dans les sols par analyse chimique

Le dosage des desherbants du type s-triazine dans les sols par analyse chimique est important pour l'évaluation de la persistance de ces composés dans les sols traités selon des pratiques agricoles normales. Les trois principaux stades de l'analyse des sols sont: l'extraction, l'élimination des substances interférentes de l'extrait et la détermination finale de la quantité de desherbant présent. Diverses méthodes publiées pour chacun de ces stades sont discutées.

Une méthode détaillée utilisée dans nos laboratoires pour les analyses de routine de l'atrazine dans les sols est décrite. L'extraction est réalisée sous reflux avec un mélange acétonitrile-eau. L'élimination des substances interférentes est accomplie par chromatographie sur alumine et le dosage de l'atrazine est exécuté par chromatographie gazeuse. Une sensibilité de 0,05 mg/kg est facilement atteinte.

L'efficacité de l'extraction de l'atrazine par l'acétonitrile et l'eau des sols modifiés par les intempéries est comparée à celle d'autres mélanges d'extraction. L'utilisation de l'eau comme un des constituants du mélange extractif paraît être essentielle.

Zusammenfassung**

Quantitative Bestimmung von Triazin-Unkrautvertilgungsmittel in Böden durch chemische Analysen

Die quantitative chemische Analyse von Böden auf Triazin-Unkrautvertilgungsmittel ist zur Beurteilung der Hartnäckigkeit dieser Verbindungen in Böden, die durch normale landwirtschaftliche Verfahren behandelt worden sind, wichtig. Die drei Hauptpunkte im Analysieren von Böden sind Extraktion, Reinigung des Extrakts und endgültige Bestimmung der vorhandenen Menge des Unkrautvertilgungsmittels. Verschiedene Verfahren, die in der Literatur für jeden dieser Punkte angegeben worden sind, werden diskutiert.

Eine genau beschriebene Methode, die in unseren Laboratorien für Routineuntersuchungen benutzt wird, wird vorgelegt. Die Extraktion des Bodens wird durch Rückfliessen einer Acetonitril-Wasser-Mischung durchgeführt, die Reinigung wird durch den Gebrauch einer Aluminium-Säule zustande gebracht, und die Bestimmung des Atrazins wird mit Hilfe der Gas-Chromatographie durchgeführt. Eine Ermittlungsgrenze von 0.05 p.p.m. wird leicht erreicht.

---

* Traduit par R. Mestres.
** Übersetzt von M. Düsch.

Die Wirksamkeit der Extraktion von Atrazin aus verwitterten Bäden mit Acetonitril-Wasser wird mit anderen Extraktionssystemen verglichen. Der Gebrauch von Wasser als ein Bestandteil des extrahierenden Systems scheint wesentlich zu sein.

References

BACHE, C. A., and D. J. LISKE: Microwave emission residue analysis of carbamate and triazine pesticides. J. Gas Chromatog. **6**, 301 (1968).
BENFIELD, C. A., and E. D. CHILWELL: The determination of some triazine herbicides by gas-liquid chromatography with particular emphases to atraton in soil. Analyst **89**, 475 (1964).
BRODY, S. S., and J. E. CHANEY: Flame photometric detector. J. Gas Chromatog. **4**, 42 (1966).
BURCHFIELD, H. P., and P. H. Schuldt: Pyridine-alkali reaction in the analyses of pesticides containing active halogen atoms. J. Agr. Food Chem. **6**, 106 (1958).
CHILWELL, E. D., and D. HUGHES: Detection of traces of some triazine herbicides by gas chromatography. J. Sci. Food Agr. **13**, 425 (1962).
COULSON, D. M.: Electrolytic conductivity detector for gas chromatography. J. Gas Chromatog. **3**, 134 (1965).
—— Selective detection of nitrogen compounds in electrolytic conductivity gas chromatography. J. Gas Chromatog. **4**, 285 (1966).
——, L. A. CAVANAGH, J. E. DE VRIES, and B. WALTHER: Microcoulometric gas chromatography of pesticides. J. Agr. Food Chem. **8**, 399 (1960).
CRANMER, M. F., and J. J. CARROLL: A comparison of electrolytic conductivity and microcoulometric gas chromatography. Nat. Meeting Amer. Chem. Soc., Chicago, Ill., Sept. (1967).
DELLEY, R., K. FRIEDERICK, B. KARLHUBER, G. SZÉKELY, and K. STAMMBACH: The identification and determination of various herbicides in biological materials. Zeitschr. für analyt. Chemie, Band 228, Heft 23, 23 (1967).
EBERLE, D.: Private communication, *J. R. Geigy A. G.*, Basle (1969).
Geigy Chemical Corp., Anal. Bull. No. 10: The determination of thiomethyltriazine residues in plant material, animal tissues, milk and water using an ultraviolet method. Agricultural Analytical Chemistry, Analytical Department, *Geigy Chemical Corp.*, Ardsley, N.Y. (1965).
—— Anal. Bull. No. 7: The determination of chlorotriazine residues in plant material, animal tissues and water using the ultraviolet method. Agricultural Analytical Chemistry, Analytical Department, *Geigy Chemical Corp.*, Ardsley, N.Y. (1965).
GYSIN, H., and E. KNÜSLI: Chemistry and herbicidal properties of triazine derivatives. Adv. Pest Control Research **3**, 289 (1960).
HAYES, M. H. B., M. STACEY, and J. M. THOMPSON: Polarographic analysis of triazine herbicides. Chem. & Ind. p. 1222 (1967).
HENKEL, H. G., and W. J. EBING: Contribution to the gas chromatography of triazine herbicides. J. Gas Chromatog. **2**, 215 (1964).
KNÜSLI, E., H. P. BURCHFIELD, and E. E. STORRS: Simazine. In G. Zweig (ed.): Analytical methods for pesticides, plant growth regulators, and food additives. Vol. IV. New York: Academic Press (1964).
McGLAMMERY, M. D., F. W. SLIFE, and H. BUTLER: Extraction and determination of atrazine from soil. Weeds **15**, 35 (1967).
MAJOR, A.: Determination of small quantities of atrazine and simazine by paper chromatography. J. Assoc. Official Agr. Chemists **45**, 679 (1962).
MARTIN, R. L.: Fast and sensitive method for determination of nitrogen. Selective nitrogen detector for gas chromatography. Anal. Chem. **38**, 1209 (1966).

MATTSON, A. M., R. A. KAHRS, and J. SCHNELLER: Use of microcoulometric gas chromatograph for triazine herbicides. J. Agr. Food Chem. **13**, 120 (1965).
MONTGOMERY, M., and V. H. FREED: The uptake translocation and metabolism of simazine by plants. Weeds **9**, 231 (1961).
NEGI, N. S., H. H. FUNDERBURK, JR., and D. E. DAVIS: Metabolism of atrazine by susceptible and resistant plants. Weeds **12**, 53 (1964).
RADKE, R. O., D. E. ARMSTRONG, and G. CHESTER: Evaluation of the pyridine-alkali colorimetric method for determination of atrazine. J. Agr. Food Chem. **14**, 70 (1966).
RAGAB, M. T. H.: Ph.D. Thesis, Univ. of Ill., Urbana, Ill. (1959).
——, and J. P. M. McCOLLUM: Colorimetric methods for determination of simazine and related chloro-s-triazines. Jr. Agr. Food Chem. **16**, 284 (1968).
STAMMBACH, K., H. KILCHER, I. FRIEDERICK, M. LARSEN, and G. SZÉKELY: Analytische Untersuchung von Triazen Herbiziden. Weed Research **4**, 61 (1961).
SHEETS, T. J., and P. C. KEARNEY: Extraction of some s-triazine herbicides from soils. Weed Soc. America Abstr., p. 10 (1964).
TINDLE, R. C., C. W. GEHRKE, and W. A. AUE. Improved GLC method for s-triazine residue determination. J. Assoc. Official Anal. Chemists **51**, 682 (1968).
THOMPSON, J. M.: Some studies of the absorption of s-triazine herbicides on soil-organic matter. M.S. Thesis, Univ. of Birmingham, Ala. (1965).
U.S. Food and Drug Administration: Pesticide analytical manual. Vol. I: Methods which detect multiple residues. Washington, D.C. Revised Jan. (1968).

Summary and conclusions

By

P. C. KEARNEY*

Contents

I. Introduction

No one can deny that the introduction of the s-triazine herbicides into American agriculture has had a profound effect on our ability effectively and economically to control many weeds in agronomic and other crops. The fact that these herbicides are active primarily through the soil has spurred additional research on the fate of these compounds in the soil environment. One result of this research has been a deeper appreciation of the complexities of soil chemistry, physics, and microbiology. By the same token, research on herbicides in soils has stimulated and given new blood to these classic disciplines in soil science. One cannot also fail to be impressed by the sophistication, diversity and volume of research produced on this one class of herbicides in ten brief years.

It is difficult for many people to appreciate the importance of the highly specialized type of pesticide research such as discussed in this symposium, but the present concern on environmental aspects of pesticides will justify more research on pesticides in soils. If we accurately portray pesticides in the environment, then we will be in a position to answer pesticide critics. The beneficial effects of herbicides

---

* Research Chemist, Crops Research Division, Agricultural Research Service, U.S. Department of Agriculture, Beltsville, Md.

far outweigh any problems associated with continued usage. We are most vulnerable, however, in those areas when we have failed to generate sufficient scientific information on the fate of pesticides in the environment. For this reason alone, research on pesticides in soils needs to be accelerated and expanded. The willingness of industry, Government, and academic institutions to cooperate can only further this objective.

What direction has this research taken and what do we know at the present time? Research on pesticides in soils has been primarily aimed at understanding how various processes affect herbicide performance and persistence.

The major processes operating on the s-triazine herbicides can be classified under three main headings of physical, biological, and chemical (Fig. 1). Each of these major headings can be subdivided into a study of the following processes. Under physical, we can list photodecomposition, volatility, leaching, and adsorption. Under biological we can list metabolism, and plant uptake; and under chemical we can list specific reactions such as hydrolysis or oxidation. Metabolism encompasses a study of the effect of the herbicide on the organism and the effect of the organism on the herbicide. Each process in some way affects the persistence and performance of the s-triazine herbicides.

Ideally, if we knew how much each process contributed to the ultimate fate of the s-triazine herbicides in soils, we should be able to draw the type of picture shown in Figure 2. In other words, at any

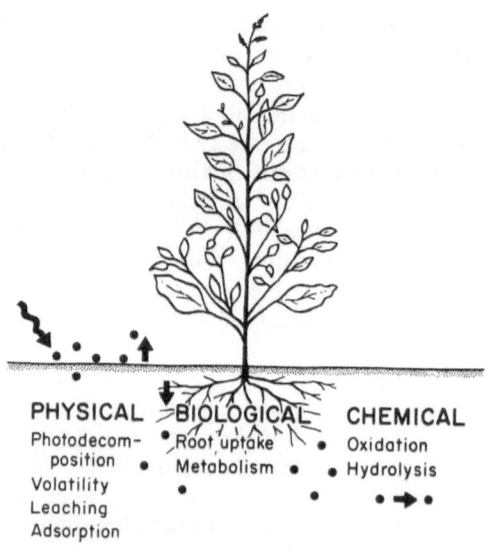

Fig. 1. Processes affecting the fate of s-triazines in soils

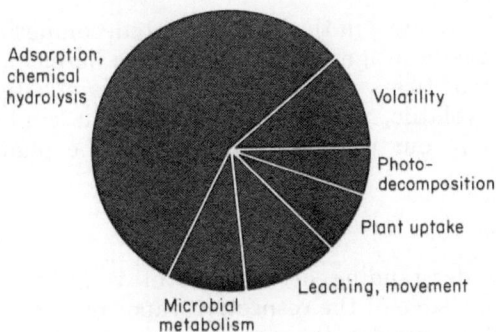

Fig. 2. Hypothetical diagram portraying significant processes affecting s-triazines in soils

one time we would like to say how much each of the processes contributed to the degradation or loss of the herbicide. Obviously, this is impossible, first since this whole scheme would change with time and second, we are unable to quantitate each process. Nevertheless, with this problem in mind, let us examine what we know and what research needs still exist.

II. Photodecomposition

In order for a photochemical change to occur, the molecule must absorb energy. Natural sunlight produces little radiant energy below 2,950 Å. Since most s-triazine herbicides absorb in the region of 2,200 Å, photodecomposition under natural sunlight would appear to be unimportant. Furthermore, once a herbicide is below the soil surface, photodecomposition must be nonexistent. An estimated 98 percent of the ultravoilet light that enters a crystal is absorbed within ca. one Å from the surface. Recently, PLIMMER (1969) irradiated simazine (2-chloro-4,6-bis(ethylamino)-s-triazine) and simetone (2-methoxy-4,6-bis(ethylamino)-s-triazine) in methanol at 2,200 Å. The chloro group was displaced and 4,6-bis(ethylamino)-s-triazine and simetone were obtained. From simetone, loss of the alkylamino side chain took place and 4-ethylamino-2-methoxy-s-triazine was obtained (Fig. 3). What

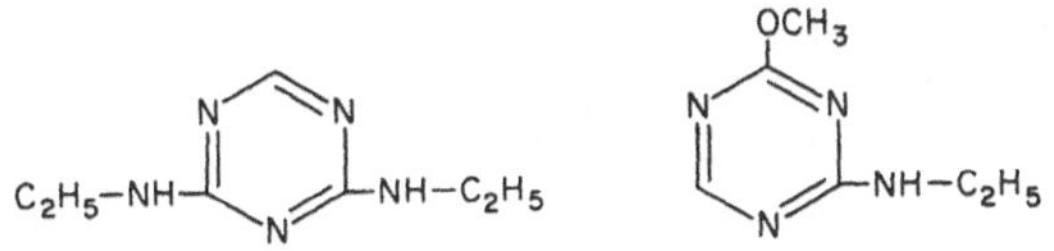

Fig. 3. Photochemical products of simazine and simetone

further needs do we have in the area of photodecomposition? Is photo-decomposition really insignificant in s-triazine herbicide degradation? What role do sensitizers play in the photolysis of s-triazines? If the s-triazines are volatile, what role does gas phase photolysis play? These are gaps in our present knowledge of the photochemistry of pesticides.

III. Volatility

Our true understanding of the role of volatility is meager. At first glance, knowledge of the respective vapor pressures would appear to be sufficient to predict the rate of loss of the s-triazines in the vapor phase. Studies from planchets and glass suggest this may be the case. In soils, however, a consideration of vapor pressures alone becomes meaningless. Subsequent research will show that the amount of herbicide lost as a vapor is a function of temperature, soil-water content, vapor pressure and adsorption (Equation 1). This last factor

$$\text{Loss} = f\ (\text{Temp., Vap. pressure, } H_2O) + f\ (\text{Adsorption})$$

probably makes the greatest contribution to the whole theoretical equation.

What research needs do we have in the areas of volatility? These questions need to be considered: What role does volatility play on the soil surface and in the air spaces between soil particles? How can we devise experiments to separate the effects of volatility from photo-decomposition in the field? Can we devise a suitable method for monitoring triazines in air? How much herbicide is being transported on soil particles moving through air and how far? How can we devise more meaningful experiments to measure volatility losses from soils and relate this to actual field conditions?

IV. Leaching

The s-triazine herbicides are not considered to be very mobile compounds in soils. A scale has been devised for representing the mobility of all classes of pesticides (Fig. 4). Those compounds with mobility numbers near one on the bottom scale are immobile or on the soil surface. Compounds with mobility numbers of six are extremely mobile and may move several feet or more in soils. Without discussing this figure in great detail, the following points are significant: the chlorinated hydrocarbon insecticides are the most immobile pesticides, while the acid herbicides are very mobile; the acid herbicides include compounds such as 2,3,6-TBA (2,3,6-trichlorobenzoic acid), picloram (4-amino-3,5,6-trichloropicolinic acid), fenac (2,3,6-trichlorophenyl-acetic acid), and dicamba (2-methoxy-3,6-dichlorobenzoic acid); the s-triazines exhibit moderate mobility, depending on their structure and the soil type.

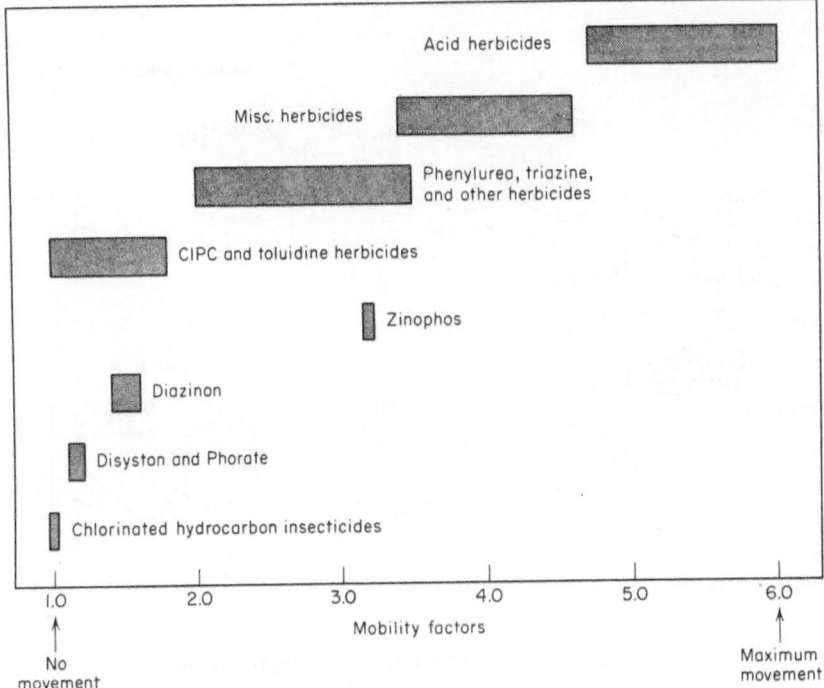

Fig. 4. Relative mobility scale for the vertical movement of pesticides in soils (from HARRIS 1969)

What gaps exist in research on leaching? Most information centers around four or five major triazine herbicides; what are the leaching characteristics of their analogs and of newly developed s-triazines? Miscible displacement studies need to be performed with a large number of the triazines. In addition, we need answers to the following questions. Is diffusion an important process? What role does the plant root play in translocating the s-triazine down into the soil profile? What role does soil particle migration contribute to downward movement? Do soil organisms contribute to the vertical movement in soils? Are s-triazines moved significantly by water or wind erosion?

V. Microbial metabolism

First, let us consider the effect of the microorganism on the pesticide (Fig. 5).

Two routes of degradation are generally recognized in soils. One route involves replacement of the 2-chloro substituent by an hydroxy group. In plants the reaction is apparently not enzymic but catalyzed

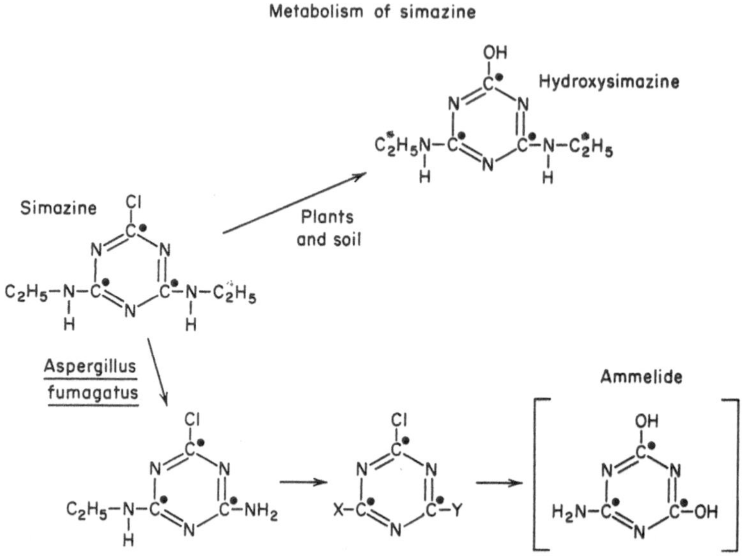

Metabolism of simazine

Fig. 5. Degradation pathways for simazine (from Kearney et al. 1965)

by a cyclic hydroxamate. A second route of degradation by soil fungi and most higher plants involves N-dealkylation to give the 2-chloro-4-amino-6-alkylamino-s-triazine. Recent evidence suggests this might be a free-radical reaction (Plimmer and Kearney 1968, Plimmer et al. 1969).

What information is still lacking in the metabolism area? The following questions need to be answered. What role do soil microorganisms play in the degradation process? What happens to the hydroxy triazine in soils? Is there a common intermediate through which all triazines eventually are metabolized? What is the nature of the triazine bound to the humic acid fraction in soils? At what stage does ring cleavage occur? Are amino acid adducts encountered in soils?

VI. Adsorption

Adsorption plays an important role on the ultimate fate of triazines in soils. We have devoted considerable time to this subject in this symposium. The observation that adsorption and subsequent protonation of the triazine at the colloid surface facilitates hydrolysis is a most significant finding. Adsorption also influences photodecomposition, volatility, leaching, plant uptake, and probably microbial metabolism. We appear to be making excellent progress on a physical understanding of the absorbed molecule. Our gaps in information relate to how adsorption affects these other processes. Does adsorption

retard or enhance microbial metabolism? How does orientation of the
molecule on the surface of the adsorption site influence the rate of
hydrolysis? JORDAN *et al.* (1964) have shown that the triazine adsorbed on filter paper is altered in the presence of light. Does adsorption enhance photolysis? How does the adsorbed molecule influence
plant uptake through the roots? What effect does this have on performance? What influence does adsorption play in the persistence of
these herbicides? Can changing their adsorptive properties change
their persistence and all of the other processes discussed in relation to
adsorption? Studies on adsorption still offer some fascinating problems
to the researcher. Obtaining answers to these questions will require
the best efforts of a team of well trained and diversified scientists.

VII. Persistence

The speed with which each of the preceding processes operates on
a pesticide will determine its persistence. How persistent are the
triazine herbicides? Recently we have completed a rather extensive
review of the literature on the persistence of 11 major classes of pesticides in soils (Fig. 6). Each bar represents one or more classes of
pesticides. The length of each bar represents the time in months
required for 75 to 100 percent loss of activity. The open spaces represent individual pesticides in the larger family of compounds. These are:

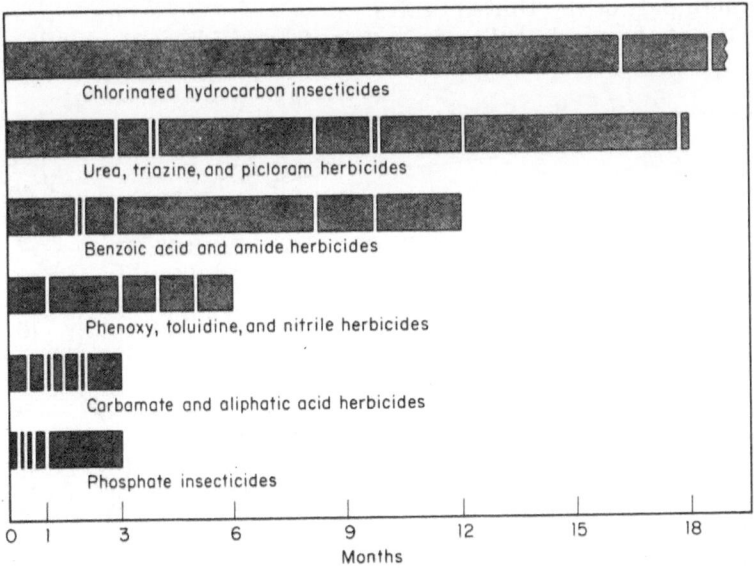

Fig. 6. Persistence of 11 major classes of pesticides in soils (from KEARNEY *et al.*
1969). (Courtesy Charles C. Thomas, Publisher, Fort Lauderdale, Fla.)

however, based on our best estimates from the literature from a number of soil types and locations. It comes as no surprise that the chlorinated hydrocarbon insecticides are the most persistent pesticides in our environment. At normal rates of application, some of these materials persist for two to five years. In contrast, the toxic phosphate esters dissipate within a few weeks. The organic herbicides exhibit a broad spectrum of persistence. Some of the urea and triazine herbicides are fairly persistent. What are some factors which contribute to persistence of the triazines? Two of the important factors are location and depth. Recently, we examined the effect of depth on atrazine degradation in a number of soils (HARRIS *et al.* 1969) at 12 sites in the United States. Samples of atrazine were placed at depths of three, nine, and 15 inches in the soil. The samples were buried at the beginning of the growing season and retrieved in the fall. All samples were returned to Beltsville for residue analyses.

The sites were located in the states of Washington, California, Minnesota, Nebraska, Kansas, New York, Maryland, Mississippi, Florida, and in Puerto Rico. Both Puerto Rico and Nebraska had two sites. The results for atrazine are shown in Figure 7. The three numbers shown for the states indicated represent the amount of atrazine remaining at three, nine, and 15 inches, respectively. These numbers represent the percent of the herbicide remaining of the amount originally applies. With a few exceptions, atrazine tended to persist longer at the lower depths and second, it tended to be more persistent in the northern part of the United States.

One of the most challenging areas of herbicide research relates to

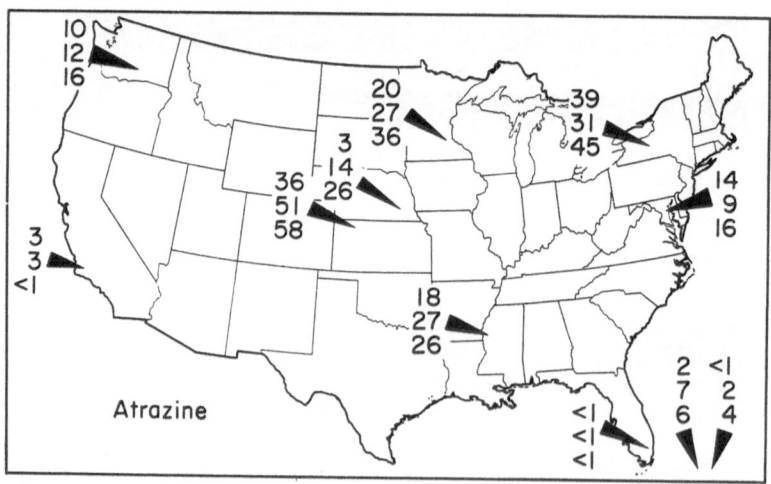

Fig. 7. Persistence of atrazine at three depths in soils (from HARRIS *et al.* 1968)

the problem of altering persistence. We have been successful in prolonging the life of a herbicide, but not in making it less persistent.

VIII. Measurement

Finally, we come to the problem of measurement.

The plant bioassay continues to play a significant role in detecting phytotoxic residues in soils. The ultraviolet method has been a fairly reliable and widely used chemical method for determining s-triazine residues in soils. What needs still exist in methodology? Can the sensitivity offered by electron capture gas chromatography be used for triazine analyses in soils? Can silanization be used as a derivatization process for the quantitative conversion of s-triazines to volatile derivatives suitable for gas chromatography? Will more specific detectors be developed for s-triazine analyses?

IX. Prognosis

I have outlined some gaps in our knowledge of triazines in soils. I don't suggest that these are urgent needs or we must take immediate action to solve some of the problems outlined. Nor do these gaps in information exist only in the s-triazine class of herbicides. Many of the questions I have asked reflect on our need to acquire fundamental knowledge on the organic and physico-chemical properties of these compounds. There are some excellent research projects ahead of us. From the scope and depth of papers presented at this symposium, I am sure we can successfully meet these challenges.

References

HARRIS, C. I.: Movement of pesticides in soil. J. Agr. Food Chem. **17**, 80 (1969).

HARRIS, C. I., E. A. WOOLSON, and B. E. HUMMER: Dissipation of herbicides at three soil depths. Weed Science **17**, 27 (1969).

JORDAN, L. S., B. E. DAY, and W. A. CLERX: Photodecomposition of triazines. Weeds **12**, 5 (1964).

KEARNEY, P. C., D. D. KAUFMAN, and T. J. SHEETS: Metabolites of simazine by *Aspergillus fumigatus*. J. Agr. Food Chem. **13**, 369 (1965).

——, R. G. NASH, and A. R. ISENSEE: Persistence of pesticide residues in soils, Chapt. 3, pp. 54–67. In M. W. MILLER and G. C. BERG (eds.): Chemical fallout: Current research on persistent pesticides, Springfield, Ill.: Thomas (1969).

PLIMMER, J. R.: Unpublished data (1969).

——, and P. C. KEARNEY. Free radical oxidation of pesticides. Weed Sci. Soc. Amer., Abstr. p. 20 (1968).

—— ——, and U. I. KLINGEBIEL: A study of the mechanism of detoxification of some s-triazines. Weed Sci. Soc. Amer., Abstr. p. 202 (1969).

Subject Index